The RIDDLE *of the* ROSETTA

THE
RIDDLE
OF
THE
ROSETTA

How an ENGLISH POLYMATH
and a FRENCH POLYGLOT
DISCOVERED *the* MEANING
of EGYPTIAN HIEROGLYPHS

Jed Z. Buchwald & Diane Greco Josefowicz

PRINCETON UNIVERSITY PRESS
PRINCETON & OXFORD

Published by Princeton University Press
41 William Street, Princeton, New Jersey 08540
6 Oxford Street, Woodstock, Oxfordshire OX20 1TR

press.princeton.edu

Library of Congress Cataloging-in-Publication Data

Names: Buchwald, Jed, 1949– author. | Greco Josefowicz, Diane, 1971– author.
Title: The riddle of the Rosetta : how an English polymath and a French polyglot discovered
    the meaning of Egyptian hieroglyphs / Jed Buchwald, Diane Greco Josefowicz.
Description: Princeton : Princeton University Press, 2020. | Includes bibliographical
    references and index.
Identifiers: LCCN 2019032326 (print) | LCCN 2019032327 (ebook) | ISBN 9780691200903 (hardback) |
    ISBN 9780691200910 (ebook)
Subjects: LCSH: Young, Thomas, 1773–1829. | Champollion, Jean-François, 1790–1832. |
    Rosetta stone. | Egyptian language—Writing, Hieroglyphic.
Classification: LCC PJ1531.R5 B77 2020 (print) | LCC PJ1531.R5 (ebook) | DDC 493/.1—dc23
LC record available at https://lccn.loc.gov/2019032326
LC ebook record available at https://lccn.loc.gov/2019032327

British Library Cataloging-in-Publication Data is available

Editorial: Rob Tempio and Matt Rohal
Production Editorial: Sara Lerner
Text and Jacket Design: Chris Ferrante
Production: Erin Suydam
Publicity: Amy Stewart and Alyssa Sanford
Copyeditor: Cathryn Slovensky

Jacket, Endpapers, and Frontispiece Credit: From the *Vetusta Monumenta* (1746),
engravings of the Rosetta Stone (vol. 4), hieroglyphs designed and engraved by James Basire (pl. 5)

This book has been composed in Adobe Text Pro, Archetype, and
The Fell Types (digitally reproduced by Igino Marini)

CONTENTS

LIST *of* FIGURES

The RIDDLE *of the* ROSETTA

INTRODUCTION

One day in the early 1820s, a traveler to London stopped at the Antiquities Gallery of the British Museum, noting that it was "the only institution in London which does not charge you for entry." Apparently even the nonexistent fee was not low enough, for the traveler soon found much to dislike about the antiquities on display. The "pillars of ancient temples, tombstones with inscriptions, and damaged statues" inspired a dejected identification. Just as he, a foreigner, felt alienated by the city, so too did these objects "lose their meaning in the modern capital of a strange country." He rallied, however, on seeing "the stone brought from Rosetta," with inscriptions in "the supposedly secret hieroglyphs" as well as in Greek and demotic. The latter two inscriptions prompted him to wonder "whether they were known to the researcher Champollion," as "they could contribute to his claim about the hieroglyphs, which have so far been considered a mystery and in his opinion are only letters."[1] That the Rosetta Stone, with its triple inscription, should rouse this melancholy witness to speculate about the relationship between the three scripts attests to the broad interest that the recent work of both Jean-François Champollion in France and Thomas Young in England was beginning to elicit at this time.

Champollion and Young are often said to have worked in broadly similar ways in their efforts to "decipher" these scripts.[2] The *Oxford English Dictionary* defines "to decipher" as "to convert (a text written in code, or a coded signal) into normal language." Both Champollion and Young did aim to read Egyptian texts, and the dictionary meaning of "decipherment" does include at least that much. But it also suggests that a script constitutes a "code," that is, "a system of words, letters, figures, or symbols used to represent others, especially for the purpose of secrecy," again according to the Oxford definition. Until well into the eighteenth century, and even later, those interested in Egyptian hieroglyphs did for the most part think of them as a "code" in the sense of a system of inscription designed to ensure secrecy. Each hieroglyphic, it was frequently thought, stood for an elaborate allegory in which sacred knowledge was concealed, knowledge that was thought to be the exclusive possession of priests.[3]

Neither Young nor Champollion subscribed to this interpretation. For Young in part, and for Champollion altogether, all three Egyptian scripts were intended to convey specific words and word phrases, which meant that the ancient texts could be understood by anyone who knew the Egyptian language. The purpose of many codes or ciphers—to maintain secrecy—was for them not at issue, except insofar as literacy itself had been limited to a select group for such a purpose. (Neither Young nor Champollion took much interest in the extent to which ancient Egyptians were literate.) To write of their work as a "decipherment" in the customary sense, though having the virtue of long-established usage, can obscure much of what Champollion considered his own aims to be. For his part, Young did see his efforts rather like the cracking of a cipher, but he also found the signs used by the Egyptians redolent of their religion, which he disdained. He insisted that, in the centuries before Alexander's conquest of Egypt, hieroglyphs always encompassed much more than the direct words and phrases of the ancient language, and what they did convey then was hardly precise. As Young noted in a remarkable letter written in 1827, although a "simple picture, for instance, of a votary presenting a vase to a sitting deity; each characterized by some peculiarity of form, and each distinguished also by a name written over him" might "be called a pure hieroglyphical representation," such a picture "scarcely amounts to a language, any more than the look of love is the language of the lover."[4] To Champollion, in contrast, the scripts were traces of the language of a venerable ancient civilization. Where for Young ancient Egypt held little interest in periods before the Greek invasion, for Champollion the Greco-Roman period marked the cultural decline of a once admirable civilization under foreign domination.

As our remarks already suggest, this book is a study in contrasts. The differences in the approaches of Young and Champollion to Egyptian writing evolved from their dissimilar circumstances, with different attitudes toward antiquity, different material cultures of printing and engraving, different ways of handling evidence, and more. Although we examine the details of the controversy that emerged following Champollion's 1822 announcement of his claims at the Académie des inscriptions et belles-lettres—the renowned *Lettre à M. Dacier*—our account begins much earlier. We concentrate on how each of our protagonists arrived at his conclusions in order to show the separate developmental arcs of their ideas. To this end, we have made extensive use of archival materials not previously thoroughly exploited: in particular, Young's manuscripts from 1814, when he first turned to the Egyptian scripts, Champollion's drafts for talks that he gave at the Académie des inscriptions before the famed Dacier letter, and manuscripts that document Champollion's passionate engagement with Coptic. All in all, Young could not take a view of the Egyptian scripts that was not also a position on the sophistication of the ancient civilization relative to his contemporary values; whereas for Champollion, the study of the ancient scripts provided much-needed *escape* from the present, especially during periods of personal difficulty. "What a distraction, indeed, for the heart and soul is a grammar that is more than six thousand years old!" Champollion wrote to a friend in May 1816, in the midst of internal exile to his sleepy hometown of Figeac.[5] To Young, Egyptian culture itself seemed principally responsible for the character of Egyptian writing, which he considered vastly inferior to Greek. For Champollion, in contrast, the scripts held out the hope of entry into a world that he wanted deeply to understand, about which he made few a priori claims. As a result, he moved fruitfully along paths that Young would not, and perhaps could not, follow.

Late Georgian attitudes shaped Young's work with the Egyptian scripts. Propriety in language, deportment, and written expression mattered a good deal to Young, as did contemporary expectations concerning the proper forms of mathematical reasoning. For Young, these expectations went considerably beyond mathematics itself. Born to a Quaker family and encouraged to value mastery of Greek and Latin, Young became expert at manipulating the written forms of both. An avid reader of natural philosophy and mathematics treatises, he mastered these skills as well. As a young man, he found favor with Georgian elites who valued ancient languages as much as he did, and who were rapidly coming to value scientific expertise. Young's path into London society was additionally smoothed by his mother's uncle, the prominent London physician Richard Brocklesby, who shepherded Young through his early medical career while welcoming him into his circle of influential friends.

Although Young came to eschew Quaker dress and practice, residues of his religious upbringing can be detected in his later attitudes. The zoomorphic idolatry he perceived as the foundation for ancient Egyptian religion rendered the entire culture suspect. This, too, shaped his attitude toward Egyptian writing. In the end, for Thomas Young the scripts of Egypt were uninteresting for anything they might have to say about a culture he viewed as barbarous, until the salutary effects of Greek culture changed the scripts in ways that he hoped might be used to illuminate the histories of the invading Greeks and Romans. Disdain for ancient Egypt strongly colored his attitude toward its writing, and the scripts interested him chiefly as a particularly resistant puzzle to solve.

Young's attitude was not unusual. Many of his contemporaries regarded ancient Egypt as useful only insofar as it illuminated the condition of the modern Briton. We see this attitude vividly expressed in Turner's early paintings on biblical subjects. One of these paintings referenced the familiar story of ancient and idolatrous civilization getting its divine comeuppance at the hands of a virtuous prophet.[6] Exhibited at the Royal Academy in 1800, J.M.W. Turner's *The Fifth Plague of Egypt* featured a desert landscape in which threatening

FIGURE I.1. J.M.W. Turner, *The Fifth Plague of Egypt* (1800). Courtesy of Indianapolis Museum of Art at Newfields.

clouds are pierced by a shaft of light that illuminates an inscrutable pyramid (fig. I.1). An instantly recognizable symbol of ancient (i.e., biblical) Egypt, the pyramid revealed nothing of its meaning or relationship to the landscape it dominated, while animals and human figures are shown prostrate in the foreground, suffering the divinely meted-out consequences of their idolatry. Such interpretations condensed religious and antiquarian themes into an easily consumable narrative of social order—who was in, who was out, and who was responsible for the present situation. As a symbol, Egypt gave form to otherwise inchoate anxieties about heritage, identity, and belonging. These anxieties were stoked not by Egypt per se but by events closer to home, such as the difficult incorporation of territories like Scotland into the body of Great Britain and the intrusion of scientific expertise into antiquarian-adjacent fields like archaeology and philology. As we will see, Young took care to distance himself from antiquarianism. But even his wariness suggested the degree to which he shared in these broader conflicts.

Born seventeen years after Young in 1790, Champollion was the youngest son in a large family headed by a domineering father, a bookseller with a taste for drink and a willingness to incur ruinous debts. Despite these unfavorable beginnings, Champollion was supported in his studies from a tender age. In perhaps the luckiest turn of his young life, Champollion's practical and worldly older brother took charge of his education. While employed by a Grenoble textile concern owned by a member of their extended family, Jacques-Joseph (known later as Champollion-Figeac) encouraged Champollion to develop his affinity for languages, notably Hebrew, as well as the usual Latin and Greek. But quiet scholarship apart from public life ultimately held limited appeal. A fiery Bonapartist during his early adulthood, Champollion narrowly avoided incarceration and worse during the Bourbon Restoration. In keeping with his Bonapartism, he exhibited a philosophe-like disdain for official religion; at the same time, he nurtured a distinctly un-philosophe-like admiration for the culture, as he then knew it, of ancient Egypt. Napoleon, of course, undertook a massive expedition to the country, largely in order to secure his bid to be remembered as an empire builder in the manner of Alexander the Great. For Champollion the appeal was different. He regarded pre-Alexandrian Egypt as aesthetically and even morally superior to what it became under Greco-Roman domination. To read the ancient scripts was to open a window onto an attractive world peopled by individuals whose lives he admired and whose beliefs he respected.

Guiding his younger brother's education, Champollion-Figeac encouraged Champollion to value some kinds of evidence over others. A librarian and scholar eventually selected to head the newly established École des chartes, Champollion-Figeac merged old-fashioned antiquarianism with the nascent sciences of archaeology and philology.[7] The result was a distinctive historical sensibility, one he shared with contemporaries who were, like him, keen to modernize French antiquarianism. Historians have noted the fingerprints Champollion-Figeac left on his younger brother's career, as he shepherded Champollion through the treacherous world of elite Parisian scholars and edited the younger man's writings in order to limit otherwise inevitable criticism and dismissal. From his older brother Champollion absorbed much about the nature and practice of history that prepared him for work on the Egyptian scripts. Among other things, Champollion-Figeac alerted him to the possibility that textual evidence could be scrutinized for uniquely persistent features such as linguistic roots and toponyms that described inalterable features of the landscape.

As we will see, the question of whether the Egyptian scripts were phonetic became central to Champollion's work in contradictory ways. The question was then a matter of considerable general import, for the advent of an alphabet was indexed to the presumed sophistication of a civilization. Though not everyone agreed on this point, Coptic—the liturgical language of the Coptic Orthodox Church in Egypt—was hypothesized to be a late descendant of the spoken language of ancient Egypt.[8] However, since Young saw Coptic as a developed language of late pre-Islamic Egypt, he was skeptical of its utility for investigations of pre-Alexandrian texts. For Champollion, on the other hand, Coptic virtually *was* ancient Egyptian, little changed from the original. As a student in Paris, he became involved with Copts recently arrived from Napoleon's Egyptian expedition who introduced him to Coptic as a living language. Champollion found in Coptic something like a fossil from remote antiquity, one that accordingly promised unique access to the country's earliest beliefs and cultural practices. The question, for him, was how Coptic could be used to elucidate the scripts. Assembling dictionaries of Coptic, Champollion even tested hypotheses about which bits of the Rosetta demotic might correspond in some manner to various Coptic roots. By the early 1820s, following a stunning reversal, Champollion became convinced that even the most ancient of the scripts had to be phonetic. In what follows, we trace Champollion's path to this conclusion.

When Young began to engage with Egyptian writing in 1814, he was an established medical doctor known for his pathbreaking and occasionally controversial natural philosophical investigations. At forty-one years old, he could boast a respectable scientific reputation even among the French savants, as scholars in natural philosophy, mathematics, and other areas were called, and this despite the acrimonious years of the Revolution and the Napoleonic Wars. Champollion, in contrast, was just twenty-four in 1814 and scarcely known outside a select group in Paris and Grenoble; his first book, *L'Égypte sous les Pharaons* (*Egypt under the Pharaohs*), was published that year after a great buildup to disappointingly mixed reviews. Unlike Young, he was uninterested in mathematics and natural philosophy, though the corpus of his published and unpublished work amply attests to his powerful ability to classify and synthesize a vast array of heterogeneous evidence. Their temperaments also differed. Until pressed by friends to counter what they saw as an emerging attempt to devalue his work on the scripts in comparison with Champollion's, Young remained, if anything, both cordial and helpful when contacted by the young Frenchman. Even the book that he produced to assert his priority in particular points was politely and carefully framed. Champollion, on the other hand, reacted vehemently to insinuations that he had merely amplified Young's positions.

The seemingly trivial matter of nomenclature emblematizes the difference between the two men, who could not even agree on what to call the intermediate Rosetta script. Young used "enchorial," a straight transliteration of the Rosetta Greek ΕΓΧΩΡΙΟΙΣ, which he took to refer simply to the vernacular "of the country." The choice was consistent with his view that the inscription's translation should adhere closely to the Greek text in both structure and meaning. Champollion, following Herodotus, preferred "Demotic," which could be construed as referring to the popular, idiomatic language spoken by the general populace and captured in the Rosetta Egyptian. Words carry a variety of meanings and connotations, and in the difference between "enchorial" and "Demotic," we spy more than an argument over scholarly possession, though it was at least that. Despite his Quaker background, Young was by training, inclination, and position hardly prone to see a reflection of Egyptian popular culture in the Rosetta inscription. The manifestly republican Champollion held a viewpoint that was altogether different. The lexical discontinuity functioned, and continues to function, as a reminder of the stark differences in the two mens' political attitudes and suggests relationships between those attitudes and their views of ancient Egypt.[9]

These differences are perhaps most apparent at the juncture where their public lives intersected with their intellectual preoccupations. Champollion's views on the nature of the scripts were matched and occasionally echoed by his political radicalism. Although Champollion's reputation as a radical did not redound to his professional benefit, outspokenness continually secured the benefits of an audience for his views. Young, in contrast, generally sought to avoid controversy, as it threatened the medical practice that provided his livelihood.[10] Such reserve could not have been more foreign to Champollion. Though his career suffered, he freely expressed anticlerical and anti-Bourbon views, counting on his brother's diplomatic talents and extensive connections to keep him from serious harm. Little wonder, then, that his and Young's paths to understanding the ancient scripts differed so markedly.

Despite their differences, Young and Champollion did share important traits. Both had an acute ability to work with unfamiliar graphical forms, that is, the Egyptian scripts themselves. Both were exceptionally able to organize these forms systematically—to such an extent, in Champollion's case, that he would be criticized for embodying the eighteenth-century grand sin of having been too committed to "system" and not enough to available evidence. To some extent, Young and Champollion also shared a literary tradition. Both absorbed eighteenth-century travelers' accounts of Egypt as well as works by predecessors who attempted to read the scripts. They knew as well the principal accounts from antiquity, those of Horapollo, Herodotus, Diodorus Siculus, and Clement of Alexandria, together with what had been written about these ancient remarks during the previous century. Both would grapple with the complexities posed by contemporary efforts to reproduce the signs on the mummy wrappings, papyri, and carved inscriptions.[11] Of course, much of what they held in common they also shared with their less well-known contemporaries, and their social and intellectual contexts were hardly mutually exclusive. People, publications, and letters flowed back and forth as freely as circumstances allowed.

We have divided our account into five major parts. Part 1, "A Quaker's Odyssey," begins nearly in medias res, with a dinner party thrown in 1803 by publisher Thomas Longman at which Young confronted attitudes toward ancient history and language, which soon pervaded his work on Egyptian writing. The rest of part 1 traces aspects of Young's early biography that were instrumental to his later work on the scripts, concluding with his first mature encounter with a physical remnant of antique writing, significantly in the form not of Egyptian hieroglyphs but the Greek of a Herculaneum papyrus. Part 2, "Antiquity Embraced," turns to Champollion's early investigations of Coptic, placing them within the history of European encounters with the language and contemporary responses to the Rosetta Stone's discovery. In part 3, "Scripts and Bones," we examine Young's earliest, and enduring, work on the Rosetta scripts, done during his sojourn at Worthing in the summer of 1814. Preserved at the British Library, the manuscript of that work permits us to follow the evolution of Young's ideas as he grappled with the unfamiliar signs. Young's subsequent studies of papyri published in the *Description de l'Égypte* cemented his views of the scripts in ways that did not significantly alter thereafter. At roughly the same time, Champollion and his brother undertook an investigation of the Roman settlement of Uxellodunum. Begun during the brothers' forced exile to the family home at Figeac, the Uxellodunum study places Champollion's labor on the scripts at the intersection of ancient history, antiquarianism, and philology.

After leaving Figeac for Grenoble, Champollion developed the startling claim that none of the Egyptian scripts were phonetic at all. When he finally returned to Paris, he detailed those conclusions in unpublished lectures given at the Académie des inscriptions, manuscript drafts of which are preserved in the Paris archives. In his *Lettre à M. Dacier*, presented only months later in the same venue and in the presence of Young himself, Champollion offered a considerably different assertion. How and why Champollion's views abruptly changed is the subject of part 4, "Reading the Past." In part 5, "Antique Letters," we examine the earliest reactions to the Dacier letter; the reasons for the church's otherwise surprising embrace of Champollion, the anticlerical firebrand; the beginnings of the priority controversy; and Champollion's efforts to clarify his system and defend its novelty. In conclusion we briefly follow Champollion's later career and assessments of the Egyptian scripts and language, published posthumously by Champollion-Figeac, who carefully tended and sometimes burnished his brother's reputation. We do not further explore the subsequent development of Egyptology into an organized academic discipline, since that is in itself a complex story.[12]

Our account is not intended to introduce readers to modern apprehension of the Egyptian scripts. Some readers may be disappointed that we do not point out where or in what ways the views of Young and Champollion differ from the present dispensation, but this book is not an exploration of modern Egyptology. In order to illuminate how an English polymath and a linguistically talented Frenchman came to their respective views, we have tried as much as possible to remain within their perspectives, attentive to what they themselves would have known and to how their views interacted with their social and cultural contexts. Of course, the present understanding did evolve out of the original work by our protagonists, and it is hardly uninteresting to see where Young's or Champollion's views diverge from present knowledge. However, to do so here inevitably risks casting an illusory light on what they actually knew and thought. It is altogether too easy to hand out marks for which of them first correctly understood this or that sign or sign sequence, but no one at the time had the benefit of two centuries' further investigation. Those who are curious about such matters can consult the many modern texts that provide Egyptian sign sounds, meanings, and grammar for comparison, if they like, with those of Young and Champollion.

Much has been made of the conflict between Young and Champollion, and between their several epigones. These accounts frequently have nationalist overtones. The two men had their differences, to be sure, but histories of their work on the scripts only took virulently nationalist forms following both of their deaths. Despite some exceptions, their conflict did not pertain principally to the military and political struggle between England and France.[13] Their concerns were narrower and more local. Exploring these interests while remaining faithful to their different points of view has constituted the bulk of our task in this book. Questions about priority can too easily mislead if the goal is to understand how each of our protagonists worked in his particular context, for the two had always kept quite different aims in mind.

We are, of course, acutely aware of the risks of adding yet another volume to the already extensive literature on the reading of the ancient Egyptian scripts. However, no published account has made full use of unpublished manuscript sources, and this alone provides sufficient reason for a fresh treatment. With this volume, we hope to strike a balance between an intimate exploration of our protagonists' different points of view and a presentation of the major turning points in the history of their conflict. Perhaps unsurprisingly for a work stemming from unavoidably biographical premises, character quickly became a central preoccupation. Stories need characters to unfold their plots, and we had both characters and plots in abundance. But even as they were surrounded by conflict and occasional deceitfulness, our protagonists lived and worked within networks of helpful others, many of them largely forgotten, who also crowd our canvas. These friends and collaborators prompted and consoled, delivered hard-to-find books and rare manuscripts, and freely opened their homes, their larders, and even their wallets when the need arose. These essential figures provide the context within which our two central figures played out a drama that was at once something more, and something quite other, than a contentious meeting of minds. We have explored the lives of Champollion and Young in order to see what happened when their separate stories intersected, to join in remarkably revealing ways.

The Young archive at the British Library in London and the Champollion fonds at the Archives nationales open a new window onto their decipherments. Young's treatment of the Rosetta scripts illustrates how Georgian beliefs concerning propriety in language, writing, and even mathematics combined with belittlement of ancient Egypt's religion and culture to pattern his approach. We follow Young as he disassembled the Rosetta inscriptions, seeking to reconstruct the meaning of words written by a people he saw as inferior to the Greeks and Romans. Champollion's intense engagement with Coptic coalesced with his early admiration of ancient Egypt to lead him in an entirely different direction as he sought to uncover a lost world through its language. Unlike his English counterpart, who, though a Quaker by birth and training, nevertheless joined the social and cultural world of metropolitan London, Champollion was ever the provincial outsider, buffeted by the currents of late Napoleonic and early Restoration France. Young avoided notoriety; Champollion courted it. His archives reveal him to have been comparably bold in his willingness to change his mind about the nature of the ancient scripts as he sought a way to make a novel impact. Young, in contrast, never markedly altered his views. And yet, different though the two were, both were riveted by the struggle to find meaning in the signs of ancient Egypt.

PART I

A QUAKER'S ODYSSEY

CHAPTER I

DINNER AT
LONGMAN'S

In May 1803, the publisher Thomas Norton Longman arranged a Saturday supper for a half-dozen guests at his offices in Paternoster Row, "that crowded defile north of the Cathedral" where stationers and booksellers hawked their wares in the shadow of St. Paul's. As the church bells tolled quarter-hourly reminders of mortality, smudge-fingered "despots of literature" toiled in the streets below—less noisily, perhaps, but fixed more strongly on the alternative prospect of immortality offered by books. Not that the secular option was easier: "Many a groan," one contemporary observed, "has gone up from authors in this gloomy thoroughfare," particularly outside "the immense emporium of Longman's," which stretched across two storefronts decorated with "little Ionic pilasters, and [an] iron crane, emblematic of the very heavy commodities in which the proprietors are sometimes compelled to deal."[1]

Any bookman could appreciate the irony. Those splendid neoclassical pilasters juxtaposed with a burdened crane neatly expressed the trouble with literate culture, at least from a publisher's point of view: a book was a durable object of determinate heft but uncertain value. If some books had the power of monuments, others were just so much dead weight. As the city's lamplighters plied their trade, a debate arose around Longman's table that reprised different facets of the same incongruity. In what did a tradition consist? To what extent were ancient sources of any tradition—sacred, secular, classical, scriptural—relevant in rapidly modernizing London? How much authority should be ascribed to ancient sources, and which ones, and why? Among the guests at Longman's dinner was the physician and natural philosopher Thomas Young. How he tried to read the writing systems of ancient Egypt, and why he only partly succeeded, are questions bound up with the table talk that night at Longman's. Young's work with the Egyptian scripts occurred at a moment when traditional forms of authority were rapidly ceding ground to secular ones, particularly modern science. Although the Egyptian controversy was still in the distance, by 1803 Young was already involved in broader but related arguments about the relevance of antiquity to contemporary problems.

"A plain man of the old citizen style," the redoubtable Longman had selected his guests with a canny publisher's eye, attentive to their overlapping interests in antiquity and science, subjects of special concern to Longman.[2] Four of Longman's guests were additionally committed to establishing a distinct literary heritage for Scotland at a time when Edinburgh vied with London for cultural dominance by positioning itself as morally and aesthetically superior to the corrupt imperial metropolis.[3] Two of Longman's guests that evening, David Irving and Thomas Campbell, were Scots poets with pronounced antiquarian interests. A third was a promising young writer by the name of Walter Scott, then a barrister and sheriff in Selkirk who had just published three volumes of folk ballads collected from the northern border; Longman would soon become his London publisher of a related work, *The Lay of the Last Minstrel* (1805).[4] Joining them was the poet, antiquary, and political satirist George Ellis and two representatives of scientific London: Humphry Davy, presently the toast of London's social scene thanks to his engaging public lecture series at the newly established Royal Institution, and Young himself, who was also lecturing at the Royal Institution and who was Ellis's longtime friend. "Such guests as these," Irving later recollected, "could not now be assembled at any table in the kingdom."[5]

Of this group, Ellis was the éminence grise. A fellow of the Society of Antiquaries since 1800, Ellis had cemented his reputation with *Specimens of the Early English Poets* (1790), a wide-ranging miscellany that brought antique poetic forms to contemporary readers; a second edition appeared in 1801, expanded to include examples of poetry in the languages of medieval Scotland and Ireland. Irving was finishing *Lives of the Scottish Poets*, the first Scots literary historiography, a work he considered no mere scholarly exercise but the first "literary biography of Scotland."[6] In contrast to Ellis and Irving, on the evening of Longman's party, Scott was only thirty-two and virtually unknown. Nevertheless, he was, according to Irving, "at all times conspicuous for his social powers, and for his strong practical sense . . . full of good humor and [with] many stories to tell." In keeping with his excellent reputation as a lecturer, the chemist Davy also showed himself admirably, "willing to talk, in an easy and unpretending strain, on any subject that was discussed." Irving was less impressed by Campbell, who was still a relative newcomer. His pretensions offended Irving, who recalled that "[a]mong these men, Campbell did not appear to much advantage: he was too ambitious to shine, nor was he successful in any of his attempts."[7]

Irving's criticism hinted at trouble simmering beneath the evening's surface while adroitly sidestepping the fact that Campbell was only half-responsible for the argument that boiled over by evening's end. New methods of textual criticism had recently arrived on British shores, threatening to supplant traditional appreciations of classical sources on antiquarian or aesthetic grounds.[8] That evening at Longman's, Young locked horns with Campbell specifically over "the Homeric Question," an ancient debate concerning Homer's authorship of the *Iliad* and the *Odyssey* that had sparked a reassessment of the authority of classical antiquity, its relevance to contemporary audiences, and the relationship of ancient texts to oral sources. Irving recalled that Campbell "was much inclined to dilate on the subject of Homer, and the poems which bear his name, but on various points was opposed with equal decision and coolness by Dr. Young." Their argument, Irving noted, turned specifically on the significance of Friedrich August Wolf's *Prolegomena ad Homerum* (1795), which Irving credited as having "introduced a new era in classical criticism." The *Prolegomena* refreshed the question, attributed originally to a group of Alexandrine skeptics known as the Chorizontes, of whether a single person—the bard, Homer—was responsible for the epics. Considered a modern approach to ancient texts, Wolf's philology carried a strong whiff of fashionable, scientific London, which eagerly sought fresh uses for antiquity in architecture, literature, and fine arts.[9]

Expressions of Scots nationalism may have been uniquely problematic for Young, who was still smarting from an anonymous attack on his wave theory of light in the January 1803 issue of the *Edinburgh Review*. The author of the attack was Henry Brougham, an Edinburgh lawyer and sometime mathematician who was also one of the *Review*'s founders.[10] The attack was understood specifically in terms of Scottish perfidy. At the end of the nineteenth century William Henry Milburn, an American Methodist preacher, characterized the affair as nothing less than an infiltration into England from the barbaric North: "No sooner had Young's 'Memoir on Light' appeared than Brougham rushed to attack him with all the fierce savagery of his cattle-stealing, house-burning, marauding forebears," he wrote, further describing Brougham as a "fierce and turbulent young borderer" with "a memory like a row of pegs to hang grudges on."[11] Young replied to Brougham in a detailed but comparatively restrained rebuttal that concluded with a vow thereafter "to confine my studies and my pen to medical subjects only," a promise soon abandoned.[12] Others more talented at public relations rallied to his defense, and the matter was dropped in favor of broadsides against the *Edinburgh Review* and its writers.[13] Nevertheless, the wound remained. All Campbell had to do was salt it.

Recalling the fight, Irving gave few specifics, so we can't be sure of the precise terms of either Campbell's position or Young's. We can infer some possibilities, however, by examining Wolf's approach in terms of what it likely implied for them both. Rather than using the Homeric texts as a pretext for

effusions over the talents of the bard, Wolf focused on the long trail of transcribed, translated, incomplete, and fragmentary documents comprising the
Homeric corpus.[14] By separating the remnant text from its origins, Wolf made
it difficult for readers to read Homeric epics literally, that is, to slip imagi-
natively from Homer's text to Homer's world. The boundary Wolf set upon
the reader's imagination would have proven uncongenial to Campbell, who
hoped to move readers to identify with collective experiences—of battle, of
nationality—through epic poems intended to suggest not fantasy re-creations
of the ancient world but its historical and social realities. Wolf's argument,
in contrast, asserted that contemporary Homeric texts not only bore little
resemblance to the original poems but also revealed more about the poems'
Alexandrine reception than about either Homer or his context. "The Homer
that we hold in our hands now," Wolf wrote, "is not the one who flourished in
the mouths of the Greeks of his own day, but one variously altered, interpo-
lated, corrected, and emended . . . Learned and clever men have long felt their
way to this conclusion by using various scattered bits of evidence; but now
the voices of all periods joined together bear witness, and history speaks."[15]

History did indeed speak—though most frequently in ways that modern
historians are not likely to applaud. As the favored historical mode of the period, antiquarianism—history as delectation, according to which objects were
valued simply for their age or aesthetic qualities—was closely allied to the
interests of virtually all the guests at Longman's party, albeit in different ways.
Antiquarianism was, first of all, a way to claim a heritage. In 1818, Scott commissioned J.M.W. Turner as one of several artists to make the sketches for his
Provincial Antiquities (1819–26), a catalog of the remnants of Scotland's remote
past that were still extant and visible in the landscape.[16] During a two-week
trip around the north, Turner filled three sketchbooks with drawings of prominent ruins juxtaposed with smaller human and animal figures, according to a
compositional principle similar to that of his *Fifth Plague of Egypt*. Although
these compositions connected viewers with an exalted heritage, they did so
ambiguously, by putting them on the scene of its decay. In Scott's *The Antiquary*
(1816), ambiguity gave way to outright irony, as the novel poked fun at both
the antiquarian's fantasies of belonging to an exalted past and the scholar who
threatened to undermine those fantasies using philology.

Wolf's historicism notwithstanding, he never claimed that close study of
the Homeric texts could offer no knowledge whatsoever of remote antiquity.
On the contrary, he believed that Homer was at least partially retrievable by
means of the separation of older material—at least as far as Wolf would identify
it—from later editorial intrusions.[17] As we will see in later chapters, this methodically comparative approach bore a strong resemblance to the talent for

sorting and classifying that Young would later bring to bear on many projects, including his work on the Egyptian scripts. In a cognate manner, Wolf had combed the Venice scholia for inconsistencies of tone, diction, word choice, punctuation, and meter that signaled minute departures from a more authentically Homeric presentation, at least by his lights.[18] The result was dry, tedious, and not especially rewarding to those, like Campbell, who doggedly sought evidence in ancient epics to underwrite claims to a very old, literate civilization with which he could identify on grounds of national heritage.[19]

Campbell would certainly have been wary of Wolf's interpretation of the Homeric epics. To him it was unthinkable that the poems as they stood did not fully convey the rich texture of antique life, and his own verses were steeped in the sentimentality he wished to attribute to Homer. In 1803, Campbell's offerings included a lyric poem about the battle of Hohenlinden. "The combat deepens. On, ye brave, / Who rush to glory, or the grave!" is an apposite example of his style. Another of his efforts, "Lochiel's Warning" (1802) effected a similarly lugubrious commemoration of the bloody defeat of Charles Stuart's forces at the battle of Culloden Moor, with consequent crackdowns on Highlanders. Just as he sought to evoke the emotions of battle in his own poetry, Campbell similarly thought to hear the cries of the Achaeans in the Homeric epics. From the very sonority of the ancient words, he opined, anyone might glean their emotional meaning. Concerning Homer, he once asked a friend, "Don't the words carry the meaning to your ear?" She replied disappointingly that the poem, "very fine [as] it is," nevertheless conveyed "no distinct ideas to my mind."[20]

Campbell's heated lyricism appealed to contemporary moods that followed the loss, in the eighteenth century, of the traditional Highlands world. This upheaval found expression in a controversy over the provenance of *Fragments of Ancient Poetry, collected in the Highlands of Scotland, and translated from the Galic [sic] or Erse language*, published by the Scottish poet James Macpherson in 1760. Purported to be translations of an ancient Gaelic saga, the *Fragments* featured a Scots hero known as Ossian and implicitly glorified an ancient Scots civilization.[21] Oriented ambivalently toward London, Edinburgh's literary establishment welcomed the *Fragments* as proof of a local poetic genius independent of England.[22] Decades before Campbell's *Hohenlinden*, Macpherson's *Fragments* sang a remarkably similar tune: "Oscur my son came down; the mighty in battle descended . . . There was the clashing of swords; there was the voice of steel. They stuck and they thrust; they digged for death with their swords . . . Here rest the pursuer and the pursued."[23]

Such strong stuff well suited a period of imperial expansion and the resistance such expansion inevitably induced. Even Napoleon admired the epic, and rumors of its inauthenticity did nothing to stanch its popularity.[24] As late as 1810, one could find echoes of Ossian in works by authors such as John Grieve, an Edinburgh businessman not overly burdened with work, who wrote *Lochiel's Farewell*, a lament for the Highlands after Culloden, that "Land of proud hearts and mountains gray, / Where Fingal fought and Ossian sung!" That these works were sufficiently popular to attract publishers attests to a profound need for recognition and solace in light of the failed struggle against intrusions from the South. The counterresistance—to Ossian, and by extension to an independent Scots literary tradition—also persisted for decades after Macpherson's *Fragments* appeared.[25] By the time Longman arranged his dinner party, the reaction had become so common as to constitute a tradition in itself—one in which Young, by temperament, training, and family connections, was firmly rooted. As a doctor and natural philosopher, he was an increasingly prominent representative of philosophical London; as the scion of a prominent Quaker family with firm ties to London's literary establishment, he was bound in yet another way to the values of the metropolis. Young would have found it hard to agree with Campbell about Wolf's approach to Homer for these reasons alone. But there were others.

Take Samuel Johnson. A frequent household guest of Young's great-uncle and major patron Brocklesby, Johnson was fiercely skeptical of claims made for the authenticity of the Ossian poems. The popularity of Macpherson's *Fragments* prompted Johnson to remark that the "Scots have something to plead for their easy reception of an improbable fiction: they are seduced by their fondness for their supposed ancestors. A Scotchman must be a very sturdy moralist, who does not love *Scotland* better than truth; he will always love it better than inquiry: and if falsehood flatters his vanity, [he] will not be very diligent to detect it."[26] His animadversions on the *Fragments*, set forth in his *Journey to the Western Islands of Scotland* (1775), led to a famous dispute with Macpherson over the work's authenticity.[27]

As a lexicographer invested in the production of a standardized English, Johnson was committed to a normalization of writing that was incompatible with enthusiasm for excursions into oral forms and vernaculars.[28] His view was also consistent with commitment to a special kind of literate culture that tallied with England's long-standing admiration for ancient Greece. Johnson may not have been as enamored of Greek as many of his contemporaries, but he had taught it as a young man, and others confirmed his considerable knowledge. Admirers of Johnson's Grecianism included Edinburgh's professor of Greek, Andrew Dalzel, who later became a good friend of Young's.[29] In keeping with this privileging of Greek literature, particularly the Homeric epics, Johnson asserted that the Ossian epic must be a sham because it was simply "too long to be remembered, and the language formerly had nothing written"—a point that anticipated Wolf's use of the Homeric epics as evidence for the centrality of writing in the transmission of any ancient language. As we will see, this point may also have proved troublesome for Campbell in his debate with Young at Longman's.

Those who, like Johnson, could credit the *Fragments* with neither authenticity nor literary value frequently presupposed that literate cultures were more civilized than oral ones, a view that raised few eyebrows in Johnson's day. Most followed John Locke in seeing language simply as a means of expressing ideas derived from experience with the natural world; languages had emerged as soon as primitive people could emit cries that went beyond expressions of pain, pleasure, or fear. To be sure, writing per se was not of the same urgent concern as language itself. But because inscription systems necessarily involved complex syntax and semantics, it followed that only linguistically advanced cultures could have developed the art.[30] This view of language's evolution and its implications for writing persisted in various guises through the 1830s in England, most famously in John Horne Tooke's well-known philosophical-philological treatise *The Diversions of Purley* (1785–1805), and it became for a time the dominant linguistic theory in Britain.[31]

Alternative views did exist, however. At least one of them, expounded by the prominent Scottish jurist James Burnett, Lord Monboddo, was substantial enough to compel Johnson's serious attention. Monboddo, whom Young would meet, rejected the idea that language resulted from the sonic expression of ideas grounded principally in experience of nature.[32] In his *Origin and Progress of Language* (1773), Monboddo instead argued for language's basis in culture and history. "I maintain," he asserted, "that the faculty of speech is not the gift of nature to man, but, like many others, is acquired by him; that not only must there have been society before language was invented, but that it must have subsisted a considerable time, and other arts have been invented, before this most difficult one was found out." Indeed, "without the closest intercourse of social life, it appears to me impossible, that an art of such refinement as the art of language could have been discovered."[33]

The advent of a symbolizing process did not guarantee that expression would be efficient. On the contrary, Monboddo remarked, primitive languages were often cumbersome. The first true words could not have been the monosyllabic cries that Johnson, following Locke, had imagined.[34] "That the first articulate cries expressed the names of things, I can no more believe," Monboddo wrote, "than that the neighing of a horse, or the lowing of a cow, is a name for any thing." Rather, these primitive utterances must have been polysyllabic—as one could see, Monboddo averred, in languages such as those of the indigenous Huron of North America.[35] He maintained that their language consisted of "words of a remarkable length" formed from smaller units in ever-longer concatenations that dwarfed those "found in the languages of civilized nations." Speakers of these syntactically unsophisticated languages simply agglutinated noises to create meanings, using tone and gesture to express fine distinctions nonlinguistically. In perfect speech, Monboddo wrote, "There should be no obscurity or ambiguity . . . otherwise the principal end of language cannot be answered, which is to convey the meaning to the hearer. In both these last articles the barbarous languages are very deficient and they supply the defect, we are told, by accents or tones of the voice, and no doubt by gestures, or action of the body." Putatively more civilized speakers, such as the ancient Greeks and even the Romans, relied on the condensing power of syntax to accomplish the same thing. Replacing concatenation and gesture, syntax made primitive theatricality redundant; rhetorically stylish expression—the skilled deployment of syntactical structures, or the "art of composition," as Monboddo called it—supplanted a primitive gestural lyricism.[36]

Touring Scotland, Johnson met Monboddo, remarking that "the magnetism of his conversation easily drew us out of our way."[37] In a 1773 letter to Hester Thrale Piozzi, Johnson noted a fundamental conflict between their views of language and civilization, relating that in conversation, "Monboddo declared boldly for the Savage and I perhaps for that reason sided with the Citizen." The dispute with Monboddo pervaded Johnson's Highlands tour. Johnson could not accept that primeval languages were constructed in Monboddo's fashion. Such a thing could not have obtained in the dawn of civilization or indeed among the contemporary barbarians Johnson identified as Highland Scots. He described their language as "the rude speech of a barbarous people, who had few thoughts to express, and were content, as they conceived grossly, to be grossly understood."[38] For Johnson the pitting of "Savage" against "Citizen" equated to the opposition of language primeval to language civilized. Monboddo's linguistically developing primitive contrasted sharply with Johnson's image of the civilized exemplar, characterized above all by the possession of sophisticated—and ipso facto written—language.[39] Unsurprisingly, Johnson was no more enamored of Monboddo's theory of language than of Macpherson's Ossianic epics. "Monboddo does not know that he is talking nonsense," Johnson remarked to James Boswell.[40]

Despite his differences with Locke and Johnson, abbreviation was important even to Monboddo, who understood the art of composition as exerting the relevant concentrating force. For all three, and their successors, the more closely a language approached the quick economy of thought, the more civilized its speakers were believed to be.[41] Abbreviation, a method of compression that made expression more efficient, played an important role in virtually all theories of linguistic development around the turn of the century. Tooke, whom Young admired, promoted a similar view just a few years later.[42] Looking back on the period and its intellectual fashions, William Hazlitt in 1835 described Tooke's linguistics in terms of the new chemistry: "Mr. Tooke . . . treated words as the chemists do substances; he separated those which are compounded from those which are not decomposable. He did not explain the obscure by the more obscure, but the difficult by the plain, the complex by the simple. This alone is proceeding upon the true principles of science." Even language might have become more efficient over epochs, or so the new philology suggested.[43]

When Longman gathered his guests, Johnson and Macpherson were long dead, and Monboddo's corpse was not so fresh, either. But their influence persisted as part of a lingering argument over the uses to which antiquity should be put. To reject the possibility of an oral tradition capable of rendering epic sagas into writing was to reject Macpherson, Ossian, and the general idea of an independent Scots poetic genius arising from a distant past. Wolf certainly opposed this idea in its narrowest form, as a principle of classical philology: Homer's poems could not stand on their own as proof of the existence of a literate culture in Homer's day. But if philologists were the only authoritative interpreters of ancient texts, then the ambitions of those like Campbell, who sought to anchor a national identity in the written remnants of an ancient epic, were insupportable. Although patriotic Scots robustly reciprocated English disdain, mere contempt proved a poor defense against Wolf's philology.

In this context, Campbell might easily have transmuted a debate about ancient poetry into a personal affront. In doing so, however, he seems to have crossed a line. Scott, who had freely interpreted the lowlands ballads he and Ellis collected, might have taken a position similar to Campbell's on the Homeric question, and surely Ellis had a stake in the argument as well. Yet, that night at Longman's, as Young and Campbell argued, neither Scott nor Ellis intervened. Campbell "began to wax somewhat too earnest," Irving reported, and, "finding that he did not attract all the attention to which he evidently thought himself entitled, he started from his seat at an early hour, and quitted the room with a very hasty step."[44] Campbell's enthusiasm for recovering a golden age of Scots heroism could not have appealed to Young, who had little interest in a people who, like the antique Picts, hardly reached what Young and others of his time took to be the high social and intellectual order epitomized by Greeks of the classical era. We shall see that the inhabitants of ancient Egypt were, to Young, little different in this respect from Campbell's Scots.

That night at Longman's, Young's reading of Egyptian hieroglyphs—the work that would secure his popular fame in England and his notoriety in France—was still in the distance. As he did with Brougham, and as he had just done with Campbell, Young would once again find himself locked in battle with a kind of alter ego. And yet, though he had vanquished Campbell that evening and would, in time, vanquish Brougham, he would not fare so well with Champollion. Young's failure stemmed from assumptions about language and writing that led him to disdain a civilization he saw as depraved. As we shall see, Champollion brought to the problem a considerably wider and more sympathetic perspective that would in the end secure his success.

CHAPTER 2

IN THE CLASSROOM
OF NATURE

Thomas Young suffered a prodigy's childhood: his intellectual reach rarely exceeded his grasp, and the adults responsible for his education routinely scrambled to find appropriate teachers and schools. Born in 1773 in Milverton, Somerset, to a strict Quaker family, Young had learned to read by the age of two; at four he had "read the Bible twice through" and was soon memorizing poetry. As he grew, so did his awareness that those around him had limitations he did not share. His voracious reading served him well in this regard, providing essential recognition his immediate family could not.[1] He spent most of his first six years at the home of his maternal grandfather, Robert Davis (d. 1785), "a merchant of great respectability," and supplemented his attendance at the local village school with private instruction by his aunt, Mary Davis.[2] When he was six, Young "learnt by heart" Oliver Goldsmith's *Deserted Village*. Dedicated to his friend Sir Joshua Reynolds, whom Young would meet years later in London as an acquaintance of Brocklesby, Goldsmith's lament for the idyllic imaginary village of Auburn decried its eventual destruction by the forces of untrammeled wealth:

> The man of wealth and pride
> Takes up a space that many poor supplied;
> Space for his lake, his park's extended bounds,
> Space for his horses, equipage, and hounds;
> The robe that wraps his limbs in silken sloth,
> Has robbed the neighbouring fields of half their growth;
> His seat, where solitary sports are seen,
> Indignant spurns the cottage from the green;
> Around the world each needful product flies,
> For all the luxuries the world supplies.
> While thus the land adorned for pleasure all
> In barren splendour feebly waits the fall.[3]

Goldsmith's sentiments would have been congenial in a Quaker household, which deplored extravagant display, though its deprecation of trade was rather out of step with Young's grandfather's substantial success as a merchant.

Rural melancholy soon gave way to Latin grammar, a curricular expansion consistent with educational expectations for male children of Young's social class, and consonant as well with his grandfather's interest in classical learning. The exchange of lyric poetry for Latin grammar was accompanied by heightened feelings of practical obligation, likely also imbibed from his worldly grandfather, who placed firm limits on the extent to which Young was permitted to absorb himself in ancient epics and dead languages. From an early age, Young's enthusiasm for classical learning mingled with an awareness of the practical advantages offered by direct study of nature. This tension would mark Young's entire career. Reflecting years later on his early education, he wrote, the "principles which I imbibed, and the habits which I formed under the guidance of these dear and excellent relatives, have more or less determined my character in future life."[4]

A closer look at Young's reading from this early period reveals a pattern of ambivalence about the value of classical learning. While the study of classical languages was acceptable, substantial engagement with the ideas of antiquity risked charges of pretension, which meshed poorly with Quaker ideals of simplicity, industry, and naturalness. Young's earliest reading assignments, completed under the tutelage of an otherwise unknown "dissenting clergyman" named Knyston, included *Gay's Fables*, a compendium of satirical verse by one of the eighteenth century's most biting ironists, John Gay, who nurtured a special aversion to scholarly pretension. In his "The Shepherd and the Philosopher," an unlettered shepherd "modestly" avers that, although he knows nothing of classical learning, "From nature too I take my rule, / To shun contempt and ridicule. / I never, with important air, / In conversation over-bear." Direct, industrious Nature contrasts with the overweening "philosopher" who disparages the shepherd by asking:

> Whence is thy learning? Hath thy toil
> O'er books consum'd the midnight-oil?
> Hast thou old Greece and Rome survey'd
> And the vast sense of Plato weigh'd?
> Hath Socrates thy soul refin'd,
> And has though fathom'd Tully's mind?
> Or, like the wise Ulysses, thrown,
> By various fates, on realms unknown,
> Hast thou through many cities stray'd,
> Their customs, laws, and manners weigh'd?

Here we see a tension between the study of the experiences and wisdom of antiquity in contrast to a "rule" that can be learned from the observation of an assiduous Nature inspired by "[t]he daily labours of the bee," which "[a]wake my soul to industry." A paean to hard work in a classroom devoted to lessons drawn from nature, the poem concludes on a distinctly worldly note: "[H]e who studies nature's laws, / From certain truths his maxims draws: / And those without our schools, suffice, / To make men moral, good and wise."[5]

Young would soon consume a great deal of classical material but, like the busy, worldly personification of Nature in Gay's poem, he never devoted himself to ancient learning per se—to learning, that is, outside of works on mathematics and natural philosophy. Although he was hardly alone in this by the late eighteenth century, he found a unique resolution to the tension. The structure of Latin and especially Greek rhetoric mirrored for him the directness of observation of the natural world. From an early age, Young understood language as one of many props in the classroom of nature, substantially separate from the culture in which it was embedded. Nevertheless, despite being one natural object in a world full of them, language also had a distinctive moral valence. Like nature itself, the purest linguistic forms were simple and governed by unpretentious rule—rather like an ideal Quaker home.

Around 1780, Young was sent to a "miserable boarding school" run by a certain King and located near Bristol.[6] There he suffered for a year and half, all the while developing the knack of teaching himself. He first learned arithmetic with the aid of Francis Walkinghame's widely used *Tutor's Assistant*, though it seems he hardly needed the help, for Young reached the end of the book before the master had managed to arrive at the middle. Essentially on his own, Young, now eight, worked his way through John Clarke's treatise on Latin composition as well as Maturinus Corderius's sixteenth-century *Colloquia Scholastica*, which illustrated by example how to carry on a correct Latin discussion.[7] As ever, Young read a great deal: two books of Phaedrus's (Aesop's) *Fables*, *Robinson Crusoe*, simplified stories from Shakespeare, and, importantly, his first substantial introduction to natural philosophy, the extraordinarily popular *Newtonian System*, published by John Newbery. This volume may have particularly appealed, because in it the small master Tom Telescope, "a young gentleman of distinguished abilities," charmed the Countess of Twilight, as well as her several "young visitors," by explaining the elements of nothing less than Newtonian natural philosophy. A direct and didactic youngster, Young would easily have seen himself in Tom Telescope, whose disdain for amusements like cardplaying would have been especially resonant in a Quaker home. "Playing at cards for money," declared Tom, "is so nearly allied to covetousness and cheating, that I abhor it." Instead, Tom preferred "solving problems and paradoxes on orreries, globes, and maps, and sometimes [to play] at natural philosophy, which I think is very entertaining." Young learned about matter and motion through evocative examples involving balls and tops, and about sound and light, based on a discussion of "the senses." For example, Tom explained that sound

> is propagated at a great rate; but not near so fast as light.—I don't know that, says Lady Caroline,—Then your Ladyship has forgot what passed in our Lecture upon Air, replied the Philosopher [i.e., young Tom]; and to confirm by experiment what I advanced, I must beg his Lordship to order one of the servants to go a distance into the park, and discharge a gun.— The gentlemen were averse to this; it being an observation they had made a hundred times; but to gratify the young people, my Lord ordered his game-keeper out; and when the piece was discharged, they had the satisfaction of seeing the fire long before they heard the report.[8]

Mercifully released from the clutches of King's school, Young at eight and a half returned home to spend six months revisiting places of happy memory. During this period, he spent many hours with a close neighbor named King-don, "a man of great ingenuity who, though originally a tailor, had raised himself by his talents and good conduct to a respectable situation in life—being at that time a land-surveyor and also land-steward to several gentlemen in the neighbourhood." In Kingdon's library, Young discovered a three-volume *Dictionary of Arts and Sciences*,[9] in which were illustrated instructions for making intriguing devices, some of which Kingdon, as a surveyor, had on hand. Because the descriptions were quite difficult even for a child as adept as Young, Kingdon's daughters and nephew explained to him the operations of the devices.[10]

In 1782, after a long search, Young joined a Quaker school run by Thomas Thompson in the town of Compton in Dorsetshire. Apart from a six-month interruption in 1784, Young flourished there for four years. As "a man of liberal and enlarged mind," Thompson "possessed a tolerable collection of English and classical books, which his pupils were allowed to make use of."[11] At Thompson's Young also learned how to turn a lathe, grind colors for drawing, and make telescopes, thanks to lessons from the mechanic, Josiah Jeffrey. When Jeffrey left, Young appropriated "some of his employments and perquisites" to sell "paper, copper-plates, copy-books, and colours to my schoolfellows." This entrepreneurialism, coupled with his parents' allowance, enabled him to buy from Thompson "some Greek and Latin books" and Arius Montanus's Hebrew Bible, whose edition was widely used to teach Hebrew, for it also contained Greek and Latin versions of the texts. Years later Young recalled that he "was at that time enamoured of Oriental literature . . . and before I left Compton school, I had succeeded in getting through six chapters of the Hebrew Bible."[12] Already an inveterate list maker, Young kept careful track of his reading, so we can follow his developing expertise in languages from his ninth to fourteenth year. He completed his reading of Phaedrus, begun at Kingdon's, and pressed on to Virgil, an "expurgated" Horace, and Cicero. He studied Greek grammar using the standard Westminster school text, a short and stringent compendium, "the greater part of which I committed to memory."[13] By 1786 he was reading the *Iliad*.

Meanwhile, he pursued his interests in natural philosophy, which were no doubt stimulated by working at lathing and telescopes with Jeffrey, who also lent books. These loans took Young considerably beyond Tom Telescope. One, John Ryland's "easy and pleasant introduction" to Newton, was already too simple for him. But the other, Benjamin Martin's "new and comprehensive system" of the same, was more appropriately challenging. For there he found, among other topics, an extensive lecture on optics, and, Young recollected, "rules for the practical construction of optical instruments." From Martin he learned for the first time about the structure of the eye, its function as an image-forming, organic device, and its most common defects. In small text at the bottom of the pages, Martin developed his account using Newtonian fluxional calculus, which at the time was beyond Young. Curious to understand the fluxional parts of Martin, Young read "a year or two afterwards," that is, probably in 1786, a "short introduction to the subject," most likely by Thomas Simpson.[14]

Young's study of mathematics, which went much beyond arithmetic, revealed a practical orientation. Reading Ewing's 1779 *Synopsis*, he focused on topics related principally to surveying (e.g., plane trigonometry) but omitting the chapter on gunnery, anathema to a Quaker. He studied as well Dilworth's sensible *Young Book-Keeper's Assistant*, replete with examples of how to balance debt with credit.[15] Young supplemented these studies with explorations of "Modern Geometry (meaning Algebra)" in the works of John Ward's somewhat outdated *Compendium*, which he complemented with a more modern, practical work by Charles Vyse.[16]

Even on vacations, Young pursued his interest in practical matters, mechanical devices, and measurement. Visiting the Davis household, he came to know "a saddler of the name of Atkins, a person of considerable mechanical skill and ingenuity." John Atkins was the first professional natural philospher Young had met; in 1784, Atkins's measurements of local atmospheric pressure, temperature, wind direction, and weather appeared in the London *Philosophical Transactions* in a paper communicated by Sir Joseph Banks, the society's president. From Atkins Young learned to use a quadrant to measure heights. Another Quaker in Young's circle, Morris Birkbeck, urged botany on Young, who thereupon combined botanizing with his enthusiasm for lathing and optical devices by constructing a microscope.[17]

Chemistry entered as well when Young read the work of Joseph Priestley, discoverer of "eminently respirable air" (oxygen, or, as he conceived it, "dephlogisticated gas"). Priestley mingled chemical discussion and description with his views on philosophical investigation. The preface to the volume that Young likely had, and which "delighted [him] greatly," included an apposite remark that insisted on two of a student's larger rights—to self-determination and to taking responsibility for error—which Young's own works in natural philosophy and, later, Egyptian writing would later exemplify:[18]

> [A] person who means to serve the cause of science effectually, must hazard his own reputation so far as to risk even *mistakes* in things of less moment. Among a multiplicity of new objects, and new relations, some will necessarily pass without sufficient attention; but if a man be not mistaken in the principal objects of his pursuits, he has no occasion to distress himself about lesser things.[19]

It is not hard to discern the sources of this remark's appeal to Young. Priestley's autonomy, for one, would have been congenial. A Unitarian minister with unorthodox religious and political views, Priestley cared little for public opinion. His political commitments included support for the "American War," as it was known in Britain, and for the French Revolution—two unpopular causes that contributed to Priestley's 1794 emigration to America three years after an angry Birmingham mob incinerated his house. In Priestley, perhaps Young saw a role model—a person whose integrity stemmed from a set of lucidly understood values that also informed his approach to natural phenomena.

At Thompson's Young had begun to study French and Italian, assisted in the latter by a schoolmate named Fox who "had made himself master" of the language. Avid to learn "Oriental" tongues besides Hebrew, Young sought the help of a dissenting minister, Joshua Toulmin, whose Baptist congregation met at the Mary Street Unitarian Chapel in Taunton, not far from the Davis home in Milverton. Like Priestley, Toulmin opposed the "American War" and was attacked for supporting the French Revolution.[20] "That most excellent man, Mr. Toulmin," Young recalled, lent him "grammars for 'Chaldee,'" the colloquial term for Aramaic, the language spoken in Israel at the time of Christ, as well as "Syriac and Samaritan," both Aramaic dialects. Toulmin was himself a prolific writer of books and essays, and from him Young also borrowed "The

Lord's Prayer in more than 100 Languages, the examination of which gave me extraordinary pleasure." And for the first time Young read a work by Gregory Sharpe, also lent by Toulmin, on the origin and nature of languages, which took an evolutionary view of their development.[21]

Sharpe died two years before Young was born; by the time Young was old enough to read him, he was already nearly forgotten. But because Sharpe viewed linguistic evolution in terms of changes specifically in orthography, his posthumous contribution to Young's intellectual development is worth discussion. Although a youthful "indiscretion" had removed Sharpe from Westminster sometime around 1729 or 1730, the scandal did little to undermine his respectability. By the time he died, in 1771, he was chaplain to George III and master of London's ancient Temple Church. Sharpe also wrote a great deal, ranging from sermons to works on classical subjects, in which he became so respected an authority that he eventually became the director of London's Society of Antiquaries.

Where Sharpe's linguistic interests overlapped with natural philosophy, they also tended to land him in the midst of religious controversy. His defense of Newtonianism extended to a planned treatise against the Hutchinsonians, some of whom, following their leader John Hutchinson, regarded Newton's concept of gravity as tending to irreligion. Hutchinson insisted that proper philosophy could be found complete in the Mosaic texts, which further led him to argue that Hebrew should be read without points (vowel marks) since these must have been added a millennium and a half after Moses by Jewish scholars, the Masoretes, and were therefore not actually used by the ancient Hebrews. Sharpe, in contrast, held that the vowel points were merely unnatural intrusions into an otherwise simple linguistic structure.

Although Sharpe's treatise against Hutchinson's anti-Newtonianism never reached print, we have remarkable visual evidence of it, for he had commissioned an etching by Hogarth that "depicts the dead Hutchinson, now transfigured into a lunar witch, dispensing a torrent of fluid nonsense—obeying the law of universal gravity—that washes away the corpses of his disciples (the mice), who died as a consequence of reading his book. The few remaining disciples continue to attack Newton's *Principia* and his telescope, but to no avail" (fig. 2.1).[22] Although Young likely never knew of Sharpe's unpublished brief, he might have heard about Hutchinson since the latter despised Quakers as exclusionist blasphemers.[23] In any case Young did thoroughly absorb Sharpe's views on languages—and linguistic representation in script.

Hebrew, in Sharpe's view, not only came closest to the earliest language but was also uniquely related to human vocal apparatus. Moreover, Hebrew syntax developed by mating the affordances of this natural apparatus with orthography. Following the sixteenth-century philologist Isaac Casaubon, Sharpe asserted that students would gain "a more perfect knowledge" of Greek by studying Hebrew because "the greatest part of the Greek language is most evidently derived from the Oriental dialects." Continuing in a romantic vein sure to capture the adolescent Young's attention, Sharpe assured readers that with Hebrew would come Arabic, with which one could "search after the remains of ancient Aegypt, Greece and Rome, which may still be found, perhaps in the manuscripts of the East." This may have been Young's first encounter with the language of ancient Egypt. Even if it were not, Sharpe's analysis evinced a way of thinking about languages and scripts that Young "studied with great diligence."[24]

FRONTIS-PISS.

FIGURE 2.1. *Frontis-piss* etching by Hogarth, intended for Gregory Sharpe's unpublished polemic against John Hutchinson. © The Trustees of the British Museum.

Hebrew, Sharpe insisted, was so simple that "one may be forgiven for calling it the language of nature, or the first language of the world."[25] The notion that Hebrew may have been the original language was hardly new, but Sharpe went on to sketch a plan for linguistic evolution in which Hebrew, stripped of points, exemplified the structure of the first languages.[26] Though he mentioned neither Locke nor Condillac, Sharpe wrote in a broadly Lockean vein by setting aside the notion that the first language was "a gift from the creator." The "creator," Sharpe contended, generated only an innate ability to assimilate the world's sounds and assemble them into meaningful linguistic structures. Instead of being a divine gift, language had its roots "in the wild notes of animals untaught, and every other sound from things animate or inanimate, at first imitated by man, originally endued with powers and propensity to imitate, and then improved by art and use, and at length diversified, by a thousand different circumstances, till all traces of its original form seem obliterated or over-run, so as not to be discoverable."[27] Language emerged out of the "necessity and convenience" required to deal with natural circumstance.

Sharpe waffled as to just how the creator communicated with Adam. Departing from the traditional view, Sharpe did not think that Adam spoke Hebrew. Sharpe allowed that Adam may have been "endowed with one tongue" but, in those earliest times before Babel, there may have in fact been "more than one original language," such as Chinese or "most probably the old Aegyptian."[28] Sharpe also conjectured a match between language and vocal apparatus. Here Hebrew provided the medium, as the language was traditionally held to be the most perfectly correlated with that apparatus.[29] "Man has natural sounds, as every other animal has," Sharpe opined. The first words for animals were therefore human imitations of animal sounds. Noting that Hebrew has clear examples of sound referents, Sharpe transliterated אדק (partridge) as *quera*, inserting the intermediary vowels, and claimed that it "happily expresses the note of a partridge, when she is calling to her young." He provided many such examples to drive the point home. Significantly, nature, and not some creator, supplied the cornucopia of sounds that could be turned into representation; "art" was necessary to produce syntactical language. And about that Sharpe had a theory.

Simplicity was the key to Sharpe's conception. In creating novel words humans combined available elements in the most efficient fashion. These combinations were arbitrary but short, "for ease of memory, and readiness of speech." Hence Hebrew's primitives "seldom or never exceed three letters." Simple combinations based on animal sounds preceded more complex ones and leaned heavily on consonants. To generate complexity, Sharpe invoked combinatorics, a move that surely appealed to Young, who was increasingly interested in mathematics and mathematical philosophy. According to Sharpe, when letters are taken in pairs, an alphabet of twenty-two letters, such as Hebrew, yielded too few primitives, while four at a time produced too many. By means of an elaborate arithmetical argument, Sharpe found that three-letter roots provided just the right number of words—245,410 by his unusual count.[30] Over time inflections and other speech forms would be added, yielding complex syntactical structures. "From this draught," he wrote, "rude and imperfect as it is, may be traced the several steps or outlines of language, from the natural notes of man in his infant state, through its progress to perfection."[31]

The human vocal apparatus played a role in language formation. Anatomical features worked in concert, rather like a "musical instrument," on which the actions of "the skillful artist" would make it "speak a language that shall command all the passions of the hearer," while the less skilled "shall offend the ear with intolerable harshness and discordance of sounds, by striking false notes, to no time or tune." Defects in the vocal organs, as well as in hearing, could alter linguistic structures. Climate also made a difference. In warm climes people spoke with open mouths, he believed, and so their "languages abound with gutturals," whereas in cold climates, closed mouths produced monosyllables. The talents of the anatomist were necessary to apprehend vocalization in all animals, including humans, according to Sharpe. Through anatomy one could learn the ways in which "the use and contrivance of the larynx, and other organs of speech . . . account[ed] for difference of voice and note in different animals."[32] Sharpe's anatomical focus would later find an echo in Young's development of the "vocal circle," to be discussed below.

Just as language resulted from the transformation of animal utterances into linguistic equivalents, so would the first figural representation of speech have been an image or "rude draught" of the object.[33] This was eventually followed by the representation of sound itself through the invention of an alphabet. "It is amazing to think," Sharpe wrote, "how the objects of one sense, by the use of letters, came at first to be transferred to another sense; how sounds could have been transferred from the ear to the eye; to think how men could contrive a method of conveying, in their own terms, their own thoughts to latest posterity!"[34] Here, Sharpe's views owed much to an important contemporary, the notoriously combative bishop of Gloucester, William Warburton, who advanced similar claims in his controversial *Divine Legation of Moses* (1738). At this point, Young had not yet read Warburton, but he certainly knew of the latter's claims, for Sharpe had recapitulated them.

Published a decade before Sharpe's treatise on language, Warburton's *Legation* aimed primarily to prove that Israel was divinely instituted and governed. To do so Warburton inverted Enlightenment claims that religion, via the promise of an afterlife, had been instituted in order to secure moral behavior through fear. Since Moses never mentioned an afterlife, Warburton argued, then Israel must have enjoyed direct divine support; otherwise, it would have collapsed into immorality.[35] The *Legation* included an account of the evolution of writing intended to further Warburton's claims concerning ancient Israel. According to Warburton, writing developed in three stages as written expression became increasingly abstract, concise, and efficient.[36] At each successive stage, particular rules governed the corresponding gain in abstraction and precision. Theological provocations aside, Warburton's account of the development of writing proved lastingly influential.

The first stage, Warburton said, was "picture-writing," in which a pictogram stands for an object or a concept, with the pictogram bearing a figurative resemblance to its referent. One familiar recent example of Warburtonian picture-writing might be the series of images one finds included with assemble-it-yourself furniture. This form of writing did not always involve, in modern terminology, logograms—that is, they did not always represent specific words or phrases. A logogram of a bird might stand for the word "bird," or one of a sun just above a line for the phrase "risen sun," regardless of how the word is vocalized in any particular language. By contrast, though a drawing of a kneeling man with upraised arms might stand for a logogram that represents the phrase "a man at prayer," it would not do so according to Warburton's notion of picture-writing. Such an image would instead signify a situation and not a specific word or phrase. A sign that looks like a kneeling person with arms upraised might stand for any of a variety of ways to speak a condition in which someone, kneeling upon the ground with arms upraised, prays to a god. Such a written sign, or sign sequence, that does not produce a specific word or phrase—Warburtonian picture-writing—is best termed a *sematogram* to avoid confusion with signs representing words or phrases.[37]

Because these notions can raise complex issues that will be important for our later discussion of Champollion's work, if not for understanding Warburton or Sharpe, they are best introduced here. The vocalization of a logogram or logogrammatic sequence obviously depends upon the language of the reader. A sign or sign sequence intended to represent the word for a feathered, flying animal would be vocalized as "bird" in English and "oiseau" in French. Moreover, the order and typology of the signs of an inscription composed of such logograms, even though nonphonographic, may nevertheless depend upon the structure of the underlying language. German, for example, places the verb at the end of a sentence whereas English places it immediately following the subject, so that the English sentence "you caught the yellow bird" would run "you the yellow bird caught" in German order. A logogrammatic sequence written to represent a complex sentence ordered according to English rules might as a result be nearly incomprehensible to a German reader even though the sign sequences do represent words or word phrases that have German cognates. An inscription composed of such signs might yield the precise meaning of the inscriber only to a reader who knows the inscriber's language. Conversely, the syntactic structure of a logogrammatic sequence written by the speaker of an otherwise unknown language may provide information concerning the language itself. This, we shall see, constituted a central, and quite early, element in the young Champollion's approach to the Egyptian scripts.

Rules entered Warburton's second stage of writing when the writing of the first stage became too "inconvenient," at which time "the more ingenious and civilized nations" invented "methods to abridge it." As these transformations occurred, the original, unregulated technique became one that involved "both a picture and a character." To explain this, Warburton divided the resulting "abridgment" into three distinct kinds that emerged successively. In the first, the earlier sematograms were narrowed in a way that brings writing to speech through the representation of set words or phrases, though not specific vocalizations, that is, the introduction of logograms. Warburton described this first transformation as making "the principal circumstance of the subject stand for the whole." He gave several instances, such as a grapheme of "an armed man casting arrows," which could represent the word "tumult," or "two hands, one holding a shield, and the other a bow" to connote the word "battle." Our previous symbol of the kneeling man might now represent the word "prayer." This might be termed a graphical version of metonym, for here a possible attribute or instance of a subject stands for the whole.[38]

Warburton's next transformation involved representation of an entity or an entity's power by a "real or metaphorical" means through which it might act. An eye might stand metaphorically for the phrase "God's omniscience," while an eye and scepter together might stand for a monarch, according to Warburton's notion of realistic representation. This is similar to a graphic form of synecdoche in that a real or metaphorical part stands for the whole. Using modern terminology, we might say that Warburton had replaced rule-less sematograms with two classes of logograms of increasing abstraction. In the first, the grapheme represents some thing or things that are normally involved in the subject. In the second, it represents a word or phrase for an entity's power by means of a part associated with the power's exerciser. The step from first to second increased the level of abstraction.

Warburton's third transformation made "one thing stand for, or represent, another, where any quaint resemblance or analogy, in the representative, could be collected from their observations of nature, or their traditional superstitions." He gave as examples a "sunrise," represented by the "two eyes of the crocodile, because they seem to emerge from the head," or "a wife who hates her husband, or children who injure their mother, by a viper." The abstraction is greater at this third stage, since the grapheme now bears little direct relation to the represented subject. Warburton believed Egyptian hieroglyphs had reached this stage. They were accordingly highly abstract and had logographic significance, but they were not phonographic in the manner of an alphabetic or syllabic system. That is, hieroglyphs did represent elements of speech and could therefore be vocalized in a particular way—but only by a reader who knew the language of the inscriber, since the precise ways in which a sequence of hieroglyphic logograms should be spoken would of course have depended on the grammatical characteristics of the inscriber's language.

According to Warburton, language developed out of a sort of sign system in which the sign consisted of a physical action that referred symbolically to an event. For support, he adduced the examples of the Hebrew prophets and the Oracle of Delphi.[39] Over time, "speaking by action was smoothed and polished into an apologue or fable," which "corresponds, in all respects, to writing by hieroglyphics, each being the symbol of something else understood." Like writing, speech had evolved through a process of increasingly abstract representation as simile was displaced by metaphor. "Thus we see," he concluded, that "the common foundation of all these various modes of writing and speaking, was a picture or image, presented to the imagination thro' the eyes or ears; which being the simplest and most universal of all kinds of information . . . we must needs conclude them to be the natural inventions of necessity."[40] Whether vocalized or graphically produced, communication developed according to a pattern of increasing abstractness from a foundation in natural signs. For Warburton this implied that Egyptian hieroglyphs had not been invented by priests "to conceal their knowledge," as many others, most famously Athanasius Kircher, thought.[41] On the contrary, hieroglyphs constituted a highly developed system intended to represent the words and phrases of the Egyptian language.

Unlike Warburton, Sharpe did not provide a detailed account of the evolution of writing, though he did think that alphabetic structures emerged through the transformation of pictograms, with their iconographic significations, into representations of sounds "as they are formed by the organs of speech."[42] Sharpe gave, as an example, the first Hebrew letter, which he imagined to have originated as a pictogram for an ox. It was intrinsically connected to an animal sound, for "the sound of the first letter of the alphabet is the first sound of animals. The name of it א aleph (alp) implies priority and power, it signifies the ox, and the form of the letter bears some resemblance to the head of that animal, which is the chief of those that man is more immediately concerned with for food."[43] Other letters had similar origins in vocalizations and representative pictograms, according to Sharpe.

Young would read Warburton eventually, but Sharpe's work prepared him to think about the origins of language in connection with vocalizations, and the earliest writing as involving pictograms of objects. The second of Sharpe's *Two Dissertations* corresponded to this way of thinking, for there he argued that Hebrew should be learned without the vowel points added for pronunciation in later centuries. In this Sharpe agreed with the otherwise objectionable Hutchinson, but for the very different reason that, in Sharpe's opinion, vowel points obscured the evolution of the earliest writing. It is worth noting that Young also read Charles Bayley's Hebrew grammar, which viewed vowel points as practical necessities, "an *essential* part of the language."[44] For, Bayley noted, even those who rejected points admitted that they were necessary for pronunciation, regardless of whether they were original to ancient Hebrew.

In any case, no one had proposed an alternative to what he asserted would otherwise be an immense number of possible vocalizations.[45]

Young's studies were coalescing into a unique amalgam that powerfully structured his developing views of both nature and language as governed by rules. Sharpe had introduced Young to the idea that language and writing evolved similarly from natural signs linked to the leading consonantal sounds of human and animal vocalizations, with point-deprived Hebrew as the primary, and perhaps original, exemplar. At the same time, Young learned from Bayley the ways in which vowels could be annexed to such a structure. Sharpe's was not the only work that led Young into linguistic origins, for he also "read through the greatest part" of Sir William Jones's 1771 Persian grammar, in which Jones hinted at a common ancestry linking English and Old Persian.[46] Young's education had mingled Greek, Latin, Hebrew, Persian, and Arabic, as well as the origins of languages and writing with devices optical and mechanical, not to mention algebra, fluxions, botany, chemistry, and, all along, intimations of the links between anatomy and speech. Young would soon encounter these rules in more developed form when he read through Newton's *Principia* and his *Opticks* and delved more deeply into mathematics and Greek literature.

*

In 1787, at the age of fourteen, Young moved away from home for another five years. Once again, Quaker connections effected the change. Young's grandfather, Davis, had good relations with David Barclay, a successful Quaker banker in London who had acquired a manor, Youngsbury, in Hertfordshire, thirty miles north of the metropolis. With his first wife, Martha Hudson, Barclay had one daughter, Agatha (b. 1753), who married Richard Gurney, also from a Quaker banking family.[47] In 1775 Agatha gave birth to a son, Richard, named for his father. After taking charge of his grandson's education in 1787, David Barclay sought a suitable companion to join Richard in his studies.

Young would be selected for the role, but not before other connections had been forged. Barclay's younger sister, Christiana, had married Richard Gurney's paternal uncle, Joseph (b. 1729), in about 1757. He died in 1761, leaving her with four-year-old Priscilla, their only child. In 1787 Priscilla (d. 1828), suffering from some disorder, was advised to spend several years in the countryside. Priscilla set out for Minehead on the invitation of Young's devoted aunt Mary, who happened to be Priscilla's close friend. At Minehead, Priscilla met Thomas, who made such a wonderful impression that she later wrote to Barclay about him, no doubt recommending Thomas as an excellent companion for Richard; her position was seconded by her stepfather William Watson, to

FIGURE 2.2. (*Left*) Hudson Gurney in later years (The Royal Society) and (*right*) John Hodgkin (the Wellcome Collection https://wellcomecollection.org/works/gfwrcaat#licenseInformation).

whom her mother had been remarried since about 1772. Barclay thought the idea a good one, and Thomas was sent to Youngsbury.[48]

Young arrived to find his new arrangement in shambles, as the tutor engaged by Barclay had, at the last moment, obtained a better position.[49] Although a new tutor would soon be found, during the interregnum—and despite being little more than a year older than Richard—Young stepped into the missing tutor's shoes. Teaching the younger boy Latin and Greek, Young returned to those two languages with his understanding newly enriched by his studies of Hebrew and the nature of language. At the same time, he pursued interests in mathematics and natural philosophy. During his first two years at Youngsbury, George Peacock records, Young read "the whole of the Principia and Optics [*sic*] of Newton,"[50] the latter in a Latin edition. He also worked through an annotated edition of Euclid's *Elements*, thoroughly assimilating its synthetic methods, and he advanced his knowledge of algebra, reading texts that took him beyond what he had learned from Simpson's short piece on fluxions.[51] He also studied Thomas Simpson's full treatise on the fluxional calculus, to which he had been introduced earlier.[52] Chemical and botanical studies rounded out the curriculum.

A new tutor, John Hodgkin, arrived in 1787. According to Hodgkin, Barclay had convinced him to join the boys by noting that the situation would offer him time to pursue his interests in classical studies and that he would be aided by "deriving some advice and assistance in them, from the extraordinary youth," that is, Young, "whose stability of conduct and intensity of application seemed to place every desirable object of literary or scientific pursuit within the reach of his astonishing mental powers."[53] Hodgkin, also a Quaker, was only seven years older than Young, which no doubt facilitated the solid relationship that rapidly developed. Hodgkin remained with the boys at Youngsbury until 1792.[54]

During this time, Young kept a journal of his studies. Although the journal, written in Latin, disappeared in the years after Peacock used it to write his biography, Peacock's comments indicate that, even at this early stage, re-encoding foreign scripts helped Young understand pronunciation and grammar. According to Peacock, the journal records that Young had "written out specimens of the Bible in thirteen different languages," including "English, French, Italian, Latin, Greek, Hebrew, Chaldee, Syriac, Samaritan, Arabic, Persian, Turkish, and Aethiopic." The "Eastern languages" were, Young effused, "beautifully written"—and, significantly, were provided with transliterations into the Latin alphabet, to assist pronunciation. In Sharpe's *Dissertations* Young had read that if "we know the sound of the Greek and Roman letters, we shall not be long strangers to that of the Samaritan and Hebrew." Sharpe had also noted that one can "derive great light and assistance from the proportion of letters in each language, when taken separately, or when one language is compared with another"—a remark that tallies well with Young's efforts years later in comparing ancient Egyptian writing with that of Coptic.[55]

Young focused on the syntax and grammar of Greek and Latin, poring over the works he had previously studied, as well as new sources, including the "Port-Royal Greek Grammar" by Arnauld and Lancelot, which to some extent anticipated their principles, subsequently published in general form, for a universal grammar—an idea whose appeal would have been irresistible to Young, who was busily compiling specimens of many different languages for the purposes of comparison.[56] Young continued to list every book he studied, including Greek and Roman plays, orations, letters, and a book on Latin prose composition that "gave him a very nice and even critical perception of the principles of the Latin language."[57] Then, from 1790 to 1792, Young, working closely with Hodgkin, cemented a connection between the exposition, structure, and orthography of Greek that would, we shall see, inflect his attitude toward the character of ancient writing, in particular Egyptian hieroglyphs. Hodgkin had begun a work on proper Greek calligraphy—on, that is, the best way to write Greek "with ease and elegance" relative to the standards of Georgian England. The volume that reached print in 1794 carried both Young's and Hodgkin's names. A title page added in 1807 specified that it was Young himself who had formulated "the method" used, to which we shall return.

Young came to his grand-uncle Brocklesby's attention when he contracted what was likely tuberculosis in 1789. Barclay's second wife, Rachel, nursed Young back to reasonable health by means of a treatment developed in consultation with, among others, Brocklesby. From this initial encounter Brocklesby developed a lasting interest in the boy, who initiated a correspondence of striking range and warmth. When Young eschewed sugar in protest against slavery, Brocklesby sympathized but advised him in October 1789 that his abstinence would have little effect since "reformation must take its rise elsewhere, if ever there is a general mass of public virtue sufficient to resist such private interests." He closed the letter "your loving Uncle"—a surprising show of affection from an otherwise hard-nosed character.[58]

Warm as it was, Brocklesby's affection for his grand-nephew was not free of prideful, quasi-parental pressure. This grooming suggests that Brocklesby nurtured an ambition to launch Young into his influential London social circle. Although Young did not read contemporary literature, around 1790 he studied Edmund Burke's *Reflections on the French Revolution*, in part no doubt because Burke was Brocklesby's friend. Brocklesby in turn made sure that Burke knew about his grand-nephew's linguistic abilities, cementing a relationship between the two. Six months before leaving Barclay's, Young produced a Greek paraphrase of Wolsey's farewell speech to Thomas Cromwell in Shakespeare's *Henry VIII*. He did so at the suggestion of Charles Burney, another prominent friend of Brocklesby's who had influentially critiqued Greek versifications according to the standards of the time. Young had apparently studied those critiques with care as he prepared his versification of the Wolsey farewell. On receipt of the result, written carefully on vellum, Barclay told Young that "Mr. Burke has taken the Greek manuscript from me, and means to show it to divers learned men of his acquaintance for their philological criticism," and he advised Young to "write frequent moral essays and keep them for my perusal and for the sight of Mr. Burke, who has taken a great fancy to you, and will be glad to aid you with his best advice in all your ways."[59] Burke, in turn, urged Young to a second task, namely, to produce a Greek version of Lear's infamous curse upon his daughter Goneril. Then, in November and December 1791, Young stayed with Brocklesby in London.

This period with Brocklesby would enlarge Young's circle of acquaintance with the leading scholars of the day, including Burney and especially Richard Porson, who was known even at the time as the greatest classicist of his generation. From 1792, Porson had been regius professor of Greek at Cambridge; Porson's 1790 attack on George Travis concerning the famously contentious *Comma Johanneum*—a treatise on a short clause of uncertain authenticity in the book of John—earned the acclaim of Edward Gibbon, who considered Porson's

FIGURE 2.3. (*Left*) Richard Porson (© National Portrait Gallery, London) and (*right*) Richard Brocklesby (© The Royal Society).

critique "the most acute and accurate piece of criticism since the days of Bentley," which was high praise indeed, since Richard Bentley had provided the high-water mark for classical scholarship in the eighteenth century.[60] On several occasions during Young's stay with Brocklesby, Porson, who would have already known something of Young's interests from the volume, coauthored with Hodgkin, on Greek calligraphy, engaged Young in elaborate exchanges concerning the proper renderings and emendations of Greek texts. Among the topics discussed were the origins of Arabic numerals, which Young at first thought came from the Greek, and other questions concerning the derivations of particular orthographic characters.

The following fall Young left Youngsbury to undertake medical studies in London. There, in the company of Brocklesby and his influential friends, he would hear contemporary arguments concerning Greek antiquity, particularly the origins, structure, and emendations of letters. Around this time, Porson published a critique of Richard Payne Knight's *Analytical Essay on the Greek Alphabet* (1791). Perhaps Porson dilated on Knight's *Essay* at one of Brocklesby's dinners. If so, Young would have heard early rumblings of the argument over Homer that would later disrupt Longman's party. For, Porson wrote, Knight "endeavours to rectify the orthography of Homer's words, by restoring the aspirates according to the directions of the metre."[61] Knight asserted, in the course of his emendations to Homer, that the Homeric sagas, despite centuries of defacement "by the varnishes of criticks, grammarians, and transcribers" were nevertheless so perfectly executed in the original "that we can still trace the minutest touches of the master's hand, and ascertain, with almost mathematical certainty, the principles upon which he wrought." To this Porson demurred in ways that anticipated Wolf's views. "Homer's poetry," Porson wrote, "however exalted and embellished by learning and genius, must partake of the rudeness and simplicity which are always incident to the infancy of language and society." Those who, like Knight, "attribute to [Homer] all possible perfection" should answer who he was, where and when he lived, and what of writing was known in his time.[62]

By the time Young arrived in London, he had read a great deal about languages and writing. He had learned to write characters according to specific rules in various scripts and had absorbed conventions for capturing sentiments in a way strictly suited to Greek prosody, at least according to the canons of his time as spelled out by men like Burney and Porson. But Young's attainments were not limited to Greek. His study of Hebrew and Persian, combined with Sharpe's disquisition on the evolution of language and of writing, permitted Young to differentiate writing systems according to the economy and precision they afforded. The earliest writing systems could never approach the precision of Greek, where the exacting formation of alphabetic characters matched the accuracy of a prosody governed in ways that approached lawfulness, or so Young absorbed from his readings and from talk around Brocklesby's table. Lawfulness also governed the other major arena of Young's studies as he grappled with the intricacies of Newtonian natural philosophy and the fluxional calculus. A moral dimension tinged much of this, for, though Young would distance himself from the rule-bound ethical universe of his Quaker parents, nevertheless the sense that both the human and nonhuman worlds were governed by morally salient law can be detected in the ways in which he attacked problems, not least when he came to hieroglyphs. The next several years in London, Edinburgh, Göttingen, and finally Cambridge would cement these apparently disparate areas into a unified way of thinking about both nature and human expression, ways that tightly governed how he approached Egyptian hieroglyphs when, years later, he submitted them to rule.

CHAPTER 3

AN ERRAND IN
THE CITY

One fall morning in 1792, Young came into an unusual possession: an eyeball from a freshly slaughtered ox. Although no receipt exists to confirm where and how he acquired this object, we may surmise that, as he made his way from his lodgings in Little Queen Street, Westminster, he stopped at a butcher's stall or perhaps even braved the fray at the huge livestock market at Smithfield in London's Clerkenwell neighborhood.[1] Unlike other shoppers on their way to market that morning, Young's intentions were far from culinary: he planned not to eat his purchase but to dissect it. The dissection would provide the experimental basis for his first scientific paper, on ocular accommodation—the ability to focus on objects at different distances. Young's investigation broke with a traditional anatomical orientation, and it is precisely here that we can spy the first formation of Young's distinctive way of thinking and working, which extended from natural phenomena to his understanding of sounds, words, and their representations.

Eighteenth-century medical London was—is—a source of enormous fascination. The nineteen-year-old Young, heading toward the dissecting room in 1792, was part of a complex amalgam of loosely affiliated doctors, institutions, and professional societies, all participants in a shadowy economy of suffering and death. Anatomy was at the center of this universe, a potent switch-point between science, corporeality, and commerce. As London's central livestock market for more than a thousand years, Smithfield would have provided the cheapest, freshest raw material for the anatomical dissection Young had in mind on that autumn day.

Before Smithfield Market became what it is today—an abattoir on the grandest scale, housed under an arcade roof of glass and steel, where the disquieting odor of blood is just perceptible beneath the soap and antiseptic—it was only a four-acre plain, a "smooth field," to use the twelfth-century phrase that originally gave the place its name.[2] This vast grazing space made Smithfield the largest outdoor slaughterhouse in London. Enormous numbers of live animals—some 30,000 on the Great Day, the busiest market day before

FIGURE 3.1. The Old Smithfield Market, from the *Illustrated London News*. Antiqua Print Gallery / Alamy Stock Photo.

Christmas—were slaughtered there, in throngs of what one observer at mid-century described as an "agitated sea of brute life" whose energies were barely contained by whip-waving drovers wearing thigh-high boots.[3] According to the physician Andrew Wynter, a visitor to the market "sees before him in one direction, by the dim light of hundreds of torches, a writhing party-coloured mass, surmounted by twisting horns, some in rows, tied to rails which run along the whole length of the open space, some gathered in one struggling knot."[4] Dickens described the market as "a stunning and bewildering scene, which quite confounded the senses."[5] Even Dickens's sturdy Pip grew disturbed at Smithfield, "the shameful place, all asmear with filth and fat and blood and foam, that seemed to stick to me."[6]

In search of his ox eye, Young would have made his way carefully over cobblestones slicked with by-products of the business that had been in full swing since the early hours of the morning. Steps away, St. Bartholomew's Hospital was already doing a brisk business of its own. Although it is unlikely that Young, still a teetotaling Quaker, stopped in for a pint, he may have passed the Fortune of War pub, located on the corner of Cock Lane and Giltspur Street, which was a favored haunt of the so-called resurrection men. These shadowy figures stole corpses from graveyards and sold them to hospitals, like Bart's, where the corpses were in great demand for the teaching of dissection—a point that was surely not lost on Young, who was then studying anatomy at the Hunterian. At the Fortune of War, these grisly wares were displayed in the upstairs room, where the anatomists came to appraise the goods in a grotesque pantomime of the bustling meat market outside.[7] Dressed in black Quaker garb and broad-brimmed hat, Young must have cut an unusual figure as he hurried away, his macabre parcel tucked under one dark-sleeved arm.

At the Hunterian, dissection had additional pecularities derived from its association with John Hunter. Perhaps the most eminent physician-anatomist of the age, Hunter collected specimens, dead and living, from the entire animal kingdom, "that he might trace the peculiarities of each."[8] His acquisitiveness in the name of anatomical knowledge brought him renown and even a degree of scientific respectability, but it also led him into some tight spots. For instance, by acquiring the corpse of seven-and-a-half-foot Charles Byrne, Hunter risked exploiting an unseemly popular interest in the postmortem fate of the "Irish giant" who had, in life, drawn crowds to London's freak shows. At his Earls Court home, Hunter's menagerie included two leopards that once escaped confinement in the barn, raising a ruckus that disturbed his neighbors—but perhaps not more than his habit of keeping such apex predators on the property.[9] His collecting proved so compelling he was frequently short of cash to cover expenses and had to delay even his marriage due to a shortfall that was directly attributable to his greed for specimens. It also made him a target of ridicule: his detractors referred to his extensive anatomical museum as "Pigs' pettitoes."[10]

For all the trouble it caused, Hunter's specimen collection did serve a larger purpose, as it gave Hunter opportunities to develop his considerable acumen in tracing obscure anatomical features and to speculate about their function. He was particularly concerned with muscular activity, including the muscles that keep the eye fixed on an object when the head moves. In his *Animal Oeconomy* (1786) Hunter wondered whether the eyeball possessed an inherent elasticity that returned it to a "natural situation in the orbit," or whether, when one muscle slackened, another contracted in the opposite direction.[11] In subsequent years, he returned to ocular anatomy in an effort to explain, by means of muscular action, the eye's ability to focus on objects at different distances. This question, of the eye's accommodation, was the very phenomenon on Young's mind when he obtained his ox eye.

The problem was hardly new. Since at least the seventeeth century, investigators had attempted to identify the part of the eye responsible for focusing on near and distant objects and to understand how it worked. In the seventeenth century, Descartes had claimed that both the lens and the eyeball changed shape, and the microscopist Antonie van Leeuwenhoek had observed "fibers" in the eye's lens that he assumed to be muscles. In 1738 the Edinburgh physician William Porterfield had asserted that the ciliary ligaments surrounding the lens (or "crystalline humour" as it was usually called) were muscular, and he thought that accommodation worked by means of these ligaments mov-

ing the lens. At Leiden, the natural philosopher Pieter van Musschenbroek—
remembered as the inventor of the Leiden jar—held a similar opinion but included the cornea in the effect.[12] Young's own preparation included familiarity
with the 1719 work of Henry Pemberton who, anticipating Hunter, thought the
lenticular fibers were muscular and used for accommodation.[13] Despite this
long and illustrious pedigree, the problem had not found a satisfactory solution
by the time Young and Hunter separately turned attention to it.

While there is no persuasive evidence that Young was aware of the de-
tails of Hunter's work on this problem, he certainly would have known about
Hunter's general approach to anatomical dissection. In the autumn of 1792, at
Brocklesby's urging, Young began to attend lectures at the School of Anatomy.
Hunter had by that time ceased to lecture there, but in the fall of 1793 his lec-
tures were given by his brother-in-law Everard Home.[14] Young never did hear
Hunter himself speak since by this time Hunter had become too ill and in fact
died that October.[15] Nevertheless, at some point during the months before
Hunter's demise, perhaps as a result of attending anatomical lectures, Young
became interested in the question of accommodation.

Young's predecessors had never thought to combine precise measurement
of the eye with optics, which is what made Young's work so original.[16] Putting his anatomical training to use, Young meticulously dissected the ox eye,
examining it under "a strong light," using either direct sunlight or possibly
an Argand lamp, an intensely bright oil lamp that Hunter himself favored for
early-morning dissections.[17] The first order of business was to remove the ge-
latinous center of the eyeball—the "crystalline" (lens)—from the surrounding
tissue, or "capsule." He dissected the capsule's "coats" or layers and examined
them "with a magnifier," finding that the "lens of the ox is an orbicular, convex,
transparent body composed of a considerable number of coats, of which the
exterior closely adhere to the interior."[18] Blowing gently through a long, thin
pipe, Young separated the layers. The lens, he saw, was not single but tripartite,
segmented along the lines of fibers. Applying "a little ink to the crystalline,"
Young added, "is of great use in shewing the course of the fibres."[19]

With respect to these fibers, Young went considerably further than Leeuwenhoek, who thought they formed a single, continuous strand. Not so, Young
claimed, for the fibers in a coat are each separate from one another, connected
at their ends to "membranous" tendons that are themselves arranged in "three
equal and equidistant rays, meeting in the axis of the crystalline [lens]."[20] No-
ticing the fibers' mutual independence, Young supposed them to be muscles,
each of which would contract against the pull of the surrounding tendons when
stimulated by "the influence of the mind" as conveyed through the nerves
and ciliary processes. But this idea—which, as we shall see, did not differ

significantly from Hunter's—did not satisfy Young. Was it possible for the lens to change shape sufficiently to focus on near and distant objects, given its refractive properties and those of the cornea and surrounding aqueous and vitreous humors? The question played to Young's strengths, for he had closely studied Robert Smith's 1738 comprehensive geometrical text on optics, which had explained how to calculate refractions through sequences of surfaces.[21] Young likely absorbed Smith's text about two years earlier, during the period in which he was immersed in Newton's *Principia* and *Opticks*.[22]

Young's combination of anatomical observation with optical calculation was powerfully productive. No prior investigator of accommodation could have shown that the optical properties of the lens were sufficient to account for the changes in focal lengths that create accommodation. Young did so. He measured the ox-eye lens's dimensions and determined its refractive index when adapted to focus rays from an object very far away. To do that he extracted the lens from its surround to determine how it behaved "in the atmosphere," where the lens's fibers, acting according to Young's belief that they were muscles, would be unstressed, allowing the lens to assume its normal shape. The unstressed lens had accordingly to be comparatively flat to bring distant objects into focus. To observe near objects the lenticular muscles would necessarily contract, forming the lens into a sphere. It was a shape to which the muscles were "admirably adapted," he wrote, "for, since the least surface that can contain a given bulk is that of a sphere . . . the contraction of any surface must bring its contents nearer to a spherical form."[23] Assuming that the humors refracted essentially like water, Young computed the lens's refractive power when in the ox eye proper. Coupling geometric optics and fluxions (that is, Newtonian calculus) to physical and anatomical measurement, Young demonstrated that the lens of the ox eye would have to shrink its effective diameter from about 700 thousandths of an inch to 642 thousandths, rendering it more nearly spherical, in order to focus on an object about a foot away.[24] Young reckoned this result more than adequate for an ox's visual purposes and well within the range that the lens's elastic properties could accommodate in any case. Even more so, then, for the human eye since "the human crystalline [lens] is susceptible of a much greater change of form." The implication was clear. Optics and calculation demonstrated that the lens's malleable shape was sufficient to account for accommodation, whereas the dimensions of the eye would have to change far too much to do so. No one before Young had essayed anything like such a calculation to demonstrate this point. Hunter certainly could not have.

In the passage from dissection to publication, Young found himself enmeshed in a cleaner and more elegant universe than the insalubrious environs of the meat market and dissecting room. His work on the eye gave him access to the Royal Society, where even the worst passions tended to smear only the pages of their publications—a far cry from Smithfield's "filth and fat and blood and foam," not to mention the more ordinary grime of the city's daily life at street level. In these years, Banks presided over this recondite haven. Impressed by the growing influence of the teaching hospitals, Banks had turned his attention to lively recruitment into the society's ranks of young medical men such as Astley Cooper, Henry Cline, and John Abernethy. Thanks to his paper on the eye, Young would soon join them.[25] Yet even the reception of this first work had a disappointing whiff of the meat market, sullied as it soon was by a scandal.

Young initially showed his work to Brocklesby, who particularly liked the material about the dissection. At his uncle's urging, Young submitted the paper to the Royal Society, of which Brocklesby and several of his friends were members. Young read the paper to the Royal Society on May 30. It was officially communicated by his uncle and then printed in the *Philosophical Transactions* for 1793, some months after Young's twentieth birthday. The response was mixed. The paper seems to have touched off a priority dispute that involved Hunter himself, at least indirectly. In a letter to Banks written either after learning about Young's lecture or shortly after seeing its publication, Hunter put himself abruptly and overtly forward, proposing to give a prestigious Croonian Lecture to cement "my claim to the discovery of the crystalline humour being muscular."[26] We'll look in more detail below at what Hunter had done on the question of accommodation, but first we will consider the events that occurred in consequence of rumors concerning Young's prior knowledge of Hunter's work. While the details remain unclear, the affair had a decided effect on Young.

On November 6, 1791, a year before his matriculation at the Hunterian and his dissection of the ox eye, Young dined at the home of Sir Joshua Reynolds. Brocklesby squired Young, then only eighteen, to the dinner. The other guests included Sir Charles Blagden, a physician and chemist; Walker King, coeditor of Burke's *Works*; and Boswell.[27] Reynolds, who had suffered for two years from progressive blindness, was obliged to wear a bandage over one eye and a green shade over the other.[28] Perhaps the great painter's loss of vision was on the minds of the assembled guests, or perhaps the science of vision came up in passing. In any event, in late 1793, shortly after Home posthumously published Hunter's claim to have been working on visual accommodation, a rumor arose that Blagden may have discussed Hunter's ideas at the dinner party. This was no trifling allegation. If true, it would seriously undermine Young's claim to originality.

Blagden himself may have started the rumor. A Scots physician trained at Edinburgh, Blagden was insecure about his lack of traditional education. He remedied the problem by acquiring "a considerable acquaintance with languages," which proved a boon as it helped him to socialize more comfortably in London.[29] He had, moreover, served as secretary to the Royal Society since 1784, publishing papers on chemistry and one on the nature of inks and ways to recover illegible ancient documents. He worked for many years as an "assistant and amanuensis" to the renowned natural philosopher Henry Cavendish before the relationship dissolved—but not before Blagden found himself embroiled in priority disputes of other sorts, functioning as a go-between among competing "originators."[30] He was in touch, as well, with Young's future nemesis, Henry Brougham, who bitterly criticized Young's 1793 paper on accommodation in a letter to Blagden. In other words, he was no stranger to this sort of conflict.

About the Young affair Blagden left only a small clue. According to Peacock, upon learning of this rumor in 1793, just after the publication of his paper, Young, deeply worried and perhaps prodded by Brocklesby to clarify the situation, wrote frantic letters to the guests present at Reynolds's dinner two years before. In these appeals, which, no doubt understandably, seem not to have been preserved, Young apparently asked "whether the subject of vision and any recent researches connected with it, were mentioned." All denied that any such thing had been discussed, so far as they could recall. Possibly fearful that he had spoken once too often, Blagden hastened to tell Young that "he was by no means so clear as to be sure that he had told him Hunter's opinion."[31] That ambiguous reply seems designed to shield Blagden from any distasteful consequences that might arise from having irritated Brocklesby, who had powerful friends. At the same time, the intensity of Young's efforts to efface the hint of plagiarism attests to his certainty that he had heard nothing about Hunter's work at the time. There is, after all, no mention whatsoever of Hunter in Young's paper as printed in the *Philosophical Transactions*, which he would certainly have mentioned precisely to avoid the kind of difficulty that did arise.

Notwithstanding Blagden's possible rumor-mongering, the immediate cause of the flap was an undated letter written by Hunter to Banks, unpublished at the time of Hunter's death but printed about a month afterward, in November 1793, by his brother-in-law, Home. The posthumous letter suggests that Hunter had been working on the problem of accommodation by the late 1780s. In this letter Hunter, who was in poor health and preoccupied with "official business on account of the war" with the French when he wrote, explained what he had intended but was unable to do with respect to ocular accommodation. In his covering letter, Home contested Young's priority, claiming that it "is now *many years* [emphasis ours] that Mr. Hunter has had an idea, that the crystalline humour was enabled by its own internal actions to adjust itself, so as to adapt the eye to different distances." Home explained that Hunter had arrived at this idea by means of dissection, which counted for more than passive observation; the actively intervening anatomist could make a stronger claim for priority in discovery. While dissecting a living tapeworm, Home reported, Hunter "was surprised" to observe contractions "in a membrane devoid of muscular fibres." This result led him to dissect the lens of a cuttlefish, whereupon he discovered that it contained "laminae," seemingly fibrous. According to his letter to Banks, Hunter next obtained an ox eye to examine whether it, too, might have such a structure. By then his health was failing, and he was unable to move ahead.

As we have seen, the novelty in Young's work was his demonstration that optical laws, when combined with careful measurement, permit the lens to change shape to account for accommodation and to gainsay the shape of the eye sufficiently to do so. Hunter assumed such calculation to be inconsequential, for "the laws of optics are so well understood, and the knowledge of the eye, when considered as an optical instrument, has been rendered so perfect, that I do not consider myself capable of making any addition to it."[32] This stance—that there was no need to add anything to what was already so well known in optics—neatly set to the side precisely what Young had been the first to accomplish. What then had Hunter himself to add? He argued that accommodation could not be due to changes in the eye's shape or the position in it of the lens. It was just Hunter's knowledge of the muscular structure surrounding the eye that had led him to conclude these muscles were inadequate to the task. That left only changes in lens shape to account for the phenomenon, in which case, the lens had to have "a muscular action within itself." This was the claim about which Young was rumored to have heard. To support the assertion, Hunter noted that the lens, when coagulated, "had a fibrous structure like muscles." He aimed to prove the claim by investigating whether, when an ox was quickly killed, the lens showed signs of contraction, for "in all violent deaths the muscles contract." Hunter could get no further than that.

Both Young and Hunter, then, agreed that accommodation could not be due to changes in the shape of the eye, and that the lens itself was striated internally with muscle fibers that produced the requisite changes.[33] Since Young thought the muscles to be part of the lens, he concluded that their relaxation corresponded to the flatter state, with the spherical shape arising on contraction. On this view, the relaxed lens accommodated for distant objects, while the stressed lens accommodated for near vision.[34] Hunter likely thought the same, for both of them imagined that a nonmuscular lens was necessarily pulled out of its intrinsically spherical shape by surrounding muscles when they contracted.

Despite their common view of the lens as muscular, a decided difference separated Hunter's mode of thought from Young's, and in this difference we can discern, in a kind of *forme fruste*, the assumptions that would later shape Young's understanding of representation, particularly writing. Hunter devoted a portion of his 1786 *Animal Oeconomy* to muscular actions, which seemed to act autonomously throughout the body, including sense organs such as the eye. "Muscles," he wrote, "are the active parts in an animal body, producing different effects, according to the circumstances in which they are placed; and most parts requiring a variety of motions, it became necessary to have a variety of muscles suited to these motions."[35] With the eye, he continued, "being an organ of sense, which is to receive impressions from without, it was necessary it should be able to give its motions that kind of direction from one body to another, as would permit its being impressed by the various surrounding objects." He would have agreed that the mind stimulated the irritating nerves in some manner, but he was not concerned with that. The activity of the mind passed without comment. By contrast, Young invoked the mind's action on the eye quite directly. "I conceive," he wrote, "that when the will is exerted to view an object at a small distance, the influence of the mind is conveyed through the lenticular ganglion . . . to the orbiculus ciliaris . . . and thence by the ciliary processes to the muscle of the crystalline [lens], which, by contraction of its fibres, becomes more convex, and collects the diverging rays to a focus on the retina."[36] The will, which is to say the principle of the mind that ultimately activated the muscles, altogether escaped Hunter's concerns, whereas Young would not leave it aside. His resistance is puzzling. Why invoke the will at all when, after all, Young, like Hunter, needed to show only that the lens could be a muscle? The answer may lie in the stark difference between Hunter's and Young's social and educational backgrounds, one that inflected Young's work whenever he dealt with phenomena that involved the relationship between thought and expression.

Hunter's lack of a classical education appears to have cast a long shadow over his life. His first biographers were particularly concerned to establish the discerning cast of Hunter's mind despite his lack of such instruction. This preoccupation seems unreasonable until we recall the obsession with learning, language, and social status that marked the late eighteenth century. Classical learning was the preserve of men of genius. When the English physician Jesse Foot, house surgeon at Middlesex Hospital and Hunter's bitter enemy, produced a biography almost before Hunter's body had cooled, he vilified Hunter for precisely this flaw. Foot, who had extensively critiqued Hunter's work on venereal disease in 1786 and 1787, wasted no time castigating both Hunter and his brother William: "Genius sits easy upon him who intrinsically possess[es] it . . . If ever there was an instance where two men have so often been disappointed, by mistaking themselves, as the Hunters, I know not where to find it. All their diligence, their art, and their contests, only prove that they struggled indeed for it, but could never obtain a reputation bearing the smallest resemblance to men of genius."[37] According to Foot, Hunter struggled particularly with writing. "The truth is," Foot charged, "that [Hunter] only furnished the images, and that the writing part was always performed by another:—he prepared the skeleton, and another covered it with composition . . . he was incapable of putting six lines together grammatically into English." This deficiency, which Foot attributed to "want of education" in grammar and rhetoric, explained why "his notions of things were so very imperfect, and his conceptions so very contracted." Foot particularly criticized *Animal Oeconomy* as exemplifying Hunter's poor rhetorical command, which he sardonically described as "the wonderful art of hanging heavy weights to slender wires."[38]

Others sought to counter Foot's critique, but not by denying that a proper way with grammar and rhetoric was to be desired. Another quality—one that Hunter was claimed to possess—compensated, namely, the perspicacity of his mind. His expression might not be elegant, but it was exact. In response to Foot, the physician Joseph Adams defended Hunter against accusations that his shaky grammar reflected a less than first-rate mind. Even if "Mr. Hunter's language was not always strictly grammatical," Adams reasoned, "it was perspicuous to those who knew the value of his facts, and were willing to explore them; and that how much soever we may lament his early inattention to school learning, he was at this time much better employed than in studying the classics." Adams stressed perspicacity, the ability to find the right words, regardless of the grammatical sophistication of the expression.[39]

To prove his point, Adams recurred to Hunter's notion of coagulation as developed in his study of visual accomodation. In his letter to Banks, Hunter had noted that, once he conceived that the lens might be muscular, he was "confirmed" in the notion on "finding that in many animals, when the crystalline humour was coagulated, it had a fibrous structure like muscles."[40] Reminding the reader that when Hunter referred to coagulation, he meant something quite specific, Adams alluded to Tooke's linguistic system as proof that Hunter's cast of mind was no less discerning than anyone else's. "Coagulable," Adams reflected, ended in "able," and that ending, according to Tooke, "included the following senses or meanings: Potential, inasmuch as it is a condition in which they *may* be found; Passive, inasmuch as they have not the power of assuming that condition, but only *suffer* such a change by force."[41] Although "[i]t may be urged that this usage has long ceased to confine this termination to a passive signification" in "looser compositions," Adams pointed out that it was "altogether inconsistent with philosophic correctness." In contrast, Hunter's use of a passive construction—the crystalline humor *was coagulated*—rigorously excluded considerations of volition. This refusal was characteristic. Although Hunter did speculate carefully about anatomical function, matters of volition remained outside his purview and accordingly merited no discussion in his *Animal Oeconomy*. He also disdained any recurrence to mental processes as explanations for visual accomodation. Criticizing earlier claims that distinct vision of an object can occur if the object's image moves across the retina, he complained that those who so thought "even explain the effects which would be produced by it *on the mind of the observer*."[42] Adams insisted that Hunter's use of the passive construction in his discussion of accomodation was "only one of numerous instances in which Mr. Hunter's accuracy of thinking induced, as it always must, a corresponding accuracy of language." To Adams, Hunter's lack of grammatical sophistication was amply compensated by his ability to choose just the right words.

Young's paper on the lens was not aimed at matters of volition, but in his scattered references to the activating power of the will, we can detect the effects of his very different background and education, and we shall see something entirely similar in his discussion of the human voice. The will for Young implicated a process of decision that, in the case of lenticular accommodation at least, worked according to rules in order to bring an object into focus. Whatever these rules might be, they must be inherent in the very operations of the mind. Language, too, was thought to be the product of a rule-governed mental process. In late eighteenth-century England, someone who had absorbed the contemporary ethos concerning grammar and rhetoric would not have thought linguistic rules to be arbitrary conventions but rather reflective of properly working, inherent mental structures. Languages certainly differed, but they were nevertheless based on mental structures generative of particular procedures. These rules could be, and were, violated among the uneducated or among those who chose deliberately to break them for social and political reasons. The rules existed nevertheless, and they reflected intrinsic processes. In England at this time (and later) linguistic rules were not merely abstract mental features but socially—and therefore, politically—significant.[43]

The mind as a distinction-producing machine had a particular position in Young's social and intellectual milieu. For Young and many of his contemporaries, language, thought, education, and social class were thoroughly entangled. The educational system, in particular the grammar schools and universities that educated the sons of the ancestrally privileged and the well-to-do, promoted an ideal of mind that rested upon the mastery of classical languages, particularly Greek. This attitude in turn presumed that the grammar of ancient Greek, understood as a model of elegance and clarity, mirrored the clarity and elegance of the Greek mind. The quality of one's expression indicated the refinement of one's intellect and sensibility. A refined intellect was very much a Greek one.

This view extended a line of linguistic thought that we mentioned above and that may be traced to the seventeenth-century's concern with universal grammar, familiar to modern readers of Locke and Condillac as well as to those acquainted with the work of the relatively more obscure Port-Royal grammarians Claude Lancelot and Antoine Arnauld, with whom Young was conversant from his time at Youngsbury.[44] Universal grammarians of the eighteenth century assumed that, while "languages were fundamentally alike in the way that they represented the mind," they differed, equally fundamentally, "in the quality of mind and civilization they represented."[45] No commoner could hope to understand the learned languages properly, let alone reproduce them. The educated, however, were trained to do so, or at least to imagine that they could. Later in the eighteenth century, practical grammarians under the sway of these ideas frequently assigned, as did Young's tutors, "construals" in Greek of English texts such as Shakespeare's plays.

The basic idea was this: only those who knew these ancient languages could hope to express themselves appropriately in English. Even Burke, whose conservatism should not obscure how important his work was to radical movements of the 1790s, found himself on the wrong end of this stick. Despite his disdain for the uneducated and the commoner, Burke's writing provided a useful example of runaway success in the vernacular. His direct style allowed readers to make use of his prose in ways that less accessibly written books disallowed. No less a radical than John Thelwall raved about Burke's *Reflections on the Revolution in France*, claiming that he had not held a conscious political thought in his head before reading it, and admired the book as having "made more democrats, among the thinking part of mankind, than all the works ever written in answer to it."[46] Unsurprisingly Burke's social superiors subjected his prose style to stinging criticism. In 1791, a correspondent to the *Monthly Review* faulted *Reflections* for being at once "sublime and groveling, gross and refined," a combination that entirely befuddled him. Similarly, the Whig Phillip Francis begged Burke to make his writing less vehement, and therefore less likely to ignite a rude pamphlet war or, what was worse, a rebellion: "Once for all, I wish you would let me teach you to write English."[47]

Taken on its own, the ire directed at vernacularly oriented writers might seem like a relatively narrow conflict between the lettered and the nearly so.

But in the broader developing struggle over social class in England, related matters of language played a surprisingly prominent role. Between 1797 and 1818, Parliament repeatedly barred from general admission and discussion petitions in favor of extended or universal suffrage that were not written in sufficiently proper English. These refusals were made not only decisively, often with a huge majority, but also dismissively; no one even bothered to change the response even as petitioners made efforts to conform to the desired style. Moreover, because no guidelines were offered beyond a vague demand for "decent and respectful language," petitioners made their revisions in the linguistic dark, and Parliament changed the standard when the petitions came too close to meeting the goal.[48] One petitioner begged for more guidance, so that "the people might know in what language the House would be inclined to lend ears to their grievances."[49] In general, conflicts over suffrage—perhaps the most fraught political issue of the day—depended on an implicit agreement about vulgarity. Vulgar people were considered incapable of participating in public life, and the index of vulgarity was one's written expression. Participation in public life was limited to those with inherited titles, status, and wealth. But it was not these inheritances that secured one's ability to participate; rather, it was the fact of being an inheritor. Even Hannah More, who was otherwise such a force for the education of the poor, stopped short of teaching more than the rudiments of grammar, and limited handwriting instruction to the "common purposes of life."[50] This instruction, codified in popular grammars aimed at laborers, perpetuated the idea that schooling of this sort prepared the student "for Trade" rather than for effective participation in civic life, which was curtailed in any event by limitations on suffrage.

Although this ideology of language, conceived as both a mirror of the mind and an index of its refinement, exercised enormous influence from its beginnings in the mid-eighteenth century, by 1790 fault lines had begun to appear. For one thing, the class division that the ideology justified and perpetuated was under pressure from political movements, such as the revolutions of 1776 and 1789, that threatened existing structures. Inspired by his experiences in 1777, when he was tried for using allegedly seditious language in an advertisement in support of American revolutionaries, Tooke made an epic study of the relations between parts of speech and acts of mind, published in 1786 under the Greek title *Epea Pteroenta*, which came to be known as *The Diversions of Purley*. Here Tooke critiqued much that was implicit in the older theories of Monboddo and his correspondent, James Harris, which, he claimed, falsely elevated obscure language as a way to perpetuate existing class divisions. Grammar, again, was the central issue.[51] According to Tooke, parts of speech should not, contra Harris and Monboddo, be said to correspond to distinct, purely volitional acts of mind. Rather, they represented elements of syntax, which was to be conceived as an independent quality, not of a particular mind or national genius but of language itself, considered to be the common property of all. The spreading debate reached even into Hunter's neck of the woods. As we have seen, Adams had used Tooke to counter critiques of Hunter by concentrating on the latter's exactness of expression rather than its elegance.

Parading one's knowledge of Greek, and to a lesser extent Latin, maintained social distinctions by reinforcing the largely unspoken moral premises from which those distinctions derived. People of quality could be identified from the refinement of their expression. To eighteenth-century grammarians and others, only classical Greek and Latin could express fine distinctions. Vernacular English was insufficiently refined for this purpose. Johnson, for instance, complained that writers ignorant of either ancient language would express themselves in coarse ways that echoed this fundamental problem. "Illiterate writers," that is, those writing in English who have neither Latin nor Greek, "not knowing the original import of words, will use them with colloquial licentiousness," by means of which they will "confound distinction," and even worse, "forget propriety."[52] Johnson's charge of "licentiousness" suggests that the word's meaning would be carelessly handled, so that the wrong words will be used in place of the right ones; the result, according to Johnson, was a loss of "distinction." Muddled prose was a failure, above all, of discernment, which could only be developed through exposure to ancient languages.

Young learned early to critique "improprieties in grammar, syntax, of the use of words and phrases" that others had produced in their examples of Greek versification. Peacock's comments on Young's Greek rendition of Lear for Burke are similarly suggestive. Young's rendering, he wrote, "is correct and scholarlike, but rather remarkable for the peculiar circumstances under which it was produced [i.e., as response to Burke], than as adequately expressing the terrible energy of the original."[53] Peacock here gestured gently, though not uncritically, toward Young's persistent, characteristic effort to force words, like natural processes, into a precisely configured structure governed by rules. "Even when a boy," Peacock wrote, in his more usual tone of perfect admiration, Young would "write his Greek and other exercises, not merely with the most minute attention to accentual and other diacritical marks, but which extended likewise to the most complicated contractions of the early Greek printers. I know of no practice which is more calculated to form those habits of accurate observation which are so essential to give the last finish to the edge of critical scholarship."[54] Young also attended closely to the grammar and syntax of the ancient languages he studied.

As much as this preoccupation may have grated on those accustomed to engaging with classical literature on aesthetic rather than philological grounds, Young's grammatical focus put him at the leading edge of a wide-ranging cultural shift. As science gained legitimacy, it threatened the advantages traditionally conferred by study of the classics, while classical study became more closely identified with philology. Writing in the 1850s, Young's biographer Peacock asked whether the purpose of instruction in classical languages was "well-disciplined training in their grammar and syntax" or the cultivation of taste. He dismissed the latter as unimportant, believing it too outdated to merit serious consideration, and opined that "the business of education is one of selection . . . much must be postponed and much must be sacrificed, in order to lay a secure basis for the more important parts of the vast fabric of human knowledge." How much had changed by 1855. In the 1790s, this "vast fabric" did not yet even include the expansion of natural philosophy, in which Young himself would play an important role. Young's ways with words, dissections, and natural processes were profoundly inflected by his understanding of Greek and Latin grammar and rhetoric. Like many others of his time and place, Young believed the principles governing these structures inhered in the mind's organization and produced both meaningful speech and the representation of these meanings through inscription.

CHAPTER 4

THE VOCAL CIRCLE

The hieroglyphics of the Egyptians were rather injurious than
beneficial to science. They converted the lively observation into
an obscure and dead image, which assuredly could not advance,
but retarded the progress of the understanding.

—Herder, *Outlines of a Philosophy of the History of Man* (1800)

Despite the contretemps over his eye memoir, Young was admitted as a fellow
to the Royal Society on June 19, 1794, a little more than a year after he had
read the memoir before the society, and six months after Home's claims on the
now-deceased Hunter's behalf reached print. Medicine, natural philosophy,
classics, and antiquities were all represented among the signatories to the
nomination, received on March 19. Taken together the group suggests the extent to which Young's leading interests matched up with those whose favor he
sought. Among the group's members, Brocklesby provided the most direct link
to Young. Home himself also signed on—which is perhaps not unexpected, despite his paper on Hunter, since subsequent events proved Home to have been
sedulous in advancing his own interests, and Brocklesby was too influential in
the medical community to be safely antagonized by, for instance, withholding
support for his promising young relation's candidacy. The list also included
classicist Stephen Weston and antiquarian Charles Townley, among others.[1]

In the months following his election, Young was free to pursue activities
that reflected his already varied interests and his widening circle of influential
friends. That summer, Young visited Burke, William Herschel, and the duke
of Richmond, who offered him a position as his private secretary. In October
Young read a paper describing a new species of the Australian plant *Opercularia*
to the Linnaean Society, which had been founded by one of his nominators,
Edward Smith; the performance resulted in Young's election there as well.[2]

The paper on *Opercularia* was not Young's only publication in 1795. The
previous December he had penned an introduction to the work on Greek calligraphy that he had composed with Hodgkin while at Youngsbury. This work,

Calligraphia Graeca, bore Young's and Hodgkin's names, and among the subscribers to the publication were Brocklesby, Burke, Porson, Dalzel, and even Young's soon-to-be nemesis, John Hunter. The work cemented connections between the exposition, structure, and orthography of Greek. It contained examples, in Peacock's words, "of the correct and elegant formation of the Greek characters" produced according to a system that, Hodgkin informed Peacock, had been developed by Young himself.[3] To do so Young had drawn what he and Hodgkin considered to be the best orthographic forms, according to late eighteenth-century English standards. They then generated a comparative table that reduced the several forms from various manuscripts to groups with the corresponding acceptable Greek capital in the leftmost column. To illustrate how the system worked, the *Calligraphia* provided a table of different alphabetic forms that were then used to produce a graphical table of Greek words in the several variants (fig. 4.1).[4] This tabulation of lists of orthographic variants transformed discrete items from different sources into canonical sets organized by common visual forms.

Here we have a first published instance of what would become an enduring characteristic of Young's working style: the production of carefully configured tables based on elaborate lists. His table is particularly significant in light of his later efforts with hieroglyphics, because it provides a graphically structured comparison of written signs. We shall see that one of Young's (and, for that matter, Champollion's) working methods was precisely to seek common visual patterns among apparently disparate orthographic structures. But there was something more to Young's presentation, for it was married to a brief specification (in Latin) of the act of writing: "μ, originating from a thin beginning, has two curves similar to the posterior part of α. The subsequent ones πωτι are drawn beginning from a thick transverse line, but for the rest they are thin . . . In order that these ideas may be more easily understood, I have appended an illustration that indicates the starting-place of each of these lettered points and that distinguishes the later strokes of the pen from the initial diversity of lines."[5] Young was particularly concerned with penmanship, to such an extent that Porson once remarked that he even had the "command of hand" over Porson himself.[6]

The emphasis on orthographic technique implicates the physical action of inscribing in the production of letters. Just as the working of the vocal apparatus produces sounded words, the working of the hand produces inscriptions. For Young, bodily actions linked speech to logographic writing, whether phonographic or not, and these actions were governed by the operations of the mind. We will see in a moment that Young would turn in Göttingen to the production of sounds by the human voice in an attempt to understand the physical

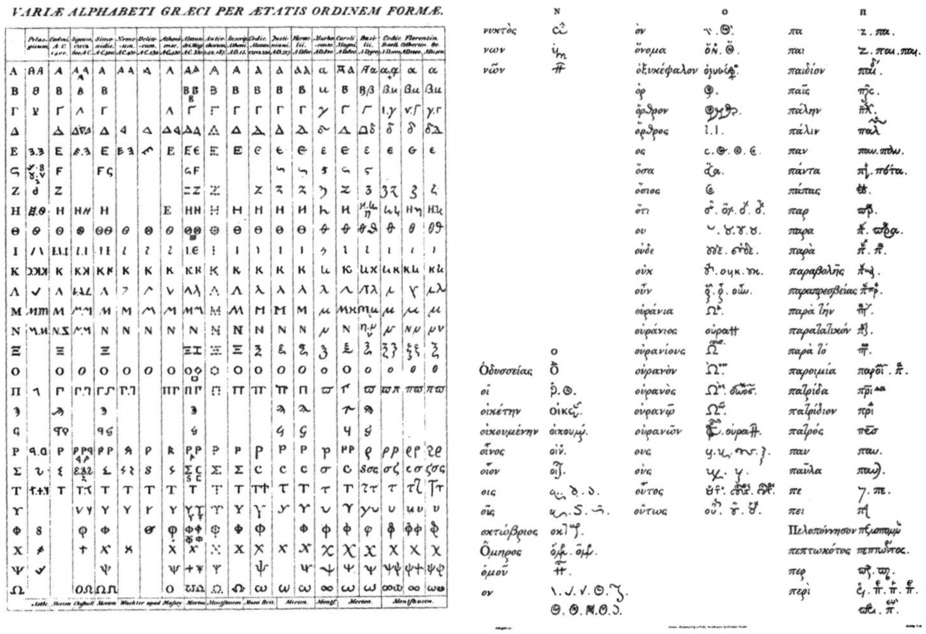

VARIÆ ALPHABETI GRÆCI PER ÆTATIS ORDINEM FORMÆ.

FIGURE 4.1. (*Left*) Thomas Young's table of variants and (*right*) an example of his and John Hodgkin's list of variant word forms.

basis of speech—and, from that years later, to the character of logographic signs considered in their origins as material inscriptions produced by hand.

All this was done, so to speak, off the side of Young's desk, for he was primarily pursuing his medical education in these years. Despite urging from his uncle and from Burke, attending Cambridge or Oxford was out of the question since Young's residual commitment to Quakerism forbid his admission. On October 20, Young arrived in Edinburgh under the assumption that recent changes to the rules for admission to the London College of Physicians as a licentiate simply required two years of university study.[7] After Young had been advised by Sir George Baker, who was president of the college at the time, that these years could be spent at different universities, Young planned a geographically dispersed course of study, attending Edinburgh and then Göttingen. In fact, the statutory revisions drawn up for the college specified that these two years must be spent at one and the same university—a surprise that Young did not discover until after his sojourns in Scotland and Germany. As a result, and with his Quakerism no longer an obstacle, he later spent an additional period at Emmanuel College, Cambridge, where he became increasingly immersed in questions concerning sound and, eventually, light.[8]

Young's move to Edinburgh was marked by both rupture and continuity. At Edinburgh Young broke decisively with the formalities of Quakerism; he learned to dance and visited the theater.[9] Meanwhile his anatomical investigations continued to bear fruit. He attended the lectures of the physician Alexander Monro, from whom he gleaned little, but to whom he gave a copy of his work on the lens, cementing the relationship. Shortly thereafter, Home's defense of Hunter, coupled to Home's own countervailing experiments, came to Young's attention. Disturbed, Young brought the matter directly to Monro, who proved himself an ally. During one of his lectures Munro spoke of Young's paper "with as much respect as it deserved" while insinuating that Home's results were essentially similar to ones that he himself had obtained. On tendering his paper Young felt that Monro would "think the better of [him] for having been in any manner opposed by Hunter," since Monro had previously noted in a lecture that the claim for the eye's fibrous structure long predated Hunter.[10]

In November Young met Monboddo, the lawyer and theorist of human linguistic development whose ideas we've discussed in connection with Samuel Johnson's hostility to Ossian. This meeting occurred at the home of Dr. James Gregory, a physician and student of Greek who had served as head of the medical school since 1790. Gregory, Young wrote, invited him to meet Monboddo "as a Grecian," a role that tapped into the part of Young's identity that was most concerned with the language of ancient Greece. Monboddo had heard of Young through Porson, who had shown him Young's "Greek writing," that is, the *Calligraphia*, to which Monboddo had subscribed. During the evening they talked of "ancient authors, and their editions." But—not surprisingly, given Young's early reading in Sharpe and the affiliations of Brocklesby's circle with Monboddo's critic, Samuel Johnson—Young was unimpressed. Monboddo was not, he thought, "a man either of the deepest learning or the finest taste."[11] Though he did not meet with many "Grecians" in Edinburgh, Young continued not only to converse about classical matters but to deepen his work in the area. Andrew Dalzel (mentioned above) was both professor of Greek and librarian at the university, as well as one of the founding members of the Royal Society of Edinburgh. He took an interest in Young, and they remained in close contact until Dalzel's death. While preparing a work on Greek poets for pedagogical purposes, he asked Young to assist with some of the selections and in writing appropriate notes. Young contributed sixty notes or emendations to Dalzel's *Collectanea Graeca Majora*, including the specific forms, meanings, and even enunciations of words and phrases.[12]

After the lectures ended, Young traveled on horseback through the Scottish Highlands, taking with him recommendations to various notables he intended to visit along the way. In late October 1795, he arrived at Göttingen. Over the next six months, Young turned from the eye to the voice, joining physiology to the sonic production of words on the basis of a system governed by fixed links between mind and body. These investigations would carry forward to structure another set of linkages, one that bound the hand and eye to meaning by means of voice, captured by inscription.

Young sat for his oral examination the following April, after which he completed a formal, written dissertation that was printed in early June. Bearing the title "De Corporis Humani Viribus Conservatricibus" ("Preserving the Strength of the Human Body"), the work discusses factors involved in injury or disease, concentrating on muscular and vascular issues. Young concluded the prescribed part of the work with a list of ten theses, including the remarkable "Thesis III"—"fibers are arranged in the same order in the human crystalline lens as in that of the cow" (lentis crystallinae in homine fibrae eodem ordine dispositae quo in boue)—which echoed his adventures in the dissecting room and the resulting paper on ocular accommodation. The dissertation had to be circulated as well as printed, and to complete the requirements Young was obliged to give a public lecture on a topic of his choice. He selected the human voice, having developed a novel account of the production of vocal sounds while working on his dissertation. Although few details remain of Young's theory, elements were printed on the dissertation's concluding sheet "with a view," Peacock wrote, "of filling up some pages."[13] This description, which severely understates the significance of the material, has virtually condemned the work to obscurity. In fact, the pages on what we will call Young's "vocal circle" demonstrated the extent to which he joined his investigations of speech and writing with anatomical knowledge gleaned from medical studies, and did so according to a method of tabulation similar to the one he'd used to great effect in his analysis of ancient Greek calligraphy.

Developing the vocal circle, Young divided "articulate sounds," meaning those produced by the human vocal apparatus, into eight classes (fig. 4.2). These groups do not separate simply into consonants and vowels but are complex structures, six of which might be said to involve both. Young defined these classes in terms of bodily actions, with each class corresponding to a particular anatomical configuration of tongue and lips in relation to the upper and lower palates. The first six groups all involved what Young called "vocal sounds," by which he meant sounds in which the vocal cords participate. The first, and major, of these six consisted of the "pure vowels," which were "formed in the larynx, not interrupted by the tongue and lips, nor passing in any degree through the nose." The second class, of "nasal vowels," passed "without interruption" through both nose and mouth. These two groups were distinguished from the remaining four among the "vocal sounds" by the absence

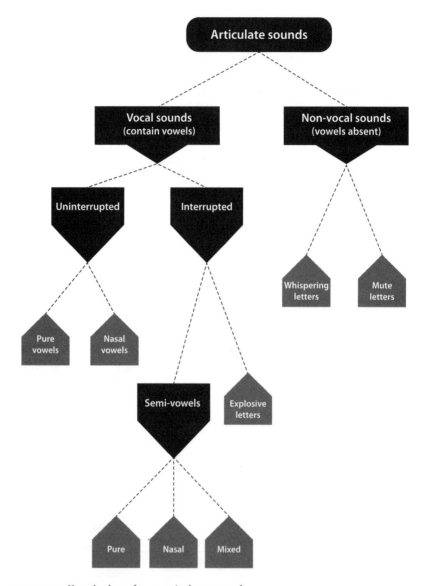

FIGURE 4.2. Young's scheme for expressive human sounds.

of any "interruption" by tongue or lips. The next "interrupted" four involved sounds produced in the flowing fashion of the first two classes modified by impediments generated either by mouth or nose. The third, fourth, and fifth of these were all subsets of Young's "semivowels," in which the impediments were limited in character, serving to modify the pure or nasal vowels; he called

these semivowels, respectively, "pure," "nasal," and "mixed." The sixth consisted of "explosive letters," which involved any of the previous "vocal sounds"
combined with a completely impeding stop. The seventh and eighth classes
differed from the others in not involving "vocal sounds." The seventh included
"susurrant or whispering letters," while the eighth corresponded in effect to
pure consonants, "mute letters" that were "incapable of being sounded alone
or continued."

Not satisfied merely to dissect expressive sounds into specific physiological operations, Young also conjectured an underlying set of operations out of
which all such sounds could be constructed. He believed these combinations
were governed by an algorithm, which he embodied in a circular graphic, his
vocal circle. This device was certainly based on one that Newton had developed
and presented in the *Opticks* for finding the resultant color when spectral lights
are compounded. Young, recall, had read the *Opticks* at Youngsbury, and he
would have been especially interested in the method since, while at Thompson's school, he had worked grinding colors.[14]

Newton's color circle worked in the following way (fig. 4.3).[15] The circumference is divided into seven regions of lengths that fit Newton's "musical"
division of the spectrum into distinct segments, with each segment corresponding to what he termed a "homogeneal" color, that is, one that, if passed
through a prism, would remain undecomposed. For each such color, place a
weight—measured by its brightness—at the center of the corresponding cir-
cumferential region of Newton's circle and then compute the center of gravity

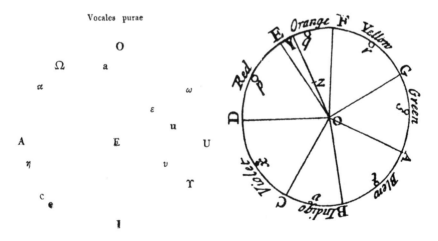

FIGURE 4.3. (*Left*) Young's vocal wheel, as printed in his Göttingen dissertation, and (*right*) Sir Isaac
Newton's color circle from the *Opticks*.

of the system. Draw a line from the center O of the circle through the resulting point, call it Z, and extend the line past Z to the circle's circumference. The region intersected by the line's extension specifies the color, while the distance from O to Z measures what Newton termed the color's "intensity," by which he meant what would later be termed its saturation. Newton thought the circle to be "accurate enough for practice, though not mathematically accurate."[16]

Young's vocal analog worked similarly, though he provided nothing beyond the diagram and a brief statement following a dissection of his first class (pure vowels) into five types, defined as always by particular bodily operations. It's worthwhile quoting him here to see just how thoroughly Young grounded his system on physiological actions that could be corralled within a geometric scheme.

E. The tongue and lips in their most natural position, without exertion.
A. The tongue drawn backwards, and a little upwards, so as to contract the passage immediately above the larynx.
O. The contraction of the mouth greatest immediately under the uvula. The lips must be also somewhat contracted.
U. The contraction continued below the whole of the soft palate.
I. The contraction formed by bringing the tongue nearly into contact with the bony palate.

From these principal vowels all others may be deduced by considering them as partaking more or less of the nature of each, accordingly as they are situated nearer to them in this scheme.[17]

The accompanying diagram (fig. 4.3, *left*) indicates how the scheme worked. Any vocal sound that could be made had to contain combinations of the pure vowels. Young positioned four pures—namely, A, O, U, and I—at the quadrant boundaries of the circle. The fifth—E—marked the circle's center.

Consider two of the sounds marked on the diagram, say a and u, respectively on the periphery between A and O, slightly above the line joining E to U, a bit closer to U. The first of these, Young wrote, represents the sound at the beginning of the French word *Âme*. According to the diagram, it must lie right at the transition between the sound for A and the sound for O, which, given Young's specifications, would implicate a transitional formation of the vocal apparatus with elements of both A and O structures. This procedure marks an apparent difference from Newton's color circle, in that Newton classified regions of the periphery into his seven spectrals and not isolated points. That, however, is likely an artifact of the incomplete development of Young's scheme, in that his intention was apparently to classify regions on the periphery by bodily configurations, with the quadrant boundaries constituting extremes. This intention is especially evident from his placement of *u*, representing the sound in *but* or *shut*. According to the diagram, it is a weighted mixture of *O* and *U* modified by the addition of *E*, presumably following a center-of-gravity calculation as in the color circle, though instead of Newtonian optical brightness, here we have degrees of physiological configuration, such as the magnitude of a particular contraction of the mouth. If we follow the implications of the diagram, the central position of *E* must represent a precise mean between the bodily conformations involved in *A, O, I*, and *U*.

From his scheme Young developed a rudimentary writing system that embodied physiology in script. His second class—the "nasal vowels"—were to be written using the same characters as the basic, pure ones but marked "with the addition of the grave accent." Young introduced a system of letters, both Greek and Latin, upper- and lowercase, to represent the sounds in the several classes produced by particular bodily configurations—for example, lowercase *n* represents a nasal vowel in which "the passage of the mouth is closed by the middle of the tongue." Incompletely specified though the scheme was, it nevertheless strikingly expressed the tight connections he perceived between expressive sounds, physiological actions, and their representation by means of an alphabetic system, all governed by an underlying algorithmic scheme designed to capture all "articulate sounds."

How the human vocal apparatus might relate to the shapes of letters had been a matter of inquiry for some time. In 1657, the alchemist Jan Baptist van Helmont claimed that the aleph took its shape from the aleph-like arrangement of the vocal organs when seen in cross section. In 1752, Johann Georg Wachter had linked Latin letters to the positions of the vocal organs, so that, for example, the letter *o* took its shape from the speaker's open mouth.[18] Though Young did not venture anything directly similar to such claims, in the course of establishing correspondences between sounds and motions of the mouth, nose, tongue, lips, and larynx, he had systematically associated vocal motions to inscriptions drawn from the Latin and Greek alphabets. Moreover, Young had insisted on correct ways to inscribe letters in the *Calligraphia Graeca*. Correctly produced writing, that is, required specific motions of the hand, while articulate sounds required specific motions of the vocal apparatus. It followed that the conversion of sound into vocalizable writing had to be governed by parallels between two sets of bodily motions, one for the voice, the other for the hand.

Though Young had worked to good effect on his dissertation at Göttingen, he found the town's society less congenial than Edinburgh's. He was particularly disappointed to have secured virtually no social invitations from professors, with the notable exception of naturalist, physiologist, and professor of medicine Johann Friedrich Blumenbach.[19] At Blumenbach's, Young met the classical scholar Christian Gottlob Heyne. Young did not join Heyne's seminar, nor did he share Heyne's guiding historicist ethos.[20] Heyne did, however, approve a Greek epigram that Young had written in praise of Johann Dominico Fiorillo, from whom he was learning to draw and who infused his lessons with remarks on the history of art. Young did attend Heyne's lectures on ancient art, and the two became sufficiently well acquainted that Heyne entrusted Young with the delivery of a letter to Dalzel in Scotland. In several paragraphs added by Young to the letter, Young remarked that Greek seemed not to be as extensively pursued in Germany as in England, and observed that "the literati" of Germany "are the slaves of the booksellers, and know little or nothing of Greek; but they are good Latinists, are immensely laborious, have the most extensive libraries at their disposal, and execute every task of drudgery infinitely better than any other."[21]

Despite his social failure in Göttingen, Young's work on vocal sounds caught the attention of Johann Gottfried Herder, the prominent Weimar philosopher and author of treatises on the origin of language and the philosophy of history. At Göttingen Young had met Herder's son, who was also studying there; he provided Young with the requisite introduction. Having become "doctor of physics, surgery and man-midwifery" on July 16, Young had gone on tour, stopping to visit Herder at Weimar on the way.[22]

Recalling the visit, Young noted that Herder "expressed a curiosity to see more of my attempt to illustrate the human voice. I wrote out for him a little specimen of the manner in which I would describe the pronunciation of the most current European languages."[23] The offering provided rich conversational fodder for the evening, as Herder had long studied the origin of speech. In 1772, he had entered a treatise on the origin of language to a prize competition set by the Berlin Academy of Sciences.[24] Contestants were asked to address Johann Peter Süssmilch's claim, which was really more of a commonplace by this point, that only divine action could have produced human speech. Herder offered instead natural development, albeit in a manner that differed considerably from the views of Condillac or Locke, both of whom had also linked speech to the establishment of conventions. Speech did arise, Herder agreed, as the human voice imitated sounds originating in the world rather than being divinely inspired. But the imitation did not lead to a conventionalized association of, for example, a humanly vocalized bleat with a sheep, but rather with the *activity* of the bleating animal. On this view, verbs constituted the first words, and speech developed from the cross-linkage of the different biological senses among one another, mediated by the uniquely human rational faculty. But, among all the senses, "the ear became the first teacher of language," for the phenomena of vision "were so cold and dumb, and the sensations of the cruder senses in their turn so indistinct and mixed up."[25]

Though Herder saw his account of linguistic development as naturalistic, his naturalism was grounded in history rather than in eternally valid, abstract rules. In that sense his philosophy of language bore affinities with Monboddo's. However, Monboddo insisted that symbolization was the primary function of language and could only have occurred substantially after the development of social intercourse. Language involved for Monboddo a type of transformative symbolization that produced the names of things. Herder, by contrast, insisted that language emerged through a different process, one focused on actions and productive of verbs. On this view, a society's language would necessarily be unique because the transformations from inarticulate cries to articulate speech necessarily depended on that society's characteristic experiences of action. Still, regardless of specific social context, Herder's position on language required language to function in a particular way: language connected sounds heard to actions perceived. Monboddo countenanced no such linkages.

Young's categorical mapping of the sounds produced by the human voice would certainly have interested Herder, who needed to explain the function of language (as connecting sounds to actions) in terms of natural structures. Perhaps their discussion touched on whether a native speaker of one language could ever truly express a thought in another. What if the very subset of sounds that defined *one* language, crystallizing it out of available vocalizations—so carefully delineated and even tentatively theorized by Young—altered the native speaker's ability to vocalize in *another* language that deployed a different subset? That narrowed ability, Herder conceived, had been exacerbated by the invention of writing. "How much does the Englishman," he opined, "torture himself to write his sounds, and how far is he who [only] understands written English still from being a speaking Englishman!" Herder contrasted English with French, Italian, and Greek, finding all of them less limited than English because speakers of these languages used their vocal apparatus differently. "The Frenchman, who draws breath up from his throat to a lesser extent, and that half-Greek, the Italian, who, so to speak, talks in a higher region of the mouth, in a finer ether, still retain a living sound," he opined. "Their tones have to remain within the organs where they were formed; however comfortable and distinctive the long use of writing has made them, as depicted letters they are forever only shadows." Poetry being closest, he thought, to the linguistic soul of a nation, a poem in one language could only imperfectly be captured in another. In a similar vein, letters could only imperfectly represent speech, for "not a single livingly resounding language can be completely captured in letters." This insufficiency only intensified the need to specify the developmental phases of phonographic writing, much as Young and Hodgkin, albeit with different motivations, had attempted in their *Calligraphia Graeca* (see fig. 4.1).[26]

Writing in the late 1760s, Herder mentioned the history of writing in several remarks on recent German literature, noting that the technique had developed long after the "art of speaking," and claiming that for centuries the earliest forms of writing "count for nothing." Anticipating by nearly fifty years Wolf's remarks on the Homeric sagas that Young would find so congenial, Herder opined that tradition, "in those days the sole preserver of historical reports, had long since shouted itself hoarse, got mixed up with lies and fables, before the remains of its legend were recorded in writing." Indeed, Herder later remarked, "in that age, when people had not yet thought of books, what was language then? Nothing but singing and speaking nature."[27] What then of the evolution of writing? What of its earliest form, before letters were invented to seize vocalization?

To Herder writing that sought to capture sound, such as alphabetic writing, was necessarily degraded because, being a partly visual medium, writing had a lower status than speech. The earliest writing had been different, albeit equally debased. It had not reproduced the sounds of speech but it was, if anything, just as inimical to the purity of lived nature. Where phonographic writing deadened the lively sounds of speech, "hieroglyphical" picture-based writing vitiated another, if inferior, sense, the visual. "The hieroglyphics of the Egyptians," Herder complained, "were rather injurious than beneficial to science. They converted the lively observation into an obscure and dead image, which assuredly could not advance, but retarded the progress of the understanding." These hieroglyphs, he continued, enclosed "secrets" such that "a series of them, must necessarily have remained a mystery to the many . . . [who] could not be initiated into the study of them, for this was not their business; and of themselves they could not discover the meaning."[28] The mystery Herder had in mind was not a complex of esoteric knowledge, as Kircher had it, but rather an intricate system of visual correspondence. For every hieroglyph, inasmuch as it was not (he thought) a representation of sound, could only be understood by means of an elaborate set of correspondences that had to be learned through the hard labor of initiates. "What poverty of ideas, what a stagnation of mind," he carped, "do the Egyptians display, in so long retaining this imperfect mode of writing, and continuing to paint it for centuries with immense trouble on rocks and walls." These obscure, imagistic references resembled, in Herder's view, the arcana of contemporary science—though he faulted modern arcana for worse. "Every hieroglyphical science of modern times," he wrote, "is a ridiculous obstacle to a free diffusion of knowledge; while in ancient times hieroglyphics were no more than the most imperfect mode of writing."[29]

Although Young and Herder probably did not discuss the Egyptian language or hieroglyphs during Young's brief visit, Herder may have been tempted to share his understanding of how a language's original sonic purity might be violated by letters upon being shown Young's scheme, which, he claimed, captured all the sounds that the human voice could make. Young had used more characters than the Latin or Greek alphabets each taken by itself could provide, a choice that aligned with Herder's conviction that no system of two dozen or so letters could fully transform the aural into the written, though one doubts whether Herder was entirely persuaded by Young's attempt to do so by using more characters. Neither would Young have been persuaded by Herder's aversion to the notion, widespread during the Enlightenment, that the relationship between sign and idea was conventional. For Young the vocal apparatus could only produce certain specific sounds, but what conception sounds might represent was not otherwise linked to them, the essential notion being that sounds *re-presented*, as it were, already-formed conceptions. Herder, however, conceived that the expression was not parasitic on the idea but that the two were dialogically related. Young's vocal scheme would no doubt have interested him because through it one might begin to see how ideas could evolve within a particular sonic matrix. Whereas for Young, the strictures of a rule-governed vocal system imposed specific forms on articulations but did not affect conventions governing what the sounds represented.

Young left for England on the seventh of February (1797), where he enrolled as a fellow commoner in Emmanuel College at Cambridge in order to satisfy the requirements of the College of Physicians. Brocklesby died on the thirteenth of December, shortly after dinner with Young and his other nephew, Robert Beeby, who inherited the bulk of Brocklesby's estate. Brocklesby left his furnished house to Young, including its library and several paintings by Joshua Reynolds, together with about 10,000 pounds.[30] That same December, an up-and-coming French general, Napoleon Bonaparte, arrived in Paris, where he presented the Directory with the Treaty of Campo-Formio transferring Austrian possessions to France. Two months later, Napoleon proposed to counter British maritime power by invading Egypt. On March 5, the Directory agreed, and the French expeditionary force, comprised not only of soldiers but of naturalists, mathematicians, linguists, skilled technicians, and artisans, set sail. Altogether four hundred ships carried more than thirty thousand men to Egypt. For the next three years, as stories and artifacts flooding into Europe brought Egypt vividly to public attention, Young stayed largely aloof from the drama, continuing to build his medical practice while pursuing investigations in natural philosophy.

CHAPTER 5

LECTURER AND PHYSICIAN

As essential as the inheritance from Brocklesby proved to Young's success as a physician in London, a degree of ambivalence had always marked their relationship. According to Peacock, Young's biographer, Brocklesby was "inclined to be somewhat querulous" in his dealings with Young, "and exacting in claims to respect." A doctor himself, Brocklesby met with mixed results when attempting to advise Young. "Though generally judicious in the course which he recommended Young to pursue," Peacock wrote, "[Brocklesby] was sometimes rather unreasonably suspicious" when Young failed to heed his wishes. As these wishes were "often very obscurely intimated," Young may not have fully understood them in the first place. Brocklesby became overbearing, so that even his kindnesses "were not altogether unmixed with some fruit of bitterness . . . Although liberal in great things, he was somewhat parsimonious in small."[1] A closer look at the world of medical London—Brocklesby's world—gives a picture of the broader professional and personal challenges facing Young as he sat down for dinner at Longman's in May 1803.

At this time, it was difficult to start a medical career without participating in a network of patronage that, it was hoped, would culminate in a credibility-enhancing hospital post or at least an affiliation. Like any other patronage system, London's hospital structure was rife with abuses, as "[s]tudents were dependent on the conscientiousness or otherwise of their masters . . . In any case brains and hard work were still a small part of the desiderata; it was far more important to pay a high fee to a good man and have his patronage when the time came to start on your own." Yet inadequate participation in this system meant ruin, as those who failed to find even inferior patrons were virtually excluded from professional advancement.[2]

The alternative route, cultivation of a private practice, demanded social skills and the energy to apply them, as well as a willingness to occupy an ambiguous position, somewhere between paid servant and independent professional. Brocklesby's notoriously prickly relationship with Johnson, for

FIGURE 5.1. Thomas Young in the 1820s. Buchwald Collection.

whom he served as physician and occasional whipping boy, illustrates the
social difficulties of eighteenth-century London physicians.[3] Wealthy and
well-connected patients needed their doctors but resented their intrusion;
the doctors, for their part, were forced by economic necessity to tolerate a
great deal of poor behavior, for no one could afford to lose a well-off patient.
Young would soon confront this dilemma.

As members of a profession whose status was already insecure, London physicians were anxious to distance themselves from the quacks and charlatans with which the city abounded. A 1797 mock advertisement promoting the services of one Martin van Butchell, "A Quack of Quacks," hints at this problem. Proclaimed to be a "Healer of Mankind," van Butchell claimed to be fitted for his task by dint of being "Briton born and bred" as well as remarkably virile at the age of sixty-two: "a Christian strong: (an entire man) having comely Beard: full ten inches long," not to mention "eight legitimates [and] another coming." Patients were advised that full payment was expected in advance, and that the doctor's goodwill could be substantially undercut by attempts to pay with counterfeit coin. At which point, and despite its prior insistence on van Butchell's "Briton" identity, the advertisement took an overtly prejudicial turn, shifting to the Scots dialect in order to inform the reader that van Butchell "do no' ken bad notes, or evil dollars, fa' we no take 'em."[4]

The fear of association with quackery was heightened by the fact that physicians could offer very little beyond interventions such as leeching and purging. The pharmacopoeia was similarly limited. In addition to a few instruments, the standard contents of the doctor's bag included a handful of nostrums of questionable therapeutic value, of which the opiate laudanum was perhaps the most familiar.[5] Even respectable medical establishments such as St. George's, where Young eventually found a post, relied on questionable nostrums such as Dr. Woord's Pills, whose active ingredients were camphor and pepper; various preparations of Daffy's Elixir, a purgative; and Sir Walter Raleigh's Cordial, a strawberry-flavored alcoholic restorative favored by Jonathan Swift as well as Queen Anne.[6]

To compound the difficulty of distinguishing comparatively trustworthy doctors from outright charlatans, both quacks and legitimate practitioners sought clients through similar channels. The walls of coffeehouses, for instance, were plastered with advertisements for medicines and medical services.[7] Some canny operators extolled the medicinal values of the coffeehouse's own fare.[8] One doctor in Young's circle, George Fordyce, touted the medicinal value of the punch at the popular Chapter Coffee House.[9] Competition from all these quarters put many obstacles in a fledgling doctor's way. Young was no exception, even with Brocklesby's bequest to cushion him. Supplemental assistance soon arrived, however, in the form of the formidably learned and well-connected naturalist Joseph Banks, president of the Royal Society, who offered Young a position outside medicine entirely, at the newly established Royal Institution of Great Britain.

Designed as "a great metropolitan school of science," the Royal Institution offered lectures on scientific, medical, and technical subjects ranging from the ordinary to the exalted, from the theory of heat to the most efficient way to make a pot of coffee.[10] In keeping with its reputation as "the workshop of the Royal Society," the institution offered chances to observe experiments made with the latest scientific instruments and invited subscribers to peruse its library of twenty-seven thousand volumes on all branches of natural philosophy.[11] Granted a royal charter in 1800, the Royal Institution was located in Mayfair at 21 Albemarle Street, a graceful Georgian building fronted by a row of fourteen Corinthian columns.[12] The interior was no less impressive: "Ascending the staircase," one observer reported, "which is extremely beautiful, on turning to the right is the apparatus room, communicating with the theatre, in which lectures are delivered" and which could accommodate nearly a thousand people.[13] In addition to the grand staircase, the laboratories, and auditorium, the building included a dining room for so-called experimental dinners during which such devices as the newly invented coffee percolator were displayed and tested.[14]

The motive force behind the institution's establishment was Benjamin Thomson, Count Rumford, inventor of the Rumford stove, a streamlined fireplace designed to push heat into the room while funneling smoke out. Born in the colony of Massachusetts, Rumford had a colorful early life: after a childhood marked by poverty and neglect, as a young man he walked to Cambridge—a journey of twenty miles round-trip—to attend the scientific lectures of John Winthrop at Harvard. During the American Revolutionary War, Rumford penned espionage letters in invisible ink while in the service of the Royal Army. Still later he organized an English garden and an efficient large-scale workhouse kitchen in Munich, for which service he was made a count of the Holy Roman Empire in 1791.[15] His *Essays*, published throughout the late 1790s, "enjoyed an extraordinary popularity," as even George Peacock, Young's biographer and no great fan of Rumford, had to admit.[16] No sooner had Rumford settled in London in 1795, than he set in motion plans to establish an institution that would permit him to pursue his scientific projects while attempting, as he put it in a 1798 letter to a prospective patron, "to make benevolence fashionable."[17]

The Royal Institution's early days were challenging. The property at Albemarle Street required substantial renovation, which Rumford managed, personally and minutely, while living in the midst of the construction. When the first professor of Natural Philosophy, the physician Thomas Garnett, proved an uncongenial hire, Banks drew Rumford's attention to Young.[18] Rumford enthusiastically approved of him but was more reserved about Humphry Davy, who was also hired at that time. Leaving for the Continent, Rumford wrote to Banks on September 21, 1801: "I earnestly request your watchful care over all that I have left behind me in Albemarle Street. Dr. Young promises to be a useful acquisition to us. Davy may do very well indeed, if he gets the better of his natural disposition to be idle and to procrastinate."[19] Despite this rocky start, the Royal Institution flourished. Lectures routinely drew such crowds that Albemarle Street was rezoned as a one-way street—the first in London—to accommodate the traffic.

In 1801, at the age of twenty-eight, Young took up Garnett's vacated post, having agreed to a salary of three hundred pounds per annum and rooms. His duties were at first limited to editorship of the institution's primary publication organ, the *Journal of the Royal Institution of Great Britain*. In the fall, Young shepherded into print a set of papers on galvanism while preparing a series of lectures to be delivered in the spring semester. This series, thirty-one in all, ran until May. The following year, he expanded the series to sixty on a broader range of natural philosophical topics. As the preparation and delivery of the lectures meant additional demands on his time, he shifted some of his editorial responsibilities to Davy, who coedited the publication with him until 1803, when the journal ceased publication due to lack of funds.

In keeping with Rumford's emphasis on maximal beneficence, the Royal Institution's lectures were aimed at a broad audience that included working people, women, and students of various subjects, especially medicine. The lectures were typically advertised in cards and printed proposals that could be obtained from the lecturers themselves, who were expected to "exemplify and systematize facts" so that, "to the practical they also add some portion of theoretical knowledge."[20] Young's lectures covered topics in natural philosophy as well as inventions, equipment, and the principles that made such things work. According to the syllabus that appeared in January 1802, he lectured on Mondays and Wednesdays at two in the afternoon and on Fridays at eight in the evening, with plans to shuffle the lectures in future years so that the Friday series would include materials previously given on Mondays and Wednesdays, ensuring that no student would miss anything on account of the pressures of a working person's schedule.

By this time Young's work in natural philosophy was becoming known. To Nicholson's *Philosophical Journal*, where Davy had also been publishing, he sent a note, which was quickly printed, on the propagation of light by waves in an ether; following this initial publication, he read a memoir on the same subject to the Royal Society in November 1802. As a lecturer, however, Young was distinctly unsuccessful. His listeners found him abstract, hard to follow, and at times dull. He understood the difficulty but saw no way around it. He remarked that although it "would be desirable [that] every syllable advanced should be rendered perfectly easy and comprehensible even to the most uninformed," still "I shall esteem it better to seek for substantial utility than tempo-rary amusement."[21] Not everyone agreed. Some, uncomfortable with Young's evident intelligence and wide-ranging knowledge, found him condescending. Even Peacock, who knew Young personally and was his most sympathetic biographer, had to admit that Young's temperament "was not adapted for a popular lecturer. His style was too compressed and laconic, and he had not sufficient knowledge of the intellectual habits of other men, to address himself prominently to those points of a subject where their difficulties were likely to occur."[22] It seems that Young's inability to make himself beloved at the lectern was related to his difficulties with clear communication, difficulties he masked with a prickly unwillingness to suffer fools. In a letter to his friend Hudson Gur-ney, Young reported that one of his critics had taken offense at his "attempt . . . to clothe the ideas of a short algebraical calculation in the language of words" rather than leaving the calculation in "its obscurity." The problem, as Young saw it, was that such translation assumed the writer had the ability to under-stand what was being communicated in the first place. "I would rather . . . be severely handled by a man who understands something of [a] subject, than be treated with affected contempt by a fool," Young sniffed. Oddly, he reflected, it was exactly this "which, by a strange fatality, has hitherto been my lot."[23]

Young's uncompromising style was the least of the Royal Institution's prob-lems at this point. Its founders held different and to some extent incompatible views of the purpose and necessity of scientific work, and these contradictions became more apparent as the institution matured. For Rumford, the Royal In-stitution existed for the production and dissemination of practical knowledge to improve the lot of London's poor, knowledge that should be made available to all regardless of what, to Rumford at least, were the petty and irrelevant concerns of social class. This stance alienated those who viewed science as a gentlemanly pastime. With an old Eton friend by the name of John Hookham Frere, future prime minister George Canning wrote a satirical "Prospectus for the Royal Institution" that suggests the mocking ambivalence with which it was met:

> Too long, alas! has human kind
> In ignorance perverse and blind
> Plodded straight on: without a care for
> Cause and effect, and why and wherefore
> . . . .
> 'Tis time at length that abstract science
> With useful art should form alliance—
> Should quit her academic leisure,

To dig, and spin, to gage and measure.
Turning the abstract law to practical,
Should teach mankind in strain didactical
The art of mending chairs and mats,
Of frying sausages and sprats:
How shining ploughshares turn the ground up,
How watches make a noise when wound up.[24]

Ginning up a list of supposedly useful items to be invented at the Royal Institution, the authors of this spoof proposal took an indirect swipe at Young while directing sourness at those, like him, whose interests extended to antiquity. Among the benefits that the institution could be expected to bestow, Canning and Frere teased, there was promised "a barrel-organ framed by Handel / a scheme for putting out a candle . . . / A jar of pickles sealed hermetically / A course of lectures read pathetically." If that weren't enough, there was also a "cast of a dying gladiator" matched by "a late-invented nutmeg grater." Lest the linguists feel left out, there was also a "dictionary, of Greek and Hebrew" and, to wash it all down, "a receipt for better beer than we brew."[25]

It is difficult to overstate the social provocation that the Royal Institution presented to well-heeled Londoners with the inclination and leisure to study natural philosophy. "This Institution of Rumford's furnishes ridiculous stories," snickered Elizabeth Fox, Lady Holland, an acute observer of the Royal Institution's struggle for legitimacy. Relating an account of an experiment with nitrous oxide, the so-called laughing gas "so poetically described by Beddoes," she reported that the substance was first tried on "a corpulent, middle-aged gentleman" who, when asked to describe its effect, responded, "Why, I only feel stupid." According to Lady Holland, "this intelligence was received amidst a burst of applause," though "most probably not for the novelty of the information." The next experimental subject, Sir John Coxe Hippisley, was a founder of the Royal Institution and one of the insiders who opposed Rumford's wish to further aims of social improvement. In a moment famously captured by cartoonist James Gillray, the gas entered Hippisley's mouth but exited his posterior (fig. 5.2).[26] "The *effect* upon [Hippisley]," Lady Holland sneered, "was so *animating* that the ladies tittered, held up their hands, and declared themselves satisfied."[27] Beyond these problems of public relations, it was difficult to reconcile experimentation for serious purposes with the institution's commitments to utility, especially in the domestic arts, and to the advancement of knowledge among working people. Fittingly, Young's experiments were limited to a single, domestically oriented contribution, recorded in the *Journal of the Royal Institution*, involving "[bones] taken from a piece of beef which has

FIGURE 5.2. James Gillray on an unfortunate effect of laughing gas. © Science Photo Library Limited.

been roasted, reduced to a smooth paste, and made into a quart of broth with proper vegetables." The broth, it was recorded, "showed no tendency to jelly when cold," "tasted only of vegetables," and there was "nothing disagreeable in its insipidity."[28] Though Young was no chef, his experiment had shown that even offal had its uses—a point of particular anxiety at a time when increased taxation made even the well-off defensive about the contents of their larders.[29]

The Royal Institution soon ran into what one contemporary delicately called "pecuniary embarrassments."[30] Its supporters had been unable to persuade London elites that it was desirable from any standpoint, social or scientific. Henry Brougham, who had attacked Young, disparaged the institution for focusing too much on "fashionable science" suitable only for ladies.[31] Rumford's personality seems to have added to the institution's woes, as he was a peculiar and intense person, with a tendency to buttonhole listeners at social events and lecture them on his latest schemes. His eccentricities included an insistence on the energetic superiority of wide wheels over narrow ones, in defense of which theory he would later drive around Paris "in a broad-wheeled chariot." Similarly, in an effort to "economize the heat of his person," he wore only white clothes during winter.[32] In 1801, there appeared "An Ode to Count Rumford," a scurrilous portrait in verse by "Peter Pindar," the pen-name of John Wolcot.[33] This "Ode," which ran to several hundred frequently scatological lines, was so pointed that Young, many years later, was moved to respond, with some heat, that Rumford "was even mortified by becoming, in common with the most elevated personages of the country, the object of the impertinent attacks of a popular satirist."[34] Eventually Rumford removed himself from the Royal Institution, fed up with London society's insistence that economic handling of resources for the improvement of the conditions of the poor was an unconscionable gaucherie.

On June 6, 1803, after two years at the Royal Institution, Young submitted a letter suggesting that he, too, intended to resign. For his pains, he was rewarded with an invitation to deliver another twenty lectures on a topic and terms of his choice. But his friends were pressing him to return to more lucrative medical practice and to follow his neglected but still substantial professional prospects in that field. Before making the final break, Young revised his lectures for publication, with the aim of producing a durable capstone to his years there. "When this work is finished," Young insisted, prematurely as it turned out, "my pursuit of general science will terminate; henceforwards I have resolved to confine my studies and my pen to medical subjects only."[35] This resolute mood correlated with significant changes in Young's personal circumstances. Having married Eliza Maxwell in 1804, Young had many fresh responsibilities—a new household to provide for, another family to integrate with his own, another person's interests and happiness to consider. Brocklesby's bequest would take them a certain distance, but the creature comforts it afforded were not unlimited.

The revised lectures eventuated in two quarto volumes of some 750 pages each, with forty plates and nearly 600 figures, maps, and other illustrations. Joseph Johnson, a prominent publisher with offices in St. Paul's Churchyard, not far from Longmans in Paternoster Row, contracted with Young for the work. A Dissenter himself, Johnson built his list around works by like-minded fellow travelers, including the natural philosopher Joseph Priestley and luminaries like Mary Wollstonecraft, Thomas Malthus, and Tom Paine; he also published works by the physicians George Fordyce, who expounded on the salutary effects of Chapters' punch, and Thomas Beddoes of laughing-gas fame. It should have been a pleasurable and beneficial engagement, as Johnson was keenly aware of the manifold benefits of creating a community from a booklist; like Longman, he held weekly suppers where his writers could mingle and relax in his intriguing upstairs dining room, known for its "odd shape."[36]

Unfortunately Young's relationship with Johnson did not prosper, and the publisher dragged his feet over the publication of Young's lectures. Although Johnson had gone bankrupt during the preparation of Young's materials, his hesitation may have had less to do with a shortage of funds than with an insufficiency of confidence in Young's abilities after his fight with Henry Brougham in the *Edinburgh Review*. In late 1804, George Ellis complained to Walter Scott, who had deep and long-standing ties to the *Review*, that Johnson had buckled under "the ridicule" that had greeted Young's work, and moreover that this ridicule "had so frightened the whole *trade* that [Johnson] must request to be released" from his contract with Young. Scott replied, sympathetically, that the story, if true, was a "very pitiful catastrophe," and that "one must always regret so very serious a consequence of a diatribe." The truth, as Scott saw it, was that Brougham had acted in an infantile way, failing to keep in mind "the fable of the boys and frogs," a reference to Aesop's tale, in which boys throwing stones into a frog pond are chastised by the frogs, whose lives are threatened by such behavior, and that "it is much more easy to destroy than to build, to criticise than to compose."[37] For Young, Scott's support was, at best, cold comfort. The *Lectures* were published belatedly in 1807, but Young never received the thousand pounds he was promised for the work.[38]

Leaving the Royal Institution also meant losing Banks's patronage. Although his connections in London's scientific and antiquarian circles would be of comparatively little help to Young in obtaining a good medical position, Banks's support had at least been reliable. Different patronage networks operated in medicine, and Young had been out of that world for several years. Between 1803 and 1809 he worked hard to catch up, taking the medical degrees at Cambridge that he needed for admission to the Royal College of Surgeons. In May 1806, Young's name turned up among candidates for the position of physician at Middlesex Hospital, but he could not muster the votes needed to secure the post; nor did a second attempt avail. He did obtain a lecturing position at Middlesex, and during the winter of 1808–9 he was much occupied with this material, as well as with preparations for an admissions examination to the College of Physicians, to which he was elected as a fellow in December.[39] He lectured on various subjects—physiology, chemistry, nosology, materia medica—and eventually compiled these lectures in another book, *Introduction to Medical Literature*, which was published later, in 1813, by Underwood and Blacks.

By dint of this effort, Young was appointed as a physician to St. George's Hospital on January 24, 1811. Young's friends pushed him to apply for the position despite his lack of significant clinical experience, and despite the unpromising fact that the favorite to succeed Edward Nathaniel Bancroft, Bond Cabbell, was a particularly beloved scion of Lord Aberdeen, the senior physician at St. George's. To Young's surprise, his appeal found an appreciative audience, and he prevailed over Cabbell by eight votes—not a particularly comfortable margin, but enough to tip the balance. "[I]t is remarkable," Young mused, upon learning of his victory, "what a variety of interests I have been obliged to bring into play; scarcely any one of my friends having procured for me more than two or three favourable answers . . . every one lamented how very little he could do; yet the aggregate was sufficient for the purpose." Winning popularity contests had never been Young's particular talent—quite the contrary—yet circumstances had forced him to develop in this direction. Drily he hinted at the private toll exacted by this period of uncertainty: "Mrs. Young has emerged from death to life by the event of this contest."[40]

Located in Hyde Park Corner, St. George's Hospital opened in 1733 after a dispute among the board members at Westminster Public Infirmary went so sour that a faction broke off to start a hospital of its own.[41] The hospital was situated neither so far from central London to fail to attract patients nor so close as to be overwhelmed by them, in a setting known for its salubriousness due to its proximity to the fresh air of Hyde Park.[42] One observer, the publisher and writer Charles Knight, not only praised "the magnificent scale of St. George's Hospital" as "worthy of the capital of a great nation" but also found the neighborhood remarkably congenial, "a place that makes [the visitor] feel out of town the moment he reaches it."[43] Lanesborough House, where the hospital was established, had been built in 1719 by James Lane, 2nd Viscount of Lanesborough who, sensible of the merits of a home that was not too centrally located, had inscribed over the door the following motto: "It is my delight to be, Both in the town and country."[44]

By the time of Young's arrival in 1809, St. George's had established itself as a leading teaching hospital. A number of famous doctors—Benjamin Brodie, Robert Keate, Everard Home—had joined the staff. The hospital itself encompassed three large floors of wards, offices, and rooms for the study and display of anatomical specimens. The wards could accommodate two hundred and fifty patients, who were treated gratis, with the exception of soldiers diagnosed with venereal disease, who were charged 4d per day.[45] Gin was forbidden, but milk flowed liberally, from both cows and donkeys (the latter was considered a potent cure-all). Patients who died were buried in the hospital's cemetery at Brompton. In the hospital's early years, a Mrs. Hoare, who was responsible for the burials, was not paid directly for this service but the hospital agreed to purchase all the vegetables she grew on the cemetery's land.[46]

Despite this efficient organization, problems plagued the hospital. The physical plant was crumbling. Patients and staff alike had to contend with a leaky roof and faulty plumbing, not to mention the rats in the drains.[47] Worse, the doctors were deeply divided over the subject of teaching, perhaps because of the income students represented, for doctors at St. George's took fees from students rather than from the patients they treated.[48] Young's experiences with his colleagues, students, and patients were disappointing but not surprising, as the fractious institutional environment naturally did little to induce Young to abandon his customary reserve. After Young's death, Brodie recollected that Young's students "complained that they learned nothing from him," and pointed out that his apparently stable clinical track record—he tended neither to kill patients nor to work miracles—could be attributed to the fact that few sought him out in the first place, as he was not often recommended by the private practitioners who referred patients to St. George's.[49]

Perhaps Young compensated for his indifferent clinical talents by turning to what he was truly capable of doing, namely, gathering vast amounts of material relevant to some subject and transforming that material into lengthy, list-like works filled with obscure details. To take just one example: his 1815 volume on consumption, from which he had long before suffered, combined direct observations of his own with references to the disease, in which it seems that nothing was too ancient or too trivial to mention. "I have endeavoured," he wrote in its preface, "to comprehend in this work every fact of importance, which I have been able to observe, or find recorded, with respect to the nature and cure of the diseases belonging to a single genus."[50] Valuing comprehensiveness over selectivity, he preferred to canvass ancient sources as completely as possible rather than using facets of that source material to advance a point of view. As we shall see, these priorities, which gave an antiquarian cast even to his medical research, would also characterize his approach to the ancient Egyptian scripts.

CHAPTER 6

THE HERCULANEUM
PAPYRI

The antique world was all around Young in these years. After the British defeat of Napoleon's forces in Egypt, ancient artifacts seized from the French began to reach England in considerable numbers, reinvigorating an "Egyptian Revival" that had been under way for some time in the decorative arts.[1] These motifs only grew more popular as images from Napoleon's Egyptian expedition multiplied and spread.[2] Reminders of Egypt appeared on monuments and in architecture. In 1808, the architect Sir John Soane famously gutted the central portion of his Holborn residence to create a dramatic, top-lit, three-story void, at the bottom of which reposed an Egyptian sarcophagus, glittering with traces of its original gilt. In 1810, a building near Young's home on Welbeck Street was renovated with an Egyptian facade, making it one of only two buildings in London with overtly Egyptianized architecture. The other, a monumental structure known as Egyptian Hall, was built in 1812 from the artist Dominique-Vivant Denon's drawings of the temple at Dendera and housed the ethnological collections of William Bullock. Young would have often passed both buildings, as Egyptian Hall was just steps from St. George's in Piccadilly.[3]

Although some reminders of ancient Egypt were prominently visible, not every artifact enjoyed the same high profile. More modest items circulated less publicly among a coterie of specialists and antiquarians. For instance, following Napoleon's capture of Naples in 1799, a cache of carbonized papyri scrolls discovered in the ruins of Herculaneum were shuttled to and from the exiled court in Palermo until they were finally forgotten at Naples just before the city became, temporarily, an English protectorate. During the six years of English rule that followed, the Prince of Wales supported efforts to unroll and transcribe the papyri. He selected the Reverend John Hayter, future chaplain to George IV, to oversee the project.[4] After traveling to Naples in 1800, Hayter, working with eleven assistants paid for by the British parliament, claimed to have opened two hundred of the charred scrolls, increasing by nearly a factor of ten the number unrolled of the eighteen hundred or so that had been discovered. Two assistants, Carlo Orazj and Giuseppe Casanova, produced apographs (scribal facsimiles) of some, if not all, of the two hundred.[5]

When the French retook Naples in 1806, the court, with Hayter in tow, absconded once again to Palermo, taking the apographs but not the scrolls. After a struggle, the newly appointed English plenipotentiary to the area, William Drummond, convinced the court at Palermo to hand over the apographs in August 1807. He passed them to Hayter, who managed to engrave only part of one. This disappointing performance may have reflected the encroachment of other interests, since a year later Drummond excoriated him for frequenting local brothels and for his apparent involvement in a scandal involving a Neapolitan baron's daughter.[6] "I am sick of the stories of your battles in the brothels," Drummond griped, not omitting to point out that even while at Naples, Hayter had opened only half as many scrolls as he had claimed. After dismissing Hayter, Drummond convinced the court to allow the apographs to be sent to England for publication under the auspices of the Prince of Wales, who soon had them transferred to Oxford, where they remain.[7]

Meanwhile, a set of similar scrolls had come to Young's attention. These six scrolls had been sent to London years before in their original, unrolled state as a present from the Neapolitan court to the Prince of Wales. A handful of interested parties, including officials from the Royal Society and the British Museum, were attempting to determine the best method of unrolling them. Young was selected to assist. His medical and scientific training were particular points of favor, for he was "known to have formerly employed himself in minute anatomy, and to be familiar with the processes of mechanics and the operations of chemistry."[8] Working on the scrolls, Young experienced firsthand the physicality of an ancient object. Solving the problems presented by the papyri drew on his varied skills. The scrolls' recalcitrant forms, their very materiality, demanded his dexterity, which had been honed from years of handwriting practice, anatomical dissection, and painstaking experimental work. He was no less dexterous in making sense of these physical remains of ancient words by linking their spatial dispositions to meaning. When, shortly thereafter, he grappled with a publication that offered a transcription and translation of one of the scrolls, he was prepared to offer a critique that joined physical structure to his understanding of Greek meaning and syntax. These methods carried forward into his earliest engagement with Egyptian hieroglyphs to combine in fruitful ways with his vision of the evolution of language in connection with the forms of human vocalization, tempered by his absorption in ancient Greek, which late eighteenth-century English classicists perceived to be the epitome of perfected linguistic structure.

To unroll the papyri, Young devised a method that drew on his anatomical expertise, leavened by the knowledge he had gained in his experiments using air to investigate the spread of sound. Breathing humid air through a tube onto the charred scroll, Young attempted to use an "anatomical blowpipe" instead

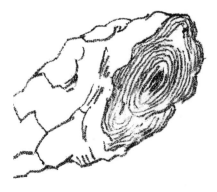

FIGURE 6.1. Contemporary drawing of a Herculaneum scroll from F.C.L. Sichler's *Die Herculanischen Handschriften in England* (1819).

of "a dissecting knife." The method was reminiscent of one recalled, years later, by an unnamed fellow at Cambridge who had "once found [Young] blowing smoke through long tubes, and I afterwards saw a representation of the effect in the Transactions of the Royal Society to illustrate one of his papers upon sound."[9] To keep unfurled leaves apart, Young proposed the use of goldbeater's skin, the fine coating of a lamb's intestine used to make gold leaf, following the practice of Antonio Piaggi, the "ingenious monk" who was the first, the most careful, and the most successful unroller of papyri scrolls. Even though the breath's humidity caused a layer of papyrus "to curl up and separate from the parts beneath where the adhesion was strong," this expedient ultimately proved a failure. Young next tried "chemical agents of all kinds," even "maceration for six months in water," to little avail. He proposed, presciently enough, leaving to the future further work with these fragile relics.

In 1810 a volume entitled *Herculanensia* appeared, containing "archaeological and philological dissertations" on the papyri, together with transcriptions of "a manuscript found among the ruins of Herculaneum." Its two authors were William Drummond and Robert Walpole.[10] Walpole, a rector in Norfolk, had graduated Cambridge with a prize for a Greek ode; his translation of Greek comic fragments into Latin and English verse appeared in 1805.[11] Drummond, of course, was the appalled administrator who had berated Hayter before leaving Naples to become ambassador to the Ottoman Porte. In 1797 his translation of the *Satires of Perseus* had earned him something of a reputation as a classical scholar. Then, in 1805, his controversial *Academical Questions* reached print. This work pleaded for ancient philosophy as a prophylactic against the baleful influence of Scottish common sense philosophers like David Hume and Thomas Reid. "The schools of Greece contributed not less to the glory of that country, than its civil institutions and its military fame," he opined. "Let not those, who would be steady in their opposition to all speculative reasoning, and who would still persevere in imposing new fetters on philosophy, ever tread the soil where the sages of Greece conversed with their disciples." Unsurprisingly, the *Edinburgh Review* responded with less than perfect sympathy. *Academical Questions* seemed to have "no other view . . . than to expose the weakness of human understanding, and to mortify the pride of philosophy, by a collection of insolvable cases, and undeterminable problems."[12] As we shall see, the Herculaneum papyri offered Drummond an opportunity to revisit points adumbrated in his *Academical Questions*. Thomas Young would take a peculiar double-edged ax to these hopes: deploying the minute source criticism that was supposedly Walpole's speciality, Young decimated their translation of the papyri and, in turn, undermined both Walpole's authority as a translator and the conclusions Drummond so eagerly drew from their translation.

Dedicated to the Prince of Wales, who had after all put Hayter to work on the papyri, the *Herculanensia* consisted of ten "dissertations." The first seven ranged from the population and topography of Herculaneum through discussions of the Greek language, Roman painting, and the materials used in writing. Some were written by Drummond, others by Walpole. The eighth, by Drummond and written "at Palermo in the year 1807," engaged the paleography of the scrolls. The ninth, also by Drummond, provided a discussion, transcription, and emended text of a "manuscript of Herculaneum" that, it was claimed, had the title περι των θεων (*On the Gods*).[13]

Drummond maintained that the manuscript in question represented "the sentiments of an Epicurean, concerning the system of theism professed by the Stoics," and that Cicero himself had drawn on it in parts of his *De Natura Deorum* (*On the Nature of the Gods*). Drummond was, however, not concerned with the transcription and emendation per se, for which he declared to have relied on the "Academicians of Portici," the group that had been working with the papyri in Naples since the 1750s. These, he assumed, were the fragmented manuscript's emending editors—a curious position since, having employed Hayter and his crew, he had good reason to think otherwise. Drummond was principally interested in finding Epicurean accusations against the Stoics in the manuscripts. Both sects, he declared, "loudly accused each other of that atheism, of which both affected to be abhorrent, and of which both were indubitably capable." Explaining this mutual finger-pointing allowed him to expatiate at length on these philosophical positions, though he well knew that his *Academical Questions* had already aroused "the censures of some critics who were angry at hearing so much of Greek metaphysics."[14]

To others, the scrolls represented an opportunity to do the more modest work of correction and emendation. Charles James Blomfield, a fellow of Trinity College and recent author of a commentary and edition of Aeschylus's play *Prometheus Bound*, reviewed the *Herculanensia* for the *Edinburgh Review*. He was unenthused by the papyrus, remarking that the "MS is in a very mutilated state; and, from the mode in which it is printed, any attempt to fill it up from conjecture is almost a hopeless task; but, in fact, it does not seem to be worth much trouble; and if all the unedited Papyri contain matter as little interesting as those already published, no good is likely to result from their being given to the world, but an accession to our stock of paleographical knowledge, and not even that, unless future editors will give accurate copies."[15] Since Drummond had so written, Blomfield attributed the effort to correct parts that could not clearly be read to "the Academicians of Portici" whose labors were "often singularly unhappy, but sometimes successful." (The author of both the transcription and the emended text was not an "Academician of Portici," however, but Hayter himself—who, we shall see, reacted hotly to the *Quarterly Review*'s critique of precisely this part of the *Herculanensia*.) Drummond offered "one or two conjectures" of his own about that, though nothing extensive, and neither did he essay a translation. Just after sending the review off to Edinburgh, Blomfield learned that Young had produced his own critique for the *Quarterly Review*, in which Young had provided an elaborate reconstruction along with a translation, going so far as to retitle the fragment altogether. As a result, Blomfield apparently tried to stop publication of his own effort, to no avail.[16]

Young's twenty-page review began on a positive note, remarking that "this highly interesting volume must ever be considered as a memorable event in the history of classical literature," since it included only the second of the Herculaneum manuscripts that had ever been printed. He commended the "profound erudition and extensive knowledge" displayed by Drummond and Walpole, and even quoted at some length Drummond's account of the scrolls' history. Brief remarks, also positive, followed concerning the first four dissertations. The fifth dissertation, by Drummond, on the names of places, attracted Young's criticism. In his view, Drummond had been "somewhat precipitate in his conjecture respecting the sense of a passage in the Bacchae of Euripides," which Young proceeded extensively to correct. He made other "observations" of this sort, his critique escalating until he reached the eighth dissertation, by Walpole, on paleography. Here Young first brought to bear his views concerning writing.[17]

In 1807, Walpole had observed that all but one of the opened scrolls were written in Greek, and that all had been written in capital letters with neither spaces nor accents. Even so, Walpole insisted that accents had existed as early as the 133rd Olympiad (248 BCE), citing for evidence the orthographic features of a Herculaneum wall inscription as well as passages from several authors. Young, in contrast, merely recurred to a "usual belief" that accents had been invented a half century later by Aristophanes of Byzantium and maintained that the passages in question referred to pronunciation rather than orthography.

This apparently minor question was of surprisingly profound significance. Its resolution would shape any attempt to restore the text of the fragment, as specialists in paleography knew from long and hard experience. Debates concerning Greek accents, particularly in ancient poetry, had been ongoing for nearly a century and a half. In 1673, Isaac Vossius had insisted that accents should be ignored altogether. The accents normally used, Vossius maintained, were late inventions that did not exist on inscriptions, coins, or manuscripts known to be early in any case. Vossius argued that the stress patterns provided by accents should be replaced by emphases in proportion to a syllable's length, and the result should be metrically appropriate.[18] John Foster, a notably ineffective upper master at Eton, where the use of accents in teaching Greek was a sine qua non, revived the matter in a 1773 diatribe against Vossius. One of Foster's principal claims was that the use of orthographic elements such as accents was closely tied to pronunciation and, in turn, to the anatomical organs of speech. According to Foster, Vossius and his followers had failed to consider that "as vocal sounds are formed by organs of speech which are essential and immutable parts of our nature, they must have been in all ages *substantially* and *formally* the same, tho' variously *modified* in their application . . . I have therefore drawn an argument from the nature of speech itself, in proof of the existence of ancient tones distinct from quantity."[19]

Young may not have read Foster—there is no evidence that he did—but he would certainly have agreed that inscription of the human voice must be connected to the nature of speech. At Göttingen, Young sought to construct phonemes out of a small set governed by a Newtonian-like rule, and he had, years before, read Gregory Sharpe's argument for a strong connection between the structure of language and the voice, where Sharpe had also argued that the earliest inscriptions to capture speech (Hebrew, of course) had necessarily omitted accents. Young would likely not have disagreed with Sharpe about that, but he certainly had no doubts about the appropriateness of supplying accents in order to capture the full sonic expressiveness of discourse. He even criticized Drummond for "disfiguring" his Greek by omitting them. What did grip Young was the variety of possible connections linking pronunciation to inscription, which was why he commented on Walpole's remarks and delved briefly into the accent-quantity argument.

Musical notation provided Young with a good example of how articulation might be captured in writing, for it was "capable of expressing length or shortness, acuteness or gravity, and force or softness, all accurately, and independently of each other."[20] Music had served Young a decade earlier in developing an understanding first of sound and then, through that and by extension, of light itself.[21] For him, acoustic expression, whether of music or speech, operated according to rules derived from the anatomical structure of the ear or the vocal apparatus, or from innate perceptual and mental processes.[22] Inscription would necessarily evolve to capture subtle intonations, with the more developed systems embodying these governing rules. Systems that did not do so would be subject to the vagaries of culture, an effect Young strongly disliked. To avoid using accents in representing ancient Greek was, for Young, tantamount to reverting to an early, comparatively primitive, and even morally objectionable form of Greek writing.

Just as Young took exception to Walpole's having overlooked the foundation of writing in the representation of pronunciation, he also objected to the transcription of the manuscript fragment that appeared in *Herculanensia* and the translation to which the transcription gave rise. Because Young held that inscribed meaning depended upon the spatial imprinting of sound in specific ways, it was essential to consider the physical placement of words in the Herculaneum papyri. Young accordingly faulted the fragment's transcription for being insufficiently attentive to location, in particular to the blank places where words might once have been. Attending to the spatial character of the scroll, Young proposed a different restitution of the passage, including a new title. As a result of this effort, contemporary English classicists began to take Young seriously. According to his first biographer, Young's review "at once placed its author, in the estimation of the public, in the first class of the scholars of the age."[23]

Although the apographs deposited at Oxford included the one discussed in the *Herculanensia*, Young had not seen Casanova's original, while the *Herculanensia* contained only Hayter's rendition. Hayter's version departed from the original in several respects. Although the original was entirely uppercase and lacked spaces, Hayter had rendered it in lowercase with lacunae. He appended a reconstruction of the text that filled in these empty spaces, substituting words and phrases where Casanova struggled to read the text. Certain that the sense of the document could only be divined from a reproduction that did not introduce such elements, Young produced, and included in his review, what he termed "a specimen of the state of the manuscript," giving "its first page, which is the most defective, as nearly as possible in the form in which we suppose it to stand." The differences between the *Herculanensia*'s fragment and Young's reconstruction reflect Young's conviction that, when attempting to determine the meaning of inscriptions of a phonographic language, the spatial arrangement of the inscription must be taken into account (fig. 6.2, first three images).

Here Young's early experiences in drawing and in calligraphic technique stood him in good stead. To restore the physical form of the fragment, Young carefully reworked the extant text as reproduced in the *Herculanensia*, eliminating spaces and using only uppercase letters. Young's restoration matches up well with the original apograph—which, again, he had not seen. In line 17, for example, Young eliminated spaces and moved the letters, now uppercase, to the right, *precisely as in the (unseen) apograph*. To six lines of the *Hercula-nensia* version, Young added a letter to make sense of a word. In one amended line (no. 26) Young omitted six letters—των ανο—from the *Herculanensia*'s transcription. "This line," Young wrote, "as printed, contains twenty-five letters, and is totally unintelligible: by leaving out των ανο it may be made to accord perfectly with the context." Comparing his version to the apograph by Casanova, we find that he has matched the spacing of the apograph correctly.[24]

Not content merely to regenerate the original on which the apograph was based, Young next recreated the entire passage. His notes apparently no longer survive, but Peacock had them to hand while writing his biography. Young, wrote Peacock, had worked "an exquisite copy of the entire manuscript—which is now before me—with the lacunae filled up in a differently coloured ink." This expedient had eventuated in an "intelligible text, which is good Greek," while allowing "the reader to judge of the propriety and probability of the restorations which are proposed, not only with reference to the space which they occupy, but likewise to the words or portions of words which precede and follow them."[25] Although Young did not include in his review more than the fragment's first column, even a quick comparison of his reconstruction with Hayter's shows substantial differences (fig. 6.2, final two columns).[26]

Young also found grounds for changing the title. Instead of the *Hercula-nensia*'s περι των θεων (which appeared nowhere in the transcription), Young turned to the final four lines of the fragment's last page and extracted ΠΕΡΙ ΕΥΣΕΒΕΙΑΣ ΚΑΤ ΕΠΙΚΟΥΡΟΝ, leading him to retitle the passage as "A Treatise on Piety, according to Epicurus." This result differed completely from Drummond's, with devastating effects on the latter's argument concerning Cicero. Having translated the title as "On the Gods," Drummond was convinced that the fragment had provided Cicero with the fourteenth, fifteenth, and sixteenth chapters of his *De Natura Deorum*, in which Cicero has Gaius Velleius represent Epicureans. Young demurred, and to prove the point he translated the passage of Cicero to which Drummond referred and listed the supposedly comparable lines. "The reports are not by any means so precisely similar," he wrote, "as to induce us to suppose, that Cicero had ever undertaken the justifiable liberty of saving himself some little trouble by making use of another author's abstract from Chrysippus, and from Diogenes the Babylonian."[27] This coup de grâce was preceded by a complete translation into English of the re-created passage.

Young concluded with a provocation directed at whomever had produced the textual re-creations. The "editors of Philodemus," that is, the Epicurean referenced in the papyrus, "are universally allowed to have succeeded admirably

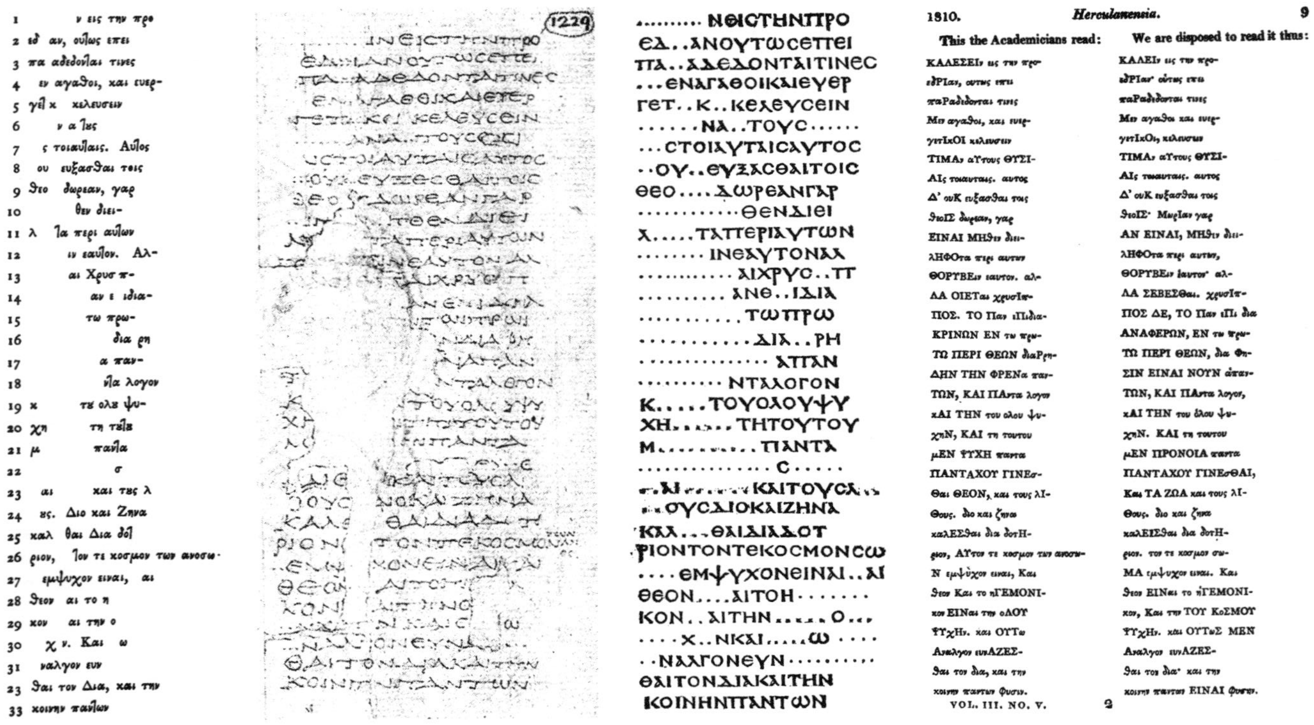

Herculanensis.

1810.

This the Academicians read:

We are disposed to read it thus:

FIGURE 6.2. (*First on the left*) John Hayter's version of the Herculaneum fragment; (*second*) Giuseppe Casanova's apograph (The Bodleian Libraries, University of Oxford MS Gr. Class. c5, 5, fol. 1230); (*third and fourth*) Young's restoration and re-creation of the fragment compared with the *Herculanensia*'s.

in their attempts to restore the genuine text of their author," he wrote. On the other hand, "we are very sorry to observe how lamentably the modern Academicians of Portici," that is, Hayter, "have fallen short of their predecessors." Hayter quickly and rather ironically sought Drummond's support as he drew up a response to Young's review, which he described as a "severe animadversion." Young's emendations to the apograph particularly irritated Hayter who, finding them entirely faulty, told Drummond that "the supposed facsimile of the Reviewer is not a facsimile of the original, neither in the distances, nor in the form of the characters."[28] Hayter may have claimed as much, but we have already seen that, "facsimile" or not, Young's proposed revision of the transcription does fairly well match the original.

Because Young's critique implicitly tarnished the diligence and competence of the apograph's transcriber, Hayter devoted much of his reply to insisting that his dimensions for the lacunae were accurate. Not only had they been "taken with much care by the copyist," but Hayter had also personally checked them against the original. He insisted that his own "conjectural letters" took account of potential sources of copying errors, and perhaps he had (though Young's point was not that the *Herculanensia*'s transcription added too many letters for the extant spacing, but that it made little sense in context).[29] Finally Hayter claimed that, unlike his own reconstitution, Young's emendations did not fit the apograph. In Young's original "facsimile"—or a copy of it, obtained perhaps from Young himself—Hayter spied an apparent inconsistency with Young's printed revision. In the ninth line of his reconstruction of the apograph, Young gave ΔωPEAN, "free," for Hayter's lowercase δορεαν, whereas in Young's proposed reconstitution of the text, the word ΜωρΙαν, "folly," appears instead. The apograph's ninth line does appear to contain a Δ in the appropriate place, but the letter is broken and was likely hard for Casanova to read. Young had substituted "folly" because the latter made sense in context and accordingly justified his emendation of a character that Casanova may well have mistaken. Hayter nevertheless fastened on this anomaly as evidence of Young's incompetence.[30]

For the third time Young was faced with a public critique of his work. Innuendo had greeted his early exploration of the eye's focusing mechanism, while Brougham's disparagement of his new theory of light had appeared, humiliatingly, in the *Edinburgh Review*. But this time, the criticism stung less. Printed privately, Hayter's response seems to have had little effect, perhaps because he had unwisely included Drummond's letter of rather lukewarm support with the publication.[31] That Young's review was published anonymously may have additionally reduced any pressure he felt to reply, which he did not. The damage, at any rate, had been done. Four decades later, Peacock noted that "no subsequent attempt was made to re-establish the character of Mr. Hayter as a scholar, or to vindicate his conduct with respect to the abandonment of the papyri and of the copies made of them."[32]

Young's foray into the words and inscriptions of antiquity reawakened his long-standing interest in languages, their origins, and the ways in which their sounds could be visually represented. Two years after his essay on the *Herculanensia*, Young produced another for the *Quarterly Review* that delved directly into these issues. The occasion was a three-volume work on the affiliation of languages entitled *Mithridates* by the prolific German philologist and grammarian Johann Christoph Adelung, who from 1787 had been principal librarian in Dresden to the elector of Saxony.[33]

Young's elaborate review, which ran to forty pages, covered the spectrum of known languages, going well beyond those discussed by Adelung. As was his custom, Young presented his argument by giving extensive lists. This time he listed languages, but the method was just the same as his prior treatises on nosology and so-called consumptive diseases.[34] This was ever Young's technique when delving into an area: examine as much of the extant literature as possible, make lists, and develop categories based either on natural affiliations or similarities found by means of comparisons. "A perfect natural order of arrangements," he wrote in regard to languages, "ought to be regulated by their descent from each other and their historical relations: a perfect artificial order ought to bring together into the same classes all those genera which have any essential resemblances, such as are not fortuitous, nor adoptive, nor imitative or derived from onomatopoeia." These "essential resemblances" permitted one to group instances by shared similarities, creating an ordered reservoir of specifics from which generalizations could be drawn. "When our ideas are once stored up in the intellectual treasury," he continued, "they seem to possess the same property which belongs to their originals, allowing themselves to be traced at pleasure according to a variety of different principles of analogy and of association." Intriguingly, Young did not expect to proceed linearly in this activity: "Wherever the human mind pursues any process of nature, it must be subjected to the inconvenience of breaking off occasionally some one train of connexion, in order to pursue another; although that system must in general be the most perfect, in which this happens the least frequently."[35]

Here we see, directly stated, Young's general method and, as it were, philosophy: the "human mind" forges "connections" and in so doing is frequently deflected from one arena into another, just as Young himself was led from the voice to sound and then to light. Eventually a system may evolve out of the mélange that is not subject to redirecting connections but has a coherence of its own. Such a mental system would be one in which the tracing of links among its ideas would replicate the linkages among the ideas' "originals," that is, among their natural analogs. One detects clear echoes of Locke in Young's rumination, which echoed as well his own experiences in effecting a new system for light as well as in elaborating nosologies for diseases and, here, for languages. Within a few years he would take the same tack with Egyptian hieroglyphs.

Beyond presenting and critiquing Adelung's schema for languages, Young provided his own affiliations. As was his practice, Young alluded extensively to contemporary and earlier literature on the subject. His references ranged from the sixteenth through the early nineteenth centuries, covering works both on languages and, especially, on the variant printed forms in which a work had been made available, such as John Chamberlayne's polyglot *Oratio Dominica*, which printed the Lord's Prayer in a hundred languages with orthography proper to each.[36]

Echoing Adelung's linguistically and geographically wide-ranging discussion, Young devoted a substantial chunk of the review to Chinese and languages "immediately beyond" in Tibet. Young opened this discussion by focusing on the curious fact that these languages were all principally monosyllabic: "[F]rom this peculiarity, as well as from the singular simplicity of their structure, they are supposed to constitute the most ancient class of existing languages, though it must be confessed that much of our author's reasoning on this subject is extremely inconclusive." Despite this unprepossessing start, Young had nevertheless picked out a detail that seemed promising. He noted "a much more marked distinction between these and all other languages" insofar as "their essence consists . . . not in sounds, but in characters, which, instead of depicting sounds, are the immediate symbols of the objects or ideas."[37] This, Young, added, "we have already explained very fully on a former occasion," namely, in a detailed review of a recent volume on Chinese, Joshua Marshman's *Chinese Languages*, that had appeared in the *Quarterly* for May 1811.[38]

The review, which was published anonymously, was in fact the joint work of John Barrow and George Thomas Staunton.[39] Barrow had been in China for two years (1792–94) as a member of the mission led by George Macartney to secure British trading rights, while Staunton had been employed since 1798 by the British East India Company in Canton and later became its head there.[40] In July 1816, he had written another piece on Chinese for the *Quarterly* about which an "excessively delighted" Young enthused to his friend Gurney that "it gives me a most accurate and perfect idea of the singular structure of that unique language; I had no conception of its nature before: I should almost like to be acquainted with the 214 radical characters, merely as a matter of curiosity." Barrow's 1816 piece went into considerably more detail concerning Chinese word "radicals" than had his and Staunton's of five years earlier—which is likely why Young, punctilious at least since the Hunter affair about when and from whom he had first truly learned something, remarked, somewhat exaggeratedly, that he previously had "no conception" of Chinese.[41]

In their discussion of Chinese writing, Barrow and Staunton speculated on the manner in which the inscription of its sounds, and not only its meaning, might have evolved. Much had certainly been written about Chinese script by Jesuits, but, according to the reviewers, Marshman was the first to pinpoint "the connection between the characters and the system of sounds." They termed the basic system of inscription an "imitative alphabet," by which they meant its foundation in the representation of what they took to be the language's monosyllabic words by means of signs (in their words, "characters"). Chinese writing, they explained, dated back about four and a half millennia and originated in signs that bore "a resemblance to the objects which they were employed to represent." Barrow and Staunton provided a representative sample of such signs that they had produced on the basis of information to the Royal Society of London by the French Jesuit missionary Jean Joseph Amiot, then stationed in Pekin (Beijing) (fig. 6.3).[42]

The system, they explained, had developed by concatenation. Characters representing words or phrases were assembled out of component elements. As an example, they offered the character for "hunger," which was compounded out of those for "I" and for "eat." This "imitative mode of expressing ideas" was "antecedent to the invention of any alphabet." Although a phonography—a set of signs for Chinese monosyllables—had eventually emerged, it had never displaced this "imitative" system. This "alphabet" of some forty-eight characters was "simple in construction, effective in operation, and capable of being extended." Thirty-six characters represented initial sounds, while the remaining twelve represented finals. Concatenation produced 432 "simple monosyllabic sounds," which, to avoid ambiguity, required modifying the finals in various ways, and then further modifying the result by extending quantity and applying accents, to produce in the end more than two thousand distinct sounds.[43]

Barrow and Staunton attempted to trace the origins, development, and spread of this system and to explain how Chinese phonography was "derived from their hieroglyphic characters." They speculated that words arose from an original need to provide a "name" for a character that expressed a complex idea or foreign word for which there may have been no specific term—and so no vocalization. For example, the sign for the concept of "splendour or brilliancy" could reasonably be compounded out of the characters for sun and moon, but how then to "name" the result? Suppose, they ventured, that "the inventor" decided to name the idea *ming*. To communicate its sound, characters would

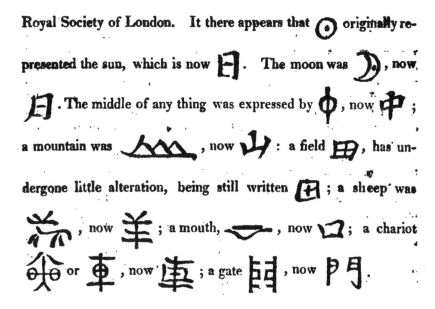

Royal Society of London. It there appears that ⊙ originally represented the sun, which is now 日. The moon was 🌙, now, 月. The middle of any thing was expressed by ⏀, now 中; a mountain was 〰️, now 山: a field 田, has undergone little alteration, being still written 田; a sheep was 羊, now 羊; a mouth, ⌣, now 口; a chariot or 車, now 車; a gate 門, now 門.

FIGURE 6.3. John Barrow and George Staunton's exemplification of original Chinese logograms.

have been sought such that the vocalization of the one might begin with the sound *m*, while the vocalization of the other might end with the sound *ing*. The characters, respectively, for *wood* and *blue* were vocalized as *moo* and *ching*. Barrow and Staunton proposed to take the consonantal beginning of *moo* and to it add the ending of *ching*, producing the desired compound *ming*.[44] Combining the characters for *moo* and *ching* would produce the representation in the script, with the understanding that the initial and terminal sounds of their Chinese names would alone be used. Expanded to a general procedure, such a process could, they theorized, have produced the Chinese "alphabet."[45] Nevertheless, the fact that the language's monosyllables also constituted its words inevitably militated against the widespread use of such a system. In Barrow and Staunton's opinion, the tight coupling of a character's meaning with the specific, univocal sound of the Chinese monosyllabic word that it represented made the replacement of an "imitative" script by a purely phonographic one highly improbable. Indeed, they thought it unlikely that any language whose words were altogether or principally monosyllabic would produce a widely used writing system that broke thoroughly from semantics.[46]

Seven years before this joint review with Staunton, Barrow's widely read *Travels in China* portrayed a country in decline. Once a comparatively advanced civilization, the Chinese had "remained stationary" for more than two thousand years, so that "compared with Europe, they can only be said to be great in trifles, whilst they are really trifling in every thing that is great."[47] We do not know whether Young read Barrow's *Travels*. He may have, as the book was extremely popular, going through three printings, two English and one American, in two years.[48] Barrow's scorn for civilizations without analogs to contemporary European material and scientific culture did not bode well for sympathetic understanding of ancient Egypt; excepting only the pyramids, the Egyptians had apparently matched neither the Romans in technology nor the Greeks in mathematics and science, at least by Barrow's lights. Barrow did distinguish Egyptian hieroglyphs from Chinese signs since he thought that hieroglyphs had never changed into graphically abstract representations with little resemblance to what they had originally been devised to represent. Nevertheless, what he had to say about the Chinese writing system applied equally well to hieroglyphics; and, damningly, neither civilization had originated an alphabet. "The invention of the Chinese character," he wrote, "although an effort of genius, required far less powers of mind than the discovery of an alphabet; a discovery so sublime, that, according to the opinion of some, nothing less than a divine origin ought to be ascribed to it."[49]

Young shared Barrow's disdain for non-Western cultures to a considerable extent. Years earlier he had absorbed from Sharpe the origination of writing in pictograms, which assimilated closely to Barrow's and Staunton's "imitative" characters. Their review, and perhaps Marshman's book itself and Barrow's more extensive discussion of China in his *Travels*, may have led Young further into the ways in which meaning could be conventionally represented by logograms, and the difficulties attendant on any attempt to transform such a system into one that could represent sounds. He would soon give a great deal of thought to this as he began to grapple with the two Egyptian scripts inscribed on the Rosetta Stone.

PART 2

ANTIQUITY EMBRACED

CHAPTER 7

WORDS FROM
EGYPT'S PAST?

In 1799, the French expeditionary force was garrisoned at Rosetta, the largest Egyptian town east of Alexandria. Located on the Mediterranean coast, it was a fine site to mount a defense against arriving British forces, and its appeal was increased by the nearby presence of a fifteenth-century Ottoman fort, known locally as Borg Rashid, that promised to serve as an excellent stronghold if properly renovated.[1] In mid-July, with construction well under way, Pierre François-Xavier Bouchard, a captain in the corps of engineers, was clearing debris from one of the fort's crumbling towers when he discovered a stone made of what appeared to be "fine black granite, with a very delicate grain and very hard to the hammer."[2]

The stone, which measured about a meter high and almost as wide, was 27 centimeters thick and weighed more than 700 kilos, as reported on September 15 in the occupation's newspaper, the *Courrier de l'Égypte*.[3] One side, well polished, was engraved with about a hundred lines of close, tiny writing, in three scripts—Greek at the bottom, a damaged hieroglyphic section at the top, and a particularly puzzling middle script. Bouchard, who had studied archaeology, translated the Greek. Its last line asserted that the decree carved on the rest of the stone was the same in all three scripts. In the initial report sent to Cairo on July 29, Michel Lancret wrote that this "second inscription, below the hieroglyphic part, is the most complete; it is composed of thirty-six lines in alphabetic characters that follow the inverse direction of the inscription above it."[4] Although the report cautiously averred that the nature of this script was "unknown," it was believed to represent some form of Syriac.[5] Under Bouchard's supervision, the stone was shipped a hundred miles south to Boulaq, on the outskirts of Cairo, and soon thereafter to the Institut d'Égypte in Cairo where it remained until the end of 1799.

Napoleon abandoned Egypt on August 22, leaving Jean Baptiste Kléber in uneasy command of the occupation's increasingly tattered remnants. As Kléber's hold on power weakened, the Rosetta Stone fell to the care of Jacques-François Menou, the flamboyant general who would succeed Kléber as army

FIGURE 7.1. The Rosetta Stone. Photo by Hans Hillewaert. https://commons.wikimedia.org/wiki/File:Rosetta_Stone.JPG.

commander after the latter's assassination the following year. Menou stored the artifact in his tent until the summer of 1801 when he transferred it, wrapped in thick mats, to a warehouse in Alexandria. When the British demanded rights to all antiquities seized by the French, Menou reported, disingenuously, that the French had in their possession only two sarcophagi. The rest, he claimed, were either already in France or the personal property of the French in Egypt and therefore exempt from seizure under the terms of the armistice. Menou argued that since the Rosetta Stone was his personal property, he was not required to hand it over.

Menou's claim to ownership came to the attention of Edward Daniel Clarke, an English collector, mineralogist, and naturalist who acted as General John Hely-Hutchinson's proxy during the British attempt to secure possession of the artifact. In his account of the negotiations, Clarke disparaged Menou: "A more grotesque figure could hardly be conceived." The short, fat Menou had stuffed himself into a "flowered embroidered waistcoat, with flaps almost to his knees," and over this ensemble he had laid "an overcoat covered in broad lace," much to the amusement of his more conservatively dressed British counterparts.[6] According to Clarke, Menou's responses to the demands of the British authorities were as hyperbolic as his couture. Standing just outside Menou's tent after one of these meetings, Clarke overheard Menou, "[his] voice elevated as usual . . . in strong terms of indignation remonstrating against the injustice of the demands made upon him. The words 'Jamais on n'a pillé le monde' [Never has the world been so pillaged] diverted us highly."[7] When Clarke asked Menou's permission to copy the stone's inscription, which would at the very least have forced Menou to divulge its whereabouts, Menou shouted that the British had "as much right to make such a demand as a highwayman has to ask for my purse!" Seeing his listeners struggle to suppress their smirks, Menou only became more enflamed: "[Hutchinson] has a cannon in each of my ears, and another in my mouth; let him take what pleases him. I have a few embroidered saddles and a tolerable stock of shirts, perhaps he may fancy some of these!"[8]

A second account of the Rosetta Stone's fortunes after Napoleon's departure from Egypt was provided by Tomkyns Hilgrove Turner, a fellow of the Society of Antiquaries who was responsible for seized antiquities in Egypt. In an account of 1810, Turner recollected that the French were less interested in preserving these antiquities than with avenging their defeat by destroying them. Although his account is prejudicial and, for that reason, difficult to credit fully, Turner did note a striking detail: "When it was understood by the French army that we were to possess the antiquities, the covering of the [Rosetta] stone was torn off and it was thrown upon its face." Turner reported that he had made "several remonstrances" against these and other desecrations, to no avail. He finally secured a sentry from the maritime prefect, one Monsieur Le Roy, who "behaved with great civility," quite unlike "the reverse [behavior] I experienced from some others."[9] After Turner applied to Hutchinson for assistance, reinforcements arrived in the form of "a detachment of artillerymen, and an artillery-engine, called, from its powers, a devil cart," a two-wheeled, single-axle gun carriage, also known as a triqueballe, used to transport heavy objects. Thus fortified, Turner arrived at Menou's warehouse, from which he "carried off the stone, without any injury," save whatever wounds he and his retinue may have incurred from the barrage of "sarcasms of numbers of [the] French officers and men" who stood by, jeering, as the cumbersome object was wheeled away through the neighborhood's narrow streets. The jibes were well worth it, however, as the Rosetta Stone was, Turner averred, not only "a most valuable relic of antiquity, the feeble but only yet discovered link of the Egyptian to the known languages," but also "a proud trophy of the arms of Britain." Indeed, he went on, "I could almost say *spolia optima*," but of course that is just what he did say, "not plundered from defenceless inhabitants" as the French had done, "but honorably acquired by the fortune of war." Which turned out to be fortunate indeed, at least for the British.[10]

According to Turner the war spoils were finally divided along natural/artificial lines: the materials in the natural history collections were returned to France as personal property while the "artificial" items, meaning virtually all the antiquities, were transferred to the British.[11] The Rosetta Stone was loaded onto a seized French ship, *the Égyptienne*, and sent to England under Turner's supervision. It arrived at Portsmouth in February 1802, and was eventually transported to the Customs House in Deptford, outside central London. By the time it reached the Society of Antiquaries on March 11 it bore two new inscriptions: "Captured in Egypt by the British Army 1801" and "Presented by King George III."[12]

A letter from Turner accompanied the artifact to the Society of Antiquaries in March 1802, explaining how it had come into British possession. In his subsequent recollection, Turner was careful to suggest that the object's missing fragments were likely to have disappeared during the French refortification of Borg Rashid, and furthermore, that "there was a stone similar at Menouf, obliterated, or nearly so, by the earthen jugs being placed upon it" and also that a comparable fragment had been "used and placed in the walls of the French fortifications of Alexandria."[13] Henry Salt, who was by 1815 the British consul-general in Cairo, obtained permission to dig there, hoping to discover the missing Rosetta fragment and other similar objects, as Young would note in 1816.[14] The Rosetta Stone remained at the Society of Antiquaries of London until mid-1803, when it was moved to the British Museum, where it remains to the present day.

Difficult as it may be to imagine now, the Rosetta Stone was not an immediate cause célèbre. Beyond a coterie of antiquarians and specialists in Greek, it generated little interest.[15] Early published reports about the stone were subdued, its significance largely lost in the general excitement over the British victory and the more picturesque and exotic spoils that were pouring into the country as a result. The arrival was belatedly noted in the *Times* on August 11: "About a fortnight since, a vessel arrived in the River [Thames] from Egypt, which contained a great number of Egyptian antiquities, collected by the French army, and become the property of the Conquerors." In addition to a sarcophagus, a gigantic stone hand, a "deified Ram's head," and a variety of other unusual objects, there was also "[a] very curious stone" that measured, according to the reporter's inaccurate estimate, "about seven feet long and five feet square." The stone, the report continued, "has three inscriptions in different languages, all supposed to be on the same subject." "The Greek has been made out," the report continued, "and it proves to be an Edict of their Priest for deifying one of the Ptolemies for his great and good deeds done for his country."[16] The *Times* writer was less impressed by the two unreadable scripts than by the sheer tonnage of the antiquities in the remainder of the shipment, which required a team of twenty horses for transport to the museum.

One of the first substantial notices of the Rosetta Stone appeared in a supplement to the *Gentleman's Magazine*, a widely read monthly compendium of news, literature, and science, for the year 1801. The notice, which described the stone as "a piece of black and extremely fine-grained granite," focused on the inscription and its potential to stir up scholarly activity as well as general curiosity. "General Dugua, lately returned from the Egyptian expedition, brought home two copies of a remarkable inscription," the report began.

> The inscription is three-fold: one portion presents a succession of hieroglyphicks in several very regular lines; another portion, which has not yet been sufficiently examined, presents a greater number of lines, in characters which yet have some uncertainty, and which require a very attentive examination; the remaining portion consists of 33 lines in Greek. One of the members of the French Institute, having undertaken to read and explain this part, thinks it a monument of the gratitude of some priests of Alexandria, or some neighbouring place, toward Ptolemy Epiphanes.

Lest readers feel unfairly tantalized, the writer noted that "Bonaparte, to gratify the curiosity of the literati in every country, gave immediate orders to have the inscription engraved; after which, it will be submitted to the examination of the learned throughout Europe." As in the report from the *Gentleman's Magazine*, the object was understood primarily as one among the spoils of the recent war. "The rare and valuable collections of plants, medals, &c. &c. made by the French Sçavans with so much toil and care in Egypt," the report continued, "[and] having been captured by the English army, will, no doubt, be brought to this country by General Lord Hutchinson. This is what Virgil would have said, *Sic vos non vobis*."[17]

Unlike other ancient Egyptian artifacts arriving on Britain's shores, the Rosetta Stone did not strike an immediate public chord. More popular objects—painted sarcophagi, glittering scarabs—were interestingly exotic even if their precise meaning and function remained unknown; these objects naturally elicited the greatest excitement among the admiring crowds who would eventually throng the British Museum. Yet, at least among specialists, the Rosetta Stone held out an important hope as a way to bring to life the sounds of ancient Egyptian. It might be made to speak—if only copies could be made of the inscriptions. After all, the artifact could only be seen in England. To bring the inscription to a broader audience, its carvings had to be represented and disseminated, increasing the odds that the scripts might eventually be read.

Years before, in Cairo, several people had attempted to copy the inscription. Three methods were used, with varying degrees of success. Nicholas-Jacques Conté, who oversaw the French occupation's workshops, coated the raised parts of the stone with a waxy substance that repelled ink, which accordingly filled the incised regions. This method produced a print with black characters on a white ground. The botanist Alire Raffeneau-Delile developed a second technique: he took a sulfur cast of the stone, which produced characters in relief. The orientalist Louis Rémy Raige and the printer Jean-Joseph Marcel, director of the press at Cairo, made copies of all three inscriptions using a quasi-lithographic method, wetting the incised regions of the stone and then applying ink to the whole. Because the water-filled crevasses repelled the ink, pressing moistened paper to the block produced a white image on a black background. After months of effort, Marcel produced a print on January 24, 1800 (fig. 7.2).

English antiquarians and specialists complained that facsimiles of the inscriptions were not rapidly forthcoming and found the delay even more galling because the vanquished French nevertheless still possessed the proofs made in Egypt. In August 1802, the antiquarian Richard Gough, writing as one "D.H." to the *Gentleman's Magazine*, warned of an unnamed French scholar, likely Antoine Isaac Silvestre de Sacy, already at work on the inscriptions who "contents himself with giving a plate of a long portion of it, where the characters appeared to him more clear and distinct." Gough anticipated the arrival of "an exact facsimile" in the near term, to be published by the Society of Antiquaries, of which he had served as director from 1771 to 1791. In the meantime, the arrival of a French competitor was all the more unfortunate since the "English Literati" were thus far not "in possession of a single copy" even of the Greek inscription.[18]

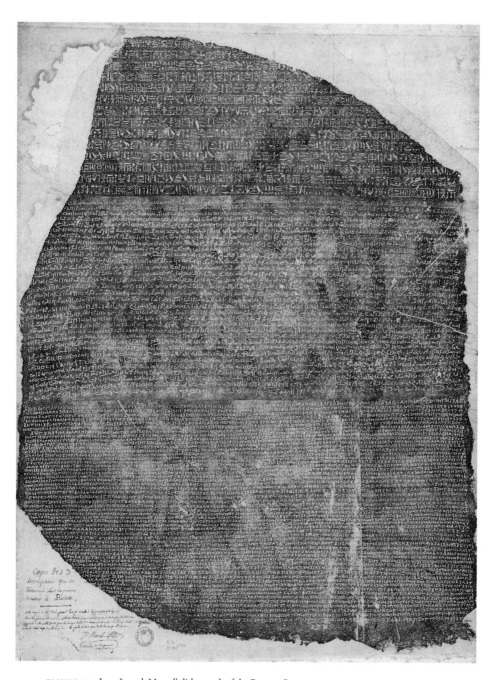

FIGURE 7.2. Jean-Joseph Marcel's lithograph of the Rosetta Stone.

Gough's complaint was not entirely justified. Access to the Rosetta inscriptions may have been limited in Britain, but the situation was nowhere near as bad as that which obtained on the Continent. An apographic engraving of the Greek inscription had in fact been printed in Britain on July 8 and was soon distributed to members of the Society of Antiquaries and to the institutions that usually received its publications.[19] The Egyptian engravings, which British scholars could have seen at the Antiquaries, were not printed until April of the following year, and then only five hundred sets were produced. Although at least one scholar outside Britain did receive an early set of the Egyptian plates, few if any others did.[20] Furthermore, four plaster casts of the Rosetta Stone were taken and sent to the universities at Oxford, Cambridge, Dublin, and Edinburgh.

Transcriptions and translations of the Greek inscription soon appeared. In Britain, the earliest public transcription of the Greek was produced by Granville Penn, a fellow of the Society of Antiquaries who had been instrumental in establishing London's Veterinary College; his transcription was printed in a pamphlet and reproduced at about the same time in the *Critical Review*.[21] The same Weston who had signed Young's nomination to the Royal Society produced an English translation of the Greek inscription that was read to the Antiquaries on November 4, 1802.[22] A Latin translation by the Göttingen classicist Christian Gottlob Heyne was read at the Antiquaries later that month.[23] On December 5 a prebendary named Thomas Plumptre, working from a copy of the society's "facsimile," sent his translation of the Greek to the *Gentleman's Magazine*. It was soon published, making his the first English version to reach print.[24] Then, in 1803, a third translation of the Greek into English, by Stephen Gough, appeared in his book on the coins of the Seleucid kings.[25]

Of all the early renditions of the Rosetta inscriptions, the most complete and perhaps the most influential appeared in the *Vetusta Monumenta*, a series devoted to antiquarian topics, featuring apographs and drawings. Although the relevant *Vetusta* volume did not appear until 1815, the plates of the Greek and Egyptian inscriptions were available at the Society of Antiquaries as early as 1803, and, as noted above, some five hundred copies of the Egyptian had been produced from engravings later used to produce the *Vetusta*. The *Vetusta* materials also included a reconstruction by Porson of the missing Greek as a separate plate (see fig. 7.5, *bottom*).[26] We will see below that difficulties in transcribing even the Greek, much less the Egyptian, quickly arose.

The plates for the hieroglyphic and Greek inscriptions were designed and engraved by James Basire, himself a member of the Society of Antiquaries and proprietor of an engraving studio located a half mile away on Great Queen Street. He also engraved the plate for Porson's reconstruction. Basire was renowned for his architectural renditions, many of which were printed in the *Vetusta*. The design for the plate of the middle script was drawn by Samuel Lysons, who had been director of the society since 1799 and would remain so until 1812. Noted for his own drawing and engraving abilities, Lysons was particularly expert with the ruins of Roman Britain. He did not himself cut that plate; that was done by L. C. Stadler.

The engraved representations of the Rosetta Stone differed strikingly from other images of artifacts in the *Vetusta*. Although engravings of other inscribed objects appeared in it, the publication's plates aimed principally to capture the object's physical qualities, not any inscription per se.[27] Consider two typical examples in figure 7.3: on the left is Roman paraphernalia found at Ribchester; on the right is a medieval English bowl adorned with Saint George in various poses. The two plates were printed in the same volume as the Rosetta inscriptions. Both were engraved with shading to bring out depth, while the bowl is also colored. Neither contained any script, though the bowl did present a series of sematograms through which the gist of the Saint George narrative could be recovered by a knowledgeable reader. *Vetusta* plates, such as those in figure 7.3, were not designed to convey information about the past; neither were they generally accompanied by texts that did more than describe the locations and circumstances of the find. They were, in effect, stylized portrayals that separated the object from its context, producing a representation consistent with the aesthetic and decorative concerns of the society's members.[28]

With the plates of the Rosetta inscriptions (fig. 7.4 and fig. 7.5), by contrast, Basire and Stadler attempted to provide some sense of the scripts' physical embodiment in the fractured stone by drawing and shading parts of the inscriptions' surround. However, the scripts are not represented as they appeared on the stone itself. Drawn in black against a white background, the scripts stand out decisively. Yet on the stone, the scripts even now appear as lightened cuts into a gray surround. In the *Vetusta* images, no attempt was made to indicate that the inscription is incised into the slab. With the exception of the shaded edges, these representations might as well have been of ink on paper, which, in a literal sense, of course, they were. The distinctly aestheticized quality of so many other *Vetusta* plates has been set aside. The object—the inscribed stone—is only vaguely indicated, and its scripts are not treated as ornaments. The *Vetusta* plates were effectively apographs, not aestheticized depictions.

Porson's re-creation of the missing Greek text (fig. 7.3) may be compared to Young's re-creation of the Herculaneum apograph. Porson himself may have designed the plate, as Basire is listed only as the engraver. Though the volume containing Porson's version did not appear until 1815, the plate of the reconstruction is dated April 23, 1810. That same year the *Herculanensia* was also

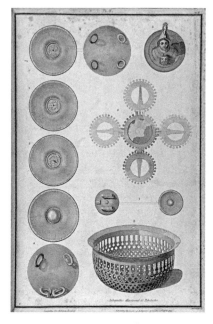

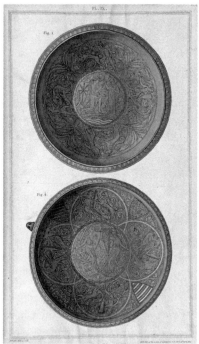

FIGURE 7.3. Plates 4 and 9 from the *Vetusta Monumenta* (vol. 4).

published, followed within two years by Young's restoration of the Herculaneum papyrus. Porson's delineation of the Rosetta's missing Greek bears a striking resemblance to Young's re-creation of the Greek on the Herculaneum material. Unlike Hayter who, as we have seen, separated words and introduced lowercase characters, neither Young nor Porson essayed such a free adaptation of their originals. Rather, both of them faithfully reproduced the original's orthographic features. Given Young's admiration of Porson, the close parallels between the restorations is unlikely to be coincidental. Young may have seen this plate shortly after its printing.

Although Young was not a member of the Society of Antiquaries, he would have had little trouble consulting its collections and prints, had he wished to. He had been appointed as foreign secretary of the Royal Society in 1802, and the Royal Society was housed with Antiquaries and the Royal Academy at Somerset House in the Strand. Young may have been attracted, moreover, to the antiquarian penchant for classification given his own interests in nosology and his habit of assembling large catalogs of previous work on whatever topic in which he was engaged. Nevertheless, Young may not have been overly keen to associate with the society. As Walter Scott would make so memorably clear in his novel *The Antiquarian*, there was something not quite respectable about antiquarianism. Its practitioners were subject to ridicule for what was seen as a fetishistic concern with antique objects (fig. 7.6).[29] An association with ill-favored antiquarianism would not have helped Young to establish his medical practice either. Other concerns may have also kept him away from the Rosetta inscriptions, as he was then preoccupied with Brougham's attack on his new theory of light. But if Young may not have been eager to examine the Rosetta inscriptions, others in France and elsewhere were, and the second script, widely presumed to be alphabetic, was a matter of particular interest.

In France, the first available copies of the scripts were the two proofs made in Cairo, either both by Marcel or perhaps one by Conté, and brought to Paris by Dugua in 1801. When the French army returned, a sulfur cast by Raffeneau-Delile was also brought to Paris. The two copies of the stone's inscription that Dugua brought back, as well as a third kept by Marcel himself, were the only ones available on the Continent until the English distributed their plates of the Greek script in the summer of 1802 and of the Egyptian ones in 1803. Until then, anyone wishing to examine any of the three scripts had to be either in Paris with good connections or else to know someone who was. News about the scripts spread quickly. In June 1800, Eichhorn, the Göttingen orientalist, wrote to Herder asking whether he had heard about it.[30] Demand for copies of the inscriptions had been growing in France since at least September 1800, when the *Magasin Encyclopédique*, under the attentive editorship of the naturalist and antiquary Aubin-Louis Millin, printed an account of the previous month's discussion of the inscriptions at the Institut d'Égypte in Cairo.[31] A major project to print accounts and engravings of the savants' findings, including the Rosetta, was soon undertaken, resulting eventually in the magnificent volumes of the *Description de l'Égypte*.

Chaos attended the project's inception. In January 1803, as the expedition's savants were returning to France, the mathematician Joseph Fourier, who had been secretary of the Cairo Institut d'Égypte, arrived in Paris along with "two or three diligences" filled with members of the Egypt commission. One of the expedition's engineers, René Édouard Devilliers, wrote urgently to another (Jean-Baptiste Jollois) to report that Fourier, acting on the orders of Napoleon, had initiated discussions about collecting the expedition's results. Meanwhile the mathematician Gaspard Monge and the chemist Claude Louis Berthollet, both of whom had also accompanied Napoleon to Egypt, hatched a plan of their own, according to which Devilliers would take charge of editing the expedition's history, which he didn't want to do. Fourier informed Devilliers that he intended to discourage this competing program by convening a meeting with Monge, Berthollet, and the doyen of French science, Pierre Simon de Laplace, in order to "show them the inconvenience of the form they

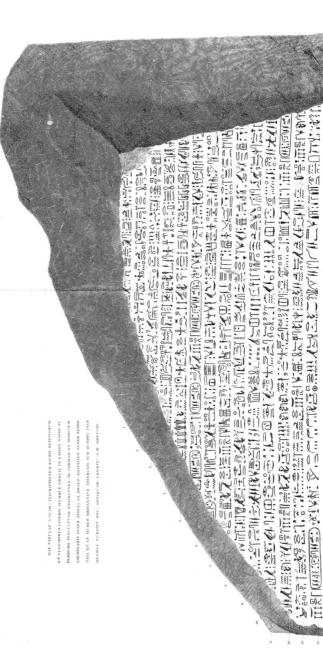

HAS TABULAS (V.VIII.) INSCRIPTIONEM SACRIS ÆGYPTIORUM

ET VULGARIBUS LITERIS ITEMQUE GRÆCIS IN LAPIDE NIGRO AC

PRIDEM INSCULPTAM EXHIBENTES AB FORMAM ET MODULUM

EXEMPLARIS INTER SPOLIA EX BELLO ÆGYPTIACO NUPER REPOR-

TATI ET IN MUSEO BRITANNICO ASSERVATI SUB SUMPTU INCI-

DENDAS CURAVIT SOC. ANTIQUAE. LONDIN. A.D. MDCCCII.

FIGURE 7.4. The *Vetusta Monumenta* engravings of the Rosetta Stone (vol. 4): (*top*) hieroglyphs designed and engraved by James Basire (pl. 5) and (*bottom*) the intermediary script designed by Samuel Lysons and engraved by L. C. Stadler (pl. 6).

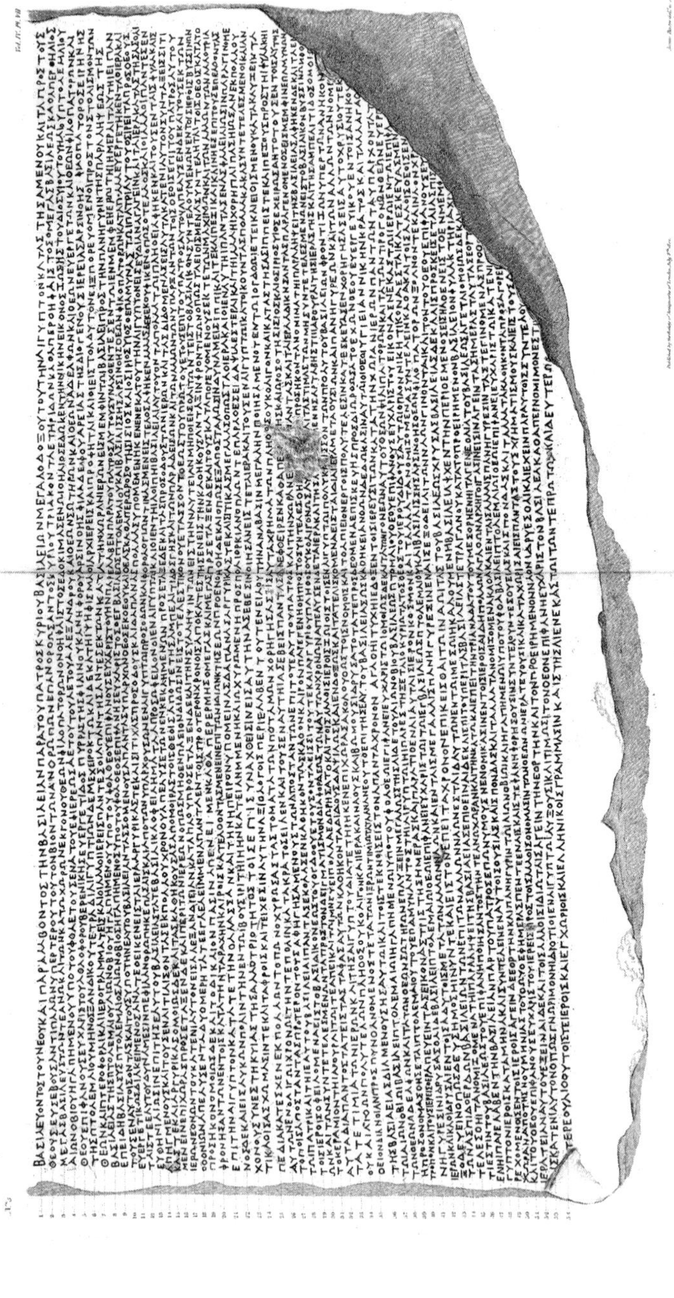

FIGURE 7.5. The *Vetusta Monumenta* engravings of the Rosetta Stone (vol. 4): (*top*) the Greek, designed and engraved by James Basire (pl. 7); and (*bottom*) Richard Porson's reconstruction of the missing Greek, engraved by Basire.

FIGURE 7.6. George Cruikshank, *The Antiquarian Society* (1812). © The Trustees of the British Museum.

wanted to give the work," and to learn "whether they hold strongly to their ideas."[32]

While the returning savants were fighting for control of the history, results, and imagery of the expedition, the artist Denon, who had also accompanied Napoleon's forces in Egypt, published an illustrated two-volume account of his experiences there, *Voyage dans la basse et le haute Égypte*.[33] Although the work met with wide acclaim in both France and Britain, Denon's volumes failed to impress the expedition's engineers, who particularly scorned his renditions of Egyptian antiquities. One of them—Jean-Baptiste Fèvre, another young member of the expedition—told Jollois that he had spoken with Gaspard de Prony, a senior member of the École polytechnique who needed little inducement to suspect the accuracy of a fine artist's drawings "of what we did in Egypt." Fèvre confessed that he had promoted Jollois's skills as an architectural draftsman and boasted that "we'll provide drawings that will convey Egyptian architecture much better than the pretty pictures of Denon, and that all the works united will form a magnificent production."[34]

By the twelfth the savants' project was afoot, with plans that called for a "historical" part written by Fourier, descriptions of monuments with drawings, and various memoirs by members of the commission. Devilliers informed Jollois that a supervisory board had been formed to take charge of the project, consisting of Fourier, the architect Jacques Guillaume Legrand, the astronomer Jean-Baptiste Delambre, the naturalist Georges Cuvier, and "an antiquarian" whose identity was not specified. In February a formal decree established a publication project to be paid for by the government, including funds for the contributors and editors. The chemist Jean-Antoine Chaptal, now minister of the interior, had overall responsibility, with a committee consisting of Berthollet as president, and members Lancret, Conté, Monge, Fourier, Louis Costaz, Pierre Simon Girard, and René-Nicolas Desgenettes reporting to him.[35] At their first meeting, on June 7, Conté was appointed as editor in chief of the project. After he succumbed to an aneurysm in 1805, following the deaths of his brother and his wife, Lancret took over, only to expire himself in 1807. Edme-François Jomard became editor in chief, a position he kept until the project ended in 1827, with Jollois as managing editor.[36]

Management of the *Description* was a full-time, and very expensive, affair. As of June 1, 1809, thirty-six people were on the payroll, and the project had burned through nearly 1.5 million francs. More than three thousand reams of paper were estimated to be necessary for the publication, and the production of the copper plates was parceled out among four printers. Late in 1806, Napoleon demanded that sheets be printed off as soon as the plates were ready, and he stipulated that publication would commence no later than 1809. The volumes in fact appeared seriatim with dates from 1809 through 1824, with a final one containing a topographical map appearing in 1828. Three volumes were devoted to plates depicting antiquities.[37]

The plates of the Rosetta inscriptions were contained in the fifth volume of the *Description*, which appeared only in 1822, though the unbound prints were produced years earlier. Before 1806 or 1807, in France only those with access to the proofs brought back by Dugua, or who could examine Raffeneau-Delile's cast, could have seen copies of the scripts. The three plates that were eventually printed in the *Description* derived from two sources: a cast taken directly from the stone by Jomard during his 1815 trip to London and the sulfur cast made in Egypt. Jomard designed the plate for the hieroglyphs himself; it was engraved by Jean-Baptiste Bigant. The plates for the intermediate script, as well as the Greek, were derived solely from the sulfur cast and were designed by Raffeneau-Delile, with an otherwise unnoted Fouquet having done the engraving.[38]

The *Description* plates (fig. 7.7) differed in important ways from those distributed in the 1815 *Vetusta* (fig. 7.4 and fig. 7.5). For one thing, less attention was paid to the Stone's physical character compared to the *Vetusta* version. Bigant engraved the hieroglyphs to show relief, though of course the characters are incised into, not raised above, the stone. In his print the ground is grayed out, with the characters blacked in and apparently raised. The plate representing the intermediary script is badly marred in several places where the *Vetusta* is not, a difference likely due to defects in the sulfur cast on which the engraving was based. There are also differences in the rendition of the intermediary

FIGURE 7.7. The Rosetta prints from the *Description de l'Égypte: Antiquités, Planches*, vol. 5 (1822).

script's characters, several of which prompted the first interchange, in late 1814, between Young and Champollion, as we shall see.

After some bureaucratic scrambling, the historian and director of the Bibliothèque de l'Arsenal, Hubert-Pascal Ameilhon, was charged with translating the Greek inscription based on the lithographed copies. Ameilhon produced translations of the Greek into French and Latin, although the latter reached print only in 1803.[39] Meanwhile copies were given to the orientalist Antoine Isaac Silvestre de Sacy and to Ennio Quirino Visconti, the latter having been entrusted with the Louvre collections that Napoleon's armies had looted from around Europe.[40]

Few could tolerate the long wait for facsimiles, transcriptions, and translations. Complaints proliferated, particularly in the German press. On September 14, 1801, Millin's assistant at the *Magasin Encyclopédique*, Théophile Winckler, wrote a letter to reassure readers of *Der Neue Teutsche Merkur* that a noted Arabist in Paris—that is, de Sacy—was occupied with the "trilingual Egyptian inscription" in order especially to examine the part that was written in the "native language," that is, the intermediate script. Tantalizingly, he continued, "the inscription does not resemble those found on the wrappings of the mummies preserved at the antiquities Cabinet of the National Library." Winckler added that Visconti, who intended to produce "a word" on the inscription, was primarily responsible for the delayed dissemination.[41]

As the copies were produced in haste by artisans who were not all equally adept at reproducing such unfamiliar scripts, even those fortunate enough to obtain scarce reproductions found fault with them. While his own consideration of the Greek was in press, Ameilhon noted that an English engraving of it had recently come to his attention, and it was of a better quality than the one Ameilhon had used.[42] That "copy," he continued, "is what we vulgarly call a *facsimile*," an appellation suggestive of attentiveness, as if "one took pride in rendering with the most scrupulous exactitude, not only the missing parts of the text, but also the mistakes of the Egyptian artist who sculpted the stone." The slipshod reproduction that Ameilhon had perforce used gave him much cause to lament: "I feel that the engraving annexed to my elucidations may not appear to have been done with altogether the same degree of exactitude."[43] He noted, in particular, that the Greek letters P (*rho*), N (*nu*), and I (*iota*) were not everywhere clearly represented, leading him into difficulties with the sense of the passage.[44]

The first transcription of the Rosetta Greek, by Penn, had replicated the English engraving's rendition. The *Critical Review*, which printed Penn's effort, congratulated him on having so nicely compensated for the Antiquaries' provision of the faulty engraving only. "From so learned a body," the critic opined, "something more might have been looked for than a bare facsimile, the accuracy of which requires to be justified." As poor as the fascimile may have been, Penn was lauded for having "favored his friends with a printed copy of it, from the stone, in a more legible form; and we flatter ourselves with the indulgence of that gentleman in venturing to annex it, with the advantages it derives from his acuteness and erudition."[45] In mid-1802 Millin hastened to print Penn's transcription in the *Magasin Encyclopédique* in order "to give it the greatest possible publicity, and also to have scholars work on this part of the inscription. It is to be hoped that the good work of C. Visconti on this Greek version will soon be published."[46]

Ameilhon did not think much of Penn's contribution.[47] As it turned out, Penn had not improved at all upon the facsimile, and Basire, the engraver, did not know Greek well enough to reproduce the letters accurately. Ameilhon had good reason to feel hard done by. While transcribing and translating the inscription into Latin and French, he had painstakingly noted orthographic irregularities such as omitted, superfluous, and confounded letters, alterations, and mutilations. On Ameilhon's reading, Penn had violated many rules of Greek grammar; perhaps, Ameilhon allowed, Penn had rushed too quickly into print. Urging caution on all sides, Ameilhon noted that Dugua's lithographs might themselves be faulty, and he hoped the English savants would consider that their engravings might be imperfect as well. As Porson's reconstruction of the missing portions of the Greek was not available, Ameilhon had not seen it; nor had he seen either Weston's English or Heyne's Latin translations since neither was printed until 1812. Deflating as it surely was, Ameilhon's cautionary finger-wagging suggested an even more demoralizing possibility: If the most familiar of the three Rosetta inscriptions had already caused so many problems, how much more difficult would the other two be?

While variously imperfect facsimiles, transcriptions, and translations of the inscriptions circulated, the stone itself remained in England, where it acquired a reputation as a quasi-talismanic object with the power to elicit extraordinary interest. The stone might be made to speak, to bring to life the sounds of ancient Egyptian. But since the artifact could only be seen in London, interested parties were limited to representations produced either by direct impression or by an engraver. Neither source was perfectly satisfactory. Engravings passed through the engraver's interpreting hand, while the material exigencies of a proof or a cast bore at best an imperfect relation to the original. Because every technique to reproduce the inscriptions was faulty in some way, issues of representation were irremediably present from the beginning, providing early commentators with opportunities to cavil over the quality of the reproductions as they laid the ground for more substantive debate.

CHAPTER 8

THE SOUNDS
OF AN ANCIENT
LANGUAGE

The first halting attempts to decipher the Rosetta hieroglyphic and intermediate scripts began while the artifact was still in the printing workshop of the French expedition in Egypt. Investigating the Greek inscription with an eye toward dating the artifact, Marcel noticed that several words, though transliterated into Greek, for example, *Pta* or deity, "are not Greek, but Egyptian," and suggested that they must therefore "indicate the epoch at which, despite the efforts of the Ptolemies, the indigenous language of the Egyptians began to mix with that of the Greeks, their conquerors." From the occasional transliterations of non-Greek words, Marcel thought to see early elements of a new language emerging from the mixture of the indigenous Egyptian with Greek, "a mixture that, successively increasing, ended towards the fourth century of the common era by forming the ancient Coptic language of which we find precious traces in modern Coptic."[1]

That language reached its conclusion, he asserted, in an "ancient Coptic" ancestral to its modern form. Marcel did not claim that the indigenous language was itself the immediate antecedent to ancient Coptic, much less the post-antique form of the language, but that Coptic must in the earliest times have been a linguistic mélange. To Marcel and the other savants, this supposition implied that the unknown intermediate script was the inscribed form of this mixed language. Moreover, since the blend evolved under the Ptolemies with written Greek widely present in Egypt, it seemed to follow that the intermediate script was likely alphabetic. But if that were so, then there was hardly any reason to think the hieroglyphic inscription bore any relation to it. After all, hieroglyphs clearly dated to centuries before the arrival of Greek with the Ptolemies, and the characters in the Rosetta inscription appeared to be no different from the ones that could be seen in the very ancient monuments at Thebes and elsewhere.

That Egyptian scripts came in several varieties had long been understood. Though Marcel did not remark the point, anyone familiar with Greek would have known as much from Herodotus, who observed that "[The Egyptians] have two sorts of writing, one called sacred and the other common."[2] Herodotus's differentiation of the "common" from the "sacred" type would certainly have reinforced Marcel's conviction, in which he was not alone, that the Rosetta hieroglyphs were effectively unaltered, or only slightly so, from their most antique form. Subsequent commentators had advanced little beyond that, however, and it even seemed that Herodotus had failed to capture anything as basic as the correct name of the nonhieroglyphic script. Herodotus's word for the common form of writing was δημοτική, "demotic." In the last line of the Rosetta Stone's Greek inscription as translated by Weston, however, the writing had a different name. Here it was decreed that the inscription "shall be engraved on a solid stone in sacred, in *vernacular*, and Greek characters."[3] The Greek on the stone for vernacular or vulgar script was ΕΓΧΩΡΙΟΙΣ, which Young transliterated as "enchorial."[4] In the second century CE, Clement of Alexandria made a tentative advance by identifying yet a third script, visible on ancient papyri, that differed from the scripts identified by Herodotus.

In the seventeenth century the indefatigable and imaginative Kircher took up the problem of the Egyptian scripts. Believing that hieroglyphs symbolically encoded ancient wisdom, Kircher effectively treated most of the signs not as logograms but as extraordinarily complex sematograms. A devout student of languages, Kircher had also decided that Coptic contained the essential forms of ancient Egyptian.[5] Although his works on hieroglyphics were appreciated through the mid-eighteenth century for their detailed plates of monuments and inscriptions, his claims about the script were widely rejected. Nevertheless, the link he perceived between Coptic and ancient Egyptian proved enduring and fruitful, and Kircher's 1643 Coptic lexicon and grammar remained, despite its flaws, the primary published resource on Coptic until 1775, when the post-humous publication of Maturin de La Croze issued a considerably corrected version.[6] Eighteenth-century scholars read Kircher selectively, salvaging what they could, especially from his linguistic studies, while remaining aloof from what Stolzenberg describes as the well-attested opinion that ancient Egyptian priests had possessed symbolic wisdom.[7]

Subsequent scholars increasingly looked beyond the hieroglyphic script, an expansion that followed and extended Clement's observations. At the turn of the eighteenth century, the collector and numismatist Jean-Pierre Rigord received pieces of a mummy wrapping from the French consul in Egypt, Ben-oît de Maillet. The borders of the wrapping were "ornamented with historical figures, with writing at their feet." Rigord could not identify the script, but he was aware of ancient writers, like Herodotus and Clement, who identified a multiplicity of ancient Egyptian scripts. If this script on these mummy wrappings could be read, Rigord opined, then "perhaps one might thereby attain knowledge of a hitherto unknown language," which was his principal interest.

To effect this understanding, Rigord reiterated the distinctions drawn by Clement of Alexandria in the fourth chapter of his *Stromata*. Clement's three forms of Egyptian writing comprised: the "epistolographic," used widely; the "sacerdotal," used by sacred scribes; and finally the "hieroglyphic," of which Rigord, following Clement, claimed two sorts, the "curiologic" and the "symbolic." The "curiologic" was written, according to Clement, "by the first elements" and was "literal," which Rigord creatively read as meaning alphabetic. Symbolic hieroglyphs came in two kinds: one referred to objects by "imitation," that is, by pictorial reference to an object, while the other conveyed meaning by allegory. In an elaboration of Kircher's original idea, Rigord added Coptic to the list of scripts, though he considered the Coptic alphabet to be "pure Greek, with very small differences in the form of the characters," and believed, like Marcel much later, that the corresponding language scarcely predated the time of Alexander. The mummy inscriptions, Rigord continued, were examples either of the sacerdotal or of the hieroglyphic curiological.[8] Beyond this, few inscriptions of either sacerdotal and epistolographic writing, to use Rigord's terminology, were known before 1800. The only others broadly available were two large but poorly drawn plates printed by the Benedictine Bernard de Montfaucon in his *L'antiquité expliquée* of 1719 and 1722. Champollion would later complain about their quality.[9]

Three decades later the Italian scholar and librarian Angelo Maria Bandini, in his account of the Augustan obelisk in Rome, distinguished hieroglyphs as symbols for concepts and objects from what he called the "vulgar" script:

> Neither [Clement of Alexandria] nor Herodotus, nor Diodorus say anything about the characters comprising the vulgar script: one may conjecture from their silence that these were the same as those of the Phoenicians, a people not merely close by, but with whom [the Egyptians] were in continual commercial relations; and moreover according to the best critics, the Phoenician letters were those used by Moses, born and raised in Egypt, and who wrote to be understood by a people in long contact with the Egyptians, and from whom they took in major part their manners and customs.[10]

If Bandini was correct, then the "vulgar" script was alphabetic. He allowed that hieroglyphs, though principally nonphonographic, might nevertheless contain a mixture of such symbols, and he held that the "curiological script" identified by Rigord had the same significance as hieroglyphs "even though [the latter] is sculpted, and the other is properly written."[11]

In 1756, six years after Bandini's publication, Anne Claude, Comte de Caylus, published his *Receuil d'antiquités égyptiennes, étrusques, grecques, romaines et gauloises*, a large volume with extensive plates; five years later, a new edition appeared, followed by six additional volumes. The fifth of these, available in 1762, contained a remarkable set of suggestions concerning the Egyptian scripts. With strong court connections and time on his hands following a military career, Caylus delved into what we would today call the material culture of antiquity.[12] His 1762 volume included an examination of a mummy wrapping (fig. 8.1) in which Caylus found three instances of a special configuration: "I perceive in the first line several characters that together form a sort of square. This configuration is the more striking, in that having been presented three times, it appears no further beyond the fourth line of the second plate."[13] Caylus claimed that this "configuration" was a group of letters, and moreover that such groups "express proper names, "on the model of those hieroglyphs which in other Egyptian monuments are joined together in ovals or squares in order to represent the names of kings and gods."[14] He thought both such "configurations," on mummy wrappings and cartouched signs on monuments, represented royal or divine names, but only the former was phonographic.

In the first volume of his series, Caylus asserted that Egyptian "cursive writing," which differed visibly from the script on mummy wrappings, comprised a phonography. Referring admiringly to Warburton's "destruction" of the idea that Egyptian priests had invented hieroglyphs to hide their knowledge, Caylus mentioned four fragments inscribed with "a kind of reasonably uniform writing" that seemed a promising avenue into the otherwise unreadable scripts. Comparing those fragments provided "a list of the characters in use among the Egyptians." The varying spatial placements of the signs promised to yield a very large set of characters, making it consistent with other languages believed to be similar.[15] "I must warn that if the resulting alphabet exceeds the usual scope of Oriental alphabets," he wrote, "the blame must be set to our continuing ignorance of Egyptian writing." The twenty-six letters of the "Ethiopian" (Amharic) script, Caylus explained by way of comparison, are deployed in such a manner that "the vowels are there attached to each letter." Since, moreover, the script "includes additionally syllabic letters, it is in effect composed of more than 202 characters." Possibly, he suggested, "it's the same with the writing of the Egyptians, and that its elements change only with position where the various strokes seem to vary the form."[16]

Reprising Warburton, Caylus suggested that the "cursive" might have "emanated from the [nonphonographic] hieroglyphs." For evidence he referred

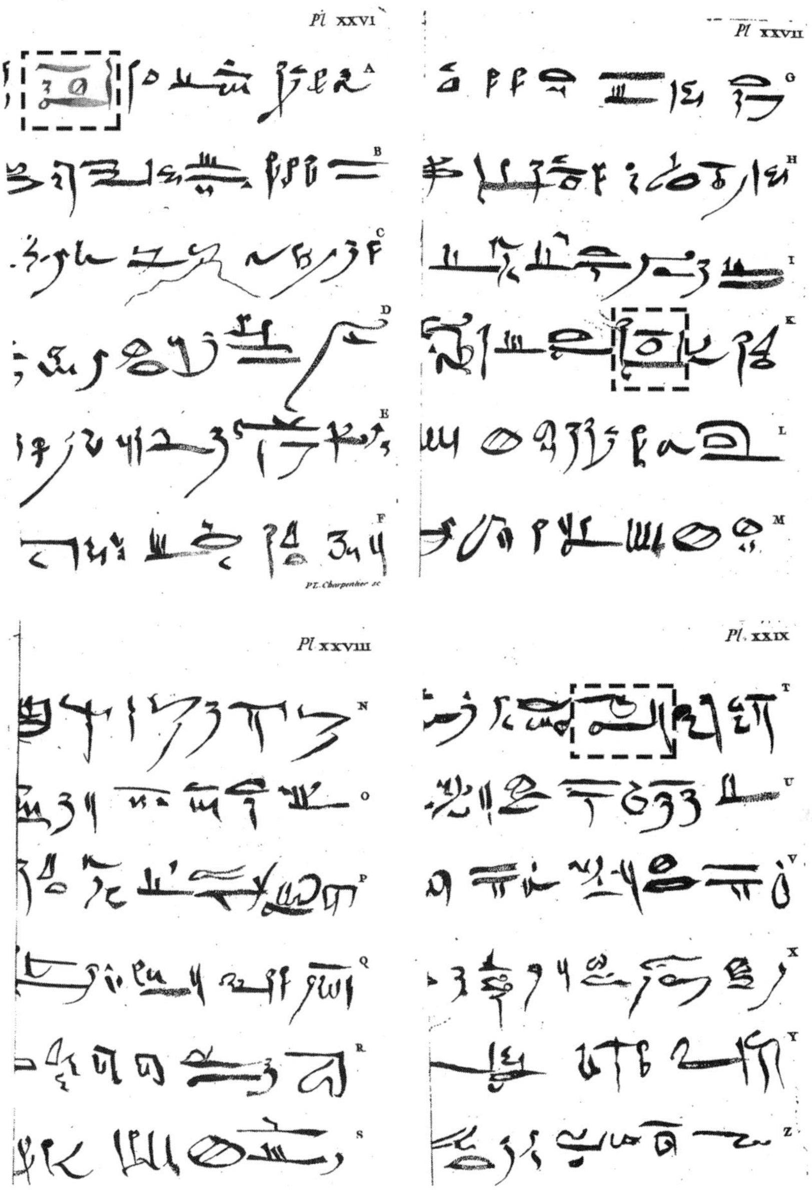

FIGURE 8.1. Comte de Caylus's plates of the mummy wrap script. Dashed boxes indicate the three "squares" that Caylus singled out.

the reader to examples drawn from one of the plates included in his volume from an example that had been (poorly) printed by Rigord (fig. 8.2).[17] For instance, the fourth hieroglyphic character, which looked like a squatting man, had been transformed into the cursive in a manner that "conserved only the outline of the original symbol." That kind of change, he continued, confirmed Warburton's claims. This cursive, Caylus decided, was an example of what the ancients termed the "vulgar" script, while the script found on mummy wrappings corresponded to their "sacerdotal." Whether the one developed from the other he could not say, but they did seem to have several "letters" in common. In his view, the presence of this commonality supported the hypothesis that both originated from the more archaic hieroglyphs. That being established, the question was whether such Egyptian "letters" had given rise to the Phoenician, which he thought likely.[18]

Doubt has surrounded the authorship of these remarks, which have long been attributed to Jean-Jacques Barthélemy. Well versed in oriental languages, Barthélemy had become keeper of the Royal Cabinet of Medals in 1753, and from 1755 to 1758 was in Italy with the French ambassador.[19] He was widely known in later years for his popular *Voyage du jeune Anacharsis en Grèce*, a vividly imagined excursion through Greek antiquity that was extensively reprinted and translated. In 1754 he produced the first reading of a dead script, *Palmyrene*; four years later, he became the first to partially reconstitute Phoenician, of which there existed very few examples.[20] Whether or not Caylus himself wrote the remarks in both the 1752 and 1762 volumes, their purport is certainly due to his connections with Barthélemy, whom he explicitly thanked in the 1762 volume for "several observations" on Egyptian writing that he would "report." Caylus was not particularly concerned with writing, except as a form of material production from Egyptian antiquity, whereas Barthélemy, a strong admirer of Warburton, most certainly was. Like Warburton, Barthélemy did not think that hieroglyphs were phonographic (or even exclusively logographic). The symbols used in later scripts, which were at least words and finally sounds, derived from hieroglyphs by Warburtonian transposition, but the ur-source itself did not represent vocalizations. "I believe with M. Warburton," Barthélemy remarked in 1767, "that the first writing was done in hieroglyphs. I further believe that the first alphabetic writing was highly defective."[21]

On the origin and development of writing, Barthélemy was also convinced by the arguments of Joseph de Guignes. A specialist in Chinese, de Guignes had been admitted to the Royal Society of London in 1752, and to the Académie des inscriptions et belles-lettres the next year.[22] He obtained the chair in Syriac at the Collège Royale in 1757 following the death of Auguste François Jault, whereupon he took the opportunity to argue, in Latin, that the French

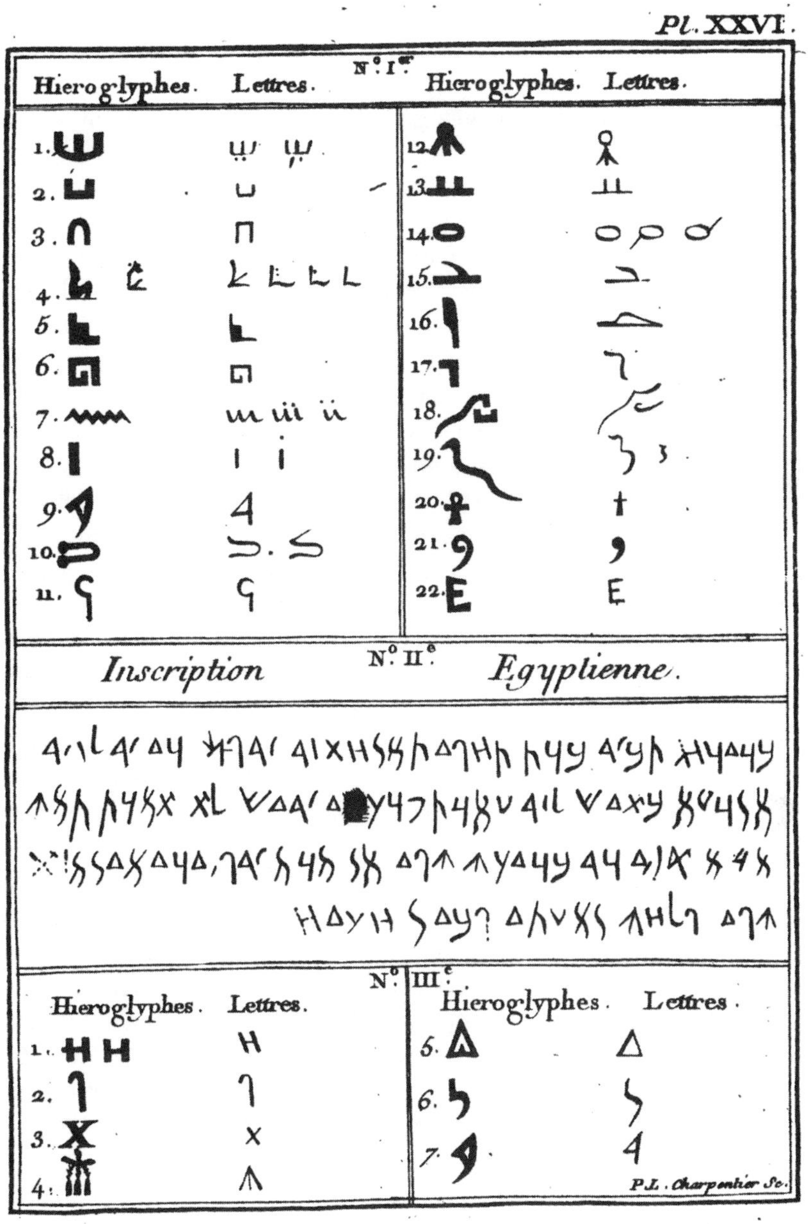

FIGURE 8.2. Comte de Caylus's plate comparing "cursive" with monumental hieroglyphs.

kings were "much better made for letters" than were the "princes of Asia," according to a nineteenth-century biographical notice that exudes the exoticism and condescension characteristic of European attitudes toward the East at the time (and often since).[23] Nonetheless, he believed that Egyptian culture had had an immense influence. Indeed, so great had it been, de Guignes argued in a lecture given at the Académie, that China had been colonized by Egyptians in remote antiquity. Chinese culture therefore originated in Egypt, and Chinese writing was itself a transformation of the Egyptian hieroglyphs. Barthélemy was thoroughly convinced: in 1758 he wrote the Comte de Saluces that de Guignes was the first "to discover, in the written language of the Chinese, the debris of the Egyptian and Phoenician tongues, and in the ancient hieroglyphs of that people, those which were first used by the Egyptians."[24]

The speculations of de Guignes and others concerning the relationship between Chinese and Egyptian contrast suggestively with scrupulous work done on the hieroglyphs by the Danish Coptic expert Georg Zoëga. Published in 1797, Zoëga's *De origine et usu obeliscorum* (*On the Origin and Use of the Obelisks*) presented all known ancient remarks concerning hieroglyphs, as well as the various theories concerning the script and the Egyptian language. He drew a distinction between Chinese characters, each of which he took to stand for a single, fixed concept, and hieroglyphs, which he considered to grant meaning by representing ideas in a connected sequence. Having examined extant materials, Zoëga distinguished nearly a thousand separate signs and divided them into categories. As had Caylus and Barthélemy, he suggested that some hieroglyphic sequences enclosed by oval "cartouches" might represent proper names.[25]

In the immediate aftermath of the Egyptian expedition's discovery of the Rosetta Stone, its two Egyptian scripts were still considered to differ completely in meaning. The hieroglyphs continued to be thought of as a form of iconography, at best loosely connected to the spoken language, while the intermediate Rosetta was considered fundamentally alphabetic and closely related to Coptic. Since, however, the artifact itself could only be seen in London, those who engaged with it were working on representations that were produced in two ways: either by direct impression (namely, Marcel's lithographs, Conté's etchings, the Delile sulfur casts, or the English plaster ones) or else by an engraver. And engravings of course reflected the judgment and skill of the artisan with inevitable differences between representations. It's hardly surprising, then, that, as we shall see, the stone's very materiality inflected the efforts made to read its inscriptions.

One scholar, de Sacy, essayed an early reading of the intermediate Rosetta script, a reading informed by his belief that its orthographic structure con-

FIGURE 8.3. Antoine Isaac Silvestre de Sacy. Buchwald Collection.

formed to the spatial configuration of alphabetic representation. His *Lettre à M. Chaptal* (1803) and the debate to which it gave rise, created the immediate terms of engagement with the Rosetta scripts. This framework proved so enduring that Champollion would eventually be obliged to adapt his own arguments to it by clothing his ideas in its idiom.

Raised by strict Jansenists, de Sacy remained religiously conservative throughout his life, as well as Royalist in sympathy. He early took up the study of Hebrew, followed by Arabic, in which he was essentially self-taught, and then Syriac, Chaldean, Persian, Turkish, Aramaic, and other languages of the Near East. During the 1780s he worked on the inscriptions at Persepolis that had been found by Carsten Niebuhr, while devoting most of his attention to biblical studies.[26] During the political chaos of 1792, de Sacy prudently retired with his family to his country house outside Paris, reemerging following the fall of Robespierre.[27] When de Sacy wrote his *Lettre à M. Chaptal*, the text with which we are principally concerned here, he headed the École spéciale des langues orientales, which had been founded in 1795 by Louis Mathieu Langlès.

At this time de Sacy possessed three copies of the Rosetta inscriptions but not the English "facsimile," which only arrived in France later, in July 1803. In 1802, Chaptal transferred to de Sacy the two copies handed over to the Institut by Dugua. One was a lithograph by Marcel; the other was Conté's etching. De Sacy also possessed a third copy, obtained directly from Marcel, the quality of which was somewhat better than the first lithograph.[28] After working for a year on the intermediate inscription (which he referred to as the "Egyptian" script, as we will hereafter), de Sacy sent Chaptal an abject progress report. "[T]he hope I initially conceived," he wrote, "if not entirely to decipher the Egyptian inscription of the precious monument found at Rosetta, at least to read enough words to know with certainty the language in which it is written, was not at all realized."[29] Despite the lack of results, de Sacy's report was printed, no doubt in order to establish whatever priority de Sacy could claim.

De Sacy's grand intention to decipher the middle script exceeded his preparation for doing so, but his more modest hope, to determine the language in which the Egyptian script was written, stemmed from significant previous work in general grammar. In 1799 de Sacy published his *Principles of General Grammar*, which was intended for students—in particular, for his son. He cited as models, among others, the "universal grammar" of Antoine Court de Gébelin, who held that writing had emerged from scripts like Egyptian hieroglyphics, but the latter's connections with spoken language were obscure if they existed at all. The first hieroglyphs, he asserted, were imitative signs that evolved through synecdoche and metonymy so that the sense of a hieroglyphic string could only be understood in gestalt fashion, by apprehending the signs as a group or "ensemble."[30] Alphabetic writing was "born out of hieroglyphs" when the desire to "pronounce" the signs arose among the so-called Chaldeans of Mesopotamia, who were then imitated by the Egyptians.[31] Although he never cited Gébelin's opinions on writing, de Sacy obliquely reprised them in his letter to Chaptal. There he averred that writing was an "ingenious method of fixing speech by painting articulated sounds, and of thereby rendering the communication of ideas independent of times and places."[32] For him such a "painting of articulated sounds" was the essence of writing. This was about as much as de Sacy was ever willing to say about the origins of writing and, in particular, about whether phonographic writing might have developed from hieroglyphs. Hieroglyphs were simply not interesting to de Sacy because they could not yield information about language, which was, after all, his primary concern.

Although de Sacy admitted to Chaptal that he had failed to reproduce spoken Egyptian, he had nevertheless succeeded in identifying fifteen letters in the second inscription. To the hieroglyphics he had paid no attention, he explained, because they "paint ideas, and not sounds" and therefore "belong to no language."[33] The copies of the Egyptian inscription that de Sacy possessed had a large portion that was a "confused mixture," though he did not know whether this was due to damage on the stone or to a failure in taking its impression. To reconstitute the script, de Sacy worked according to a simple "rule of proportion" derived from his use of a compass to find the locations at which words in the Greek inscription could be found in the Egyptian. He based his method on the assumption, which he admitted to Chaptal did not in the end work altogether well, that the Egyptian and the Greek were literal versions of each other. Although he did not describe his procedure, it clearly involved careful spatial comparisons of the positions of sequences in the scripts.[34]

The fundamental assumption guiding early investigations of the intermediate Rosetta derived from the presumed tight relationship between the Egyptian and Coptic scripts. The latter were assumed to contain signs for all of the sounds represented by Greek letters; several of these signs were visually similar to the corresponding Greek, though most were not. The full Coptic set moreover included six signs representing phonemes that, because they did not exist in Greek, were assumed to exist only in Coptic and perhaps in the more ancient Coptic projected to have been Egyptian. This writing system, with its six "extra" signs, was assumed to have appeared after the Alexandrian conquest, that is, relatively late, but the earliest investigators of the Rosetta scripts believed that the extant Coptic script retained the fundamental structure, and to some extent the signs, of the earliest form of the writing system. There would of course be important differences in word assembly between the earlier and later forms of Coptic. Nevertheless de Sacy presumed that closely related phonographic scripts should maintain closely similar positional configurations for letters in corresponding words. To capitalize on that conviction, de Sacy worked primarily with the proper names in the Greek inscription, expressing surprise at one point at having thereby identified a Ptolemaic title.[35]

The name "Alexander" appeared only once in the Greek inscription. Using his method of spatial comparison, de Sacy found the name at the end of the second line of the Egyptian. Having located the beginning of the word, he proceeded to identify its other letters; to show how he had worked, de Sacy redrew the Egyptian. The identification was complicated, as we can see from figure 8.4.[36] De Sacy relied in the first instance on Hebrew, considering that both it and this Egyptian script would share at least some common characteristics. Reading from right to left, and combining the first two letters in the script, de Sacy thought he saw a figure resembling the Hebrew aleph (א). Believing that other letters ought to follow, he identified a fourth block of signs as "incontestably" representing the Greek ξ. Given these two characters, the second and third blocks from the right must be λ and ε, respectively, which, he concluded, must be in the form of capital letters. De Sacy worked in similar fashion with other names, making adjustments along the way. For "Arsinoe," he identified the same aleph that he had for "Alexander," though we can see from the reproduction in his letter that the second sign in the two that comprise the supposed alephs are at best mirror images of each other. He additionally asserted an identification for "Ptolemy" based on corresponding positions, which, given his sign equivalents, de Sacy eventually transliterated as *Astoloma*. He explained the extra *A*s on the grounds that, as he put it, "the Orientals almost always had recourse to such an addition when they borrowed from Greek, or from some other language, words that begin with two consonants."[37]

Perhaps because de Sacy had already acknowledged his effort to be a substantial failure, he did not explain in detail just what he considered to be the successful correspondences, of which there were a handful. Among these were the figures we have just noted: the Egyptian characters for the cognate to aleph and for the Greek ξ, λ, and ε. In the sign that he identified as representing the name "Ptolemy" (fig. 8.4, *bottom*), de Sacy also identified what he took to be a single character as the aleph, again uniting the first two distinct signs; a second apparent character as a cognate for the Coptic *fei* ϕ, which he did by uniting the third and fourth distinct signs and which, he remarked, required reversing each sign to effect the identification; and a third as the Hebrew ט (*tet*). He deemed a fourth, fifth, and sixth as essentially equivalent to the Hebrew ל (*lamed*), with the equivalence between the fourth and fifth based on Coptic doubling of diphthongs, while the resemblance of both to the sixth he grounded on a similar resemblance between letters supposed to represent the same locutions in Sassanian medals that he had examined. Inconsistencies plagued his attempts, however, and so he confessed overall failure to Chaptal. The problem, he felt, was that there were too many apparent letters in the Egyptian to effect thoroughgoing correspondences. He attributed this excess to a variety of possible causes: the presence of capital letters, of letters borrowed from Greek, of variation in letterforms, and of abbreviations.[38]

While de Sacy worked on the Egyptian inscription, he frequently met with Johan Åkerblad, who had rather different ideas but shared his fundamental assumption that the Coptic and Rosetta intermediate scripts were closely linked. After graduating from Uppsala, where he had studied Turkish and Arabic, Åkerblad worked in Stockholm on royal collections of ancient coins while writing a dissertation on the Phoenician inscription that had occupied Barthélemy and an Englishman named John Swinton years before (cf. fig. 8.5). In 1783, Åkerblad took a position in the Swedish foreign service that brought him to many places in the Near East, including a seven-month sojourn in Egypt. Åkerblad visited Paris for the first time in 1789, where he conceived a durable affection for republicanism while making the acquaintance of de Sacy. Nine years later, in Rome during the brief republic established

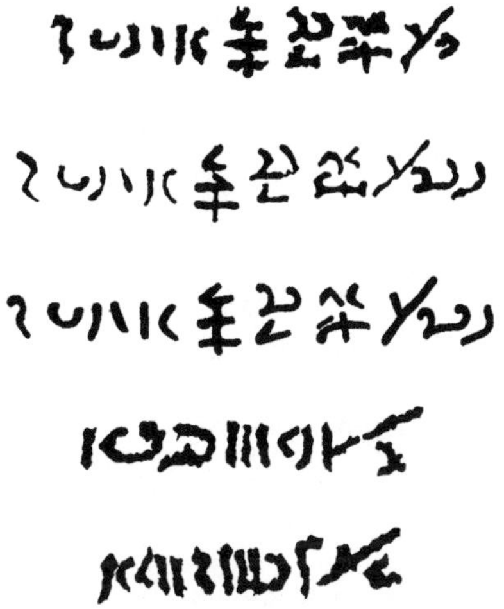

FIGURE 8.4. (*First line*) De Sacy's reproduction of "Alexander" in the "Egyptian" script; (*second*) the same signs from the English *Vetusta Monumenta* plate; (*third*) the same from the first separate plate for the Egyptian in the *Description de l'Égypte*; (*fourth*) de Sacy's "Arsinoe"; (*fifth*) de Sacy's "Ptolemy."

there by the French, Åkerblad met Zoëga, then working on a catalog of the Coptic manuscripts in the collection of Cardinal Stefano Borgia.[39] By 1801 Åkerblad was posted to Paris, where he began work on Coptic manuscripts while revising his dissertation into a publication in which he identified variant forms of Phoenician letters, going considerably beyond Barthélemy.[40] De Sacy published a laudatory but critical review of this work, which appeared in 1803.[41]

Åkerblad had identified the Phoenician letters by means of a comparison with the Hebrew script and its cognates. Working on Richard Pococke's transcription of a Phoenician inscription from Kition in Cyprus (fig. 8.5), Åkerblad had aimed principally to reach the language that the Phoenician script expressed. Like Barthélemy and Swinton, Åkerblad essayed translations of passages and not merely transcriptions. The former two had concentrated on a Maltese version of the inscription, on the basis of which Barthélemy had formed his Phoenician alphabet as a Hebrew cognate and had offered a translation based on the conviction that the language represented was, as Swinton agreed, "a mixture of Hebrew and Syriac." But there were difficulties, as the following passage from Swinton's critique of Barthélemy's reconstruction (fig. 8.5) illustrates:

The first word, לאדנו, must be looked upon as Syriac; as may likewise the sixth term, נדד, on account of the sense wherein it is used. The seventh is Hebrew, as well as Syriac. The ninth, however we render it, is undoubtedly Hebrew; and the eleventh, however this may have escaped M. l'Abbé, is certainly Syriac. The last two words are manifestly Hebrew, though in the last syllable of the former of them *Jod* is suppressed. But this is intirely [*sic*] consonant to the Phoenician form, the coins struck at Sidon generally exhibiting לעְרנם, for לעְרנים, with the *Jod* expunged.[42]

Åkerblad encountered similar problems with the Cypriote inscription since various words seemed to be missing letters essential for establishing meaning via the comparison with Hebrew and Syriac. He accordingly inserted several where he thought it necessary.

Åkerblad disagreed with both Barthélemy and Swinton concerning the cognates of several letters, and he found that some of the letterforms differed between the Cypriote and Maltese inscriptions, as well as from a third (fig. 8.5).[43] Barthélemy and Swinton had provided translations that differed substantially from each other; Åkerblad's differed from both. Juxtaposing the three, translated here into English, suggests the difficulties of extracting meaning from a dead language's script:

Barthélemy:
I sleep (I in an eternal rest) Abdassar, son of Abdssissem, son of Chad (of the village) of Tsabeth. Having passed my life in tranquility, I rest myself for the following centuries. Mathrath my wife, daughter of Tham . . . son of Abdmelee, erected this monument.

Swinton:
The marble of Abdasae, son of Abdsesam, son of Ahur. The tombstone of Lemebi, who lived 20 years of an age of pain. GO down to the eternal prison of the grave, those who died at Amathuns. This monument is the building of the house (or family) of Tain, son of Abdmelec.

Åkerblad:
I Abedasar, son of Abedsusam, son of Char, this monument to her who in my life-time, departed from my tranquil bed into an eternal one, have placed—to my wife Astarte, daughter of Taam, son of Abedmilec.[44]

These strikingly different translations illustrate the difficulty of recovering meaning even when both a strongly cognate script and language are presumed

FIGURE 8.5. (*Top left*) John Swinton's reprinting of Jean-Jacques Barthélemy's table for the Phoenician alphabet based on the Maltese inscription; (*top right*) the Phoenician inscription from Kition in Cyprus as reproduced by Richard Pococke; (*bottom*) Johan David Åkerblad's Phoenician.

to be available. All three agree on the cognate but differ on the transliteration of particular letters, and even more so on meaning. Åkerblad's translation relied upon his conviction that the presumed Hebrew cognate was nearly identical with the Phoenician and not merely related to it. All but one of the words in the inscription, he insisted, were to be found in the Hebrew Bible.[45] Reviewing Åkerblad's work, de Sacy demurred on several points, though overall he approved Åkerblad's rendition of the Kition inscription.

Meaning also lay at the core of Åkerblad's contemporaneous work on the Egyptian inscription of the Rosetta Stone. Though, like many others, Åkerblad thought an ancestral form of Coptic to be the language of the inscription, in this situation there was neither a fully cognate alphabet with which to work, nor an ancient, but known, language that, like Hebrew or Syriac to Phoenician, was presumed to have a very tight relation to that of the inscription. Modern Coptic would be the only way to reach the original Egyptian, but nothing beyond a presumption, however widely accepted, could sustain such a conjecture until a persuasive rendition of the inscription could be developed. In the case of the Kition Phoenician, none of the translators doubted its very close connection to Hebrew and Syriac both in script and meaning. Even there, as we have just seen, great variances in translation, and lesser but still significant ones in orthographic interpretation, were inevitable.

For de Sacy and for Åkerblad, experience with the Phoenician carried over directly into work on the Egyptian, albeit with significant differences. De Sacy apparently considered the orthographic structure of all alphabetic scripts to be univocal: Written either from left to right or from right to left, with or without points or accents, such a script would proceed linearly, with one letter following or preceding another in a straight line. This was how he had interpreted the Egyptian letters in his missive to Chaptal. Though Åkerblad's views concerning the fundamentals of alphabetic writing were likely akin to de Sacy's, he held a more liberal position on the spatial structure of orthography. As we will see, this basic difference would have consequences for the views each would develop in relation to the scripts.

The two met frequently while de Sacy was working on the Egyptian, but Åkerblad did not accept much of the Frenchman's work, and the suspicion was apparently mutual. De Sacy did not mention the Swede in his published letter to Chaptal. However, in letters to Friedrich Münter, a professor of theology in Copenhagen who had studied Coptic in Rome, de Sacy admitted that Åkerblad thought him to be "wholly mistaken." Although Åkerblad's effort had "not yet convinced" him, de Sacy confessed that "it has shaken me greatly." He would be "the first to applaud" if the attempt proved to be convincing, though this gesture was likely more rhetorical than sincerely appreciative.[46] Nevertheless, de Sacy lent Åkerblad one of his three copies of the inscription.

Åkerblad had added extensively to La Croze's Coptic dictionary while examining Coptic manuscripts in Paris. To that end he compiled an alphabetical list of words on cards that he stored in a shoebox. Turning to the Rosetta inscriptions, he began by seeking the analogs of proper names, just as Barthélemy had with Phoenician and de Sacy with Egyptian. Åkerblad went further, however, by searching for other kinds of nouns as well.[47] Though he relied extensively on his and La Croze's dictionary to effect correspondences with the Egyptian text, there were, he noted, many instances of words that he thought should be explicable from the Coptic but that nevertheless escaped identification. Recourse to Coptic, however, came only after a spatial comparison. Like de Sacy, Åkerblad worked by comparing physical locations between the Greek and the Egyptian, paying particular attention to repetitions of sign sequences in the Egyptian that might correspond to a repeating word in the Greek. "I recognized," he wrote by way of example, "by comparison of all the places in which temples are mentioned in the Greek inscription, that this group [of signs] must necessarily signify temples; and only after that did I try to apply the Coptic word that signifies the same thing."[48] Once he had the Coptic word, Åkerblad then tried to locate in the Egyptian the sign that represented the corresponding letter in the Coptic script. Having obtained a set of Egyptian letters in this fashion, Åkerblad could then proceed to identify other Egyptian words directly by means of his Coptic-based alphabet.

Unlike de Sacy, Åkerblad was not committed to the notion that the Egyptian analog of a Greek word would have to place its physical letters in the same manner as the Greek (or Coptic). Consider his critique of de Sacy's major claim for the name "Alexander," which he found "altogether inadmissible." How so? Because the letters that comprised the name, and that Åkerblad had found elsewhere as well, were not laid out spatially as in Greek or Coptic orthography, and because, Åkerblad claimed, the name did not begin with an aleph equivalent but rather with a sign that Åkerblad had twice identified as signaling the genitive, similar to the Coptic *mei* (ⲙ).[49] Åkerblad further opined that, though de Sacy might judge that "the contours of the letters, as I represent them, are too neat, too well-terminated," in fact "the fault lies with the artist you [de Sacy] employed, who has greatly exaggerated the inequalities of the contours."[50] Apparently de Sacy's copies were sufficiently obscure that the artist, and presumably de Sacy himself, could not distinguish the first two distinct signs that Åkerblad set as separate characters, having identified them elsewhere as well. Åkerblad in the end listed thirty-four sign equivalents (see fig. 8.6, *bottom*, including three numerals);[51] of these a significant subset was maintained over the years.

De Sacy admired Åkerblad's effort and was especially impressed by his identification of a broad spectrum of proper names. Were that all, he

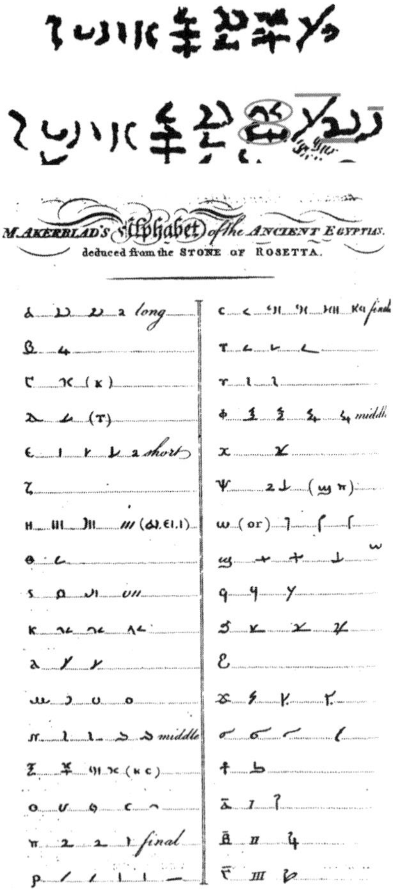

M. ÅKERBLAD's *Alphabet* of the ANCIENT EGYPTIS.
deduced from the STONE OF ROSETTA.

FIGURE 8.6. (*Top*) De Sacy's artist's rendition of three of Åkerblad's letter signs (*first row*); the same signs (*second row*) as they appear in the Vetusta Monumenta print, marked by ovals and lines above and below; (*bottom*) Åkerblad's table of Egyptian letters with Coptic equivalents in the leftmost columns. Note his inclusion of many vowel cognates.

would accept the result. But Åkerblad had gone further. There was the question of vowels to which Åkerblad had assigned signs—a "weak place of your analysis" and rather "arbitrary," de Sacy remarked. But, above all, there was the issue of "Alexander." All this must, de Sacy insisted, await a more accurate copy of the inscription, thereby undercutting in a backhanded way Åkerblad's claim that de Sacy's "artist" (and de Sacy as well, by implication) had not carefully scanned even the copies that did exist. So many "bad problems" impeded this work. "I say bad," he continued diplomatically, "because I have a presentiment that they will disappear, as the stars of the night vanish before the light of day (if you will give me this Oriental phrase), when your patience and sagacity will have triumphed over the several obstacles that impede your progress, and for me there will remain only the merit of having so long defended a bad position, and to obtain an honorable capitulation."[52]

Åkerblad's efforts rapidly displaced de Sacy's in the eyes of interested scholars. In France, Ameilhon averred that Åkerblad's work might assist in overcoming difficulties presented by the Greek inscription. He congratulated Åkerblad, hoping that the Egyptian might compensate for all that "the Greek had lost," but he demurred concerning Åkerblad's restitution of the missing Greek text.[53] Millin dedicated the 1802 volume of his *Magasin Encyclopédique* to Åkerblad, "savant Orientalist, restorer of the Egyptian alphabet" as "a tribute of esteem and friendship."[54] On April 1, 1803, Åkerblad's friend, the classicist Jean-Baptiste d'Ansse de Villoisin, enthused about his "immortal letter on the Egyptian inscription." But Villoisin, interested mostly in the Greek, devoted the bulk of his report to disagreeing with Heyne's Latin translation and interpretation. He looked forward to the appearance of other versions by Weston, Visconti, and Ameilhon, and for good measure added his voice to the chorus of complaints about slow publication of copies of the inscription, opining that "the heroes of Egypt took less time in conquering it than the antiquaries have in publishing the precious monuments that are the fruits of their victories."[55]

On March 25, 1802, the Peace of Amiens was signed, and hostilities between France and England temporarily ceased. The brief, peaceful interlude was notably productive as people, books, and correspondence flowed freely across the Channel. Åkerblad's work, which had appeared only in French, received considerable attention in England, as did de Sacy's. A letter dated August 11, 1802, from "D.H."—again, Richard Gough—to the *Gentleman's Magazine* was the first to announce that the "former possessors" of the Rosetta inscription had "anticipated us" in shedding light on the problems posed by the Egyptian scripts.[56] Five months later Gough reported at some length on Åkerblad's interpretation. He summarized what Åkerblad had accomplished and, like Ameilhon, ventured to hope that "a more perfect acquaintance with the Egyptian inscription [would] fill up the mutilated parts of the Greek"—which is principally what Gough and others thought the primary interest of the Egyptian script to be. Just as "almost all hope of recovering the antient alphabet of Egypt was nearly given up, as that of ascertaining the meaning of the hieroglyphicks has long been," Gough enthused, Åkerblad's work had "reanimated their hopes, and led them to think that the veil which covered the ancient monuments of Egypt was about to be removed." The anonymous author of the equally keen report in the *Critical Review* included Åkerblad's alphabet table.[57]

Both English reporters agreed with Åkerblad's novel view that the Egyptian script was in the form of Clement's "sacerdotal" (later hieratic) and not his "epistolographic" (later demotic), and they did so despite the presence, in the Greek version, of a term for intermediate script that meant "rural, in or of the country." Apparently they were convinced by Åkerblad's reasoning from Clement, who had remarked that the "sacerdotal" was used by sacred scribes, who were "clearly designated by our inscription as forming part of the Egyptian hierarchy, from whom the decree emanated."[58] As we shall see, the question of the relation between the two scripts, if indeed they were not essentially the same, remained contentious throughout the period during which Young and Champollion worked on the inscriptions.

Åkerblad left Paris for Holland in November 1802 to take up the position of secretary to the Swedish Legation in the Hague. There he remained through the rest of the brief peace, which collapsed on May 3 of the following year. In

October he returned to Paris, where he stayed for another six months; he was elected to the history and ancient literature (third) class of the Institut as a foreign correspondent. During this period Åkerblad was generally too pressed by diplomatic responsibilities to devote much attention to the Rosetta inscriptions beyond noting that, in the last line of the hieroglyphs, three horizontal signs have beneath them strokes that must denote the numbers I, II, and III. Recalling Clement, who had claimed that sacerdotal texts contain hymns of praise to the gods, Åkerblad pointed out that such things could not have been written in hieroglyphs because by their very nature such signs "could only imperfectly express poetic language, where the nuances of language play such a large role." Again we see that hieroglyphs were not thought to represent speech. That they might indicate the presence of an ordered series, such as a numbered list, remained within the realm of possibility, however.

Little more was done with the Rosetta inscriptions for a decade, as war disrupted scholarly work and correspondence. Åkerblad left Paris altogether in 1804, and de Sacy, as we will see, had much to distract him from this work. Even Young evinced little interest in the scripts. Visiting Paris during the peace, Young attended a lecture at the Institut and met Napoleon, yet he sought out neither de Sacy nor Åkerblad, and he seemed unconcerned with the Rosetta Stone's nearby presence in London.[59] Young would not become directly involved with the scripts until a decade later, in the spring of 1814, when he was contacted concerning a papyrus covered with Egyptian characters recently found near Luxor. In the meantime, however, a young man recently arrived in Paris from Grenoble was undertaking studies of ancient languages. We turn next to Champollion's formative years in order to understand his ways with the customs, scripts, and language of ancient Egypt, and to characterize the distinctive qualities of his attitude and approach.

CHAPTER 9

PARIS ATMOSPHERES

The parish of Saint-Roch nestles on the Seine's right bank, running parallel to the river along rue Saint-Honoré. A block and a half from the church, rue de l'Échelle, a vestige of the medieval city, sweeps between rue de Rivoli and avenue de l'Opéra. It was here, on September 13, 1807, that Jean-François Champollion first set foot in Paris. Just sixteen years old, Champollion already possessed a promising reputation as a scholar, having successfully presented a series of papers on ancient geography, religion, and literature to the Académie des sciences et des arts in Grenoble while still a student at the lycée.[1] Napoleonic reforms had militarized the educational system, and in Grenoble these changes threatened to restrict Champollion's freewheeling investigations of ancient history and languages.[2] Seeking better opportunities, he made the seventy-hour trip to Paris in a heavy diligence, a swaying "compound of the English stage wagon and coach" typically crammed with up to a dozen passengers traveling with their luggage suspended more or less securely overhead.[3] As relieved as he was to be free of Grenoble's stifling lycée, his risky and uncomfortable trip to the capital anticipated significant features of his sojourn there and hinted at the difficulties of its conclusion.

For the next two years, Champollion attended lectures at the Collège de France and the newly established École spéciale des langues orientales, pored over manuscripts in libraries and archives, and pursued private language instruction. If these were riches enough to sate the intellectual appetite of any young student of languages, they only seemed to whet Champollion's. Although the dull Grenoble years had constituted an improvement over his childhood, made chaotic by his father's recklessness with money and drink, in Paris Champollion was still making up for lost time.[4] He threw himself into his daily round, immersed in a variety of influences that wove themselves, in different ways, into his first substantial publication, *L'Égypte sous les Pharaons*. In this book—published in abbreviated, introductory form in 1811 and in the extended, two-volume treatment in 1814 that, we shall see, Young eventually possessed—Champollion invented a geography of ancient Egypt by studying

FIGURE 9.1. (*Left*) Jean-François Champollion, and (*right*) Jacques-Joseph Champollion-Figeac. Jean-François Champollion drawing by Eugène Champollion, courtesy of the Musée Champollion, Figeac (inv 03.03.1). Photo by N. Blaya.

its toponyms, linking language and history in a novel way that laid the foundation for his work on the ancient Egyptian scripts.[5]

The very geography of Champollion's daily existence was saturated with links to the past. Mingling old and new, foreign and familiar, Paris at the turn of the nineteenth century reverberated with echoes of his work on ancient Egypt. To have an address on rue de l'Échelle was already to live in conversation with historical onomastics, with the ways time and language tend both to erase and to preserve traces of human activity.[6] While many of Paris's medieval byways were slowly eliminated by Napoleon, especially near Champollion's lodgings in the first arrondissement with the creation of the avenue de l'Opéra, vestiges like rue de l'Échelle persisted.[7] Before the fifteenth century, the street had been known more prosaically as the "chemin qui va de la Porte-Saint-Honoré à la Seine," the passage from the Saint-Honoré gate to the river.[8] After the street's name changed sometime in the fifteenth century, its precise function would never again be so toponymically clear. According to some historians, the later name may have referred to the site of one of Paris's several "échelles patibulaires," pillories or stocks, erected at the behest of either the bishop of Paris or perhaps merely a chaplain of Notre Dame. Others claim that the name referred to a scale, known as a *pont-à-bascule*, for weighing livestock passing through the area on the way to market; the platforms of such scales, examples of which may still be viewed at the Musée des Arts et Métiers in Paris, were equipped with an *échelle*, a ladder-patterned surface that (presumably) helped stabilize the animal.[9] Either way, the name echoed a basic requirement of city life: to strike a balance between urgent individual needs and the equally exigent requirements of social order. The names of streets like rue de l'Échelle, which quietly but persistently evoked the past, surely resonated within Champollion, whose sensibilities had been honed by years of historical and philological study at Grenoble. What could such vestiges say about the lost world toward which they, in the face of sweeping changes, stubbornly continued to gesture? As it turned out, quite a lot—and ancient place-names provided the intellectual thrust behind the work that Champollion accomplished during his first Paris sojourn.

Hurrying between appointments, Champollion would have had many opportunities to observe these changes—not only the ways in which the old persisted amid the new, but also the ways in which the extremely old, such as remnants of ancient Egypt, could be recontextualized to yield new meanings. Champollion's Paris seethed with transformation, especially construction of public monuments and improvements to services vital to the needs of a modernizing city.[10] Symbols of French glory cropped up everywhere, ornamenting projects intended to rationalize the practical supports of urban life. The new Pont des Arts, a sturdy pedestrian bridge, linked two centers of power, the Institut d'Égypte and the Louvre, and the Passage des Panoramas, a glass-ceilinged shopping arcade, transformed ordinary marketing activity into a distinctive urban drama of seeing and being seen. In 1810, the city's food supply was modernized by means of the construction, by Bonaparte's decree, of a large new abattoir at Les Halles that distanced Parisians from the unsightly by-products of butchery and "thereby remov[ed] a nuisance long complained of by the inhabitants of Paris."[11] Nor was spectacle neglected. A number of Egyptian-themed monuments, architectural details, and even theatrical performances reinforced a rosy view of Bonaparte's occupation of Egypt.[12] In this context, the fact that modernization did not fully eliminate streets like rue de l'Échelle may have had a particular resonance. Recently arrived from the neater, smaller, and more familiar Grenoble, Champollion would have been especially sensitive to the persistence of the vestiges of an older, smaller town within the dynamic metropolis.

Champollion's biographers have emphasized his Parisian curriculum, especially the catalogs of lessons that he sent to his older brother, guardian, and confidant, Jacques-Joseph. These dispatches conveyed the contents of his studies and chronicled his absorption in a program of study that embraced both ancient and oriental languages. They also described his eminent teachers.

Louis-Mathieu Langlès, familiarly known as the "Tatar," taught Champollion Persian at the École spéciale des langues orientales. At the Collège de France with de Sacy, Champollion learned Arabic. A third official teacher, Prosper-Gabriel Audran, instructed him in Hebrew, Syriac, and Chaldean (Aramaic) later on.[13] By late winter 1809, Champollion had also immersed himself in the study of Coptic, the liturgical language of Egyptian Christianity that, we have seen, was long believed to contain vestiges of ancient Egyptian. His letters to his brother describe hours devoted to reading and copying materials in all of these languages, leavened by excursions into translation and grammar. Hustling between rue de l'Échelle, the Collège de France, and the Langues orientales, Champollion traversed a rich linguistic universe in which it was possible, in a single day, to view the same text through the lenses of multiple languages, getting a sense of what was lost, and what was retained, in different translations. Exploring Paris, Champollion had many opportunities to observe constants in the midst of change, important outward benchmarks for a person on the cusp of adulthood engaged in a parallel task—to separate, within himself, the essential from the superfluous.

Compared to Grenoble, Paris was more exciting but also more challenging. Now far from home, Champollion had more to do than simply follow his intellectual interests. He incurred debts and needed to pay them; he fretted about staying housed, clothed, and fed. The area around his rooms, near the warren of streets near the Louvre, was "gloomy and desolate," as Balzac described it, "shrouded in the perpetual shadow cast by the high galleries of the Louvre," a neighborhood of "miserable hovels" and "vile shacks" set incongruously against palatial government buildings.[14] If the mingled grime and splendor of Paris inspired Balzac, it had precisely the opposite effect on Champollion, who experienced his new city as a physical and moral trial. He complained of migraines and hot flashes.[15] Cloaking frustration with hyperbole, he claimed that even the Parisian air "wears me down. I slobber like an *enragé* and I've lost *my vigor*." With this reference to the *enragés*, the most radical of the sansculottes, Champollion tinted his criticism with politics, a habit that would intensify in the coming decades. "This place is horrible," he continued, coming down to soaked Parisian earth, "one always has wet feet. *Rivers of mud* (no exaggeration) run in the streets, and I'm bored to death."[16]

A talented caricaturist, Champollion reserved his most pointed criticisms for those nearest him, freely and wittily criticizing his lodgings, paid for by his brother, and his colorful landlady, who came in for special scorn. Reading his letters, one easily imagines Champollion enduring the indignities of a seedy public dining room like the one described by Balzac in *Le Père Goriot* (1835), with its shabby tea service, "its gilt decoration half worn away," and revolting "food-stained" and "wine-spattered" napkins assigned to lodgers and "kept in numbered pigeonholes in a box standing in a corner." Like the unpleasant and miserly Madame Vauquer, proprietess of Balzac's fictional flophouse, Champollion's landlady, the apparently formidable Madame Mécran, stood for much that Champollion despised about Paris. Like "all *women who have had a peck of troubles*," Balzac's Madame Vauquer had "the glassy eye and the innocent air of a procuress ready with a show of virtuous indignation to squeeze a client for some extra payment."[17] In letters to his brother, Champollion, too, complained of being pressured for rent by a powerful woman: "Madame Mécran torments me," "Madame Mécran hassles me every day for rent," "Send me next month's rent right away so I'm not forced to deal with a thousand harassments from Madame Mécran." As the impasse continued, Champollion's tone turned arch: "I trust you'll have the goodness to rush me the money to pay for four months' rent that we owe Madame Mécran, for I begin to perceive that I am becoming truly Parisian in the easy way that I spurned her, for the hundred thousandth time, when she came to invite me to release my eight francs."[18] As a person on whom no linguistic detail was ever lost, Champollion had surely noticed that his landlady's family name—which referred to a region of Persia (modern Makran) known to the West from at least the time of Cyrus—was itself a toponym.

That Madame Mécran, who was both foreign and familiar, should loom so large in Champollion's Parisian existence is perfectly unsurprising in view of a historical fact that even Balzac's crowded canvas omits: a recent influx of refugees from the Levant.[19] Greeks, Egyptians, Syrians, and Copts settled in Marseilles, in the suburb of Paris known as Melun, and in central Paris, where the poorer of these refugees clustered around the Hôtel des Invalides while the better off lived around the Palais-Royal and Tuileries, not far from Champollion.[20] Most were former members of the auxiliary units of the Armée d'Orient. In Paris they often worked as translators for Napoleon's growing roster of orientalists affiliated with the recently established École spéciale des langues orientales. Prominent among this group was Dom Rafaël de Monachis, an Egyptian-born monk trained in Rome who had accompanied Napoleon in Egypt, with whom Champollion translated the fables of La Fontaine into Arabic.[21] Champollion was also introduced to Mikhaïl Sabbâgh, a Damascene scholar who joined the French occupation of Egypt as secretary to General Jean Reynier and was working as a copyist at the Bibliothèque when he met de Sacy and his student, and eventual successor, Étienne-Marc Quatremère, with whom Champollion would soon come into protracted conflict, as we will see.[22]

The nearby Church of Saint-Roch provided a meeting place for Copts among the local émigré population. Located two blocks north of Champollion's rooms

FIGURE 9.2. Charles Monnet etching, *The Journée of 13 Vendémiaire, Year 4*, Saint-Roch Church, Rue Saint-Honoré, Paris. Bibliothèque Nationale, Paris, France / Bridgeman Images.

on rue de l'Échelle, Saint-Roch rises sharply from rue Saint-Honoré, fronted by a steep bank of steps. By the time Champollion arrived in 1807, the Baroque-style church was already an icon. Not only did the church contain the tombs of Diderot and the Baron d'Holbach, but the Marquis de Sade, that most unlikely of husbands, was married there in 1763. And it was from the church's steep front steps that Bonaparte had subdued a Royalist uprising on 13 Vendémiaire Year 4 (October 5, 1795; fig. 9.2). The battle became the stuff of legend: he had prevailed despite being massively outnumbered, fighting even as his horse was shot out from beneath him and dispersing opponents with a hail of gunfire, the "whiff of grapeshot" that for many marked the start of Napoleon's rise to power.

In the early years of the nineteenth century, Yuhanna Chiftichi, an Egyptian priest and former member of the Coptic Legion, began saying mass in Coptic according to the Alexandrian rite at Saint-Roch. Despite being harried, broke, and absorbed in his studies, Champollion now began attending these masses as part of an unofficial tutorial in Coptic with Chiftichi.[23] He rapidly identified with his teacher and the larger community. Turning off Saint-Honoré onto the tiny Passage Saint-Roch, which snaked along the side of the church in the shadow of its fragile campanile, was like slipping into another world.[24] On November 2, 1807, he wrote excitedly to his brother: "I am going to see a Coptic priest at Saint-Roch, rue Saint-Honoré, who celebrates Mass . . . and who will instruct me in Coptic nouns and the pronunciation of Coptic letters." [25] A little over a year later, on March 7, 1809, he felt sufficiently certain about the identity of Coptic with ancient Egyptian to conclude that his "days and hours" at the *École* spéciale had become "useless": "I am devoting myself entirely to Coptic. I want to know Egyptian [*l'égyptien*] as well as I know my French."[26] Styling himself "Saghir"—"little one" in Arabic—he grew out his beard in imitation of his mentors and cultivated a refined Arabic speaking style.[27] Dom Raphaël went so far, at one point, as to number Champollion among his children, to call him a son, a benediction that surely resonated since Champollion's father, alcoholic and running up debts in Figeac, had never had much time for him.[28]

Champollion's relationships with Dom Raphaël and with Chiftichi should not surprise us. Champollion was energetic, ambitious, and by all accounts gregarious. It would not have taken much to enrich his Parisian context in just this way. Nevertheless it remains difficult to specify the precise channel through which these connections were first made. Although Chiftichi knew Joseph Fourier from the Egyptian occupation, it is unlikely that Fourier alerted Champollion to Chiftichi's presence either directly or through the usual intermediary of his brother, who provided him with a spectrum of contacts. Both Fourier, now prefect of Isère, and Champollion-Figeac were far from Paris, in Grenoble. Since 1804, Champollion-Figeac had been working with Fourier on the introduction to the *Description de l'Égypte*—indeed, it was Champollion-Figeac who delivered Fourier's introduction personally to Napoleon. Rather, it seems more likely that Champollion learned of Chiftichi through Langlès, who had attempted to secure translation work for Chiftichi in conjunction with the production of the *Description*.[29] Alternatively, Champollion may have met him through Dom Raphaël, himself a refugee from the occupation who also enjoyed a close relationship with Fourier, who had secured him a position in Paris, translating Arabic documents; it has been suggested that Champollion met him for the first time not in Paris but in Grenoble some years earlier.[30]

Though he was likely the only representative of scholarly Paris at Chiftichi's Coptic services, Champollion was neither the city's sole Coptic enthusiast nor the only one who saw Coptic as a key to the language of ancient Egypt. As we have seen in our discussion of de Sacy and Åkerblad, some prominent Parisian orientalists had also trained their sights on Coptic and were testing the hypothesis, first presented in Kircher's *Lingua aegyptiaca restituta* (1644), that Coptic might contain remnants of the language of ancient Egypt. Reprising the general view, de Sacy in 1802 went so far as to describe Coptic as "égyptien moderne" and asserted that Coptic was "incontestably formed from remnants of ancient Egyptian."[31] Champollion nurtured a similar view. "All those who know Coptic," Champollion wrote, "and who are studying the language are deeply convinced of its identity with the language of the ancient inhabitants of Thebes and Memphis."[32] Poring over ancient Coptic manuscripts housed at the Bibliothèque nationale, these specialists sought moments when the sources seemed to vouchsafe purely Coptic words that appeared to retain no trace of influence from Greek, Roman, and Arab conquests, that is, from later periods of Egyptian history marked by domination by outsiders. These rare words or even sequences of letters were believed the most likely to preserve vestiges of ancient Egyptian.

This approach to ancient Egyptian by way of Coptic owed something as well to the latest exciting newcomer on the intellectual front: comparative philology. The field was in its earliest days, having developed from Romantic linguistic nationalism and fresh interest in Sanskrit as evidence of an Indo-European family of languages. The two strains converged, both ideologically and in terms of the people who expounded the central ideas of each one. In 1772 Herder, whose ideas we discussed in chapter 4, gave voice to Romanticism's "central linguistic dogma" in his *Treatise on the Origin of Language*, which claimed that language gave a collectivity, such as a nation, a unique identity, a position that leaned on an implicit concept of biological race, though even Herder believed only in cultural-linguistic, rather than biological, entities.[33] Years later, in 1823, the linguist and educator Wilhelm von Humboldt would make a similar point, privileging the exploration, in his own work, of the "innere Sprachform" or deep structure of a language—the formative principle that, he believed, gave rise to language, and from which speech and, secondarily, grammar were derived.[34] Following on the groundwork laid by the Sanskritist William Jones, Alexander Hamilton exercised, by historical accident, an important practical influence on Parisian orientalists. Hamilton was reviewing the Sanskrit holdings of the Bibliothéque nationale when the Peace of Amiens collapsed. After securing privileges as a prisoner of war, he found employment in the library cataloging those same manuscripts while also teaching Sanskrit and Indo-European comparative philology to the leading orientalists of Paris, including Champollion's teacher, Langlés. Shortly thereafter, in 1814, the first chair of Sanskrit was established at the Collège de France; no other European city could boast similarly generous institutional support for its study.[35]

This Sanskrit vogue inspired influential new ideas about linguistic kinship. The German poet, philosopher, and philologist Friedrich Schlegel, with whom Hamilton shared rooms, emerged as a leading spokesperson for these ideas with the publication in 1808 of his *Language and Wisdom of the Indians*. The fruit of his study with Hamilton, the book asserted that kinships between languages could be identified by looking at the structure of words (later called morphology), for example, noun inflection.[36] Significantly, the study of Sanskrit itself stemmed from Anglophone studies undertaken in India, notably by Jones and Hamilton, with a group of pandits, scholar-priests who spoke the ancient language, in Calcutta.[37] These scholars, in turn, drew partly on the insights of eighteenth-century comparativists like Siraj al-Din Ali Khan Arzu, who had earlier discovered correspondences between Persian and Sanskrit.[38] This is not to say any Sanskrit grammatical system was fully taken up, or was even very well understood, by Europeans at this point. Rather, the more available bits of Sanskrit and Persian were "classicized," made to fit into a preexisting scheme, and were interesting to Europeans principally to the extent that they could be related to more familiar languages, such as Latin and Greek.[39] While Champollion was skeptical of the first wave of excitement over Sanskrit, he nevertheless remained intermittently preoccupied with the subject as an important point of comparison with ancient Egypt until at least 1813. Champollion published an essay on Persian literature for the Annales du Département de l'Isère in 1808, followed in 1813 by one seeking links between Persian and Sanskrit and exploring parallels between philosophical and religious ideas to see "if the religion and philosophy of Egypt truly originated in India."[40]

If Sanskrit was the ancient language of the moment, interest in Coptic among Parisian orientalists nevertheless remained lively, abetted by an unusual shift in the material conditions of research. Napoleon's conquests filled not only French coffers but also Parisian libraries, and many important Coptic manuscripts numbered among the spoils. By the time Champollion arrived in Paris, books and manuscripts had been pouring into the city for a decade, in shipments of war booty from Napoleon's victories all over Europe and at least for a time in Egypt.

The removal to Paris of manuscript and printed materials from the Vatican and the conquered Italian states had been under way since Napoleon's June 1796 armistice in Bologna, which resulted in five hundred manuscripts being sent to the Bibliothèque nationale along with other items of aesthetic, scholarly, and historical interest. These were the first rivulets in what would soon become a flood. Many manuscripts, incunabula, and rare volumes came from Rome and the libraries of northern Italy. Still others rolled into Paris on convoys from libraries throughout Prussia, as well as Vienna and the Tyrol. Toward the close of the eighteenth century, those responsible for processing these shipments reported the arrival of hundreds of stuffed crates; the items within numbered in the thousands. In 1809 Napoleon seized the holdings of the Vatican Archives. All in all, his forces moved twenty convoys' worth of materials to Paris. Each convoy consisted of 4,500 crates, and each crate weighed more than two hundred pounds; the whole project required hundreds of wagons and nearly a thousand horses and oxen to drag them. The Bibliothèque nationale received materials until space ran out; Napoleon then ordered the convoys to be unloaded at the Hôtel de Soubise, a palace in the third arrondissement. Among the materials removed were manuscripts, volumes, and other Coptic-language items in the Barberini and Borgia collections, which would soon become vitally important to Champollion.[41]

Although texts written in the languages of the Near East had been scattered in libraries around Europe by the early modern period, Coptic materials were relatively rare. Perhaps the most significant was Jean-Baptiste Colbert's collection, which numbered more than six thousand items when his grandson sold it to the Bibliothèque royale in 1732. Colbert's Coptic manuscripts were collected mainly through the exertions of his librarian, the mathematician Pierre de Carcavy, who was in turn a client of Johann Michael Wansleben (better known as Vansleb), a traveler and adventurer who was also one of the most assiduous participants in the early modern book trade. He brought more than three hundred manuscripts and other Coptic materials to Paris in 1676, after years in the Levant, which he described in his *Nouvelle relation en forme de journal d'un voyage fait en Égypte* (1672–73)—a work that, despite its age, nevertheless remained current enough to provide Champollion with an important source for his *L'Égypte sous les Pharaons*.[42] The Coptic collection in the Bibliothèque nationale was further enriched by holdings seized during the Revolution from local Paris scriptoria, such as the library at the abbey of Saint-Germain in the sixth arrondissement.[43] The abbey had boasted more than fifty thousand items, many of them medieval scribal manuscripts if not more ancient, before gunpowder stored in the abbey's thirteenth-century refectory exploded in late August 1794, destroying virtually all of the abbey, including the library.[44] The materials that escaped the fire included volumes of Coptic Old and New Testaments; hymnals, liturgies, and other religious texts; and Coptic and bilingual grammars and dictionaries.

As rich as these Parisian collections were, they could not compare to those of Rome, where numerous Coptic manuscripts had been available since at least the middle of the fifteenth century when delegates to the Council of Florence (1438–45) began providing materials in Coptic, Arabic, and other Eastern languages to the Vatican Library, usually as part of efforts to forge a durable union between the Eastern churches and the Vatican. A delegation of Egyptian Copts offered the richest of these gifts, a collection of fifty manuscript codices, of which six were in Coptic. The Vatican's Coptic holdings also included polyglot materials, that is, items that were bi- or trilingual. Although many of these texts were religious, the holdings eventually also included grammars and dictionaries, secular materials that could support extensive investigation of the linguistic aspects of sacred texts. In addition, the powerful Italian noble families of the Renaissance—the Borgias, the Barberinis, and the Medicis—also collected Coptic manuscripts and printed volumes, and they sometimes bequeathed them to the Vatican. Finally, in the sixteenth century, the Vatican's holdings in Eastern languages were strengthened by the addition of materials sent back to Rome in the aftermath of different conflicts around the Mediterranean.[45] In short, Champollion and his Parisian colleagues were uniquely fortunate beneficiaries of Napoleon's European conquests, which among other things greatly enriched the Parisian pool of materials in the languages of the Near East.[46]

In addition to Kircher, whom subsequent scholars could neither accept nor fully do without, Champollion's early modern sources for links between Coptic and ancient Egyptian included works by eighteenth-century scholars such as Maturin de La Croze, who used Coptic to interpret Egyptian names for gods and kings.[47] The next generation of Coptic scholars centered on Zoëga, who worked in the Vatican Library, especially in the collection of Cardinal Borgia, who had been collecting Coptic materials since the 1770s and for whom Zoëga was cataloging Coptic holdings.[48] Zoëga's *Catalogus*, published posthumously in 1810, made an immediate, positive impression on Champollion, whose "Observations sur le catalogue des manuscrits coptes du Musée Borgia" praised Zoëga's work and introduced him to the readers of the *Magasin Encyclopédique* in October 1811, shortly after the catalog itself appeared.[49] The most current work on Coptic as a late form of Egyptian was being done in Paris, however. Thanks to their recent occupation, the French had a monopoly on the latest views of Egypt and, of course, they had access to a large number of manuscripts and antiquities unavailable elsewhere at the time. Champollion's new Coptic sources included liturgies, hagiographies, and martyrologies, Coptic versions of Old and New Testaments, Coptic and Arabic grammars, vocabulary lists, and polyglot materials. He also had a copy, though not a particularly faithful one, of the Rosetta inscriptions to compare with whatever he discovered in the mass of new acquisitions.[50]

Champollion's sources for his *Pharaons* also included materials published in the *Décade Égyptienne*, the journal of Napoleon's Egyptian expedition. These included important extracts by Marcel on "géographie arabe," which would prove useful for confirming locations mentioned in documentary sources, as well as other publications by the savants involved with operations in Egypt.[51] Among them were some familiar scholars—de Sacy, Åkerblad—as well as the up-and-coming orientalist Étienne-Marc Quatremère, whose research agenda would soon collide disastrously with Champollion's. Champollion developed a less antagonistic relationship with Åkerblad, whom he acknowledged specifically for his attention to and command of Coptic.[52] Åkerblad, we have seen, had reasoned paleographically in order to demonstrate Coptic equivalents for several royal names already identified in the

Egyptian text of the Rosetta inscription, namely, Arsinoe, Ptolemy, Alexander, and, most importantly for our purposes here, Berenice. His discussion of this name—in particular, his formulation of the nature of its hard-to-pin-down vowels—shaped Champollion's deepening engagement with Coptic as a late form of ancient Egyptian.

The name Berenice had proved challenging to de Sacy, and it was on precisely this vulnerable point that Åkerblad's argument turned. When Berenice appeared in the Egyptian, he stated, it was preceded by a particle that indicated the genitive case.[53] The consonants could also be relatively clearly discerned, notwithstanding some uncertainty over whether the first letter corresponded phonetically to *B* or *P*. Signs that might correspond to vowels, however, gave Åkerblad pause, partly because in written Coptic, vowels and diphthongs were especially subject to orthographic variation.[54] One mark in the Egyptian, consisting of three parallel vertical lines, was of particular concern. A similar three-line sign in Coptic represented an article expressed by combining the sounds of two Greek letters. However, the sign bore "no relation" to those two letters at all. This suggested to Åkerblad that it might "have its origin in the ancient writing of the Egyptians."[55] Later that year, in his *Lettre sur l'inscription égyptienne de Rosette*, Åkerblad claimed that the Egyptian original could be given a phonetic value of "o" or "i," as these options corresponded to appearances of the sign in different royal names whose Egyptian transliteration had already been established (e.g., Ptolemy, Alexander). But Åkerblad remained unsatisfied. To clarify the niggling ambiguity, Åkerblad suggested that a three-line sign signified the presence of a vowel, averring that its own meaning was however "entirely empty . . . which confirms my supposition that the sign is nothing other than *a fulcrum* that may be adapted to multiple vowels."[56] The mysterious sign, in other words, functioned like a railway switch point: the correct vowel could be supplied by a speaker who, like the train conductor, knew how to choose the right direction for the switch. On this point Åkerblad's argument departed most significantly from the main line of contemporary research into the decipherment of ancient scripts, which did not conceive of flexibly significant vowel placeholders. Its oddness was also its strength. The most promising affordance of Åkerblad's fulcrum hypothesis was that it allowed for a variable orthographic element. Such a language accordingly contained at least one placeholder element that could be specified on the fly, as context demanded. Åkerblad, we saw above, compiled his results in a chart that suggested linkages between Coptic letters and signs found in the Rosetta Egyptian (see fig. 8.6). The analysis stopped there. But the idea of identifying and tracking variable elements of a script was something Champollion would soon pursue, with better results, by focusing less narrowly, not on letters or paleography but on whole words, in particular on common nouns rather than just proper names.

Åkerblad's point about vowels was not the only one Champollion would take up and elaborate as he developed his geography of Egypt. Åkerblad had worked on manuscripts in Rome with Zoëga, and he noted with frustration that most Coptic sources were too stuffed with religious terminology to be useful to scholars like himself, who sought references to ordinary objects, place-names, and so on.[57] In contrast, he singled out texts in the Borgia collection that were, unusually for this body of material, not principally concerned with liturgical matters or patristics. Such texts could be usefully mined for "the names of towns, villages and nomes of Egypt, or other remarks that could be interesting for the geography and history of the country."[58] Åkerblad did not make use of this insight in his efforts to decipher the Rosetta Egyptian, focusing instead on paleography by forging links between it and the letters of the Coptic alphabet. Here Champollion seems to have spied an opportunity.

Champollion's thirty-six-page review of Zoëga's catalog and anthology of the Coptic manuscripts in the Borgia collection, which appeared in October 1811 in the *Magasin Encyclopédique*, extended and amplified Åkerblad's remarks about the value of ordinary words to the project of apprehending the ancient language of Egypt, raising hopes that many such might be discovered in the Borgia collection. The cache even contained a few leaves from a medical text, "a collection of prescriptions for the cures of several diseases." Champollion cautioned that although "we shouldn't expect to find learned theories in this material," nevertheless "the fact that the fragment contains only popular remedies should not be a reason to neglect it. Rather, since these [prescriptions] are usually the result of experience, we can find in them certain things that are worth knowing."[59] Such materia medica could be expected to include, for instance, references to local plants and familiar anatomical details, providing a source of common nouns. Echoing Åkerblad's position of nearly a decade before, Champollion continued: "So this interesting fragment is made all the more precious by being one of the few Coptic manuscripts that does not deal with religious matters." Ever since Coptic had been relegated to a liturgical language, "it would appear that the books dealing with anything other than asceticism have been entirely neglected," which was a great pity, "since manuscripts of a different type might have provided a basis for a wider knowledge of the general state of ancient Egypt."[60] Coptic words for ordinary things could provide clues to significant aspects of ancient Egyptian culture.

Champollion's suggestion was part of a larger framework, for he had set his sights on a most ambitious target. While the *Pharaons* was conceived, organized, and promoted as a geography, Champollion aimed to inhabit, as far as possible, an ancient Egyptian point of view. In the brief 1811 publication that would evolve into the 1814 two-volume *Pharaons*, Champollion described the work as conceived and written in terms that ancient Egyptians themselves might have recognized: "Our primary objective has been to introduce the country *on its own terms*: we have aimed to produce *an Egyptian geography of Egypt*, which has not existed until now."[61] In a statement transmitted from Champollion to Antoine-Jean Saint-Martin, a friend who was also an early reviewer of the work, Champollion's brother reprised this important theme: the work was not merely a superficial summary of the existing documentary evidence but an accurate geographical representation of ancient Egypt, "une géographie *Égyptienne*."[62] Champollion explicitly framed his Egyptian geography in terms of an "intérieur mise en oeuvre," a hard-to-translate phrase, reminiscent of Humboldt's "innere Sprachform," which suggests Egyptian civilization had unfolded according to a logic of its own. If this "interior implementation" (to translate the phrase most literally) belonged solely to Egypt, its workings could nevertheless be inferred from extant remnants, including linguistic ones, of the civilization itself. Ordinary language played a key role here, as we will see, for these residues, which persisted over remarkably long spans of time, could be found in the names used to refer not only to cities and towns but also to prominent geographical features such as the Nile. In other words, the work's originality rested not in geographical insights per se but in the way Champollion used contemporary ideas—what one might anachronistically call early nineteenth-century French linguistics, coupled to a sympathetic effort to inhabit another point of view—to unearth the ancient language of Egypt.

Coptic provided Champollion with precisely the time-travel mechanism he needed: through the intermediary of the language, Champollion could work backward to ways of speaking—and, as we will see, thinking—characteristic of the ancient past. According to him, Coptic was distinctively suited for this purpose. First, the language was, Champollion wrote, "altogether monosyllabic," governed by "fixed, constant, immutable" rules of combination. Thanks to these basic qualities, the language could not easily be corrupted; essential elements would persist regardless of intrusions from other languages. This linguistic inalterability was, in turn, culturally pervasive. Not only did it preserve rudiments of the ancient Egyptian language but it also reflected "all the institutions of primitive Egypt."[63] The latter part of his position on Coptic—that its inalterability reflected ancient Egyptian institutions—was not especially original. Wolf, for example, thought Greek especially excellent because, uncontaminated by foreign elements, it had not changed over time. Champollion took this position a step further with Coptic, however, by looking to its presumptively unchanged words for clues to Egyptian culture itself.

Champollion elaborated his idea by using a particular tool of linguistic analysis: the primitive root. The persistence of certain monosyllables—a "root," a *racine primitive*, as it was called—in words with similar meanings, or with the same meaning in different languages, had previously been remarked and used, in the latter case, to conjecture a shared linguistic origin to account for these similarities. But the focus before 1800 was on the interaction of roots with meanings, not on the persistence of roots in the unfolding history of a language. In 1662, the Jansenist logicians Antoine Arnauld and Pierre Nicole, in their *La logique, ou L'art de penser*, familiarly known as the Port-Royal *Logic*, claimed that although the form of a word—its spelling, its morphology, and/ or its sound—may persist, its meanings may change; in this way, denotation gave rise to connotation.[64] A half century later, César Chesnaux Dumarsais explored the ways in which the same word could give rise to multiple, albeit related, meanings beyond its simplest and most literal one. Meanings might be absolute, relative, collective, distributive, moral, allegorical, abstract, concrete, equivocal, or louche, to name a few.[65]

By the middle of the eighteenth century, when Jacques Turgot took up the problem of etymology in a lengthy article in the *Encyclopédie*, the idea of a durable monosyllable, persisting through gradual shifts in meaning, had a fundamental function within general grammar. Shifts in meaning could arise from multiple causes, but orthography conserved what was otherwise easily modified in speech. Writing, therefore, had a new function, as a reservoir of vestiges of roots.[66] To illustrate this point, Turgot considered translations of the same word from one language to another. As speakers of different languages met for trade or other purposes, they left traces of their languages in one another's, especially in words that tended to remain the same over long stretches of time, like toponyms. Celtic, for instance, left so many roots in French one could find "scattered remnants in the Irish, Welsh, Bas Breton in the ancient place-names of Gaul, &c." Similarly, Turgot wrote, "in Saxon, Gothic, and the different ancient and modern dialects of the German language, we will partly find the language of the Franks."[67] Foreign words entered according to the nature of the cultural links between speakers of different languages: trading cultures brought mercantile words into a destination language.[68] Etymological analysis was accordingly "the source from which flows the rules of general grammar," of which, "all our languages' grammars are only partial and incomplete instantiations." Despite the ubiquity of the system of general grammar, which "governs all languages," speakers of any given language remain unaware of it, "believing they are only following the whims of use."[69]

A decade later, Abbé Nicolas Sylvestre Bergier took up the question of how language learners could make practical use of this feature. True roots, he said, are monosyllabic and, since each syllable had a meaning, a person who has understood a few, small, primitive elements of a language may derive the rest from them. Even "syntax" or "the arrangement" of the language could be inferred from word use alone: "to master a language, especially a dead language, involves nothing more or less than knowing the meanings of all the words that comprise it." Moreover, the more ancient the language, the fewer words it contained, and the more of them would be roots rather than derivatives, as the latter required time and civilizing processes to develop. A perfectly primitive language would consist, therefore, only of roots, and these would be mere gestures and cries indicating simple things. To account for differences, Bergier suggested environmental influences, and he claimed that different languages were analogous to different physiognomies.[70]

Charles de Brosses's *Traité de la formation mécanique des langues* of 1765 literalized Bergier's physiognomic perspective, focusing on the anatomy of speech organs in order to assert mechanical links between those organs and the sounds humans made with them. Linguistic roots "produced" language just as botanical roots produced plants: they were the "original kernel" (*premier germe*) from which "branches and branches" developed.[71] Going considerably beyond Bergier with this physiological metaphor, de Brosses claimed that any given word has an essential primary (*premier*) and true (*véritable*) form (*physique*). Nevertheless, he claimed, while each word may have an essential basis, all words are susceptible to many different renderings, nuancings, and shadings, a concession of broad plasticity that echoed the linguistics of Dumarsais a half century earlier.

The promise of etymological analysis had advanced so far by the last quarter of the eighteenth century that Court de Gébelin, in his 1772 *Histoire naturelle de la parole*, was moved not to define or explain but to defend it. He did this by advancing a metaphysics of language, claiming that the most primitive words have a distinct vitality that it was the proper function of etymological analysis to recapture, an original "énergie étonnante."[72] Like his predecessors, he believed that "primitive words" were monosyllabic and referred in the first instance to literal objects. Only later did they become figurative or abstract. Words change "like rubbed stones," he wrote: they wear away with use.[73] It was for this reason that modern languages differed from ancient ones. Although he agreed with his predecessors on the special preservative quality of writing, he conceded that orthographies varied widely, making ancient languages even more difficult to understand; nevertheless changes in orthography were regular and could, therefore, be systematically identified and traced: vowels changed according to one scheme, consonants another.[74] Analyzing these changes might be complicated, but the task was not hopeless.

Comparing semantically related words from different languages in order to find morphological similarities had a separate but related history, stretching back a half century or more before Champollion's time. Earlier scholars listed words that repeated in biblical or ecclesiastical texts, typically with the goal of arriving at the original language of Jesus. One example of such a comparative practice was produced by Barthélemy in a report read on April 12, 1763, to the Académie royale des inscriptions et belles-lettres (an earlier name of the Académie des inscriptions et belles-lettres). "In the language of a people," Barthélemy wrote, "one may find an abridged version of the history of their opinions, and sometimes also in that of other peoples, the history of their colonies and their commerce. It is therefore essential to compare ancient languages with each other."[75] The point of the newer work, however, was not to delve into a single language or civilization; rather, even when their arguments touched on morphological similarities between words, the aim was to support claims about the affiliations of different languages and civilizations. Champollion arrived in Paris just in time to read the first reviews of the initial volume of Adelung's *Mithridates* (which Young himself reviewed, as we saw in chapter 6) as it appeared in four volumes over the course of more than a decade (1806–17). Adelung's massive project compared texts of the Lord's Prayer and other sample texts in all languages known at the time, seeking similarities between grammatical structures that could, he hoped, indicate historical or genealogical relationships between different groups, even those that were widely geographically dispersed.

Finally, there were precisely contemporary developments in the study of the grammars of oriental languages, particularly the study of Persian and Sanskrit, mediated, as we have seen, through the work of Jones and the immediate presence of Hamilton in Paris. Study of these ancient and recent grammarians alerted scholars to special inflected forms in Sanskrit, such as those identified and systematized by Pāṇini, whose catalog of grammatical rules, verbal roots, and other linguistic elements appeared in the fourth century BCE.[76] Of particular interest for our purposes are Paninian linguistic primes, including verbal roots and noun stems, especially the *karakas*, the various roles assigned to one class of these primitives, that is, the verbal root. These roles are implicit in the inflections attached to the root; for instance, they may specify the subject but only implicitly. In Sanskrit, notes a modern specialist, "[a] sentence is seen as a little drama played out by an Agent and a set of other Actors, which may include a Goal, Recipient, Instrument, Location and Source."[77] In this way, syntax was shown to govern semantics, at least in a limited way, an asymmetry we will reencounter as Champollion painstak-ingly sought to wrest insights into the grammar and lexicon of ancient Egyptian from his work on Coptic roots.[78]

Although Champollion, who by the age of seventeen was already an accom-plished linguist, was surely aware of these ideas about roots and derivatives, in his thinking about specifically Coptic roots he was closely influenced by at least two contemporaries apart from the Copts who tutored him in Paris. The first was Tommaso Valperga di Caluso, a Turinese orientalist who first trained as a mathematician with Joseph-Louis Lagrange; he was the author of *Literaturae Copticae rudimentum* (1783), an influential introduction to Coptic that drew from the insights of La Croze. Champollion's teacher, de Sacy, was a second source of encouragement toward Coptic studies. De Sacy advocated for the application to Coptic of analytical methods already used in so-called oriental languages, subsuming Coptic under that umbrella. The most advanced way to approach Coptic, de Sacy later wrote, was "to apply to the Coptic language the etymological system used in the dictionaries of Hebrew, Syriac, Arabic, etc., a system that has the advantage of collecting all of a root's derivatives under that same root." Not coincidentally, this method perfectly described Cham-pollion's working procedure, abundantly displayed in his Coptic notebooks, which we will discuss in chapter 10. "Without this method," de Sacy continued, "the members of a given family of words are dispersed, and each member is deprived of the support that root words [*mots radicaux*] and their derivatives reciprocally lend to each other."[79]

Champollion internalized several mutually reinforcing standards against which he anxiously measured and remeasured his results. First of all, he wor-ried about creating false etymologies. To do so was to be like Kircher, whose investigations were widely scorned as fantasies.[80] Inferences and conclusions had to follow in obvious ways; they should not seem forced or labored. In a let-ter to his brother written on October 10, 1808, Champollion expressed concern that his project might be too redolent of "system," the familiar and dreaded shibboleth of the eighteenth century—a criticism with which, we shall see, his later decipherment would indeed be accused. This concern seemed to belong at once to Champollion and to an imagined, and rather skeptical, interlocu-tor: "If mine is a young intellect that creates imaginary systems," he asked his brother, "which have no basis except in subtleties, etc. etc., why would you want [me] to bring into print an Egyptian geography, which is full of the spirit of these same systems?"[81] Finally, Champollion feared that the effort would be disparaged as "a minor work, in which it is shown that Egyptians, those bestial and rusticated [*oignonicole*] people," could not be credited for having made a civilization.

Yet the presumptive descendants of the inhabitants of ancient Egypt, descendants who had so recently frustrated, disappointed, and repelled the French during the occupation, were showing up in Paris and Marseille. They were teaching Napoleon's savants to read the vast number of otherwise impenetrable texts now swelling the holdings of the Bibliothèque, and drawing military pensions for their service in a nightmarish occupation, enduring privations—attacks of blindness and outbreaks of plague and leprosy, to name a few—most ordinary Parisians could scarcely imagine. Thanks to his friendships with these émigrés, Champollion had a clear idea of their sufferings and particularly of the unjust and demeaning attitudes toward them held by many Parisians, including his teachers. His intimacy with the Egyptian community in Paris fired his commitment to the study of the language, history, and culture of ancient Egypt. When his brother raised questions about his investigation of etymologies as key to that goal, Champollion replied:

> Call me crazy, etc., etc. This will not prevent me from studying my ancient languages and the relationships between one people and another, to love etymologies and the like . . . I'll also tell you that since I've returned to my Egyptian geography, I've only felt more confirmed in my discovery *and even my system.*[82]

His self-confidence notwithstanding, and despite the consolations of the community provided by Chiftichi and Dom Raphaël de Monachis, Champollion's patience was fraying. The longer he remained in Paris, the more alienated he became from its institutions and from those who wielded power within them—particularly his teacher de Sacy, who had already put Åkerblad forward, ahead of Champollion, and who would later support Quatrémere against him as well. For his part, Champollion called his Parisian colleagues "vampires" and faulted their hypocrisy. He complained that they behaved as if partisanship were the only legitimate source of authority, as though, having staked their position in advance of any inconvenient facts or reasoning, they could all more easily "arriver à la chaise curule," that is, occupy the scholarly equivalent of the Roman curule seat of judgment.[83] Threats and slights accumulated. Perhaps Champollion had even been blackballed. He failed to secure a position he sought in the Bibliothèque nationale as a copyist.[84] Then the orientalist Langlès, in a sudden move, proposed him for a consular post in Persia, part of a project to put Parisian students at the École spéciale des langues orientales to military and diplomatic purposes abroad. The offer held little appeal: to take the position would have completely derailed Champollion's Egyptological career. He stopped going to classes and turned for help to his brother and to

Fourier. The matter was soon dropped, but Langlès apparently neither forgave nor forgot Champollion's refusal.[85]

Discouraged, Champollion looked wistfully back to Isère. Grenoble's placid familiarity promised respite from the ambiguous excitements of Paris, but to abandon the metropolis had obvious costs—loss of professional momentum, restricted access to crucial research materials. But even as he sank into dejection, geographical matters were never far from his mind. Even his homesickness had a geographical cast. "The more I think about it," he wearily told his brother, "the more attached I feel to our mountains."[86]

CHAPTER 10

ROOTED IN PLACE

Our mountains. On October 15, 1809, the Lyon coach to Grenoble arrived at dusk, after a meandering journey through the extraordinary landscape of the Dauphiné, with its walnut groves and tobacco fields draped in mist, its dense woodlands punctuated by sheer cliffs and fed by swift-flowing rivers. Champollion was among the passengers who disembarked at Place Grenette, the town square Stendhal described as he observed the same coach arriving amid the tolling bells of Saint André and the click of children's sabots as they played hopscotch, their voices rising as they counted *un, deux, trois!* At mid-October, the leaves were already falling through air seeded with the first icy hints of winter. "Pitched in the lee of towering hills" was how one writer later characterized Grenoble, which was more or less how Champollion, gripped by homesickness, had remembered it, a place where "the Alps nudge the sky at the end of every street."[1]

Much else was nudging Champollion at this moment. These forces both protected and oppressed. Deeply engaged in his reinterpretation of Egyptian geography through the lens of Coptic, Champollion was eager to produce a definitive treatment before a competitor did. His worldly and ambitious older brother was also eager for this result. Since 1800, Champollion's life and work had become ever more thoroughly intertwined with the interests of Champollion-Figeac, whose efforts on his younger brother's behalf included everything from the securing of paid employment, to the correction and editing of his manuscripts, to the distribution of dollops of encouragement and occasionally stronger exhortations, not to mention infusions of cash.[2] His fingerprints are everywhere visible during this period, and his protective activism secured Champollion the freedom to lay the intellectual foundations of his later work. Years later, Champollion told his brother that he had long been reconciled to this surrender of autonomy: "It has been clear to me for a long time that *what is mine is yours. . . .* The present, the past, all that I have been, all that I am and will be, I owe to you."[3]

Champollion's focus on geography for the first significant publication of his career, and his first major work devoted to ancient Egypt, was a deliberate and even strategic choice. Napoleon himself placed a high value on geography, for its obvious military value as well as for its utility in arenas closer to home, where reliance on the outdated and flawed maps of the French interior by the Cassini family had led to a series of humiliating domestic catastrophes. Even the empress Joséphine had once become lost and finally stuck in the Ardennes mud while following a route laid out by Napoleon according to a Cassini map that bore an unexpectedly distant relationship to reality.[4] Though a military survey would not provide updated maps until 1818, the need was already acute.

The geographically minded, like Champollion, could turn to history—but there, too, special difficulties lay in wait. Since so little of the ancient world remains apart from what has been written about it, to produce a geography of antiquity necessarily required painstaking study of the ancient names for places. But names are not geological formations. Like any other human pro-duction, place-names are subject to sudden modification, to drift and, not infrequently, sheer obliteration by conquerors and others hostile to their pres-ence. Moreover, it has always been difficult to precisely match a toponym in a textual source to an actual location.[5] Compared to royal names, however, which could be (and were) effaced from monuments, place-names enjoy a special, if limited, persistence. For practical reasons, the phonetic instantiation of toponyms tends to be conserved even as the names are written in differ-ent languages with different scripts. The special longevity of toponyms there-fore made them ideal vehicles for the preservation of cultural information through language.[6] The modern poet Gustaf Sobin has made the same point in more lapidary terms: "Nothing's older, in many places, than the toponyms themselves. Not even the archaic sites they still, occasionally, designate: the ruins, the earthworks, the beheaded remainders, say, of some oppidum lying perched on its long since abandoned outcrop. *Nomen*, quite frequently, ante-dates *habitatio*."[7]

As early as 1811, when he published a substantial introduction to what would eventually become his two-volume study of ancient Egyptian geography, Champollion was working out connections between linguistics and geograph-ical reality, between ancient Egyptian place-names and the real-world places to which those names had been and often continued to be attached. Echoing a standard view, Champollion believed that Egyptian place-names were es-pecially long-lasting because of what he considered the closed, conservative character of Egypt's cultural institutions.[8] Citing Iamblichus, a third-century Neoplatonist from northwestern Syria, Champollion wrote that the Egyptians, like other "Asiatic peoples" tended to "persevere in their customs, keeping their manners unchanged, and . . . the names of places or other [things] they had adopted [also] remained the same." He contrasted this conservatism to the Greeks, whom he described as "friends of novelty" who "changed whatever struck them, in order to present it in a new form."[9]

Although a complete elaboration would have to wait until the 1814 publication of the *Pharaons*, the novelty of Champollion's framework was already evident in his 1811 *Introduction*. This novelty consisted in the link he made between these familiar positions and Coptic—which, he claimed, had a special incorruptibility that made it particularly likely to have retained residues of ancient Egyptian. For Coptic, Champollion asserted, "is one of those [languages] that, thanks to an inherent construction [*contexture*], cannot corrupt themselves." As we will see, the durable linguistic integrity Champollion identified in Coptic stemmed from its reliance on a discrete number of primitive phonetic elements that could not only be identified in written Coptic as it appeared in the oldest available extant texts but that were also distinct from intrusions of Arabic and Greek, the languages of Egypt's conquerors. "One might therefore argue," he wrote, "that despite the long period during which it was spoken," the Egyptian language "could not be altered in any way." Thanks to its continuation in Coptic, the ancient language "has reached us, so to speak, in its purity," a claim that he promised to develop in a separate work on the grammar of ancient Egyptian.[10] What is important to note here is Champollion's idea of *contexture*, which signaled his implicit belief that some languages were internally coherent and therefore enduring. Coptic, in particular, was distinctive for its resistance to processes of linguistic colonization.[11]

Although Champollion believed toponyms preserved residues of Egypt's long history, it did not follow that the links between them as recovered from extant sources were free of external, non-Egyptian influence. Indeed, because Egypt had been repeatedly conquered, mutations were to be expected. Champollion sought to identify encrustations of Greek, Latin, and Arabic within the toponyms that he studied and to separate those residues from Coptic influences, hoping to detect traces of ancient Egyptian in the latter. As persistent features of the landscape, toponyms were especially good for this purpose. "It was while studying and reading the written remains," he wrote, "that we happened on the idea of introducing Egypt to readers in Egyptian terms, and it is in the writings of the Copts that we find collected the names of the most ancient cities of this interesting country . . . the following reflections will prove the names appearing in the Coptic sources were the true Egyptian names."[12] Analyzing Egyptian place-names in a variety of documents in many languages and dating from many periods, Champollion sought toponymic opacities; like a person staring at a pebbly riverbed through a fast-moving stream, he looked for what visibly persisted beneath a surface roiled by alien influences.

In order to show how a language like contemporary Coptic might preserve remnants of antique speech, Champollion deployed a distinctive method, discernible from the overall plan of the 1814 *Pharaons*. Delving into his Coptic sources—fragmentary and scattered remnants of a relatively late period in Egyptian history—Champollion hoped to discover information about an even more distant past. His nearest predecessors, de Sacy and Åkerblad, had, we have seen, attempted to recover a handful of significant words, typically royal names, from the Rosetta Stone, a strategy Champollion faulted for being overly narrow.[13] He intended to base his 1814 *Pharaons* on the considerably different assumption, not shared by any of his contemporaries, that very early Egyptian words could be phonetically recovered, in whole or in part, through a careful study of Coptic words, particularly those referring to ordinary things. For Champollion, the best hope for recovering Egypt's lost language was to focus on the mundane rather than the unusual. Words with a double significance— words that were pointers to the simple objects of everyday life and expressive of cultural, often religious, perspectives—were of particular interest, as were the names of important places, which might themselves contain references to common, obvious features of the landscape. While toponyms might seem like proper names, akin to the names of pharaohs and gods, Champollion believed Egyptian place-names had developed from ordinary words describing natural and physical objects, often with a divine significance. Place-names would endure and therefore might evince vestiges of the ancient vernacular.

Champollion's extensive consideration of toponyms in the *Pharaons* accordingly depended on his presumption that they evolved out of linguistically enduring words for aspects of the landscape and nature. Extant Coptic words of that sort would necessarily be related to their remote ancestors given what he thought to be the enduring character of the language. His intense study of Coptic had prepared him for just such a search. For years, he had been reading, translating, annotating, organizing, and cross-referencing Coptic and related materials available in the libraries of Paris and Grenoble. Sifting these documents, he had systematically sought new words to fill out dictionaries for all three then-recognized Coptic dialects.[14] Entries in each dialect's dictionary were followed by lists of related words and their definitions, cross-referenced where possible to corresponding entries in the dictionaries of the other dialects. Significantly, the words Champollion selected for inclusion in his dictionaries were not invariably place-names or the names of kings, two categories that had particularly interested earlier scholars. Rather, Champollion paid special attention to the names of ordinary things, such as plants and animals, a focus reinforced by the first, highly anticipated volume of the *Description* (1809), which contained much fresh firsthand information about Egyptian flora and fauna and included detailed engravings.[15] As we will see, his results suggested a very different picture of Coptic than the one of a dead liturgical language that prevailed in his time. In his hands, the language seemed more alive than dead, more a vernacular than strictly a religious language; rife with double and triple entendres, it had a marked potential for polysemy that Champollion would eventually discover in Egyptian hieroglyphs as well. Above all, the dictionaries provide clues to Champollion's earliest efforts to link an Egyptian script— the Rosetta intermediary, or what Champollion would, following Herodotus, eventually term "Demotic"—to Coptic by means of the linguistic root. This conceptual technology permitted Champollion to move a step beyond Åkerblad's narrow focus on letter forms. Moving freely between orthography, phonetics, and semantics, Champollion hypothesized novel connections between Coptic and demotic script that extended beyond narrowly linguistic concerns to embrace elements of ancient Egyptian religion and culture.

Champollion's notebooks permit us to observe him at this work. In assembling his Coptic materials, Champollion paid special attention to words that seemed to share common roots; between 1811 and 1814, and sporadically thereafter, he developed notebooks devoted to their systematic presentation. He organized one such notebook thematically, sorting his entries by type (words for insects, plants, birds, and so on). Each entry dealt with a single root and explored the root together with its dialect variations and derivatives, which could be quite abstract, even poetic. For instance, an entry for the word "finger" contains three manuscript variants in Memphitic dialect (ⲧⲏⲃ ⲑⲉⲃ ⲑⲏⲃ) and one for the plural, "fingers" (ⲛⲉⲧⲏⲏⲃⲉ) in Theban. The ensuing list of Latin significations suggests the mundane activity of pointing and less literal variants related to putting ideas and things in order, as well as the power of doing so.[16] Uniting these elements, Champollion reached for a sense of what such a grouping might suggest, as if the Egyptian language was not only grounded in ordinary, concrete things (such as fingers) but was also pregnant with the potential for sophisticated abstraction (such as indication or ostensive signification). To count, to enumerate, to set in order—to, as it were, *point out*: what emerges is a picture of a way of thinking, a mode of thought derived from ordinary objects. Champollion proposed links between the Coptic word ⲧⲏⲃ ("finger") and a "racine primitive" (ⲧⲱⲃ), or primitive root, and several additional "racines derivés" or derivative roots. These roots were morphologically and phonologically similar to the original word for "finger" in the different dialects (fig. 10.1). Champollion, then, conceived that the same Coptic root could develop into a variety of words not only with different meanings but also with different degrees of abstraction. In his Coptic materials Champollion identified words, usually common nouns or verbs describing simple actions, in which he could discern a fundamental root. From these identifications he extrapolated a specific meaning based on the general significance of the root, taking into account the semantic and grammatical context in which the word

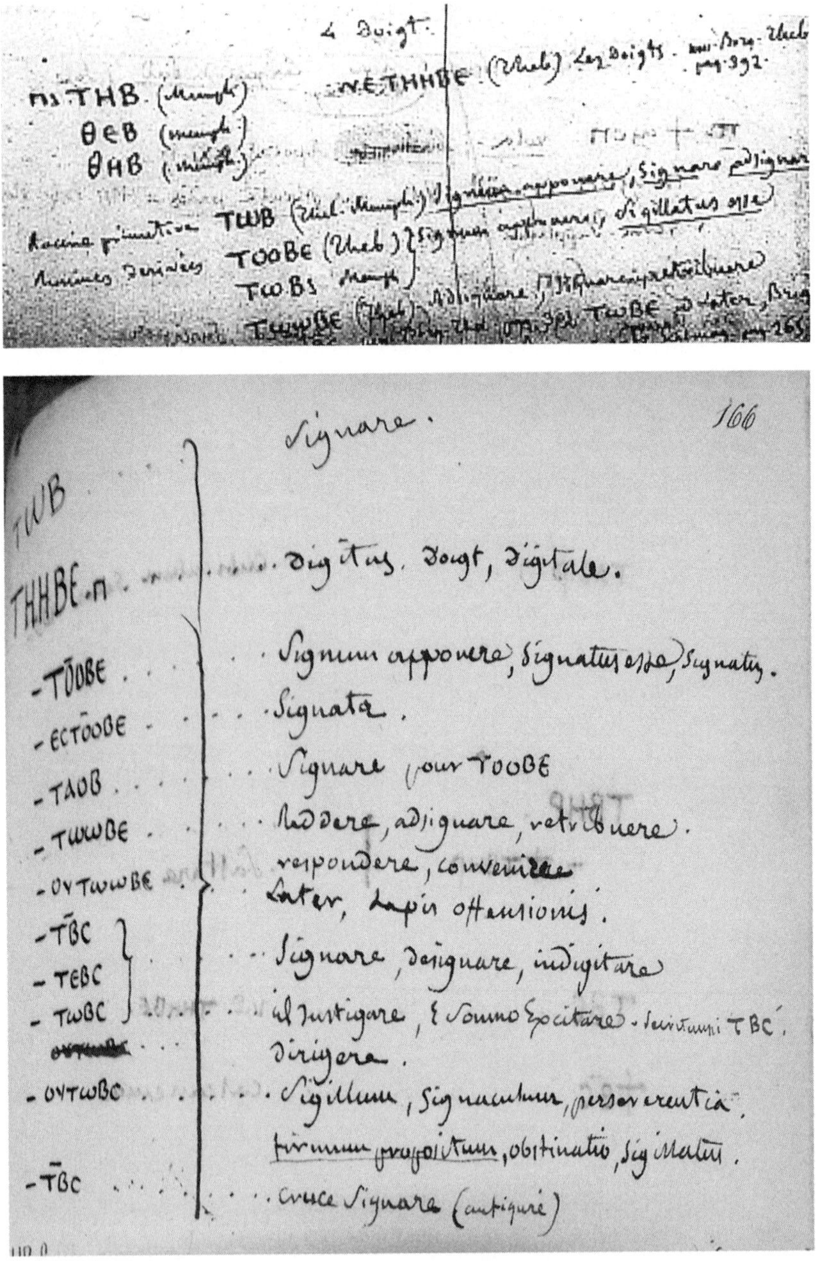

FIGURE 10.1. (*Top*) Entry for "Le Doigt" (the finger) in one of Champollion's Coptic dictionaries and (*bottom*) a related entry in a different Coptic dictionary. Bibliothèque nationale de France.

appeared. The same root could accommodate the expression of both the simple and the complex, the abstract and the concrete. Thanks to the root's ability to animate all of its derivative words, every word beyond the simplest and most literal could be built up from an original, simple meaning, which was still present within the morphology and phonology as a kind of sediment. Words were like toponyms in that sense; they contained their own histories.

The organization of these materials tells us something about Champollion's open-ended working method and, in particular, the relatively limited number and kinds of assumptions he brought to his task. Interestingly, he tended to work only on the recto side of the pages in his dictionaries of roots, leaving the verso blank. This procedure suggests work in progress, as he routinely allowed room on the left-hand pages for expansion of material entered first on the right-hand ones. In other words, he planned to be surprised. One entry, for the root ⲏⲉ, extended backward, onto the blank verso of the previous page, as Champollion discovered ever more information that could be inserted in association with it. His entries usually treated single roots (typically monosyllabic though multisyllabic roots also appear with some regularity). Now and then, however, he made more complicated entries. One of these exceptional entries exposes a particularly rich juncture in which it is possible to see Champollion going beyond Åkerblad in his search for links between words with no similar letters but related roots, exposing a rich vein of semantic possibilities.

This exception, which appeared in a different Coptic notebook, concerned the root ⲏⲉ (fig. 10.2).[17] Here, Champollion made two subentries. The first was ⲏⲉⲧ, which Champollion defined as "veritas, verité" and to it he associated six words having to do with truth, including adverbial and adjectival forms ("veridical," "truly") and an idiom ("in truth"). In the second subentry, which retains the root's orthography, Champollion defined the root ⲏⲉ quite differently, as "Amare, diligere, chérir," a matrix of words for love that, taken together, suggests that passionate affection might also include opposing elements, such as duty (having no choice) and preference (having a choice).[18] What did these two ideas, love and truth, have to do with each other?

The variants ⲏⲉⲧ and ⲏⲉ would seem unrelated despite similarities in their orthography and presumably also phonography. As Champollion developed this section, however, their meanings began to converge. The next entry, on the verso side of the previous page, associated the root ⲏⲉ to the word ⲧⲏⲉ. With the addition of the ⲧ, the result merged not only the orthographies of the root's first and second variants but also their meanings. In doing so, the word captured an image from Egyptian nature lore. It was, Champollion said, a bird name (*nomen avis*), denoting "a type of bird, the bird *of Justice*," likely a

FIGURE 10.2. From Champollion's Coptic notebooks. Bibliothèque nationale de France.

reference to the ostrich, whose feather appeared in representations of Maat, the goddess associated with justice in ancient Egypt.

In a final subentry, Champollion explicitly linked the root ⲎⲈ to a character on the Rosetta Stone. (It is one of two such entries in his dictionary of Coptic roots. The other appeared in the entry devoted to the root for "human" or "man.") He gave three examples of this character taken from the Rosetta intermediate script along with suggested Coptic and Latin renditions of them: "Who love their fathers," "Who loves his brother," and "Beloved of Ptha." Here it is possible to see the whole arc of Champollion's progress as he reasoned from roots to the Rosetta intermediate. Joining each piece to the others was an idea, derived from the root, of love-as-respect, that is to say, of piety, an esteem that flowed through hierarchically organized channels from parents to children and from deities to humans.

Although Champollion never fully explained his notion of an etymological root, he did make notes about it during this period. On a series of scraps, he essayed several definitions, as though he were attempting to clarify the idea for himself. On his first try, he offered an expansive definition, but his emendations suggest a struggle to arrive at precisely what he wanted to say: "What we call 'roots' are nothing other than words invented ~~arbitrarily and accoerding to~~ [illegible] in order to ~~express~~ represent our ideas and our thoughts."[19] In a second note, Champollion capitalized the word "root," as if to suggest it occupied a category of its own (like a proper name), and clarified the nature of its relationship to its referent: "We call [a] 'root' that which represents an idea or abstraction that makes a connection to people or things."[20] In a final set of notes, he hit on yet another way of understanding the notion by connecting it in appropriate circumstances to the infinitive in Latin: "This exposition gives the root the value of the Latin infinitive."[21] Interestingly, Champollion presented the root as functioning rather like what Åkerblad termed a "fulcrum," with an important difference. Åkerblad used the fulcrum metaphor to indicate places where a series of letters might surround an indeterminate vowel, with the missing element being deterrmined by the context of the passage. In Åkerblad's conception different words that employed the "fulcrum" had no semantic relations with one another. Champollion altered the notion fundamentally by taking it over from orthography and assigning it to the linguistic "root." In his conception, words with different *but related* meanings all balanced on one and the same "root." It was precisely here that Champollion merged two strands of contemporary linguistic thought—the idea of a root as a point of morphological continuity in time as well as semantic contiguity, that is, as an intersection of multiple words, all derived from the same root, with similar or related meanings.

Champollion used the linguistic root to probe the ways in which meanings were produced in Coptic on the implicit presumption that the language was built up from principally fixed, essentially meaningful monosyllables that could be concatenated to now form wholes whose meanings were related to those of the component parts. This assumption had a clear operational consequence: it permitted Champollion to organize his Coptic research according to these same small and simple units. It also allowed him to suspend, for the moment, questions of grammatical structure in favor of investigation of syllables.[22] Indeed, the syllable was such an important unit of analysis that Champollion organized his dictionary of Egyptian roots entirely in accordance with this priority. "Of all the methods of classifying in one and the same series the simple and compound words of the Egyptian language," he opined, "alphabetical order is the least suitable . . . classification with respect to the radical and primitive word in each, according to a certain order itself governed by their composition, is the only one appropriate to the genius of the language."[23] Since the roots presumptively originated as references to ordinary things—a river, a person, a finger, and so on—a dictionary arranged by these semantically significant structures could provide hints about ordinary life in ancient Egypt, as well as a sense of how those ordinary words gave rise to more abstract and complex expressions.

If, as de Sacy, Åkerblad, and others were certain, the Rosetta intermediate were itself alphabetic, then it might be possible to reconstitute the primitive, monosyllabic roots out of whose meanings the language had been constructed through Coptic, its final descendant. By 1808, Champollion was certain that he could accomplish that goal by divining the letters that comprised the intermediate script. At this juncture, his methods, and the assumptions informing them, lifted him from the safe harbor of relatively uncontroversial ideas (in France at least) into more novel, hypothetical, and dangerous waters. Could

Champollion reuse the method he applied to Coptic—identifying its simplest elements and building up from them—by applying it to ancient Egyptian writing, and if so, could he leverage his knowledge of Coptic roots to assist with this task? For this to be possible, it certainly had to be the case—in itself not so controversial—that the alphabetic structure of late Coptic writing had close analogs in the Rosetta intermediate, which at this point Champollion began to call "Demotic." While he would change his mind about this possibility more than once in subsequent years, on August 30, 1808, he felt confident enough to write as follows from Paris to his brother, referring to a manuscript copied by Denon in Egypt:

> On first view of this manuscript, I thought it impossible ever to decipher because I took for so many letters the frequent groups that one notices. I soon reflected that the Egyptians had doubtless followed the same path in that as other oriental [writers]; so, setting aside [the idea of] groups, I set about collecting the simplest elements. And I soon noticed, not without much pleasure, that all my groups were nothing but the reunion of 1, 2, 3 or 4 of these same simple elements. Some have so far resisted the tenacity of my ocular analyzer and scrutinizer, but I'll prevail.[24]

Youthful confidence notwithstanding, it turned out to be quite difficult to reduce Egyptian writing to its constituent parts. The boundaries between parts proved hard to identify, for one thing. Especially in running scripts like demotic, there were no obvious boundaries around an expression like the one signaled, in hieroglyphs, by the occasional cartouche. When even letters could not be distinguished from one another, it was even more difficult to know what counted as a syllable or a word. In a follow-up note to his brother, written between August 30 and October 10, Champollion reported a dismaying lack of progress on other (unspecified) papyri as he tried different methods for subdividing texts. He felt stymied by sign groupings:

> The papyri. Here is another embarrassing and obscure issue. I read a line and a half, I made an alphabet based on a well-known monument; I found a clear meaning consistent with the situation and written in a suitable style. And even so I am none the wiser. I can't go further; *the groups stop me* [emphasis added]; I have studied them, meditated whole days, and have understood nothing.

Not all was lost. "It's true that, without flattering myself, I have still gone further than all previous antiquarians in this matter, since I am able to prove that all the papyri (in cursive writing) are all in the same very similar alphabet," he wrote consolingly, "which no one had suspected until now." From one point of view, this was something of an exaggeration. Åkerblad, who had gone the furthest in this direction, had not actually advanced very far. He never devoted much attention to papyri that bore evidence of a script similar to the one on the Rosetta Stone. If he had, it's not clear that he would not have perceived a similarity. On the other hand, even Åkerblad was not prepared to light out for the territory Champollion now had dimly in his sights. If even papyri devoted to native messages could be read in the same way as documents intended to represent the edicts or statements of Greco-Roman conquerors, then considerable elements of the demotic script would likely have been developed long before the Alexandrian invasion. As this was an astounding implication, perhaps it was just as well that Champollion did not yet command the full attention of the Parisian establishment. Even at this early point, Champollion was aware of the hostility such wild ideas were likely to evoke. "Besides, I know our age and the scholars nowadays!" he exclaimed to his brother. "If I'd had the misfortune to push my discovery further, I would have had every present and future scholar all over me," he continued, condemning him to "criticisms, censures, and no more rest."[25]

CHAPTER 11

HIER POUR DEMAIN

Champollion had good reasons to feel harried. Working in the fulminant matrix of what would, within a generation, become modern philology, he was surrounded by many serious and enthusiastic competitors. Although the hub of this activity was Paris, which abounded with manuscripts written in the languages of the Near East as well as specialists who could read them, the hub had several spokes. Particularly in the German-speaking states, the philological transformation of antiquarianism formed one branch of this flourishing speciality.[1] The shift greatly interested Champollion-Figeac, whose body of scholarly work participated in the transformation, and left a substantial mark on Champollion's own thinking as well.

Historians have long agreed that the history of modern philology began across the Rhine and spread to Paris from there. Despite Napoleon's incursions, intellectual commerce between Paris and centers of learning in the German states remained lively through the early years of the nineteenth century. Indeed, when Champollion-Figeac was trying to find a suitable educational context for his brother after the Grenoble lycée, he asked his friend, the Parisian naturalist and antiquary Aubin-Louis Millin, to advise him on a course of study. His precocious younger brother had already exhausted the not inconsiderable resources of Grenoble. If not Paris, the dutiful elder brother wondered, then where? Strasbourg, perhaps? In reply, Millin starkly laid out the contours of this scholarly world: either send the young man away to Göttingen, where he could study with the renowned classicist Christian Gottlob Heyne or bring him to Paris to study with de Sacy.[2]

Millin's recommendation reflected Göttingen's status as the center of a new approach to textual and material vestiges of the ancient world. The undisputed leader of this transformation was Heyne, who trained a generation of scholars in his method by means of his philological seminar, which, by the early years of the nineteenth century, was already the stuff of legend. Heyne promoted the historicizing of antiquity, particularly Greek antiquity, as a coherent whole. That is, to understand ancient Greece fully, it was necessary to locate and analyze all of Greek antiquity's scattered remnants, whether textual, sculptural, or architectural, and present them as expressive of a consistent and coherent whole of which the present fragment was merely an extant evidentiary suggestion. After Heyne, the goal was no longer simply to edit, translate, and comment upon the texts of classical antiquity, nor was it to emulate those texts, as Alexander Pope had, by creating translations intended to be valued for their aesthetic qualities rather than their fealty to the original text, so far as one could be identified. Of the many effects Heyne's philological seminar had on classical learning, perhaps the most far-reaching was his division of ancient sources into two categories: poetic or, more broadly, literary work, that is, products of the imagination, and sources that were only meaningful in reference to objects and to processes in reality, such as bodies of law, folk practice, and ritual. By his example and his teaching, Heyne urged others to privilege the latter type of source material. Because the new philology was understood to be powerful enough to make sense of sources in both categories, it was raised above this divide, considered the premier tool for analyzing textual remnants of all sorts, literary and otherwise. Meanwhile, literary sources were valuable not for moral uplift or aesthetic appreciation but for what they said about the historical context of their production. Toward the end of the eighteenth century, Friedrich August Wolf, Heyne's student at Göttingen and later professor at Halle, had, as we have already seen, outlined a recognizably philological approach to antique texts in his *Prolegomena ad Homerum*, which analyzed the fragmented, incomplete, and relatively late body of Homeric texts to show how they had been cleaned up and packaged as coherent works of epic poetry by a single ancient author named Homer.

The contributions of Wolf and Heyne found a ready embrace among the scholars who clustered around Fourier, Millin, and Champollion-Figeac. Preparing a course on Greek to be offered at the university at Grenoble in 1810, Champollion-Figeac praised Heyne and Wolf for their contributions to classical philology and attributed the poor state of Greek studies in France to the absence of comparably sophisticated work. Following Heyne and Wolf, Champollion-Figeac carefully separated philological analysis from literary appreciation of ancient texts, and he heaped particular scorn on insufficiently rigorous literary translations of ancient works—Pope's floridly inaccurate but popular editions of Homer were, perhaps, on his mind—because all that such works could give were "skeleton" outlines of the original.[3] Despite the genuflection toward literature in its title, his course emphasized elements drawn directly from Wolf's treatment of Homer: analysis of "the use of writing" among the Greeks, specifically including its historical development and the alphabet; attentiveness to orthography, inscriptions, writing implements, and material remnants of Greek antiquity that pointed not to realms of myth and literature but to the world of everyday practice; and analysis of the history of writing or protowriting implements that might have been used in ancient Greece, a topic Wolf also took up.

The new philology was not transmitted solely by means of publications. Human mediators were also crucial. Of these, perhaps the most familiar was Germaine de Staël, the Geneva-born writer whose *De L'Allemagne*—completed in 1810 but not published, due to political concerns, until 1813—traced the contours of the new approach to art, literature, history, and science flooding into France from across the Rhine. Another early mediator, more oriented toward scholarship and the role of the university than de Staël, was Wilhelm von Humboldt, eventual founder of the University of Berlin, who was a constant presence among Parisian linguists around the turn of the century. During this time, he lived in central Paris and took an apartment in the busy winter of 1800 on the rue Saint-Honoré not far from Millin's offices and Champollion's rooms.[4] While there, he contributed an essay to a late volume of Adelung's *Mithridates* and promoted his good friend Wolf's philological ideas, even going so far as to articulate a number of the premises and commitments implicit in Wolf's neologism *Altertumswissenschaft*, the scientific study of antiquity—an effort that, ironically, meant Humboldt would be remembered for defining the term more than Wolf was for coining it.[5]

Marquee figures provided interest and gravitas, but the spadework of founding a new way of doing history fell to others, and Millin in particular did a great deal of it. Since 1796, the indefatigable Millin had run the *Magasin Encyclopédique* out of offices in the rue Saint-Honoré, steps from the Church of Saint-Roch and Champollion's rooms in the first arrondissement.[6] Among the many magazines devoted to literature, the arts, belles lettres, and the latest scientific findings published during this period, the *Magasin* was unique in the sheer breadth of topics covered, as befitted a publication aiming to keep readers up-to-date in a variety of scholarly and scientific fields. Consistent with his attentive scrutiny of the new philological methods pioneered by Heyne at Göttingen, Millin was particularly alert to developments in the German-speaking states, an interest that profoundly shaped the *Magasin*'s editorial priorities.[7] Thanks to Millin's efforts, the *Magasin* became the premier channel for the transmission of the new philology from the German states to France. In the pages of Millin's magazine, Heyne was a frequent touchstone, his name preceded by laudatory epithets: the illustrious Heyne, the renowned Heyne. His work, and that of his students, was closely followed.

Millin had been corresponding with Champollion-Figeac since at least Year 10 (1801–2). Their friendship developed while Millin was reorganizing the Cabinet des médailles at the Louvre according to the scheme laid out in his *tableau analytique*. As we have seen, it was none other than Millin who had urged Champollion-Figeac to send his younger brother to Paris, where Millin befriended him too.[8] Stendhal famously found Millin dull, but the German dramatist August von Kotzebue, visiting Paris in 1804, was more favorably impressed, describing Millin as "a lively, spirited person, with flashing eyes."[9] Visiting Paris in 1804, Kotzebue reported on Millin's Wednesday *thé littéraire*, held in Millin's offices in the rue de Richelieu, where guests milled about, conversing or perusing the latest publications from around Europe, piled high on a table.[10] A regular attendee, Champollion shared an account of these Wednesday salons with his brother. Proceedings began at eight, followed by tea and punch served a half hour later. A handful of bored notables, "princes, ambassadors, dukes, etc.," clustered in a corner while savants talked shop and enjoyed the refreshments, showing that their "throats work[ed] as well as their tongues."[11]

In these years, antiquarianism was under pressure to modernize, to become more systematic, and, in doing so, to cast local histories in national terms. This shift marked the beginning of the end of the *rêve particulariste*, in Odile Parsis-Barubé's evocative phrase—that is, the end of local antiquarian history as a pursuit that could be safely kept separate from a national narrative, the story of France.[12] During this shift, which was slow, uneven, and subject to strain, antiquarianism fragmented politically, methodologically, and in terms of topics taken up and perspectives developed. The struggle wasn't only French. In the German states, the jurist Friedrich Karl von Savigny's *Treatise on Possession* (*Das Recht des Besitzes*; 1803) critically reimagined ancient Roman law using the tools of philology and historicism before turning to its medieval reception, rejecting the creation of new civil codes in the Napoleonic fashion in favor of gradualism and sentimental ideas of Volk. For Savigny both law and language reflected a "common consciousness," that is, a national one.[13] His views had parallels in Napoleonic France as well, where Chateaubriand famously evoked the medieval past to similar ends. There, too, antiquarian-minded conservatives resisted the progressivism of colleagues, such as Champollion-Figeac and Millin, who understood the past as having value primarily as a wellspring of the future: *hier pour demain*.[14]

That view of the past developed within an institutional context recently invigorated by administrative authorities prepared to exercise their powers in the name of values dear to Napoleon and his savants. Chief among them was Fourier, who shared with Millin a keen interest in the interpretation of the ancient past, especially ancient Egypt. Installed as prefect of Isère in 1802, Fourier facilitated Champollion-Figeac's election to the Société des sciences et arts (formerly the Académie delphinale) at Grenoble in 1803 and arranged his post as secretary. Champollion-Figeac was meanwhile rising through the provincial ranks of the world of letters, starting in 1808 as an adjunct librarian;

by 1812, he had been promoted to conservator in chief of the Grenoble library. After Fourier advanced Champollion-Figeac for a faculty position, he was named professor of Greek in 1809. By 1813, he had become the director of the faculty. Altogether, this group—anchored by Millin in Paris but extending to Fourier and Champollion-Figeac in Grenoble—developed a view of the past requiring new methods of apprehending it, important elements of which appeared in Champollion's Egyptological work, as we shall see.

Although historians have traditionally situated "the discovery of the provinces" in France in the middle of the eighteenth century, the postrevolutionary liberation of royal, ecclesiastical, and noble properties substantially fueled the need to document and preserve these possessions, now conceived as property of a culture.[15] Antiquarians continued to value the local and the particular by attending to insular traditions, partly in order to extract maximum public utility from such knowledge and partly in the service of a search for national origins, which of course might themselves be turned to some practical purpose such as boosting a provincial economy through the encouragement of tourism.[16] To be sure, some antiquarians did seek ways to link local stories to a national narrative, occasionally going so far as to deploy modern but impersonal social scientific methods such as statistics. Others asserted an opposing pressure, seeking to rescue local history from incorporation into larger collectivities, such as the state. Such incorporations threatened to diminish their authority over the local past, to which they often had deep loyalties through family connections.[17]

In the first part of this book, we traced a cognate impulse—to discover and promote stories from the periphery rather than the center of social and political power—in Britain, where it was expressed by enthusiasts for the tales of Ossian while also being pilloried by satirists. In France, the impulse began with attention to local history but evolved into a somewhat different form, especially outside the capital, in places with strong local traditions and an independent regional identity. For Champollion-Figeac and others, the past was a resource for showcasing new methods—the tools of the new philology, the comparative method in particular, as well as archaeology.[18] Discourse about ancient history focused on how archaeological finds intersected with the latest philological methods, which specialists then applied to local languages, the various dialects of provincial France. Insights gleaned from these investigations propagated into public life through the organization of archival collections and the presentation of objects in museums.[19] The new methodologies also exerted an influence on Champollion who, thanks to his close ties to his brother and their friends in Grenoble and, during the years of exile, in Figeac, adapted elements of this approach to French ancient history to his Egyptian studies.

For a time, one focal point of this activity was the Académie celtique, a short-lived but influential antiquarian society that counted Millin, Fourier, and Champollion-Figeac among its earliest members. Founded in 1804 and, after a period of uncertainty, headed by the controversial Alexandre Lenoir—Freemason, apologist for the fanciful view of the Egyptian origin of all civilization, disciple of Charles-François Dupuis, and founder of the Musée des Monuments Français—this organization set out to collect, publicize, and organize evidence of an independent and relatively remote French history centered on Celtic progenitors rather than, and predating, ancient Greeks and Romans in Gaul.[20] Facts and analysis, however, were to provide the bases for conclusions, as the British example, which posited Ossian as a homegrown Celtic Virgil, proved a poor foundation for claims to a nonclassical European past. These claims, which stemmed from revolutionary contempt for the classicism of the ancien régime, were reinvigorated by Napoleon, whose admiration for Ossian was matched only by his enthusiasm for science; Napoleon even allowed himself to be styled "Emperor of the Celts" by the Italian poet Melchiorre Cesarotti. Napoleon's campaign in Egypt, promoted as the site of another golden civilization anterior to the classical age, further reinforced French Ossianism.[21] Of course, Napoleon was no Thomas Campbell, hoping to recover a lost Celtic golden age in France. But Ossian was a popular success, and Napoleon's campaign in Egypt also proved evocative. Some members of the Académie celtique even hoped to establish significant links between the ancient cult of Isis in Egypt, a similar cult found among the Celts in pre-Roman France, and perhaps early Christianity, though this last connection withered in the secular atmosphere of Napoleonic France.[22]

The idea that all religious belief in non-Greek and non-Roman ancient civilizations was nature-oriented, manifesting a direct link between man and the cosmos, had a long history. But the linkage of this idea with a French provincial tradition was new. More broadly, the Académie celtique recontextualized French provincial traditions and folklore, staking out fresh territories of meaning and recovering the givens of ordinary life, distinct from the prevailing view of theologians who viewed such folkways as mere superstition and humanists who saw them as bizarre curiosities.[23] In both goals and methods, the new antiquarianism owed much to Heyne's holism, if not to his philology. The strange, the grotesque, and the bizarre were no longer to be judged as expressions of a primitive irrationality that persisted, unchanged, through time; rather, these elements were to be viewed as fragmentary vestiges of a broader, organized social milieu. No matter how bizarre it seemed to modern eyes, every fragment of folklore or tradition was to be considered part of an ordered whole, the coherence of which could be reconstructed through historical analysis.[24]

Heyne's influence is evident here, in the assumption that historical realities were sufficiently coherent that larger social and cultural wholes could be inferred from a sufficiently comprehensive collection of evidentiary fragments.

The first volume of the Académie celtique's *Mémoires*, tucked within an extensively footnoted discussion of Ossian and MacPherson, insisted on the value of these folk documents: "What we despise today as mere popular tales, as coarse monuments, are precious vestiges of the wisdom of their ancient legislators." It was the task of the Académie celtique to "avenge" these authorities who, after all, were only the first editions of the present inhabitants of France, and who, through a process of cultural reproduction whose details it was the task of the society to explore, represented "the history of the people we are."[25] Readers were exhorted to keep open minds as they sifted and reflected on folktales and local religious beliefs, lest they trivialize the sources of their own histories:

> How many practices have been abandoned, just because they have seemed strange; and have only seemed strange because their sources and objects are unknown! How many monuments have been relegated to the world of fairy tales, attributed to the devil or to giants, rather than being seen as vestiges of an extensive, civilized people whose whole codes inhered in their customs, just as their religion and history inhered in their monuments and traditions.[26]

The Académie celtique's membership closely mapped to the Champollions' network of friends, particularly those with ties to Grenoble. In addition to Millin, Champollion-Figeac, and Fourier, the Académie celtique's first members included Champollion family friend and fellow Dauphinois, the professor of mineralogy Faujas de Saint-Fond; Valperga di Caluso of Turin, whose Coptic studies were of great importance to Champollion; and Jacques Cambry, who served as the Académie celtique's president after Lenoir, who was an ardent, if uncritical, Egyptophile.[27]

Like Millin, Cambry sought out Champollion in Paris. But when Cambry invited him to join the Académie celtique, Champollion responded with trepidation. After mistaking Champollion's powerful linguistic abilities as evidence of an equally powerful interest in the local history in which he, Cambry, was himself engaged, Cambry attempted to plunge Champollion into his own work on ancient agriculture and natural religion in the provinces.[28] Champollion-Figeac advised his younger brother to be patient with Cambry while carefully contriving not to offer him anything for publication from the forthcoming work on Egypt.[29] Given his own extensive activity with the organization,

Champollion-Figeac was eager to avoid the appearance of impropriety: "It is necessary that your *Mémoire* [on Egypt] fly with its own wings."[30] The conflict came to an unexpected and abrupt resolution when Cambry became ill over dinner with Champollion and died later the same evening, sending Champollion into a spiral of depressive guilt: "This unfortunate event has affected me very deeply and cast a long shadow over my horizon: nothing gives satisfaction anymore, the loss has poisoned all my pleasures, my black thoughts are back with a vengeance. Perhaps it is weakness. The idea of death pursues me." The ever-phlegmatic Champollion-Figeac prescribed the distraction of work: "Your work provides the means [to feel better;] use it and calm yourself."[31]

Although he did not share Cambry's enthusiasm for the agricultural version of the *rêve particulariste*, the local histories of the farms and orchards of rural France, Champollion-Figeac nevertheless retained a number of commitments that were held widely among the members of the Académie celtique regardless of specialization. The approach downplayed or avoided analyses of ancient texts and artifacts in terms of abstract and universalizing ideas of taste, favoring an approach that could reveal what each source could say about the context of its production. Scholars mined forgotten manuscripts cached in monasteries and local libraries, and sifted debris at archaeological sites, on the lookout for written materials, like inscriptions, that pointed toward reality rather than being wholly products of the imagination. The earlier the text, the more primitive and perfunctory the literary expression. Wolf was the first to call attention to the existence of such primitive forms of reference in the Homeric corpus. He noted that the verb δειξαι (*deixai*), "to show," was consistently used at moments when "to read" would have been expected. This pattern begged the question of whether written texts existed at all at the time that the poem was composed. Perhaps there were only prototexts, "some sort of marks or symbols, not true written characters." Wolf continued: "We do not know what the ancient grammarians conjectured, except that a few of them, as more have done in modern times, seem to have reckoned [these primitive signs] among the hieroglyphs"—a striking Egyptian reference to which we will soon return.[32] For now, it's enough to note that these mysterious signs *meant* almost nothing; they had virtually no semantic content—except that they *indicated*, they *pointed to* the real world beyond the text.

In the process, these signs deepened the distinction between reality and imagination that Heyne, Wolf, Humboldt, and others had drawn so sharply, separating aesthetic experience and artistic production from other spheres of human activity, including philology, history, archaeology, and linguistics.[33] This shift necessitated different ways of parsing ancient sources, many of which were considered works of art rather than, or in addition to, being pointers

to how people actually lived. Rather than focus on the best, most aesthetically pleasing way to render a translation, the philologically oriented scholar
should focus instead on what the text suggested about the world in which it
appeared and the reality toward which it gestured, even if the gesture was
oblique. Perhaps nowhere was this evidentiary preference more apparent than
in the earliest history of the École des chartes, where Champollion-Figeac
would eventually accept a prominent position. Among those who spearheaded
the founding of this grand école was Joseph Marie Degérando, whose vision
for the institution was inspired by the question of whether ancient textual
sources might be as revealing as the languages of "savage peoples" and the
deaf, two groups whose (apparently) rudimentary linguistic productions he
had also studied in detail. The "chartes" that were the ostensive objects of study
were not literary productions in the imaginative sense, not flights of imaginative fancy, but records of the establishment of relationships between people
wishing to codify rights and obligations in order to advance real-world aims.[34]

As a result of this shift in emphasis, the historical timeline opened up. Late
classical and medieval objects and texts came under renewed scrutiny as part
of a general vogue for those eras. Inscriptions were especially important when
they were explicitly concerned with the details of public life, for example, contractual matters suggestive of a body of legal practice. In 1807, Champollion-
Figeac published *Antiquités de Grenoble*, in which he inventoried the extant
vestiges of pre-Roman and Roman settlement in the city. In this inventory, he
cataloged virtually every ancient textual remnant in Grenoble—inscriptions
found on tombs, temple vestiges, and so on. For the most part, he described
these monuments in paleographic terms, noting the heights of letters and the
presence of unusual characters indicating different historical periods. But this
work went an important step beyond merely cataloging and describing inscriptions. Rather, he intended to use this evidence to document specific, local, and
ordinary historical realities.

Where possible, the study emphasized the sacerdotal significance of the
cataloged inscriptions, making special note of Egyptian vestiges—for example,
worship of Isis, and of the equivalents of Thoth (as Mercury) and Ptah (as
Vulcan or Hephaestos)—that had traveled to the city by way of its Roman conquerors. Not surprisingly, such inscriptions also often included references to
facets of local life. In this way, generic Roman deities took on local significance,
so that, for instance, Mercury, the god of writing, was especially revered in the
province of Viennes due to relatively intense interest there in belles lettres.[35]
The study also made extensive use of toponyms and other proper names.[36]
According to Champollion-Figeac, the settlement at Grenoble originally belonged to the Allobroges, a Celtic tribe whose name meant "foreigner." Like

so many of the apparently proper names in Champollion's geography of Egypt, it originated as a common noun, a simple indication of the name's status in context.

Champollion-Figeac's inventory included several entries dealing with funerary monuments. These he also treated somewhat differently relative to the main line of antiquarian study at the time. The difference reflected the larger shift in antiquarian methods and priorities we have just described. For instance, Champollion noted the presence of an ascia, a sign shaped like an ax, on a number of Roman tombs, and suggested that it was there to indicate consecrated ground; the sign was significant because it pointed to a social, perhaps even contractual, reality.[37] Champollion-Figeac similarly described an unusual find, a sarcophagus, which provided direct and unambiguous evidence of inhumation—a rare practice among Romans, who preferred cremation, but common among the Greeks. Upon this evidence he built a case for the existence of a Greco-Roman colony in Grenoble.[38] Although the work's title made explicit reference to the increasingly outdated discourse of antiquarianism, Champollion-Figeac dedicated *Antiquités de Grenoble* to Fourier, a move that positioned the work within the context of a modern approach to the past. His explorations of the social and religious practices suggested by the evidence further marked his departure from established methods. In a letter to him from the same period, Champollion-Figeac admitted that despite its clear utility, even venerable paleography had limitations: "Of all the subfields of archaeology, paleography is surely the most conjectural."[39]

Elements of comparative philology also found their way to Grenoble. What Adelung had begun with *Mithridates* was extended by investigations of the Académie celtique into the structural relationships that could be discerned between modern languages and lost or ancient ones such as Bas-Breton, Basque, and Celtic. That the comparative impulse was generally strong in Paris is clear in letters from Champollion to his brother, who observed the vogue and dabbled in it. On November 16, 1807, Champollion asked his brother to look up the word "Aour," cited in the Latin Bible of Robert Estienne, in order to authenticate a suspicion that the "city which Aboulphedah calls Haur," corresponded to "*Hour* in Coptic" and the Egyptian god Horus.[40] Champollion-Figeac responded with disappointing news. He couldn't find any place named Aour but he did find a place called Auran, which, if it existed at all, was to the north and east of Egypt: "I found in Robert Estienne *Auran*, near Damascus, [in] Hebrew *hauram*. I believe this is the same city for which you asked me for a note."[41]

Champollion, undaunted, reported to his brother a few weeks later on his submission of a draft of this work to Constantin François de Chassebœuf,

Comte de Volney, "who will weigh in on the grammatical relationships between Arabic and Egyptian," that is, Coptic. He intended to display these relationships in a table, with which he hoped to facilitate comparison of the phonetic and orthographic expressions of different words in Arabic and Coptic. Just such a comparative table did in fact appear in his 1811 introduction to the *Pharaons* and was reprinted in the full, two-volume publication that followed in 1814, along with a note acknowledging Volney's comparative textbook for the teaching of Arabic, Persian, and Turkish, as well as William Jones's work on Sanskrit and Langlès's help in uncovering these texts. Adelung's influence did not go unfelt either. Intriguingly, when Fourier proposed a comparative project to Champollion-Figeac in Grenoble, to complete a study of the dialects of the Dauphiné, Champollion-Figeac worked through 1808 on the project, and in November 1809 he published *New Researches on the Patois or Common Idioms of France, and in particular on those of the Department of Isère*, to which he appended, in an echo of Adelung, a selection of samples of "littérature dauphinoise"—proverbs, poetry, a sermon—and capped the collection with a glossary of dialect words. *Antiquités de Grenoble* also included a comparative element. Just as remnants of ancient Egyptian were presumed to persist in modern Coptic, Celtic vestiges were believed to persist in Bas-Breton. Champollion-Figeac believed that "Cularo," the ancient name for Grenoble, was undoubtedly of Celtic origin.[42]

In 1803, Fourier wrote to Champollion-Figeac regarding the demolition of the tower and walls of the old bishopric in Grenoble. The bishopric, which had been built over ancient Roman walls, was of special concern. Fourier asked Champollion-Figeac to lend a hand with conservation, specifically "to see to the preservation of ancient inscriptions that may be found on the stones from the walls of the bishopric currently being demolished."[43] Champollion-Figeac responded with a brief catalog of inscriptions in the area, including items marked "Cularo," suggestive of their relatively great antiquity, and the following year, in 1804, he identified and described two important third-century inscriptions found on the ancient city's gates, showing how the inscriptions provided a window into the local administration of the Roman colony.[44] In 1814, Champollion-Figeac again returned to the subject of Grenoble's identity in antiquity. With the publication of *New Clarifications Regarding the Town of Cularo, Today's Grenoble* (*Nouveaux éclaircissements sur la ville de Cularo, aujourd'hui Grenoble*), he intended to settle a handful of questions that had long befuddled students of Grenoble's history, leaving them to flounder in a "shoreless sea."[45] These questions involved the location of the city, which in turn had implications as to which Celtic tribe it had originally belonged. If it was located on the left side of the Isère, it belonged to the ancient tribe of the

Voconces, but if it was located on the right bank it was, therefore, a property of the Allobroges. (That the Isère marked the boundary was not in dispute.) Whatever was correct, he said, tradition would not provide the sole support for his claim; rather, he would use inscriptions and other artifacts for evidence. With this essay, Champollion-Figeac closed the circuit he and Fourier had opened years before.

Like any question of influence, it is, of course, difficult to gauge the extent to which Champollion-Figeac's investigations of the provincial archaeology of Grenoble affected Champollion's developing views of ancient Egypt. Nevertheless, we have one bit of suggestive evidence, and it involves the church of Saint-Laurent, located across the Isère, just beyond the ancient Roman city walls of Grenoble. The church, a mostly Romanesque mélange, was erected on the site of an ancient burial mound that became the Merovingian crypt of Saint-Ovand, discovered by Champollion-Figeac and described in his *Dissertation on an Underground Monument at Grenoble* (*Dissertation sur un monument souterrain existant à Grenoble*; 1804) in an edition of twenty copies.[46] This work aimed to demonstrate that the crypt was an originally Christian rather than pagan structure.

Champollion-Figeac first noted architectural details, such as the placement of columns and the cross-shaped footprint of the crypt, which were consistent with other examples of Christian rather than pagan sacred sites during the early centuries of Christianity in Europe.[47] He observed that areas devoted to prayer were oriented toward the East, again consistent with the early church, as opposed to pagan prayer conventions, which oriented prayer toward the West.[48] Nor was philology neglected. He analyzed a set of stairs found at the site in light of an older argument, set forth by Jules César Boulenger, that the Greek word BHMA was derived from a more ancient phrase meaning "to ascend stairs," and from this Champollion-Figeac inferred evidence of a sacristy where liturgical objects, such as chalices, might be stored.[49] The implicit presence of these latter items led him to confidently propose a ninth-century date of construction since Nicophorus of Constantinople, "who flourished during the ninth century," had decreed that sacristies be used to safeguard such items from abuses to which they were otherwise not uncommonly subjected.[50]

Champollion-Figeac noted the presence of details suggestive of a pagan religious structure on the site, but he posited that the mere presence of such a pre-Roman vestige did not mean the sculpted cornices and other details found in the crypt also dated from the same period. In fact, these elements seemed to refer to elements of nature that were unusual if not wholly absent from early church dogma. Those details therefore conveyed a different meaning, one that had important resonances with the argument Champollion would soon make in the *Pharaons* work. A group of "sculpted heads of the most bizarre design" included one that was horribly disfigured but unambiguously human. A handful of objects were incised with mysterious and perhaps mythologically freighted symbols—a "right hand, a kneeling figure, and a snail shell"— as well as a remarkable relief sculpture of a snake and a dragon, "covered in scales, with wings and claws of a griffon," with ambiguous outcroppings that might be pointed ears or perhaps horns. Champollion-Figeac observed that one of the heads was turned to the East, in the fashion of the period of the Crusades, toward Jerusalem, and was therefore suggestive of early Christian rather than pagan symbolism. "These *ornamens*," he wrote, referring to the ambiguous little images, of the shell and so on, "may not be very useful for identifying a pagan temple, but the line of heads, which has no hint of such ancient style, seems to be very modern in relation to them." The double-dragon with the human face provided an additional clue, but only in the light shed by modern philological and local historical analysis. "For reasons purely local and particular to the city of Grenoble, we may determine the meaning of the snake-dragon. This hieroglyph, if one may use this expression," he continued modestly, echoing Wolf's description of the Homeric use of δειξαι (*deixai*) as a kind of hieroglyphic, "can be explained in this well-known proverb on the city of Grenoble," which he gave, appropriately, in dialect: "Lo serpein et lo dragon / Mettron Grenoblo en savon." This bit of doggerel, he said, alluded to the position of the city at the mouth of the Drac river (*draco*, dragon) while the Isère was represented by a serpent, as it had similarly sinuous curves.[51]

Here was an apparently complex symbol, seemingly freighted with mythological significance, that close analysis—using local knowledge of the landscape, dialect, and culture—pointed to an obvious feature of the surrounding geography. It did so, moreover, by means of the names of the rivers—that is, through the medium of toponyms. How familiar: Champollion would make a similar argument throughout the *Pharaons* volume. More than just evidence of a shared project or even just a moment of good-natured, fraternal recognition, the convergence reflects an implicit agreement about the selection of details as well as their analysis. Champollion-Figeac emphasized the elements of the crypt that pointed to the real world—*deixai*, indeed. Choosing these fragments, he was engaged in a practice that implicitly valued certain words, often the simplest and most basic, over others. Such words, which gestured toward reality without signifying anything beyond it, were the historian-philologist's ideal quarry, the proper object of inquiry for anyone seeking access to the lived world of antiquity through its textual remnants.

Champollion was hardly unaffected by his brother's research, as his letters attest. His brother was among the vanguard, but many scholars were working

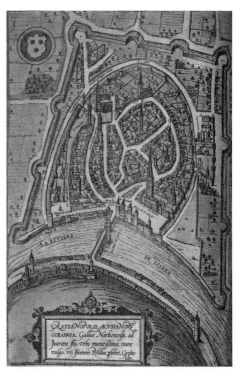

FIGURE 11.1. (*Top left*) Champollion's map (Bibliothèque nationale de France); (*top right*) Georg Braun and Franz Hogenberg, "Gratianopolis Acusianoru Colonia" (1581–88); (*bottom*) Jean Boissart, "L'ancienne ville de Grenoble" (1648).

to modernize contemporary discourse about the ancient world, shifting the familiar forms of antiquarianism toward new methods and questions in philology and archaeology. In his notes on Zoëga, made during his Paris sojourn, Champollion even embellished a copy of what is apparently a bit of Coptic materia medica, a recipe for an emollient, with a bird's-eye view of the northwestern corner of the ancient Roman boundary of Grenoble, including the Isère as it flows past Saint-Laurent (fig. 11.1).[52] Beside this map, Champollion made a lighter sketch, suggestive of the rambling Saint-Laurent structure, along with an even lighter sketch of a tower, likely of the nearby Musée de l'Ancien Évêché, the demolition of which, in Year 12 (1803), excited Fourier to urge Champollion-Figeac to search the rubble for artifacts and sparked a series of publications that opened up the late antique and medieval history of Grenoble.[53] We have juxtaposed Champollion's sketch with two views of the same area: a map of Grenoble printed in 1581, in which the same corner of the city is visible, and a panorama of Grenoble from 1648, in which the tower is visible, as are the city's ancient walls (though the church of Notre Dame may be misplaced, and the Saint-Laurent complex, across the river, is obscured by the Tour du Pont). Unlike the earlier images, Champollion's doodle gives both bird's-eye and panoramic views of the place where the Isère runs past Saint-Laurent and tower, suggesting an attempt to integrate the two perspectives. Selection and integration of relevant yet heterogeneous data characterized his approach to the ancient Egyptian language as well. Coupled with the Coptic materia medica, which brings Champollion's philology of the vernacular into the picture, this remarkable remnant suggests how closely Champollion's concerns comported with the most modern currents of research of his time.

CHAPTER 12

L'AFFAIRE POLYCARPE

Among other things, Champollion-Figeac provided his brother with a friendly alter ego with whom he could safely explore and refine his ideas. Other colleagues proved more challenging. The so-called *Affaire Polycarpe*, which pitted Champollion against his orientalist rival Étienne-Marc Quatremère, has typically been understood as a priority fight. The argument began quietly, in 1808, with the publication of Quatremère's *Critical and Historical Researches on the Language and Literature of Egypt*, and flared at intervals for the next four years, becoming markedly hostile after the publication, in 1811, of Champollion's *Introduction* to his projected Egyptian geography. Historians agree that both men simultaneously attempted to be the first to glean, completely and authoritatively, clues about the ancient Egyptian language from available Coptic sources, using ancient Egyptian geography as a focal point for these explorations.[1] But the opposition between the two protagonists' works was never starkly clear; rather, each participant in the controversy produced a considerably different kind of interpretation of similar sources. At stake was not, or not merely, who came first but whose methods and approach could supply valid results in the first place. The conflict reflected differences in the ways both men imagined the geography of ancient Egypt, and particularly to what extent each was willing to credit ancient Egyptians with sophisticated cultural forms—including humor and irony—as well as modes of apprehending themselves and the natural world.

The most significant difference between their approaches concerned the linguistic root. Champollion's analytical use of the linguistic root differed sharply from Quatremère's approach, which was more consistent with the main lines of research then being conducted. Whereas Quatremère preferred to catalog instances of place-names and other geographical references, Champollion analyzed these instances using his knowledge of word roots. Champollion's result, a detailed and comprehensive genealogy of Egyptian toponyms, went well beyond Quatremère's simple list. As we will see, Champollion's approach also closely aligned his work with the modernizing ethos of Napoleonic France.

With an understanding of just what Champollion was doing with Coptic in the years leading up to the publication of *L'Égypte sous les Pharaons*, we can better appreciate the stakes of this episode and its relationship to Champollion's work on decipherment.

Primarily a scholar of Arabic, Quatremère hailed from a distinguished bourgeois Parisian family. Both parents were devout Jansenists. His father, Marcel-Étienne, a prominent judge, was guillotined during the Terror, when his son was twelve years old.[2] At the École des langues orientales vivantes, Quatremère devoted himself to the study of languages, particularly Arabic with de Sacy, supplemented by Hebrew, Coptic, Armenian, Persian, Turkish, the different dialects of Aramaic, and even Chinese. By 1807, he was working in the manuscripts section of the Bibliothèque impériale, an appointment that brought him into contact with leading orientalists like Langlès, the conservator, while exposing him to the wealth of materials pouring into Paris, including the Coptic manuscripts that would prove so useful as he developed his position on the ancient language of Egypt.[3]

Quatremère did not go out of his way to make himself beloved; he seems to have struck others as gratingly sanctimonious. During his years at the École, he earned himself the nickname "Polycarpe" because of what Aimé-Louis Champollion-Figeac (Champollion's nephew) decades later called "antisocial habits" while discreetly failing to offer further specifics about the behaviors that had proved so intolerable.[4] Quatremère's failures at sociability may have stemmed from excessive religiosity, a quality that irritated many people, including Champollion. In any case, the unflattering sobriquet referred to the original Polycarp (80–167 CE), the traditionalist first bishop of Smyrna who was martyred toward the end of the second century during a phase of growing confidence in analyses of sacred texts that asserted a degree of independence from religious authority.[5] The scholars who swelled the ranks of Napoleon's schools and institutes were often overtly hostile to religion as well. The use of Polycarp's name to demean Quatremère may have reflected the low status anticlerical French scholars accorded any religious authority.

Quatremère's reputation was not enhanced by his position as one of de Sacy's favorites. Throughout this period, de Sacy proved more loyal to Quatremère than to Champollion, a diminishment that Champollion found difficult to forgive despite its near-inevitability, given de Sacy's personal history. According to Aimé Champollion-Figeac, de Sacy's loyalty to Quatremère stemmed from a preexisting friendship situated one generation back in time, in de Sacy's affection for Antoine-Chrysostome Quatremère de Quincy (1755–1849), Étienne-Marc's cousin, who, with de Sacy, was a member of the Institut d'Égypte.[6] De Sacy's sympathetic attitude toward Quatremère may have also been rooted in a similar religious outlook, as de Sacy, like Quatremère, came from a Jansenist family. As Quatremère's career developed, he followed de Sacy by working increasingly with Arabic sources. Although his focus on Arabic was anticipated in his work on Coptic, the latter was never of primary importance for Quatremère. On the contrary, his arguments in these early works suggest that knowledge of the language of ancient Egypt would require a substantial detour through Arabic sources, particularly those de Sacy was then enthusiastically translating.

Beyond de Sacy's favoritism and the rancor it predictably engendered, the more fundamental conflict pertained to the appropriate use of Coptic in the study of ancient Egypt. At particular stake was the evidentiary status of the different Coptic manuscript sources. De Sacy was skeptical of comparative linguistics, a position with which Quatremère found it convenient to agree.[7] If the comparative use of materials in different languages could only say so much about any of the languages under study, then Coptic would have a correspondingly limited value for anyone hoping to approach ancient Egyptian. But comparative analyses of botanical, chemical, and natural historical subjects had emerged in the last decades of the eighteenth century and were now appearing regularly. Champollion would himself soon produce a deeply comparative study of ancient Egyptian geography, about which we will have more to say in chapter 13. These works carried a distinct whiff of modernity, of which the fusty and reactionary de Sacy would certainly not have approved.

The new comparative studies frequently included charts, known as *tableaux analytiques*, that identified comparable data, grouped them, and ordered elements of the group according to a given principle or perspective. Ideally the resulting tabular take on the data permitted the meaning of the data to be apprehended at a glance: it gave a synoptic view. One of the earliest examples, the tellingly titled *Coup d'oeil général sur la France* (1765), by Louis Brion de La Tour and Louis Charles Desnos, bound maps of governmental divisions with tables of financial and population data to create a sophisticated atlas that allowed readers to compare different ways of carving up the map of France. Antoine-Louis Brongniart's *Tableau analytique des combinaisons & des décompositions de différentes substances* (1778) and Chaptal's *Tableau analytique du cours de chymie* (1778) attempted to set chemistry on a similar footing, charting the ways in which various operations on substances decomposed them into constituent parts or turned them into more complex substances. Jean Verdier's *Tableau analytique de la grammaire générale appliquée aux langues savantes* (1803) did the same for language, decomposing Latin and French into their most basic elements and attempting to build from them a system of moral education.

On June 8, 1795, the minister and politician Jacques-Antoine Rabaut-Pomier, then deputy of the Département du Gard, informed the Commission of Public Instruction that such analyses were a boon "to the eye, the spirit, the memory, the imagination," and not least, "to industry." He recommended these methods as a means of bringing order to existing collections, particularly the Cabinet des médailles in the Bibliothèque nationale, which had recently lost its conservator, the legendary Abbé Barthélemy.[8] The following year, Millin joined this short-lived project, replacing Barthélemy as conservator. Responsible for imposing order on a chaotic collection, Millin did just what Rabaut-Pomier had advocated, applying principles of systematic categorization to the coins and other materials in his care. In his *Introduction à l'étude des médailles* (1796), Millin opposed his method, which was based on specific characteristics and similar to botanical and natural historical modes of categorization, to simple alphabetized lists of holdings, which he described as "extremely bad" for scholarship. His method, in contrast, was "the thread of Ariadne that will guide us through the labyrinth of the sciences" and "as useful to numismatics as it is to botany and other areas of natural history."[9]

Millin's modernizing impulses met with significant resistance. The antiquary A. L. Cointreau, whose employment in the Cabinet des médailles stretched back to the ancien régime, felt himself particularly marginalized and took special offense at Millin's promotion of *tableaux analytiques*. In 1803, Cointreau took his revenge by publishing the provocatively titled *Tableau analytique de la réligion des anciens Égyptiens*. As one might guess from the way Cointreau's title poked sarcastically at the ambitions of Napoleon's savants, his "tableau" was not an analysis at all. Rather, Cointreau's point-by-point rebuttal showed how Millin's descriptions of the cabinet's holdings illegitimately Egyptianized them, implying that the modernizing trend represented by Napoleon was rewriting ancient history in terms that reinforced his recent claims to imperial glory. By dint of sheer repetition, references to ancient Egypt as a locus of ancient imperial prestige were hoped to produce a positive identification with Napoleon's own imperial ambitions, or so Cointreau suggested. If Millin had his way, Cointreau feared that even antiquity might become, like botany and natural history, the province of upstart savants, and scientific matters would become scrambled with antiquarianism to the latter's detriment. Cointreau fumed: "An antiquary among the naturalists, a naturalist among the antiquarians—one must not take as Gospel what is said by cit[izen] Millin."[10] The dispute between Champollion and Quatremère erupted within this context, and its salient aspects soon reappeared in their fight.

Their rivalry began, ironically enough, with announcements of competing works that did not actually exist. In June 1808 Quatremère published his

Critical and Historical Researches on the Language and Literature of Egypt, in which he proclaimed the imminent publication of a different work, his *Geographical and Historical Essays on Egypt.*[11] Although this second work was not published until January 1811, after Quatremère had moved to Rouen to take up a professorship of Greek, the 1808 announcement put the two men on a collision course, as Champollion had himself announced plans to complete a similar work to the Society of Arts and Sciences at Grenoble just before coming to Paris in 1807.[12] Moreover, Quatremère had, perhaps unwittingly, foreclosed Champollion's avenue of advancement by taking a strong position against etymological reasoning—which was more or less precisely what Champollion was accused of doing—by characterizing it as a faulty method that inevitably led to serious errors. Conjuring the image of the scholar who had succumbed to the temptations of spurious etymology, Quatremère warned that by following such linkages, "we find ourselves in a vast labyrinth in which we have only the most delicate thread to guide us. At any moment the thread may break and we will find ourselves lost in a world of shadows."[13] This point, so delicately made, was the sharp tip of Quatremère's knife: as we have seen, Champollion was all too aware of the danger of being miscast as a latter-day Kircher, seduced by specious etymology. In September 1810, a salvo from a different front intensified the pressure. Åkerblad's *On the Coptic Names of Several Towns and Villages of Egypt* arrived at the Institut, and shortly thereafter de Sacy read at least part of it to the membership.[14] A panicking Champollion-Figeac wrote from Paris on September 22: "Everybody's eyes are turned to Egypt. Åkerblad is making attempts in your field . . . Yesterday Sacy read his report . . . which treats the Coptic names of Egyptian cities and villages; rush your print!"[15]

Champollion-Figeac's anxious pressure notwithstanding, little would come of the Parisian debut of Åkerblad's essay. Åkerblad, who had not published anything since 1803, was by this time permanently established in Rome and therefore unable to prosecute his own case on the ground in Paris. Åkerblad's French colleagues proved unhelpful as well. His admiring friend Millin was traveling in France and Italy, absorbed in a quest for fresh antiquities to identify and describe, and de Sacy, after reading part of Åkerblad's essay to the Institut, simply set the work aside, perhaps to protect his protégé, Quatremère.[16] Meanwhile, Quatremère's second work on Egypt, the projected volume on geography that he had taken care to mention in the preface to his 1808 publication, appeared in January 1811.

In the latter study, which canvassed available ancient and medieval documents for references to more than a hundred Egyptian towns, Quatremère relied heavily, though not exclusively, on Arabic sources, consistent with his linguistic interests and those of his patron de Sacy. From these toponyms

Quatremère gleaned what he could, remaining silent about the names of other geographical features, such as mountains and rivers. Champollion's projected study, in contrast, aimed to treat more than two hundred towns, as well as topographical features such as mountain ranges and, of course, the Nile; moreover, he had consulted a broader range of source materials, particularly in Coptic. Perhaps for these reasons, on February 20, 1811, Champollion resolutely told his friend, the young orientalist Antoine-Jean Saint-Martin, that Quatremère's publication gave him "more fear than pain."[17] His confidence increasing, Champollion continued sarcastically: "[Quatremère] has no idea of my plan, and his lack of [knowledge] of Egyptian characters prevents him from rolling out his vast Copto-Egyptian erudition."[18]

By this time, as we have seen, Champollion had already extensively developed his Coptic dictionaries and his notebook of Coptic roots; moreover, he had begun trying to match demotic characters in the Rosetta inscription with Coptic roots and to make hypotheses about the six Coptic letters that were believed to have phonetic equivalents in the demotic. Although Quatremère had many resources at his disposal, including de Sacy's support, he did not have this depth of experience with Coptic, nor had he seen the Borgia Coptic manuscripts before his work went to press. It was in this uncertain but promising context that Champollion hastened to complete his *Introduction*, which was published in March 1811 in a (very) limited edition of thirty copies.[19] The work appeared after a delay of almost six months while his printer, Peyronard in Grenoble, apparently struggled to locate a suitable Coptic typeface, prompting Champollion to call him a *lambin* (slowpoke) in a letter to his brother.[20] That same year, Champollion brought out two or three additional essays—on Thebes, Memphis, and the different names given to the Nile—extracted from what would eventually become the 1814 edition of the *Pharaons*.[21] Taken together, these publications represented the kernel of what was most interesting and original about the 1814 edition and so amounted to a preview of that much larger work. Although they met with little fanfare, Jomard praised them, remarking rather self-referentially that the plan of the *Introduction* came as an "agreeable surprise," and one, moreover, that was "absolutely consistent with what I have written." He added that the claims advanced there were "likely to be helpful in completing" the *Description*, the early volumes of which he was just then readying for press.[22]

Any hopes Champollion may have nourished for winning the approval of de Sacy—his former teacher and most powerful patron—were dashed on August 12, 1811, when de Sacy damned the *Introduction* with faint praise in a five-page review published in the *Magasin Encyclopédique*. Worse, de Sacy explicitly and negatively contrasted Champollion's projected work with Quatremère's recently published volume. Though their rivalry predated de Sacy's review, this public humiliation hardened Champollion's opposition to Quatremère and heightened his suspicion of de Sacy.

A veteran of the battlefields of scholarship, de Sacy knew how to damage an adversary without marring his own credibility by allowing the attack to seem excessive or gratuitous. He skillfully arranged a handful of rhetorical fig leaves over the review's assault, the gist of which insinuated that Champollion had failed to acknowledge a substantial debt to Quatremère. De Sacy conceded, first of all, that Champollion's proposed work was "analytical," perhaps more so than Quatremère's. He also approvingly noted Champollion's inclusion of a *tableau analytique*, a seventeen-page chart of the familiar names of Egyptian towns and villages and their equivalents in Coptic, Greek, and Arabic (fig. 12.1). De Sacy recognized the intellectual payoff promised by the table and accepted Champollion's assertion that the remainder of the work, projected but not undertaken in this volume, would substantiate the insights at which the chart could now only hint. He also noted Champollion's suggested identification of Coptic letters in the Rosetta demotic, a point that helped to justify Champollion's central presumption that Coptic, as a late form of ancient Egyptian, really was useful as a gateway to the latter language.

Having carefully selected these items for praise, de Sacy then deftly split Champollion's interpretations from his treatment of evidence and proceeded to undermine the former by means of misdirection and attacks on the latter. "Champollion has gathered and put in order everything that the ancients wrote and that modern critics have to say about an interesting topic while also attaching his own observations," he wrote. This framing minimized Champollion's accomplishment, for Champollion had not merely "attached observations" to his data but had drawn original inferences from them. By severing Champollion's handling of his data from the imposition of a point of view, de Sacy effectively insinuated that Champollion had not done much beyond collecting and ordering. It was at this vulnerable point that de Sacy raised the alarming specter of specious etymological reasoning. While he congratulated Champollion on having "avoided getting tangled up in the spirit of system that too often fills the silences of history with hazardous conjectures," the apparent praise, read closely, was hardly a tribute at all. De Sacy did not laud Champollion for anything he had actually done but solely for having avoided calamity. De Sacy then made sure to douse any enthusiasm he may have inadvertently roused: "Champollion deserves credit for having written a useful book," de Sacy opined listlessly, "and he will see his efforts crowned with success and the approval of the learned."[23]

TABLEAU ANALYTIQUE

DE la Géographie de l'Égypte sous les Pharaons.

Nom Égyptien ou Copte.	Nom Grec.	Nom Arabe.	Nom Vulgaire.

I. Frontière méridionale de l'Égypte.

| Ⲡⲕⲁ ϩⲏ̄ⲛⲟⲟⲟⲩ.
Ⲛⲉⲟⲟⲟⲩ.
Ⲉⲃⲁⲣⲱ. | Ⲁⲓθⲓⲟⲡⲓⲁ. | El-Hhabbesch. | L'Éthiopie. |
| Ⲧⲁⲛⲟⲩⲃⲁⲧⲓⲁ.
Ⲧⲛⲃⲏ.
Ϯⲗⲓⲃⲏ. | Ⲛⲟⲩⲃⲁⲓ. | El-Noubah. | La Nubie. |

II. Nom de l'Égypte.

| Ⲭⲏⲙⲓ.
Ⲕⲏⲙⲉ. | Ⲁⲓⲅⲩⲡⲧⲟⲥ. | Missr. | L'Égypte. |

III. Nom du Nil.

| Ⲫⲓⲁⲣⲟ ⲛ̄ⲧⲉⲬⲏⲙⲓ. | Ⲛⲉⲓⲗⲟⲥ. | El-Nil. | Le Nil. |

FIGURE 12.1. First page of Champollion's "analytical table."

If the first part of de Sacy's review was dismissive, what followed was devastating. Having first singled out Champollion's *tableau analytique* for notice, he proceeded to diminish its significance by claiming that it was neither novel nor particularly special in other ways; that, in fact, it was entirely at one with Quatremère's results. De Sacy attributed the consistency to the fact that both Quatremère and Champollion relied on the same body of source materials. The "analytic principles" Champollion had applied did not accordingly originate with Champollion but were, or would be, self-evident to *anyone* who examined the same sources. De Sacy added that unlike Quatremère, Champollion was fortunate enough to enjoy special access to the Coptic manuscripts formerly owned by the Borgias and transported to Paris after Napoleon's sack of the Vatican Library. These were the manuscripts cataloged by Zoëga that Champollion had reviewed. De Sacy begged the reader's indulgence on behalf of Quatremère, who had recently transferred to Rouen, and who had not yet had the same opportunity of consultation.[24] If Quatremére's work did not exhibit the same analytical sophistication, that was to be attributed to Quatremère's limited access to sources rather than to anything particularly original or penetrating about Champollion's approach.

Once he had established the use of specifically Coptic source material, rather than anything Champollion had done with it, as the central value of the *Introduction*, de Sacy focused sharply on priority. He noted, first of all, that Champollion had not fully digested current work in his own field, for he had admitted ignorance of Åkerblad's 1810 manuscript on the Coptic names of towns and villages in Egypt, which de Sacy had announced and read to the Institut in the early part of 1811. This was disingenuous: as the only person in France known to possess a copy of the document, de Sacy had complete control over its dissemination. Although he had at least asked his friend Saint-Martin about the *Mémoire* in Paris, Champollion was not among the auditors to this presentation, and there is no indication that de Sacy ever sent a copy of the document to Champollion. Nor would de Sacy publish it for many years.[25] In addition to holding Champollion to this impossible standard, de Sacy complained that Champollion had failed as well to acknowledge Quatremère's *Mémoires géographiques*, published toward the end of 1810 and subsequently reviewed by de Sacy himself. De Sacy claimed that Champollion had evaded responsibility for this failure of acknowledgment by defensively insisting that the piece had been in print since "last October" and blaming the rush to publish for any incoherence in the *Introduction*, "which was not destined to be separated from the work of which it forms a part."[26] Since this latter point hinted at a larger project that was still in preparation, de Sacy had "no doubt that Champollion will make a point of consulting [Quatremère's] excellent work" in the promised subsequent study. De Sacy additionally hoped that Champollion possessed "enough respect for the public to avoid uselessly repeating what Quatremère has said on the topic" and that he would "make the requisite sacrifices" by reshaping his materials to be more respectful of Quatremère's priority.[27]

Champollion did not respond publicly either to Quatremère's publication or to de Sacy's review, but privately he was incensed. Letters to his friend Jean-Antoine Saint-Martin attest to his disturbance. He frequently mentioned Quatremère's posturing and de Sacy's interference, and he let Saint-Martin know that he understood Quatremère's efforts as directed specifically against his own. Concerns about intellectual property weighed heavily on Champollion's mind. In addition to dealing with de Sacy's betrayal and Quatremère's incursions, he was also smarting from his erstwhile supporter Fourier's failure to acknowledge his assistance with the historical preface for the *Description de l'Égypte*.[28] Champollion's sensitivity to Quatremère's activities may have stemmed from anger he could not express directly to Fourier about being excluded from the *Description*. But de Sacy had certainly acted provocatively by obscuring the originality of Champollion's work while staking a claim for Quatremère's priority. Strictly speaking, de Sacy's support of Quatremère was unnecessary. There was no priority to establish: Quatremère had published first, after all. If Champollion's work contained little of value and had come later in any event, why trouble to critique it in the first place?

The answer is obvious: Champollion's work, sketchy and incomplete as it was, posed a serious threat. The question was not whether Quatremère had arrived first on the shore of an unclaimed intellectual property. The question was whether he had arrived at all. Champollion's defenders pointed out that because Champollion's approach differed so sharply from Quatremère's, the two studies could not but arrive at markedly *divergent* conclusions. This point contrasted with de Sacy's aim, in his review, to demonstrate that the two works could not be anything but similar. A letter written by Champollion-Figeac, which has been preserved among Saint-Martin's correspondence with Champollion, described the fundamental difference between the studies in a two-page, densely packed rebuttal to de Sacy's review.[29] After excusing Champollion for not writing directly, being indisposed by illness, Champollion-Figeac offered his general position:

> M-Q [uatremère] has aimed to give an alphabetical collection of the towns and villages which he has found in the Coptic manuscripts that he has consulted; but he has not given any details about these towns . . . my brother, on the other hand, is giving a geographical description of Egypt

by locating the Egyptian names within their natural and political classification, indicating the latitude and longitude of the principal places, and
attending to their most important monuments and the different names
that these places have been given by the different peoples who have spoken [of them].[30]

Champollion-Figeac's description of his brother's achievement quietly
insisted on its originality and even modernity. Where Quatremère had been
content to list instances of toponyms mentioned in the source material—to
present the results of tedious archival work, the mechanical sorting of a textual
corpus that could be done by any sufficiently motivated antiquarian pedant—
Champollion had gone further. First of all, he had included the most recent
findings, for example, of latitude and longitude, associated with each toponym,
using results published in the *Description*. With this information, his assertions
about the precise locations associated with each toponym could be brought
up-to-date using the best firsthand reports from Egypt. Second, Champollion
had identified and described the linguistic and geographical contexts for his
toponyms, showing how they fit into a larger framework of ancient Egyptian
understandings of politics and nature as connected realms. Quatremère could
lay claim to none of this.

Champollion-Figeac supplemented this broad condemnation of his
brother's critics with a handful of smaller objections. He faulted Quatrèmere
for failing to provide a complete list of toponyms, whereas his brother's list was
comprehensive. Quatremère had also omitted geographical coordinates, many
of which were available in the recently published volumes of the *Description*,
for the places whose toponyms he did investigate; by pointing out this omission, Champollion-Figeac positioned Quatremère's volume even more firmly
as irrelevant pedantry. Third, Champollion-Figeac claimed that Quatremère
misinterpreted certain Coptic toponyms from manuscripts in archives with
which he alleged to be familiar. Finally, although Quatremère complained that
he'd not had the chance to consult the Borgia archive, Champollion-Figeac
argued that this misfortune was not so great as Quatremère and de Sacy had
made it out to be, since the archive was the source of only a handful of toponyms in Champollion's account. "No one could see my brother's work as an
imitation of Quatremère's," Champollion-Figeac insisted. On the contrary,
Champollion had proposed the work to the Académie de Grenoble in 1806 and
circulated copies of that lecture to Langlès, Millin, Jomard, Fourier, and even
de Sacy himself, as Champollion-Figeac was pleased to remind him.[31] If anyone
should be suspected of appropriating ideas without proper acknowledgment
of their source, it was Quatremère—or so Champollion-Figeac implied.

In a review of Champollion's *Introduction* published in the July 30 issue of the *Moniteur Universel*, Saint-Martin restated many points made by Champollion-Figeac in his closely argued letter (without referring, no doubt intentionally, to Champollion-Figeac's participation). To start, Champollion's particular use of Coptic provided a powerful tool for uncovering the past and contrasted with the less perspicuous efforts of others who knew Coptic:

> Other [scholars], highly versed in the knowledge of Coptic, a very valuable and important reservoir of the language of the ancient Egyptians, have attempted to use *hazardous and sometimes false etymologies* to explain the religion and beliefs of these people, while the less bold are content to produce editions of books that Christians of Egypt had written in the language of their ancestors and whose interest to us would be absolutely zero, if they did not also contain the remnants of that language.[32]

With these words Saint-Martin evoked the specters of *two* forms of pedantic irrelevance, hazardous etymology and tedious martyrology. Both were sure to be perceived as painfully quaint compared to the sharp-edged, new productions of Napoleon's savants. Echoing Champollion-Figeac and the modernizing ethos of savants like Millin, Saint-Martin downgraded Quatremère's contribution as a pedestrian "collection," a mere list of place-names. Where Champollion had offered a powerful chart that gave a digest of the analysis of toponyms followed by a comprehensive synthesis of geographical knowledge about Egypt organized according to his unique understanding of the linguistic history of place-names, Quatremère had settled for merely listing the place-names and other geographical markers that he'd found in extant manuscripts and the secondary sources that had made prior use of them. Champollion's *tableau analytique* "gives an idea of the plan and analysis of the work," while Quatremère had merely given "an alphabetized collection of Coptic names of cities and towns found in works by Coptic writers," a list that was, moreover, "far from being complete." Worse, "among the names on which he does report, many lack detail and the author could not fix the precise location either."[33]

With these criticisms Saint-Martin attacked Quatremère's work at its central vulnerability, on the very ground that de Sacy had with the faintest of praise remarked in his report on Champollion: unlike Champollion's work, Saint-Martin averred, Quatrèmere's was not "analytical." The result was more a concordance to European libraries' ancient manuscript sources containing geographical information about Egypt than a guide to the geography of ancient Egypt itself. Although Quatremère's stated aim was only to canvass those sources and to create what was essentially an elaborate index of them, Saint-Martin held the work to another standard. Coptic sources per se were to be used as vehicles for a larger project of the analysis of geographic data, particularly place-names, to gain a purchase on ancient Egyptian language and ways of making sense of the physical world. Seemingly narrow geographical projects in fact had critical implications for the study of ancient Egypt more broadly; it was, he implied, the writer's responsibility to trace these implications for the reader.

Although Saint-Martin's review amounted to little more than criticism of the weak principles of order evident in Quatremère's work, Champollion was sufficiently buoyed to write ebulliently to his friend on November 1, 1811, that "the miserable Quatremère sees the sceptre of the East trembling in his hands."[34] Quatremère himself responded at vigorous length in a substantial pamphlet, published in 1812, titled *Observations on Several Points in the Geography of Egypt*. Rather than take up the broad criticism that Saint-Martin, following Champollion-Figeac, had made about the quality of his analysis, Quatremère read the criticisms narrowly, focusing on specific flaws that he had identified rather than addressing the larger questions of selection, emphasis, and inference that had made Champollion's work so distinctive, at least according to Saint-Martin and Champollion-Figeac.

After "losing sight of Egyptian geography" for a while, Quatremère claimed to have happened upon Saint-Martin's notice concerning Champollion in the *Moniteur* and found the piece distinctly mean-spirited. Saint-Martin was, he wrote, "not content to give the work of his friend [Champollion] the most pompous *éloges*" but wanted also to tear down Quatremère. "According to him [Saint-Martin], my work presented two essential flaws: one, my list of the cities and towns of Egypt was substantially incomplete, and two, I failed to establish the location of a great many places."[35] This narrowly accurate gloss of Saint-Martin's criticism missed the larger point: Quatremère's work was a mere *receuil*, a collection. It was not organized analytically. For Saint-Martin, Quatremère's research method amounted entirely—as Quatremère himself pointed out—to a survey of extant Coptic sources for place-names.

Quatremère rationalized his choices in an effort to defang Saint-Martin's criticism. He insisted that some of the gaps in his presentation were more apparent than real; they were simply artifacts of the book's organization. Other Egyptian places just did not seem that important for various reasons.[36] Quatremère conceded that Champollion's assertions may have been "bolder" than his own, but this boldness was just what had led Champollion astray. With his criticisms, Quatremère hoped to "correct, or at least put within reach of correction, some mistakes [Champollion] has made."[37]

Quatremère devoted the rest of his rejoinder to nitpicking.[38] His focus on minutiae permitted him to sidestep the question, so central to Champollion, of how toponyms—some of which were still in use—might relate to ancient Egyptian religion and culture. Avoiding larger questions of that sort had been an essential part of Quatremère's modus operandi since at least 1808, when he claimed that invasions of Egypt had expunged ancient remnants from the language so thoroughly that one simply could not expect to find residues in more recent textual sources, thereby dismissing the possibility of any relationship between extant place-names and ancient Egyptian culture.[39] In passing, Quatremère raised the question of plagiarism on his own part only to dismiss it, somewhat disingenuously, by claiming that while he had seen Champollion's 1811 *Introduction*, he had not taken it seriously as it was merely a plan of work to be done and not the finished product—a position that repeated de Sacy's, who had praised Champollion's ambition while withholding reassurances about Champollion's ability to deliver.[40] His 1811 critique accordingly seems aimed less at Champollion's work than at his ambitions for it, as if the vision of identifying elements of ancient Egyptian culture in such remnants as persistent toponyms was too far-fetched to be countenanced given the current state of the evidence. For Quatremère, there was simply no route between extant place-names and the distant world of ancient Egypt.[41]

CHAPTER 13

AN EGYPTIAN
GEOGRAPHY OF
EGYPT

The two-volume edition of *L'Égypte sous les Pharaons* published in 1814 greatly extended the line of inquiry that Champollion had opened with the 1811 *Introduction*. He again began by summarizing the relevant scholarship and cataloging points of phonetic equivalence among the Coptic, Arabic, Greek, and Latin alphabets. The remainder of the work presented a series of essays in which he attempted to link Egyptian toponyms to salient features of the natural and human landscape as well as elements of Egyptian culture. The two could not be easily separated, a point that Champollion made implicitly throughout his 1814 geography and explicitly in his unpublished notes and other manuscript materials from this period. He analyzed Egyptian toponyms and developed arguments to show how they expressed fundamental elements of ancient culture. These elements were mainly religious but also had to do with important geographical features such as the Nile—a source of meaning as well as of food, water, transportation, and so much else that governed Egyptian life. The whole project required a most specific tool that, we have seen, he had carefully developed: the concept of a persisting linguistic root.

Champollion's discussion of the origin and meaning of Syène (Aswan) exemplifies the approach. He gave the town's coordinates using the latest published information, taken as usual from the *Description de l'Égypte*, in this case from Jomard's entry on "Syène." He then summarized the related information available in the manuscript and published sources at his disposal in Paris and Grenoble, focusing on the history of the toponym. Where manuscript sources diverged, he noted the differences and evaluated them. Having dispensed with these preliminaries, he linked the toponym with a Coptic root, a procedure he followed repeatedly in the *Pharaons*. "Coүxn," he wrote of Syène, "is derived from the root oүн (*ouèn*), oүeн (*ouan*), which is the same as oүωn, *ouôn, aperire, ouvrir,*" that is, various forms of the verb "to open." Next he derived various Coptic words by combining the root with the particle "*sa,* ca, indicating the attribution of ability or power to do something." Indeed, he continued, "a great number of Egyptian words are formed in this way, such as самаωι, *samaschi,* [or] weight that has been attributed to [something] by means of a scale, and саменθоүх, *samethnousj,* the power to lie," a juxtaposition that gestured toward a familiar world in which scales were routinely in need of calibration.[1] The toponym Syène, he suggested, derived from a similar combination of root and particle: "In the same manner that the Egyptians say соүхn, *aperiens,* the Copts also say есоүнn, *asouén, aperta,*" that is, "open." Contracting the Coptic yielded the Egyptian toponym for Syène, соүаn. "The signification appears to us to be in relation to the military position of Syène, considered throughout the country to be the leading frontier town in Egypt, *the key to the realm,* on the border of Ethiopia," he explained.[2] A key, of course, opens things. The toponym accordingly carried a historically useful residue, a vestige that appeared as an identifiable monosyllabic root that referred, in the first instance, to the power to open something. With this root analysis, Champollion suggested that the name of Syène encoded features of its geographical and strategic situation: Syène was to Egypt as a key to a door. The *Pharaons* bursts with derivations of this sort.

The most apparent remnants of ancient Egyptian in Coptic adhered to words and names for such natural sites as rivers and mountains.[3] Champollion argued that the ancient Egyptian word for "Nile," for example, must stem from the contemporaneous word, discoverable in Coptic, for "river."[4] Place-names could also be more abstract and symbolic, as long as they kept a degree of of connection with ordinary life. The metaphorical link between Syène as the "key" to Egypt and the Coptic word for "open" was an example of one such ordinary toponym. For a second example, consider Champollion's gloss on the name of the mountain chain near Libya:

> Memphis was situated at the foot of the mountain *Psammi-us,* a name that connoted in Egypt the part of the Libyan chain near this ancient capitol. *Psammius* is obviously an Egyptian word, that may derive from the root хоm, *Sjom, fortitudo.* Perhaps хоm, combined with mх or mоι, *dare,* might have formed, through the ordinary mutation of vowels, пхоммоι, or a word approaching the *Psammios* of the Greeks, a word that would have had this value: *dans fortitudinem, défense, boulevart.*[5]

Here we see Champollion working with the assumption that the place-name persisted in modified form from the ancient Egyptian. The toponym accordingly reflected its ancestral progenitor, a simple root and modifier that named the spot by announcing its strategic value: *Psammis*, a stronghold.

Although most entries in *Pharaons* were no longer than a page, Champollion's entry for Thèbes, which had also been prepublished as a stand-alone pamphlet, ran to twenty pages that detailed the history of the city and its monuments. Not only was it the most important of Egypt's ancient cities but, he argued, the very name announced the fact. Admitting that "the origin of this premier capital of the Egyptian empire has been lost in the night of time," Champollion nevertheless dismissed various Greek influences, claiming that the name was of Egyptian origin, and "therefore it is in the language of the Egyptians that we must seek its meaning."[6] The Greek "Thèbes," he argued, was "nothing other than" a Hellenized version of "the Egyptian [Coptic] word ⲧⲁⲡⲉ, *Tapé*, which, in Theban dialect, is used to say *tête, chef*; it therefore would naturally apply, and rightly so, to Thèbes, the *capital* of Egypt, the most ancient and important city in the realm, the foremost seat of Empire and of the [priestly] *hierarchy*."[7] Framed in this way, the term's history met Champollion's standards of naturalness and rightness. In the ancient Egyptian cosmos, he thought, orders of nature and of man were hierarchical, and natural and human laws were tightly correlated. In deriving Thèbes from ⲧⲁⲡⲉ, he believed he was applying the standards of ancient Egyptians themselves. He concluded by recurring, again, to the simple and obvious, inviting the reader to imagine ancient Egypt through the eyes of an ordinary inhabitant whose frame of reference was resolutely local, even provincial: "We must note here that, in every circumstance, the Egyptians thought about matters of place only in relation to themselves, and so they called the Nile by the generic name *River*, because it was the only one in Egypt; just so, they called Thèbes *capital*, because Thèbes was indeed the only capital of Egypt."[8]

Of course, words that sounded the same—that shared similar phonemes—might also have quite different meanings, as Champollion, whose letters are full of puns, often in several languages, was well aware. We should not be surprised, therefore, to find that an attentiveness to wordplay was characteristic of Champollion's analyses.[9] His discussion of the town of Toum/Pithoum and the neighboring Silsilis/Sjolsjel illustrates this aspect of his approach.[10] Here again, the etymological root provided essential information linking the toponym to an easily recognizable feature of the landscape. This example permitted Champollion to take a further step by showing how these roots could encode facets of a geography not merely on their own but when juxtaposed, repeated, or otherwise combined.

According to Champollion, the Roman name for the area, Toum, was "an alteration of the ancient Egyptian name, which derives from ⲑⲱⲙ, *empêchement*," an obstruction. The town, he noted, was situated in a narrow passage between the nearby mountain and the Nile—that is, in a site that could be easily cut off. In Arabic, he continued, the town was called El-Hhassir, derived from the Arabic *hhassara*, which similarly connoted a place to which access could easily be blocked. The root, he continued, also "corresponds to the Latin *obturare, boucher*," to obstruct or stop up. Champollion cited several variants of the same root that meant "to conjoin, to adhere, to loiter," that is, to stick around, and "to turn back upon oneself." Taken together, these meanings suggested a remote, inaccessible place where one might indeed feel trapped because the narrow passage connecting the town to the larger world was vulnerable to being cut off.[11]

As isolated as Toum/Pithoum seemed to be, its toponym, at least, was related by analogy to its neighbor, Sjolsjel. In both cases, the idea of adhesion—sticking, or being stuck—was at the center of the town's toponymic identity. Sjolsjel had two additionally relevant topographic features: it was situated on the west bank of the Nile as it flowed between two mountain chains, and it was built with stone from neighboring quarries. Its toponym came from the Coptic ϫⲟⲗϫⲗ, which Champollion translated as "Sjolsjel" or "a wall, that is, something that obstructs passage." Near that town, Champollion observed, a mountain is situated so near the Nile that it "in effect nearly blocks the route" out of town and so "the name is given, justifiably, to the place that we've tried to describe." He noted that the Coptic had a further connotation as an encircling or surrounding enclosure. But, Champollion continued, the word also appeared in a different form that meant "adhaerere" (to cleave, to adhere) in Latin. At this point, he observed that the first monosyllabic root, ϫⲟⲗ (*sjol*), in ϫⲟⲗϫⲗ meant something like "surrounding wall" but with a redoubled significance given the identical root that followed. The pair of roots seemed not just to repeat but to stick to each other, to *adhere*. By showing how the literal mountain chains that encircled the town reappeared figuratively in the town's toponym, Champollion suggested the close relationship between Egyptian words and the things they described.[12]

Having discovered in this marvelous example a linguistic phenomenon broadly illustrative of his point about the close connection between language and culture, Champollion seized the opportunity to expand upon a general feature of the language. "The Egyptian language abounds with words of this type," Champollion declared. Repetition changed the word's basic meaning, captured by the root, through abstraction and metaphor. To illustrate, he produced a series of vivid examples of Coptic words with a repetitive structure: ⲗⲁϣⲗⲉϣ (*lashlesh*), humility; ϫⲟⲙϫⲉⲙ (*sjomsjem*), to examine by touch; ϭⲓⲗϭⲱⲗ (*kilkoul*), to yoke; and, most noticeably, in the onomatopoetic ϩⲉⲣϩⲉⲣ (*kerker*), to snore.[13] While he observed the obvious similarity between the toponym Ϣⲟⲗⲱϫⲉⲗ and the Arabic town name of Selséléh, he rejected Denon's report, in the *Voyage*, that the Arabic name referred to the way the Nile was blocked there by mountains, a geographical supposition that Champollion considered "a complete impossibility." Rather, the name must refer to the mountains themselves, and the Arabic was merely a transliteration of the Egyptian.[14]

These and many similar excursions through the genealogies of Egyptian toponyms convinced Champollion that it was possible to work backward in time, starting from a Coptic word presumed to contain the vestiges of an ancient root that could be linked to some feature of the terrain. These concrete, ordinary words and roots nevertheless provided the foundation for more abstract concepts, including religious ones. That supposition contrasted markedly with Quatremère's approach, for he had simply listed the toponyms with cross-references to their appearances in Coptic sources and other documents. In contrast, Champollion went so far as to link place-names to elements of ancient Egyptian culture and religion. On two occasions in the *Pharaons*, Champollion made an especially careful, step-by-step argument to illustrate how a name could even reveal how a deity could be consecrated to a place.

The two occasions concerned the towns of Panopolis and Hermopolis Magna. Panopolis was also known as Schmin and Chmim, while Hermopolis-Magna was also known as Schmoun.[15] Although the towns' Hellenized names seemed to suggest that they were unrelated to each other and had simply been named for a deity (Pan ruled erotic life; Hermes, transitions and boundaries), the presence of alternate names alerted Champollion to the existence of a common root. He argued that the names—variously ϣⲙⲓⲙ, Schmim, or ϣⲙⲉⲙ, Schmem—derived from the root ϧⲙⲓⲙ, *kmim*, which he defined as "*heat [incalescere], seethe [fervere], warm up [calefieri]* in the Memphitic dialect," finding similar meanings in Bohairic and the Greek. Moreover, he noted, the relevant consonants—the Coptic *hori* and *khei*, the Greek *chi*—were frequently confounded. From all of this it followed that the town names were late developments from the root, whose meaning was easily assimilated to each town's ruling deity: "The name ⲭⲙⲓⲙ, [*Kmim*] *incalescens, fervens*, is perfectly appropriate to the god of generation and fructification." He adduced, as supporting evidence, a word list found in the Siwa Oasis containing "a word that is without doubt Egyptian, and which is derived from the root *Qmom, Khmom.*" This word is "Akhmoun, that is to say, penis, *membrum virile*, principal attribute of the god of ⲭⲙⲓⲙ, *Khmim*, who appeared in statuary throughout the area of Panopolis. Among the Greeks, Pan and Priapus played very similar roles, representing the multiplication of beings and fertility of the land. We derive the etymology of the name of the city of *Chmim* from these various circumstances."[16]

Echoing these points, Champollion developed a similar argument with respect to Hermopolis Magna.

Even at this early date, Champollion took care to connect his philological investigations to ancient Egyptian culture, particularly its religion. Champollion's unpublished notes include two pages on what he termed the "symbolic theology of the Egyptians" filed with material from 1808 to 1810.[17] This material—organized so carefully as to seem nearly an essay draft—appears alongside notes on a major source for Coptic literature, the works of Zoëga, and several drafts of a "Religion of the Egyptians," which appears to be a manuscript of a short book. In his notes Champollion admitted uncertainty about any attempt to probe religious matters: "It is necessary," he wrote, "to address the doubts that follow when an educated person dips into divine things," even religions of the deep past. Nevertheless, he aimed to show "how the Egyptians approached knowledge [*la science*] of divine things." The first step required an understanding of the relationships that Egyptians believed to exist between themselves, their gods, and the material constituents of their world. These were hierarchical, subtle, and indirect, expressed in a language of images: "In imitation of the very nature of the entire universe and the *modus operandi* of the gods, people hide mystical, hidden and secret precepts in plain sight, in their images. In this way, nature symbolically expresses itself in forms suggestive of hidden secret precepts."[18] These images, moreover, were the product of "the creative activity of the gods." Proceeding along these lines, Champollion attempted to reconstruct the Egyptians' theory of divine mind:

> Having realized that superior beings loved to see themselves mirrored in inferior ones, and would want, in this manner, to suffuse all things with goodness (that is, imitations of superior things), [the ancient Egyptians] established their Symbolic Religion in conformity with divine (superior) things.[19]

"Now listen," Champollion urged his reader, "to the sense and interpretation of symbols according to what the Egyptians themselves say. Take care, however, not to let your imagination or your ears get carried away by the material." To uncover "the meaning of symbolic interpretation" required that one work according to a specifically Egyptian point of view, one that remained alert to the possible presence of multiple meanings. Here Champollion seemed to warn against the temptation to see one's own meanings in the hieroglyphs—for instance, to understand them on one's own terms rather than theirs, or to view them as referring unambiguously to whatever they depicted. To do so was to allow oneself to "be carried away by material images," in reference to which Champollion also used the classical term *eidolon*, rather than staying focused on what he called "symbolic things." In order to keep track of those, he suggested, we must "bind our intelligence to matters of meaning."[20]

By "meaning," of course, Champollion meant "as construed by ancient Egyptians," for whom he insisted that meaning derived preeminently from a fundamental belief that "all bodily, material things, have the ability to nourish or generate." This generativity stemmed from the pervasive presence of divine power. "God," he wrote, in reference to the Egyptian supreme divinity, is "the first cause of Nature, of Generation, and of all the powers of the Elements." This divinity, who is "superior to all things, and contains all things within himself," is also "immaterial, without a corporeal body . . . hidden by himself and within himself." Because this invisible entity "contains all things in himself and communicates himself to all parts of the world," he is all-pervasive, yet he nevertheless remains apart from earthly matters, "separate, exempt, elevated, and deployed at a distance over the powers of the world and the Elements."[21]

Having established an all-pervasive generativity as the fundamental divine quality, Champollion argued that material reproduction per se was neither the only nor even the most significant aspect of divine presence in the world of ancient Egypt. Ideas could propagate as well. Champollion offered an example, a familiar "symbol," as he called it, "of a god sitting on a lotus." The symbol, he wrote, "expresses in a mysterious way the power and holiness of a god who has no affinity with matter itself but nevertheless exerts spiritual governance over all things." This governing principle pervaded creation and was expressed by the god sitting atop the lotus.

> The <u>round form</u> of the Lotus affecting all things from the leaves to the fruit, [what is] that but God, who is also himself and by himself, is neither his government nor this operation, being [illegible] Holy. And holding within himself that which is expressed by his seated representative. But when, in the Symbols, <u>God is taken as the pilot of a ship</u>, [this] power governs the world.[22]

The lotus, ubiquitous in ancient Egyptian decoration, grew abundantly in Egypt's wetlands and provided an important source of honey. A symbol of life, it was associated with Horus, the falcon-headed sky god and child of all earthly and cosmic forces, whose powers suffused the being of the pharaoh, making him omnipotent. No more efficient condensation of these elements could be imagined than the image of Horus blooming from a lotus flower.[23]

Champollion had by now developed a singular understanding of the structure and dynamics of ancient Egyptian cultural production. This understanding went beyond language to hint at an underlying mode of thought. As the Nile fructified the land, some vital flow animated all life for ancient Egyptians, and this generativity could be observed even in the structure of the language. Hints that Champollion had come to Egyptian linguistics through his study of the civilization's culture (and not vice versa) are scattered throughout the 1814 *Pharaons*. Perhaps the clearest example appears in his discussion of Panopolis, the town whose Egyptian name, recall, derived from notions of vital heat and the deity associated with fructification. The discussion had orbited about the question of whether the Egyptian variant of Panopolis stemmed from ϣⲙⲓⲛ, *Schmin*, or ⲕⲙⲓⲙ, *Chmim*. After locating variants for *Chmim* in dialects of Coptic and Arabic, Champollion pointed out that the "aspiration" of the initial vowels were "very easily confounded" with each other as well as with the Greek. "Just as the letters ⲭ and ϣ are interchangeable in Coptic, and their pronunciation is very often the same, so ϣⲙⲓⲙ will have been written for ⲭⲙⲓⲙ, which in turn will have been corrupted to ϣⲙⲓⲛ." It was therefore possible, he thought, "that the ancient Egyptian language possessed the root ϣⲙⲓⲛ as a synonym for ϩⲙⲓⲙ or ϩⲙⲟⲙ, and, like the others, it signified *incalescere, calefieri*," that is, heat and the quality of being heated. As we have seen, these qualities were affiliated with generativity and its representatives among the gods.[24]

The conflation of pointing with signification that we noted in Champollion's analysis of Coptic words related to the "finger" (ⲧⲉⲃ/ⲧⲱⲃ) encapsulates his connection of religion with language. The argument involved three steps. First, to point is to indicate: one does not look at the finger but at what it points to; similarly, one should look through the proper name of a place to its function or purpose among the Egyptians. Second, the Coptic-Egyptian root suggested that *to point* was not only *to indicate* but also to *symbolize*. Finally, the symbolism was inherently *generative*. Just as the Nile gave rise to an abundance of living *beings*, the symbolic range of the root could generate not just one but many *words*. In his notes and published writings, Champollion suggested that Egyptians saw language as a natural resource, much like the Nile. Like the Nile, the Egyptian language possessed a generative power. This language was, as it were, *sieved* by roots, as the Nile was sieved by canals, while words, the mental equivalent of the Nile's agricultural products, were generated through symbolization.

Despite his elaborate development of linguistic and symbolic reasoning, Champollion's argument remained fragile. Every point he made could be recast as an exception, acknowledged perhaps as correct but limited in significance—a weakness that was not lost on Quatremère, as we have seen. Still,

Champollion had laid a claim to Coptic as a substantial key to Egypt's past in a manner quite different from de Sacy or Åkerblad. And he had begun to work out how ancient Egypt's culture could be more fully apprehended through the scraps available in documents and on monuments. Because those scraps had symbolic weight, they were not isolates but parts of a larger whole that could only be understood by taking into account religious beliefs, daily practices, geographical constraints, and specifically Egyptian modes of thought. Incomplete though it certainly was, Champollion's project brimmed with promise.

Finally, there was the question of publication. The timing was not propitious. After Napoleon formally abdicated in April 1814, Champollion-Figeac joined a delegation from Grenoble to assure Louis XVIII that he had the city's support. In return, Champollion-Figeac was decorated by the king, invited to wear the royal fleur-de-lis, an invitation he did not, and perhaps could not, refuse. This quick change of allegiances suggests something of Champollion-Figeac's political realism, but it also likely reflected the king's own sense that moderation, including the cultivation of Napoleon's former close advisors, would be needed to effect a secure political transition. Louis, too, feared the Ultras and was seeking to ally with more moderate voices and standpoints. For a moment it seemed possible to temper royal power with a modest stab at liberalism, with the hope of stabilizing the monarchy as state power expanded beyond the king to a group of his appointed ministers, according to a new constitution.[25] Champollion-Figeac's malleability had surprised even himself, as evinced by the tone of anxious wonder with which he described the fleur-de-lis ceremony: "Here I am," he marveled, reporting from Paris in a letter to his brother, "with a white ribbon and a little fleur-de-lys in my boutonniere."[26] The Treaty of Paris was signed on May 30. In Grenoble, the cessation of hostilities was greeted with uneasy relief.[27]

Thanks, no doubt, to his political flexibility, Champollion-Figeac won a series of professional advancements that put him in a better position to assist Champollion who, for his part, did not always approve of his brother's willingness to cultivate the powerful. In the middle of June, Champollion-Figeac was elected as a corresponding member of the Institut, and by July he was chasing a position as royal censor. To this end, he produced a pair of pro-Royalist publications: *The Utility of Censorship and the Need for Censors in the Authors' Interest* and on *Hereditary Monarchy and Constitutions*.[28] Despite these efforts, he did not get the position. What he did get was support, permission, and funding to publish the *Pharaons* work. As was frequently the case with the Champollion brothers, Grenoble connections smoothed the way. Permission to publish was secured through a circuitous appeal to the Ministry of the Interior. On the advice of his friend Eugène d'Arnauld, Baron de Vitrolles, a member of an ancient Dauphinois family who was presently reorganizing the army, Champollion-Figeac wrote Louis d'Aumont, duke of Aumont and Pienne, requesting authorization to dedicate the work to Louis XVIII.[29] *L'Égypte sous les Pharaons* was published in July, prefaced by a suitably fawning royal dedication. This outcome was less than ideal. Champollion's long-awaited book had finally reached print, but only because Champollion-Figeac had ingratiated himself with the king's men. Now he and his brother were in their debt.

CHAPTER 14

INDICATIONS

> You must know that, in our part of the country and at that time, there was nothing wrong with receiving money from [a mistress], provided you spent it *hic et nunc* and didn't hoard it. *Hic et nunc* was a phrase that Grenoble owed to its *parlement*.
>
> —Stendhal, *The Life of Henri Brulard*

There was nothing essentially new about the impulse to make sense of antique fragments in terms of their relationships to one another and to practices lost to history. Indeed, Champollion-Figeac had for the past decade made a career out of interpreting the ancient artifacts of Grenoble through this lens. But the impulse took shape within a specific context and drew upon the intellectual, social, and institutional resources available in that context. Antiquarianism provided Champollion-Figeac—and, to a different degree, Champollion—with a powerful methodological point of departure, as they and their colleagues interpreted ancient textual sources as pointers to the past rather than as works of art, objects of purely aesthetic value.[1] The position drew its coherence from the implicit assumption that the past must somehow be made to serve the needs of the present, Stendhal's hic et nunc—the here and now.

Somehow: for there were many ways of going about this. Sketching his own fitful history as a young Grenoblois during these years, Stendhal had even turned valorization of the hic et nunc into fodder for his satire. To be sure, the distinction between textual sources with links to reality and those with links to the imagination could be viewed merely as a shift in emphasis. Historically in-clined scholars could write about texts cast in various literary modes—ancient epic poetry, the lyric flights of the troubadours, the earliest dramas—or they could examine documents from the more prosaic realms of contracts, laws, and governance, statements and agreements written in a mode that might be called *indicative*. The latter had a special value, at least among Champollion's coterie. For them, such documents pointed to realities beyond themselves, to historically specifiable and perhaps even archaeologically verifiable practices;

as traces of the hic et nunc, they participated in lived experience and derived authority from their relationships to action, decision, and practice.[2]

In 1806, Champollion-Figeac published a short essay on a Greek inscription found on the temple of Dendera. This was not his first exposure to the monument, as his close collaboration with Fourier during the latter's tenure as prefect of Isère ensured that the treasures of the site, cataloged by Napoleon's savants, including Fourier, remained objects of sustained attention for their potential astronomical significance.[3] The 1806 essay concerned a damaged Greek inscription, transmitted by Denon in his *Voyage* (1802), on a lintel over a door in the temple's south wall. The damage, as well as the inscription's sloppy execution, made accurate transcription difficult, and there was a further problem with the translation given by Denon. His version seemed to consecrate the temple to Isis and other deities "in the Year 31 of Caesar," followed by a phrase Denon translated as "the School of Priests of the Emperor." Champollion-Figeac took issue with the latter. He used stylistic analysis, the coin record, his knowledge of Egyptian calendrical conventions, and Fourier's earlier work on the dating of the Dendera zodiac to argue that the inscription had been misconstrued. According to Champollion-Figeac, the inscription ought to have been translated, "In the year 31 of Caesar, on the 18th of the month sacred to Thoth."[4] The error, he averred, stemmed from the translator's failure to appreciate the author's original conservationist intention, which kept posterity foremost in mind. "How could these provincial magistrates—who, in the name of the people, solemnly swore their devotion to the sovereign—how could they denominate that sovereign in such a way as to make him unrecognizable," Champollion wondered, "and, while recording the date of this event, leave the date itself uncertain!" Behind his astonishment glimmers the practical historicist orientation we have just described in connection with Stendhal's hic et nunc. In deliberate contrast to the aesthetic aims of antiquarians, Champollion-Figeac focused on elements that pointed directly to lived experience—to the realities of place, time, and sovereign.

Six years later, Champollion was also probing concrete elements that might provide access to the practices of the past. On November 12, 1812, he submerged an ancient alabaster jar in a pot of boiling water. A little over a foot tall, the jar contained a thick, hard substance, like dried pitch; it was this that Champollion hoped to loosen by means of the hot soak. The jar's lid, carved in the shape of a jackal's head, was nearby. Hunched over the bubbling pot, Champollion jotted these notes:

A water bath
Finely chopped straw
The object is wrapped in a cloth

The object is four inches by two
Its animal nature is obvious
Fibrous tissue
Near the flame: an animal smell
And boiling balm, carbonized residue
Found at the bottom of the vessel
The object is impregnated and coated with a thick layer of unguent
Wrapped simply in cloth
Two little splinters of sea green Egyptian porcelain
There's a liver, a brain or brainstem[5]

Canopic jars had long been considered simple representations of either Egyptian gods or holy elements of Egyptian life. Champollion's discovery suggested something entirely different. Rather than being representations of "little gods, or symbols of the Nile's flooding, of water, of the origin of all things" or having any other purely sacerdotal significance, they were "simply vessels, part of the embalming process." That is, they were practical, functional objects that played various roles within an integrated system of meaning. The jars were part of a ritual, put into practice by ancient embalmers and understood by their clients, that made sense of the soul's separation from the body at death.[6] "These spirits," as represented by the different figures on the lids, "are emblems of the four great qualities of God [de Dieu]: the woman expresses the *goodness* with which the soul is received by his court; the hawk [is] *the god who gives life*; the jackal [is] *the god who bestows death* and the baboon [is] the god of *divine justice*."[7] The figures on the lids functioned as nonliterary textual remnants that gestured beyond themselves to a system of meaning. The idea of what one might term the communicative vestige—the trace that, interpreted correctly, opened up a world of meaning and practice—applied to writing, too. Texts could point to the practice of writing as a vestige of speech, and to the systems of rules that governed inscription.

In these years, speech was in vogue. New political and accompanying social forms demanded a language that could represent them fully and accurately in the real time of conversation. Louis-Sébastien Mercier's *Néologie* (1801), a catalog of neologisms inspired by contemporary controversies about the status of the vernacular, reflected an understanding of how social and political disruptions could inspire new forms of speech. Conceived explicitly in conversation with the 1798 edition of the *Dictionnaire de la langue française*, the fruit of the Institut's language standardization project, Mercier's was an effort to redress the Institut's "imprudence" in failing to adapt to new linguistic realities, communicative situations, and expressive needs. Mercier warned that such efforts to "nail down the language" would "crucify" vulnerable new forms of speech.[8] The fashion for neologisms soon emerged in Grenoble, where the multifaceted lexicographer Claude-Marie Gattel—who also served as a lawyer in the parliament of the Dauphiné, a professor of grammar at the École centrale, and headmaster of the Grenoble lycée—was assembling his *Nouveau dictionnaire portatif de la langue française.*[9] Following Gattel's death in 1812, the Champollion brothers collaborated on an updated edition, published in 1813, for which Champollion-Figeac also wrote a preface, while Champollion supplied information for words with "oriental etymologies."[10] Following Mercier, Gattel's dictionary contained a list of words added to the language after the Revolution. Primarily concerned with additions related to the new government, its calendar, and its new system of weights and measures, the lexicon also included such evocative gems as *fanatiser* (to radicalize), *galamathias* (confused, unintelligible talk), *chouan* (a kind of insurgent), and *enragé* (another kind of insurgent), a word for which Champollion, as we have seen, had already conceived an affection.

Linguistic variation took on a fresh salience as well. In 1807, Napoleon's interior minister commissioned the prefect of every *département* to record the parable of the prodigal son in the local patois. These could vary considerably. The parable began: *Un homme a deux fils* (A man has two sons). In the Catalonian Pyrenees, the locals said: *Un home tingue dos fills*; in Marseilles: *Un homo avié dous eufans.*[11] Despite these differences, inhabitants of the French countryside living within fifty to eighty miles of one another could shift easily between different dialects; one report claimed that the range of mutual comprehensibility extended for 250 miles.[12] In 1809, Champollion-Figeac prefaced his study of the dialects around Grenoble with a suggestive remark by the seventeenth-century historian Pierre Bonamy: "It is from the provincial vernacular that the French, Spanish and Italian languages are formed." Champollion-Figeac opined that "the truth of this statement, which one could see as paradoxical, will finally be recognized with respect to the French language." He agreed with Bonamy that the origins of French were not to be found in any dictionary or grammar but in the history of the various dialects spoken by "the peoples of the southern parts of Europe," including swathes of Spain and Italy. "These considerations," he continued, "suffice to suggest the great importance of these truly ancient languages," variants of which were still spoken in remote areas. This is the speech of a "celebrated people," he wrote, alternately "victorious and vanquished," who "never allow[ed] anyone to subjugate that Gallic freedom which made Rome tremble."[13]

It is difficult to say precisely how Champollion viewed the Egyptian scripts in relation to phoneticism at this time. Even his most direct statements on this matter tended to be equivocal, and he did not make many of them. To uncover the ancient Egyptian language, and through it Egypt's culture, was ever a principal goal for Champollion, but the only route to that goal lay through inscriptions. Whether they could be assimilated to some phonetic system remained an open question. At the start of 1813, Champollion apparently believed the cursive script found on papyri was phonetic; he further claimed that hieroglyphs were syllabic and that at least some were "alphabetic," though he had a particular meaning for the assertion. On February 11, 1813, in a much-reprinted fragment of a letter to Saint-Martin—one that, we shall argue, has been misunderstood— Champollion separated a handful of hieroglyphs he characterized as "alphabetical" from a larger set, which, whatever they might represent, were clearly images of familiar "natural objects": "In hieroglyphics there are two kinds of signs: 1. The 6 alphabetic signs indicated; 2. A considerable but definite number of imitations of natural objects."[14] By October 1813, Champollion had gone further. "I am more and more convinced," he wrote his brother, "that our bewigged-heads were befuddled by the claim that a hieroglyph expressed an idea. As for me, I already refused them the capability of expressing an entire *word*."[15] But if neither "ideas" nor "entire words," then what might hieroglyphs represent? A clue lay, he apparently thought, in the six "alphabetic" signs.

In the letter to Saint-Martin, Champollion had emphasized the representation of grammatical elements in Coptic, which he continued to believe was close to the ancient language: "Egyptian nouns, verbs and adjectives have no ending or rather a particular ending . . . everything is done by additions [augments] or prefixes." Grammatical inflections of nouns and verbs, he continued, "all revolve around the [six] letters" in question, and these "same letters are found in the hieroglyphs in their Egyptian alphabetical form."[16] If, then, these six had more or less the same function in ancient Egyptian that they had in Coptic, then they were not simply "alphabetic" in the same sense as the Coptic letters repurposed from Greek, in fact, quite the contrary. Champollion believed them to be syllabic in the particular sense pertaining to the language's semantic monosyllables since, as he remarked, the "hieroglyphic system is like that of the Egyptian language, which is entirely syllabic." Hieroglyphs accordingly must group together to produce sign sequences built out of the Egyptian language's monosyllables, with meanings that depend in various ways on those of these constituent elements. In this they differed profoundly, he thought, from alphabetic systems like those of Hebrew, Arabic, Greek, or Latin. "I shall begin," he promised, "by proving that the words of two syllables are composed of two other [words]. This analysis of the Egyptian language undoubtedly gives me the basis of the hieroglyphic system, and I shall prove it."[17]

Although Champollion had access to two copies of the Rosetta inscriptions, the *Vetusta* engraving and the one that would eventually appear in the

Description, he did not have perfect faith in either copy. To make further advances, he needed to secure his data, to be sure he was not making claims about passages from the inscriptions on the basis of faulty copies. On November 10, 1814, Champollion wrote to the president of the Royal Society of London with a request for "a plaster cast from a mold made on the original" artifact and confirmation of the accuracy of several passages in the engravings by means of direct comparison with the artifact (fig. 14.1).[18] He listed these passages on a separate sheet, now lost. With these materials he also enclosed the volumes of his *Égypte sous les Pharaons*.

It is curious that Champollion did not direct this packet to the Society of Antiquaries, which, after all, was still in possession of the Rosetta Stone. We have no evidence to explain the lapse. Several explanations seem plausible. Perhaps Champollion was motivated by prestige alone and wished to contact the most respected scientific organization in England. Perhaps he wanted to avoid the Antiquaries because he was reluctant to associate his work with the outmoded antiquarianism of the *Vetusta*'s primary audience. Perhaps he was unwilling to solicit an organization so directly involved in the British seizure of French antiquities after France's defeat. He may have heard that the Royal Society's president, Joseph Banks, nurtured an interest in Egyptian matters and so might be especially sympathetic to his cause.[19] Or perhaps Champollion had no idea where to send his letter and simply took a guess. Since he was in Grenoble, Champollion would not have known about Young unless de Sacy had written him, which was certainly unlikely given the political turmoil of the past months and Champollion's own reputation as a firebrand. In any case, Banks passed the letter on to Young, then the Royal Society's foreign secretary, whom Banks knew to have been at work on the Rosetta inscriptions for several months.

Champollion asked for a direct comparison of the passages "done from the two engravings" that he had transcribed on the now-lost attached sheet "with the monument itself." Young copied phrases that were apparently included on Champollion's now-lost list into the draft of his reply. It seems that he had asked for confirmation of one particular line in the intermediate script for which he had already found an equivalent, a Coptic root noted in his Coptic dictionaries.[20] There he equated the first character of the Rosetta Egyptian (read right to left, fig. 14.1, *bottom*) with the Coptic preposition NE and assimilated it to the phrase "beloved of" preceding a deity's name, for example, "beloved of Pta." Although Champollion did not report this connection in his letter, he did explain, in general terms, the use he had made of Coptic in writing his *Pharaons*. The central aim, he wrote, was to establish the ancient Egyptian toponyms, its rivers and towns, and in so doing to go beyond "the scattered and confused memories of the antique glory of this country." He had been able to accomplish this via "the books of the Egyptians from the Middle Ages, written in Egypt and in its ancient language." These Coptic texts formed the basis of Champollion's denominations, which he had then compared "with those that we owe to the Greeks and the Arabs." At no point did he provide an explicit justification for his interest in the specific passages from the Rosetta inscriptions that he listed. Leaving this much unsaid, Champollion closed his letter by boasting that if he had only had a cast of the inscriptions taken directly from the stone itself then he would already "have fixed the reading of the entire inscription."

On February 26 Napoleon escaped from Elba; three weeks later, on March 19, the Bourbons fled to Belgium, and the next day he once again took control of the government. When Napoleon stopped in Grenoble on his march back to power, the town's mayor recommended Champollion-Figeac to him as a secretary. Champollion-Figeac was made responsible for publicizing Napoleon's positions and was appointed editor of the *Annales de l'Isère*. Although Champollion-Figeac's administrative talents, connections, and experience were surely enough to secure him the position, Napoleon rather charmingly found an additional, etymological indication of Champollion-Figeac's worthiness. Recalling that *léon* appeared in the old spelling for *Champoléon*, Napoleon quipped, "It's a good sign; he has half my name."[21]

Egypt continued to preoccupy Napoleon who, through the mediation of Champollion-Figeac and Fourier, sought to cultivate Egyptological studies and Champollion's work on the subject. Likely assisted by his brother, Champollion drew up a proposal seeking Napoleon's support in the creation of an Egyptian grammar and dictionary.[22] Written between March 8 and 15, 1815, the proposal covered six intensively marked-up pages in draft. It opened on a mixed note, invoking philological as well as religious justifications for the proposed work, emphasizing the dual value of the planned Egyptian study for both the study of the Bible and for emphatically secular philology. The former justification was, perhaps, a nod to de Sacy, whom Champollion correctly expected to be among the proposal's evaluators. Pursuing the biblical angle a bit further, Champollion suggested that ambiguous words appearing in the Bible and in classical texts were possibly Egyptian in origin that had either been transliterated into Greek or written in Hebrew letters. He also suggested that some Arabic roots were also Egyptian, that the two languages were related idioms, in which case their relationship would be "of special interest to orientalists," that is, to de Sacy.[23]

Champollion made a number of familiar philological points as well. First, he praised the Coptic specialists of a generation earlier, such as Christian Scholtz,

À Monsieur le Président de la Société Royale
de Londres

Monsieur le Président,

Grenoble le 10 Novembre 1814.

J.F. Champollion le jeune

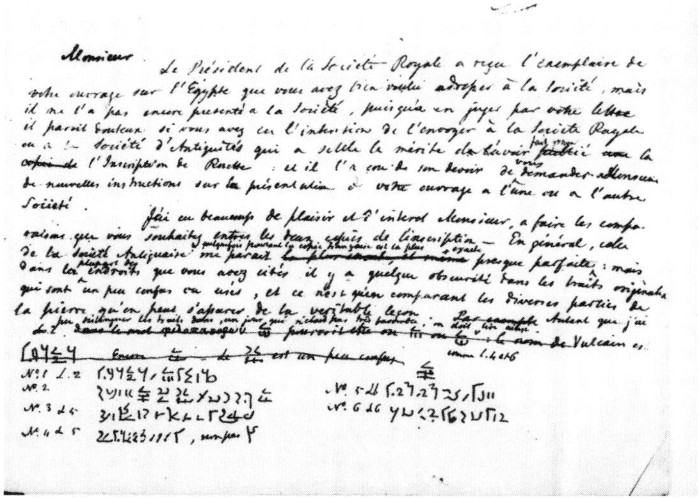

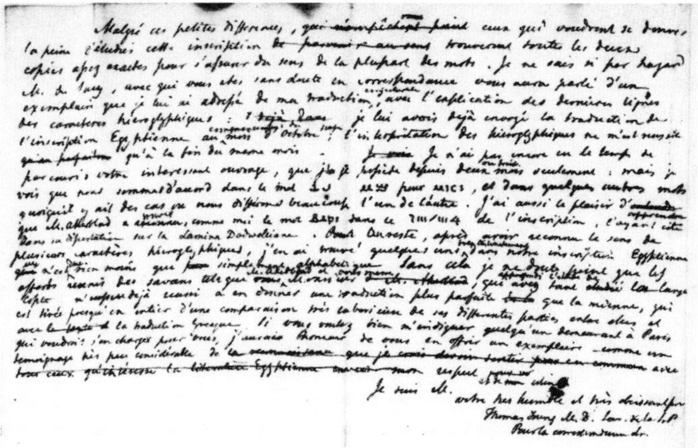

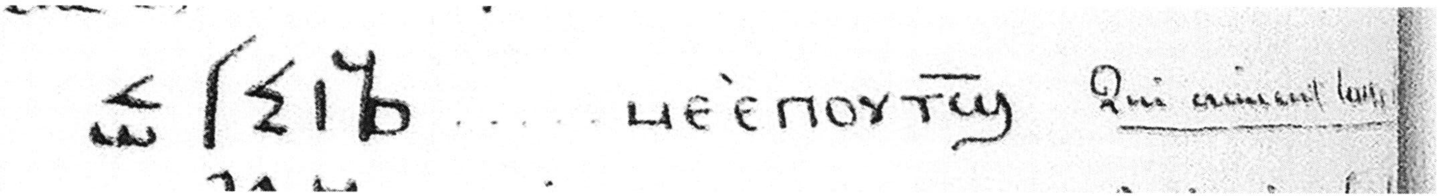

FIGURE 14.1. (*Top*) Champollion's letter of November 10, 1814, from Grenoble to the president of the Royal Society of London (© The British Library Board, ADD 21026 f15-f16); (*center*) Young's draft of a reply (© The British Library Board, ADD 21026 f15-f16); (*bottom*) from Champollion's (1812) notes (Bibliothèque nationale de France).

Carl Woide, and Valperga di Caluso, for their efforts to create comprehensive and useful dictionaries and grammars.[24] He then criticized Scholtz's work for being insufficiently attentive to the "particular genius and true development of the Egyptian language." In contrast, Valperga di Caluso had recognized its "regular," "philosophic," and "analytical" quality, but even he failed to develop the insight. Breaking with them both, Champollion intended to build up his dictionary from the simplest element, the linguistic root.[25] First, Champollion remarked, Egyptian was made of monosyllables and these monosyllables constituted the fundamental roots. "The Egyptian Language is monosyllabic," he wrote. "Each of these monosyllables is a root, and each root gives rise to derivatives and compounds; [a]ccording to this principle our Dictionary is a Dictionary of Roots."[26] He repeated the point, made in the *Pharaons* volumes and the 1811 *Introduction*, that Coptic sources provided the best route into ancient Egyptian: "The Egyptian writings, written in Greek characters, and called Coptic, preserve the Egyptian Language in all its purity, because the Elements of this language are so much combined that it can neither change nor corrupt itself."[27]

By separating Coptic source material from later influences, Champollion intended to reveal the language's deep structure in its essential simplicity and regularity. "Our Egyptian grammar . . . is free of the foreign formulas introduced into the Egyptian by the conquest of Egypt under Alexander," he declared. Eliminaton of these later elements yielded a result that represented "the genius of the Egyptian Language," one that was constructed "from the simplest" elements and that "gradually builds up to words composed according to the most abstract rules." Champollion claimed that thanks to this organic coherence, his grammar "presents the Egyptian Language in its primitive state of simplicity and regularity, that which can only be found in the MSS of Thèbes."[28] The grammar was as much a cultural object as a linguistic one. Just as an animal head sculpted on a canopic jar and an ascia carved on a tomb near Grenoble gestured toward systems of funerary practice, an understanding of the grammar of the Egyptian language could help would-be decipherers forge informative links between the written traces of ancient Egypt and the practices—religious, medical, agricultural—toward which those traces endlessly and mysteriously pointed.

Champollion applied to the Académie des inscriptions et belles-lettres for support for the proposed Egyptian dictionary on May 18, only to be rejected in July on many grounds. The evaluators struggled to understand the purpose of organizing the dictionary according to Coptic roots, and the proposed grammar seemed to them unlikely to surpass what Scholtz and Woide had already done.[29] An explanation of sorts was provided by de Sacy, who appreciated Champollion's focus on ordinary words but expressed deep reservations about everything else, for he was not ready to accept the distance Champollion sought between written signs and whatever they might stand for, whether phonemes or ideas. "It is not that we think, with M. Champollion, that the hieroglyphics are representative signs only of the sounds of the spoken language and not of the signs of thought," he wrote. "Without contradicting the testimony of all antiquity by a similar assertion, which seems to us very risky, we nevertheless believe that a profound knowledge of the synthetic progress of the Egyptian language and of the relationship between physical and sensible objects, such as the mouth, the hand, etc, and the spoken language, would help us understand the problem more clearly."[30]

Champollion was always sensitive to rejection, and this disappointment cut him to the quick. Privately, in a letter to his brother, he fumed about de Sacy's betrayal: "The Jesuit's report is just as I expected: the venom is concealed in sugar, one cannot attack the foundation, he will reject [it] on a formality. This is a caterpillar that, unable to bite and tear at a plant, contents itself with covering it with slime. None of the malice in the report has escaped me." In his distress, Champollion reprised his suspicions concerning de Sacy and Quatremère. "[De Sacy] is the master now," Champollion raved. "The Institut has welcomed my mortal enemy and you know very well how disastrous Polycarpe's success is to my prospects. I think it's useless to continue a struggle in which one must succumb sooner or later anyway; the spirit of faction will reign henceforth with more power than ever; the color of your cap will decide the contents of your mind."[31] By July 19, Champollion was alternating between resignation and despair: "I'm back to my usual work, but my heart's not in it. I'm distraught over the news. I firmly believe this is it for me. Hail Sylvestre [de Sacy] the all-powerful, [and] Polycarpe at the Institut: my path has been cut off. I'll work again, but without pleasure, without hope."[32]

Champollion's rejected proposal was hardly the worst of the troubles that now beset him and his family. The preceding spring had been much taken up with anti-Royalist political activity that exposed him to reprisals as Napoleon's hold on power continued to slip in the face of ferocious counterpressure from aligned forces of church and throne. With Joseph Rey, Champollion had founded a secret Grenoble outpost of the *fédérés*, a loose affiliation of pro-Napoleon paramilitary political associations active during the Hundred Days of Napoleon's return from Elba.[33] On June 1, 1815, Champollion was holed up at city hall with a handful of fellow dissidents, including the mayor Charles Renauldon and the law professor Jacques Berriat-Saint-Prix, uncle to Zoé, wife of Champollion-Figeac. From this redoubt, they issued a proclamation against the menacing coalition of monarchy and church. "Our country is threatened.

The heroes chosen by the French people are resisting the ambitious designs" of the coalition. "It falls to us," they railed, "to disrupt their operations" in order to prevent the "dismemberment" of "our *belle France.*" Meanwhile, Champollion expressed private doubts to his brother: "Our Éteignoirs [priests] always raise their heads; I'm afraid we won't be able to mow them down."[34] Despite having worn the royal fleur-de-lis just a year before, Champollion-Figeac was decorated for his service to the empire, an honor that now spawned new worries. Champollion, for whom the memory of the fleur-de-lis was still fresh, warned: "You will never be forgiven for your meetings with the Emperor or for your ribbon," the decoration bestowed upon him by Napoleon. "Your ribbon will become a *corde* around your neck."[35] This knowing failure to specify the precise nature of this *corde* suggested ominous possibilities: it might become a leash, perhaps a noose. Champollion was not above such punning gallows humor, and it was a dangerous time. But his warning would fall on deaf ears, as he well knew. Having by now grown accustomed to his brother's chameleon-like shifts, Champollion could only sigh, "It is a new misfortune."[36]

Inexorably defeats piled up. By June 22, Napoleon had abdicated for the second time. When news reached Grenoble on June 24, Champollion urged his brother, who had been apprised of the loss at Waterloo while attending dinner with Lucien Bonaparte at the Palais-Royal, to take any measures necessary to save himself.[37] Two weeks later, pro-Royalist Austrian fighters arrived on the outskirts of Grenoble. The city surrendered after a brief struggle and on July 9, the authorities began to round up the rebels.[38] Grenoblois lost the positions to which they'd been appointing during the Hundred Days, and Champollion-Figeac's university post was suppressed. In addition to suffering the rejection of his proposed Coptic dictionary and grammar, Champollion was now punished for his involvement with the *Annales*—considered dangerously liberal and which his brother had edited at Napoleon's request—for a pamphlet he'd written against the "calotins" (priests). By the end of the month, the Royalist flag had unfurled over Grenoble, and the brothers were under state surveillance.[39]

In early August 1815, Casimir de Montlivault was installed as the new prefect of Grenoble. He immediately set about eliminating the remaining anti-Royalists from public life by means of house searches, arrests, and imprisonments. Joseph de Barral, the eminent seventy-four-year-old former mayor of Grenoble and vice president of the Federation, was summoned to the prefecture to defend his involvement with the fédérés. Even as anti-Royalist forces rallied in the face of these affronts—by mid-September, Grenoble was awash in anti-Bourbon propaganda—the resistance only inspired fresh suppressions. In the following month came yet another purge, as Montlivault's Commission of Public Instruction dismissed seven Grenoblois from their university and government posts for being suspected fédérés. Among those dismissed were Berriat Saint-Prix and Champollion himself.[40]

As the year turned, Champollion's outlook darkened. On October 12, 1815, he wrote his brother, "This is not a good time for literary dreams. For your courage, I give you my compliments; as for me, I'm so disgusted with everything that I don't give two farthings for literary glory and all that flows from it."[41] Threatened by recent defeats and demoralized by the criticisms of his prospective father-in-law, who wanted his daughter to marry someone more gainfully employed, Champollion even briefly considered a new vocation—as a notary. The bitter irony: History in the indicative mode had triumphed so completely that Champollion entertained the idea of closing himself completely from a career in letters; the best he could do was notarize, creating the documents that later historians would examine and make sense of. They would have the privilege he could not himself attain; as a notary, he could, at least, provide future historians with their raw material.

One month later, just before he and his brother were remanded to internal exile at Figeac, Champollion had a premonition of the coming upheaval. No one could be sure of its precise contours, but one thing at least was clear. He wrote to his brother: "In any event, you, me, Bilon and Berriat Saint-Prix will be driven out." In an excruciatingly precise inversion of the violent fate he had imagined visiting on Grenoble's priests, he described their vanquished group of anti-Royalists as the ones who had finally been mowed down: "Nous sommes rasés."[42]

PART 3

SCRIPTS AND BONES

CHAPTER 15

SUMMER AT WORTHING

According to his later recollection, Young first became interested in the Rosetta inscriptions while working on the review of Adelung's *Mithridates*. "In reading [Adelung's] compilation," he recalled a decade later, "my curiosity was excited by a note [asserting] that the unknown language of the Stone of Rosetta, and of the bandages often found with the mummies, was capable of being analysed into an alphabet of little more than thirty letters."[1] This avowal notwithstanding, Young had long been aware of the Rosetta Stone, as he had years before subscribed to Thomas Walsh's 1803 account of the Egyptian campaign, in which Walsh had listed the antiquities extracted by the British from the French in an appendix.[2] Nor was this the first time Young had thought about Egyptian scripts, for he owned a copy of Denon's *Voyages*, which contained a plate depicting signs that seemed to be "manifestly compendious imitations" of hieroglyphs, with perhaps an admixture of alphabetic characters.[3] It seems he made no more of these early moments of engagement with the scripts at the time of the *Mithridates* review. The inscriptions only became objects of his lively interest somewhat later, when he had the opportunity to examine "some fragments of a papyrus, which had been brought home from Egypt by my friend Sir William Rouse Boughton, then lately returned from his travels in the East. With this accidental occurrence my Egyptian researches began."[4]

Young's friendship with Boughton was occasioned by the latter's chance encounter with an evocative remnant of ancient Egypt. Having graduated from Christ Church, Oxford, in 1811 Boughton embarked on a variation of the grand tour in which he avoided Italy, then under French control, and visited Egypt instead. "During my travels in Upper Egypt," Boughton later wrote, "I had the good fortune to meet with a mummy." This mummy, found in a catacomb near Thebes (present-day Luxor) was, Boughton recalled, "in a perfect state of preservation." Mummies were not rare, but what Boughton found with it was: a papyrus in excellent condition, covered in signs. After procuring a tin box in which to keep his find, Boughton proceeded overland to Constantinople, sending his baggage, including the box, ahead by sea. Unimproved by the marine journey, the box corroded, and the enclosed papyrus split into fragments, some of which luckily remained legible enough for Boughton to copy (fig. 15.1).[5] On April 24, 1814, he sent the Society of Antiquaries a letter presenting the "fragments" together with his "accurate copies" of them. Three weeks later, Weston read the letter to the society. The letter was accompanied by "remarks from the hand of a learned friend."

Between 1804 and 1820, Young and his wife Eliza summered at Worthing, a seaside town located about sixty miles from London.[6] At Worthing in 1814 Young was asked by Macvey Napier, an editor of the *Britannica*, to contribute a series of articles on topics of his choice to the *Supplement*. Young declined, fearful as always of damaging his medical reputation by appearing to be distracted by nonprofessional interests. "I feel it a matter of necessity," he wrote to Napier, "to abstain as much as possible from appearing before the public as an author in any department of science not immediately medical."[7] Another matter was on his mind as well. Sometime in the early spring, Boughton had persuaded Young—the "learned friend" mentioned in the letter to the Antiquaries—to add his remarks to the paper on the papyrus. He had agreed while choosing to remain anonymous, perhaps to avoid stealing Boughton's thunder.[8] Over the summer Young further developed his remarks, which he sent to the *Archaeologia* for publication the following November. The final result contained a section on the Rosetta hieroglyphic inscription. The "observations" that had accompanied Boughton's letter of the previous spring, Young wrote, "may be considered as preliminary to an attempt, which has since been made, to compare the three inscriptions of the stone of Rosetta minutely with each other," indicating that he first came fully to grips with the Rosetta inscriptions during the summer of 1814.[9] His expanded paper now included a "conjectural translation" of the Rosetta intermediary inscription together with an "interpretation" of portions of the Rosetta hieroglyphic inscription. A second publication, in May 1816, reproduced the *Archaeologia* remarks and "translation" of the intermediary inscription, but not the hieroglyphic "interpretation," in the recently founded *Museum Criticum*.[10] The *Criticum* material also included extracts of letters from Young to de Sacy and from Åkerblad to Young, all dating from 1814 through August 1815.

Nowhere in these published materials did Young explain how he went about creating his translation and interpretation. Nor did he privately explain his methods to anyone else, at least in letters that have been preserved. Although Young had notified de Sacy and others about his efforts by the end of that productive Worthing summer, he did not detail his procedures to them either. He would not provide such details until the publication in 1819 of a lengthy article on Egypt that he agreed to write four years later for a *Supplement* to

FIGURE 15.1. William Rouse Boughton's copies of Egyptian papyrus "fragments."

the *Encyclopaedia Britannica*, and even then his explanation was incomplete.[11] As we shall see, while Young did outline his methods in this article, he did not provide any examples of how he had gone about his work. Fortunately, the British Library holds an extensive manuscript begun by Young on July 13, 1814, that reveals the details of his methods. Entitled *Inscriptio Rosettensis: Memorandums of an Attempt to Decypher the Egyptian Inscription of Rosetta*, this document is the first concerted attempt by anyone to work on both the Egyptian and the hieroglyphic inscriptions of the Rosetta Stone. Young's methods, and the conception of the Egyptian scripts that inform them, remained substantially fixed thereafter.

Young's *Memorandums* betray an increasingly skeptical engagement with Coptic, as he had recently begun to mull over recent reprisals of claims, familiar since Kircher, of a special relationship between this language and the Egyptian scripts. In his review of Adelung's *Mithridates* the year before, Young had touched for the first time upon the possibility that Rosetta inscriptions were related to Coptic. He had done so because the "Copts and Egyptians demand the priority in treating of the inhabitants of Africa, from their early connexion with ancient history." Coptic, he went on, has a good deal of Greek in it but "the rest is probably old Egyptian." By that time he was superficially acquainted with the relevant work of Åkerblad who, as we have seen, was among the first to study one of the stone's scripts in ways that contemporaries found persuasive, and who was a Coptic specialist to boot. Even then, Young doubted that the the Rosetta scripts could be useful in uncovering the ancient language, remarking that although the "old Coptic" might have "some light" thrown on it "by the attempts of future investigators to decipher the inscriptions of Rosetta," they would have to do so "more completely than Mr. Ackerblad [sic] has done."[12]

Young later wrote to de Sacy that he had not read Åkerblad's work carefully before embarking on his own attempt at "decyphering [sic]" in the summer of 1814: "I had read Mr. Åkerblad's essay but hastily in the course of the last winter, and I was not disposed to place much confidence in the little that I recollected of it; so that I was able to enter anew upon the investigation, without being materially influenced by what he had published."[13] Though Young may not have carefully read Åkerblad by early 1814, he had seen enough to generate unease. This feeling, which intensified during the following summer, spilled into the remarks that accompanied the publication of Boughton's letter. The trouble had its roots deep in Young's much earlier reading on languages and scripts, in his extensive practice with orthographic representation, including his reconstruction of the Herculaneum papyri, and in his understanding of the nature of human vocalization and its representation.

Simply put, Young thought that Åkerblad had gone too far in his attempt to produce a thoroughgoing alphabetic transcription of the Rosetta Egyptian script. Some such signs might be present, he agreed, but "there is great reason to believe that these characters were more or less mixed with hieroglyphics on different occasions." In Boughton's papyrus he spied "two varieties of the old Egyptian character" in the "badly fragmented" pieces, though the difference seemed to be of size and not form (see fig. 15.1).[14] Comparing examples from the few extant papyri and "bandages of mummies" that had been printed, he found that "these manuscripts exhibit a greater diversity of characters than could be expected from the use of any one alphabet." Åkerblad could be entirely correct, Young reasoned, only if the letters in the Rosetta's intermediate script "have been combined and diversified in such a manner, as to present appearances of a much greater number." That is, if the Egyptian characters did correspond to letters, then the letters had been written in various ways—in various forms and with various phonetic values—within that one inscription. Other complications soon arose. For one, though Åkerblad's methods did work well for a spectrum of proper names, and even for "a variety of words closely resembling some which are found in the later Coptic," Young found it "extremely difficult to account for the non-occurrence of Coptic words, which must unquestionably be in the inscription." Moreover, words that should have appeared did not. For example, the Egyptian month Mechir, which in Greek has the synonym *Xandichus* or *Xantichus*, and which would refer to January in Coptic, was not present in the Egyptian script where Young expected to find it, given the Greek.[15]

Young looked to predecessors. Kircher was useless, "founded on nothing more than gratuitous and even improbable hypotheses." Zoëga "has not even attempted to enter into any investigations" concerning the hieroglyphs.[16] Horapollo provided symbolic significations in his *Hieroglyphica*, as of a lion for the Nile, "but none of these symbols is to be found" on the Rosetta, even though "the corresponding ideas occur several times in the Greek."[17] To use "external" sources to decipher these inscriptions was simply "hopeless," regardless of whether one focused on the Egyptian script or the hieroglyphic. But, Young hinted, internal evidence could be efficacious through "a careful comparison of the different parts" of the hieroglyphic inscription "with each other, with the Greek version, and with the Egyptian, when it shall have been sufficiently deciphered."

For Young the signs that formed the Rosetta scripts—both hieroglyphic and Egyptian—were best understood by retrieving their meaning without resort to phonetics. The sign sequences were to be treated as meaningful logograms— as words—regardless of what the corresponding sounds in Egyptian speech might have been. Though Young did think that the ancient language had a final descendant in Coptic, he was largely unfamiliar with the language, and, as we shall see, he doubted whether extant Coptic was similar enough to the language spoken, even in the centuries following the Alexandrian invasion, to be a reliable guide to the meanings encoded in the ancient scripts. At this point Young had little to say about the vocalizations of the Egyptian words that he identified, though we shall see that he did eventually reproduce a table with Åkerblad's alphabetic designations, and that he even added to it. However, in the absence of a way to translate sound into meaning—in the absence of a securely known language to which a putatively alphabetic script could be linked—Young felt that the only way forward would be to work with sign groups that constituted fully meaningful words, using a contemporary Coptic dictionary as, at best, a helpmate in so doing. At just about this same time, recall, Champollion had had much the same notion, though he had bound it to the idea that the signs formed concatenated groups of the language's semantic monosyllables—its fundamental "roots"—that were essentially the same as those in Coptic.

Even if the Egyptian script were alphabetic, which Young doubted, and even if the sounds of some of its characters could be divined by comparison with Coptic, Young believed that work should proceed by attempting to group the Rosetta signs in ways that make sense given the Greek lines, not least because he thought that the Greek was a translation of the Rosetta. Only then were comparisons with Coptic worthwhile. If the goal was to extract meaning, as it was for Young, then any possible phonographic significance of the signs, even if such assignments were found to be partially correct for certain words, was virtually useless in the absence of secure knowledge of the contemporaneous spoken language. Extant Coptic therefore likely differed too much from the ancient language to be relied upon in the first instance.

On July 13, Young began his *Memorandums* by comparing a fragment of the Rosetta intermediary script with the script found on mummy wrappings. "From the resemblance of some of the characters" in the Rosetta intermediary inscription "to those of the manuscripts found with the mummies, a chance is afforded that it may afford [*sic*] a key to the interpretation of these highly interesting remains of a remote antiquity," he wrote. For this reason, Young now believed that "complete knowledge of this remarkable inscription" was "an object of considerable importance." He claimed that the script, "which notwithstanding the apparent success of Mr. Åkerblad in the first steps of the investigation," nevertheless remained "almost wholly unintelligible." After ten years of stagnation, "it seems to be most eligible to consider Mr. Åkerblad's attempt as having altogether failed." Setting Åkerblad's work to one side, Young proclaimed his intention "to begin the process anew, without any regard to the partial interpretation which his labours have afforded, and without his remarks to hand to consult."[18] Still, Young knew that Åkerblad and de Sacy had worked with Coptic, and so during the previous spring he had learned the language as best he could from La Croze's dictionary and then used it to read the century-old edition of the New Testament in Coptic produced by David Wilkins.[19] What Young would do with the Coptic over the next days and weeks remained to be seen.

Young began by making "a correct copy of the Egyptian Inscription with more space between the lines than the original affords." He did so because he intended to write below the corresponding lines of the original and because the *Vetusta* plates did not "exhibit the circumstance which is very obvious in the original, that the terminations of the lines on the right are much more regular than on the left," a detail "which plainly shows that the writing proceeds from right to left, as all the Egyptian manuscripts published in the French collection [the *Description*] and elsewhere very clearly exhibit."[20] Knowing the sense and order of the Greek inscription, Young hoped to divine the meaning of the Egyptian inscription by breaking the Greek into sections and essaying fits to the Egyptian, going line by line.

The *Memorandums* enables us to follow Young's procedure. "I have divided," he wrote, "the 54 lines of the Greek as nearly as possible into 32 parts, this being the number of the Egyptian lines: so that in comparing the respective parts of both inscriptions, I shall be able to refer at once from one to the other, and shall consider the number as referring only to the division into 32."[21] The resulting division suggests that he constructed each of the thirty-two Egyptian lines by sequential selection from the Greek. He divided the fifty-four lines of Greek into thirty-two segments, each of which held together well as a group, that is, without breaking at places that made little sense. Such a method presumed the Greek to have been produced from the Egyptian since the latter's lines would have provided a template for a Greek version that captured, in reasonable sequence, the purport of the original.

The procedure recalls Young's experience with the Herculaneum papyrus. In the case of the papyrus, the original inscription was admittedly difficult to reconstitute, while the Rosetta Egyptian was comparatively intact. The incompleteness and material defects of the Herculaneum Greek had made it necessary to reconstitute missing sections by inserting words that fit the sense of the passage, or by reading a sign in a way that made sense. Young's knowledge of Greek had informed his reconstitution of the Herculaneum text. That same knowledge now aided him in redividing the fifty-four Greek lines into thirty-two in ways that, he thought, also made sense.

"After this preparation [of an initial division]," he explained, "the first step of the investigation must be to collect all those groups of E.[gyptian] characters which occur twice or more, and to search for some Greek words which are found an ~~corresp~~ equal number of times in corresponding parts."[22] At this stage, he attempted to identify a sequence of signs that might correspond to the Greek word βασιλειας (king), because that word, with its several derivatives, occurs in the Greek text some forty times.[23] Here a difficulty arose: Young could not find a group occurring with *exactly* the same frequency in the Egyptian script. In figure 15.2, a page from the *Memorandums* written on or shortly after July 13, 1814, the text adjacent to the two columns of numbers reads:

> Hence it seems probable that the word (I) must answer to "βασιλειας," unless another Egyptian word can be found that occurs at least as often— But it is impossible that the words can answer to each other in every part since the term does not occur in the Greek for L.8 to 17, while it is found 4 or 5 times in the Egyptian.[24]

The leftmost column of numbers in the figure is Young's identification, which he seems to have refined in the second column, yielding a total of "24 to 30" incidences. That result suggested that the sequence was reliable enough to use in places where the match seemed unambiguous. Young returned to the Greek in order to correct his initial division: "Having written over each quarterline

from one to the other, and shall consider
the number as referring only to the division
into 32 —

After this preparation, the first step
of the investigation must be to collect all
those groups of E. characters which occur twice
or more, and to search for some Greek
word which are found as ~~correct~~ equal number of times in corresponding parts —

Βασιλευς or its derivatives. 40+

Ι. υ1ʃ4ϸ 2. 1	(ι)		Hence it seems
1	1		probable that
	2		the word Ι υ
3	3	37	must answer
5	5		to Βασιλευς,
5	5		unless another
5	6		Egyptian word
6?	6		can be found
8?	8		that occurs as
8	8		least as often
10	17		— But it is
13	17		impossible that
14	17		the word can
15	18		answer to
16?	21		each other
17	21		to correspond
17	22		since the term
20	23		does not occur
21	24		in the Greek for
21??	24		L. 8 to 17, while
21	25		it is found by
22	26		its terms in
24	26		the Egyptian
24?	26		
25?	27		
26	27		
27	28		
28	28		
28	29		
29	32		
32			

occur from 24 to 30 times

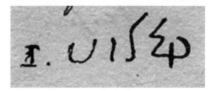

FIGURE 15.2. Young's first reconstitution of an Egyptian word, for the Greek βασιλειας (basileías).
© The British Library Board, ADD 27281 f3.

of the Egyptian text the portion of the Greek which appears to correspond most nearly to it in place of the words which had first been ~~compared in an approximate manner~~ placed by a similar approximation over the words, it may now be possible to examine somewhat of the structure of the language and the relations of its words."[25]

In these notes, as well as in his early letters to Gurney, de Sacy, and Åkerblad, Young referred to the inscription as the "Egyptian," but he would eventually insist on calling it "enchorial" rather than "Demotic," which was Champollion's preferred term. The nomenclature is significant. Young justified his choice on broadly economic grounds, claiming that it simply meant "the characters of the country" and was consistent with the expression used in the inscription itself.[26] After all, the last line of the Rosetta Greek contains the word (enchorial) and not δημοτικὰ (demoticà), meaning "rural, in or of the country." Since Young was determined to stick as faithfully as possible to the inscriptions alone, without recourse to anything beyond them, "enchorial" was obviously preferable. That Champollion chose otherwise has much to say about his very different conception of how to work with the available evidence: "Demotic" linked the script to Herodotus's terminology, and this, in turn, permitted Champollion to differentiate the Rosetta intermediate from a very different script, used on mummy wrappings, that would prove important for the eventual decipherment. Young tended to conflate the two scripts.

Having established a method, Young rapidly isolated further sequences. His second identification proved especially important, for it represented the Egyptian for Πτολεμαιος, the proper name "Ptolemy," for which the corresponding Egyptian sign sequence was already contested. De Sacy had suggested sequences that might correspond to this name, and although Åkerblad did not explicitly provide his own version, his Coptic-based alphabet might have yielded a different result (though such a version would be difficult to reconstruct since he had several variants for each Coptic letter). Although he was using a more accurate copy of the Rosetta inscriptions than had de Sacy, Young's sign equivalent nevertheless did not differ altogether from de Sacy's (fig. 15.3).[27] Young identified eleven instances of the sign sequence, of which eight closely followed the one that he had identified for βασιλειας.[28] Of this he remarked:

> This word certainly exhibits a most satisfactory coincidence occurring so nearly in the same parts of the inscriptions, and so constantly following the former [βασιλειας] in the proper place; this coincidence adds also much weight to the interpretation of I [βασιλειας][29]

FIGURE 15.3. (*Left*) De Sacy's sign sequence for "Ptolemy," followed by (*right*) Young's version.

That same day, Young tried to find a sequence corresponding to a Greek phrase that occurred several times, though he was less certain about the match. On the next day he pressed on with the same phrase, finding the "agreement amply sufficient to identify the sense of the corresponding phrases and to allow us to write a part of the Greek text under the corresponding lines of the Egyptian, although the order of words seems to vary too much to afford much assistance from this mode of comparison." Apparent variations in the phrase's repetitions provided a way to distinguish the separate words.[30] Further identifications followed, all based on pattern recognition in light of frequencies in the Greek. Young soon felt himself ready to produce a rudimentary glossary of the Egyptian cognates of Greek words and phrases (fig. 15.4, *left*), all generated without recourse to Coptic.

Over the following week, Young sought to understand variant forms of particular words. At this juncture he explicitly recognized the need to tolerate a degree of uncertainty as to sense. "It is necessary," he wrote on July 19, "to compare the different ways in which the same word and phrase is written in order to obtain all the forms of the same letter, and this may be done without understanding the exact sense of each any further than is necessary for ascertaining their identity."[31] Between July 19 and 23, he identified a total of eighty-six sign groups, almost all of which had several variants, usually slight, in orthography (fig. 15.4, *right*).[32]

Young tested his results by redividing the Greek once more in the light of new identifications. His first division must have been done entirely as to sense, since he did not initially have any way to isolate the Egyptian before identifying multiple incidents of the same sign sequence. Young apparently cut up his "correct" copy of the inscription once he felt able to link its signs at least reasonably well with Greek cognates. He then pasted the Egyptian lines onto a sheet of paper, separating them with blank spaces into which he wrote the corresponding Greek above and Latin below.[33] He also included paste-ups of the hieroglyphic inscription. In figure 15.5 (*top*) we have marked the *Vetusta* plate with Young's correspondences for the Egyptian lines, while the bottom figure is the first page of Young's cut-ups.[34]

As the summer ended, Young produced a Latin translation of the entire Egyptian script; an English version followed in the fall. These translations

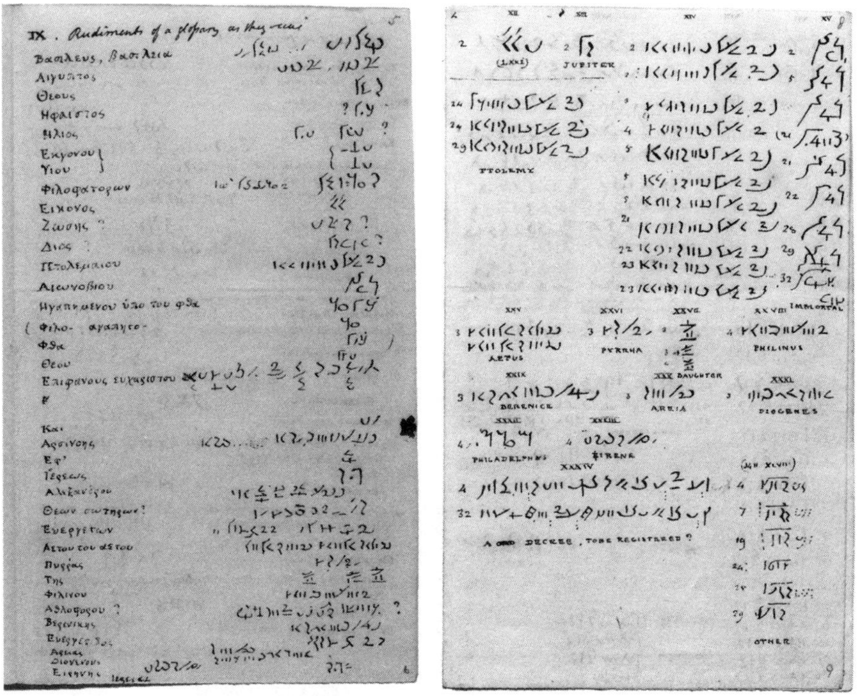

FIGURE 15.4. (*Left*) Young's "rudiments of a glossary" on July 15, 1814, and (*right*) an example of his July exploration of sign sequence variants, including the sequence for "Ptolemy."© The British Library Board ADD 27281 f8.

were not intended to be word-by-word or even phrase-by-phrase, because Young did not think that the script consistently conveyed grammatical or even syntactic information. To "translate" such a script accordingly meant to render its meaning in light of the identified sign sequences by using the Greek text to infer grammatical structure, imprecise as any such inference must necessarily be. His English rendition of the Egyptian appeared, as we remarked above, in both the *Archaeologia* and in the *Museum Criticum*; the Latin was never printed and remains only in manuscript form. The volume of the *Archaeologia* was compiled in 1817, but Young had copies of his piece available to distribute by the early spring of 1815. The issue of the *Criticum*, with its extracts of letters that Young had sent and received from August 1814 through August 1815, was available in the spring of 1816.

The *Criticum* paired Young's English translation with Gough's of the Greek as modified by Porson. It's worth having a brief look at the fifth and sixth Egyptian lines, with Young having placed his translation of the Egyptian to the

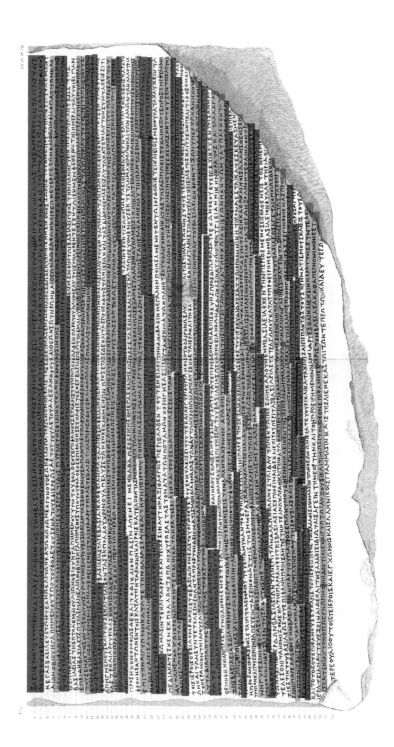

FIGURE 15.5. (*Opposite*) The *Vetusta Monumenta* marked to represent Young's divisions corresponding to each of the Egyptian lines, and (*above*) the first page of cut-ups taken from his "correct" copy of the Egyptian (© The British Library Board ADD 27281 f82).

left of what he took to be the corresponding Greek from Gough and Porson.[35] We can see from table 15.1 that he had adhered fairly closely to the Greek, principally inserting additional or modified words and phrases identified in the Egyptian sign sequences. Occasionally Young decided that the Egyptian cognate of the Greek should be expressed differently, and in doing so he revealed his conviction that the Egyptian text was not essentially alphabetic.

TABLE 15.1. Young's translation of the Egyptian script's fifth and sixth lines with his choices of the corresponding Greek from Gough's translation

Young's Egyptian (E) Line	The Corresponding Line(s) in Gough's translation of the Greek (G)
(E5) [to meet the king, at] the assembly of the lawful power of king Ptolemy the ever living, beloved by Vulcan, the god illustrious, munificent, succeeding his father; and who entered the temple of Memphis, and said: Whereas king Ptolemy, the ever living, the god illustrious, munificent, (son of) king Ptolemy	(G7) to Memphis, to the king, to celebrate the receiving of the (G8) kingdom of Ptolemy, ever living, beloved of Phtha, the god Epiphanes, gracious, which he received from his father, they being assembled in the temple in Memphis, on this day having decreed that (G9) as king Ptolemy, ever living, beloved of Phtha, the god Epiphanes, gracious, descended from king Ptolemy
(E6) [and queen] Arsinoe, the parent living gods, has given largely to the temples of Egypt, and to all within his kingdom, being a god, the offspring of a god and a goddess, like Orus the son of Isis and Osiris, who fought in the cause of his father Osiris; and being pious and beneficent towards the gods, has bestowed much silver and corn, and much treasure, on the temples of Egypt	(G9) and queen Arsinoe, gods Philopatores, has been in many things kind both to the temples and (G10) all in them, and to all placed under his government, a god descended from a god and goddess, as Orus the son of Isis and Osiris, assisting his father Osiris, well disposed towards (G11) [the worship of] the gods, has brought to the temples supplies of money and corn,

One example is particularly instructive. In the first Egyptian line, Young inserted the phrase "lord of the asp bearing diadems." A recent translation of the Egyptian gives "the Lord of the Uraei," while Gough's has "Lord of the Kings." The corresponding Greek refers neither to a Uraeus (the asp or cobra head ornament of a king), which would require ουραιος, nor to a diadem, which would require διαδημα.[36] Nevertheless, the Uraeus is certainly a cobra- or asp-bearing diadem.

Why did Young go beyond the Greek at this point? The Greek refers to "crowns" (κορώνες in modern Greek, ΚΥΡΙΟΥ in the Rosetta inscription),

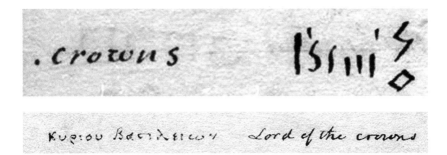

FIGURE 15.6. (*Top*) Young's translation of the Greek into "Lord of the crowns" on July 16, 1814 (© The British Library Board ADD27281 f6), and (*bottom*) his identification of the corresponding sign sequence in the Egyptian (© The British Library Board ADD 27281 f57).

and Young had in fact originally translated the Greek phrase as "Lord of the crowns" on or about July 16 (fig. 15.6).[37] But he knew that the Egyptian crown involved a diadem-like ring topped by a snake, for that was common knowledge. Furthermore, a similar sign appeared in Egyptian lines 25, 26, and 27, and in those lines the Greek has the word "asp" in close proximity.[38] Since Young considered the sign sequences to represent words or phrases but not speech per se, he seems to have rendered them as he saw fit within the limits he had already established. "Crowns" might accordingly become "asp-bearing diadems." The goal of such a "translation" was to find limited meanings rather than to render sounds or provide allegorical renditions.

Reflecting on this work, Young reckoned it a partial success. He wrote Gurney in 1814 that he had succeeded in identifying "pretty satisfactorily all the words that occurred [in the Egyptian script] more than once, and to ascertain their meaning." Even so, he could "read very few of them alphabetically," despite having had "so much reason," given Åkerblad's claims, "to expect a very general coincidence with Coptic." On that score he remained agnostic, explaining that he "made so few attempts to obtain an alphabet" that he "was not yet much discouraged." Since he had not worked an alphabetic line, he had "considered the whole as hieroglyphical"; that is, he saw the scripts as consisting of intricately formed logograms or even sematograms. On the other hand, Young recognized that if the scripts really did work that way, "I should have succeeded more rapidly than I have done, because the characters could be easily recognised when they occurred the second time."[39]

That notion naturally carried over into Young's work with the Rosetta hieroglyphs, which took place almost immediately after his grappling with

FIGURE 15.7. Young's strip for Egyptian line 32 with hieroglyph line 14. © The British Library Board ADD 27281 f100.

the Egyptian. He eventually pasted cut-ups of the hieroglyphs over what he thought to be the corresponding parts of the Egyptian (fig. 15.7).[40] The quantity of hieroglyphic script was considerably smaller than the Egyptian since much of it had broken off (see fig. 15.4). It consisted of fourteen lines, enough for Young to attempt an "interpretation" corresponding to the final sixteen lines of the Egyptian, which, as we have seen, he included in the *Archaeologia* piece together with his "translation" of the Egyptian (table 15.1).[41]

From the outset Young did not think that either of the two Rosetta scripts could be read by means of direct comparison to any writing system consisting of characters with stably fixed significations. Some hieroglyphic sequences might correspond directly to words and were consequently logographic, but not all were. Young may have seen these labile Egyptian signs as rather like

TABLE 15.2. Young's "interpretation" of the final two hieroglyphic lines with his "translation" of what he took to be the corresponding Egyptian

Young's Hieroglyphic (H) "Interpretation"	The Corresponding Egyptian (E) from Young's "Translation"
(H13) The priesthood of the god illustrious and munificent: and it *shall be lawful that the festival of the* king be *celebrated by all private* persons *disposed to honour him:* they may consecrate likewise a shrine to the king illustrious and munificent, and keep it in their houses, performing all manner of sacred rites both monthly and yearly: in order that it may be manifest that all the inhabitants of Egypt	(E30) with sacrifices, libations, and other honours: the priests living in the temples of Egypt, in every temple, shall be called *priests* of the god illustrious and munificent, besides the other sacerdotal names which they bear, in all edicts, and all acts belonging to the priesthood of the god illustrious and munificent: and it shall be lawful that the festival be celebrated . . .
(H14) . . . [w]ith due respect: and they have resolved to engrave on a column of hard stone, in sacred characters, in the characters of the country, and in Greek, the present decree; and to place it in all the temples under the dominion of Egypt, of the first, and second, and third order, wherever, shall be the image of the young king Ptolemy, the ever living, beloved by Vulcan, the god illustrious and munificent.	(E31) as it is just to do: and this decree shall be engraved on a hard stone, in sacred characters, and in Greek, and placed in the first temples, and the second temple, and the third temples, wherever may be the sacred image of the king whose life is forever.

algebraic ones. He expressed a stereotypically English dislike of algebraic reasoning in a famous and often quoted passage:

> [T]he strong inclination which has been shown, especially on the Continent, to prefer the algebraical to the geometrical form of representation, is a sufficient proof, that, instead of endeavouring to strengthen and enlighten the reasoning faculties, by accustoming them to such a consecutive train of argument as can be fully conceived by the mind, and represented with all its links in the recollection, they have only been desirous of sparing

themselves as much as possible the pains of thought and labour by a kind of mechanical abridgment, which, at best, only serves the office of a book of tables in facilitating computations, but which very often fails even of this end, and is at the same time the most circuitous and the least intelligible.[42]

The "enlightening strength" of geometry, as Young saw it, derived from the fact that geometric deductions refer to magnitudes whose mutual relations are fixed by virtue of their spatial representations. These qualities, for Young, endowed a geometric system of representation with "intelligibility." Algebraic expressions were not similarly "intelligible." An x might, for instance, represent a duration while a y in the same equation might represent a distance, whereas two lines in a given geometric diagram necessarily stand for the same kind of entity. The relations among algebraic signs consequently lack the clarity that spatial form provides to geometric expression. Such sequences cannot sustain, in Young's words, "a consecutive train of argument" that "can be fully conceived by the mind."

Though he never explicitly drew the connection, Young's view of algebra bears some resemblance to his conception of hieroglyphs. For him perhaps the most confounding feature of Egyptian hieroglyphs was their semantic lability. For example, a sign resembling an object might, in one sequence, refer to such an object, while in another, it might represent an aspect of a person associated with the object (as a scepter might signify such an object or, alternatively, the power of a sovereign). In that case, the relation of a sign to other signs in one sequence may differ from its relation to signs in another, making interpretation strongly dependent on context. Such signs were inferior to those of an alphabetic system for much the same reason that algebraic expressions seemed inferior to geometric representations.

In a related vein, the gap between hieroglyphic and phonographic systems seemed nearly unbridgeable to Young. Once even a limited phonography had infiltrated an originally hieroglyphic scheme—once sound replaced a sign's inevitably variable referent—the result was complex and unstable. Some signs might retain their labile referential values, while others would newly represent fixed sound sequences. Indeed, the same symbol might be used in either way depending on context. The primitive simplicity of semantical signs, however limited in their syntactical power, would accordingly be lost without, he thought, any true benefit of the precision that a thoroughgoing alphabetic system could provide.

The principal example of such a mixed scheme was, for Young, Chinese—a language that, like Coptic, was based in monosyllables, thereby rendering its original system of inscription semantic. Chinese, Young later wrote, as "the principal, and probably the most significant of the monosyllabic languages, is distinguished from almost all others by a more marked peculiarity, which is, that its written characters, instead of depicting sounds, are the immediate symbols of the objects or ideas . . . And in this point of view the Chinese would require to be classed with the old Egyptian only, since we know of no other language which was habitually expressed in hieroglyphics and their immediate derivatives."[43] In their essay on Marshman, Barrow and Staunton argued that although the introduction of phonography into the Chinese sign system depended directly on the language's monosyllabic structure, this shift occurred only partially, and only as the necessity of representing words foreign to the language became pressing. Such a transformation would never have otherwise taken place, or so their argument seemed to imply. As we have seen, Young studied this essay. To the extent, and to Young it seemed a very considerable extent, that the Egyptian language was structured similarly to Chinese, the same would have held for it. There could have been no phonography before the Alexandrian conquest.

A script that, like Young's Egyptian, represented words and not sounds had another advantage inasmuch as it might retain meaning through transformations in orthography even if the sign's referents changed. If the sign had not mutated so thoroughly as to make its visual connection to earlier versions unrecognizable, then the sign's semantic power might persist: despite its changes, the sign retained its semantic clout. Alphabetic scripts like Greek differed from Egyptian scripts because any significant orthographic changes would easily produce confusion. Garbling a sequence, for instance, robbed it of its semantic import because the sequence was only phonemes to begin with—a claim that was implicit in Barrow and Staunton's review of Marshman on Chinese. Years later, Young praised Barrow for "his clear and concise explanation of the peculiar nature of the Chinese characters," which had "contributed very materially to assist[ing] us in tracing the gradual progress of the Egyptian symbols through their various forms."[44] Although the symbols may have changed over time, they had always preserved their semantic values. Nothing similar, he thought, could happen with an alphabetic script.

All of this echoes Young's conviction that the Greek language and its writing system constitute a perfected exemplar of the links between sound, script, and vocal physiology. His early collaboration on Greek orthography with John Hodgkin had trained Young to think of well-formed writing not merely as elegant penmanship but as an expression of the tight connection between language as spoken and written and thought itself. Clear thinking required a proper language to express it, while the physical actions in speaking and in writing inevitably reflected the character of thought. The human vocal apparatus, Young had argued, produces sounds according to a precise, almost geometrical system akin to Newton's color circle, while the inscribing hand effects a similarly precise transformation of such sounds into script, provided that the language being sounded and written is well formed. Greek, for Young, certainly was well formed, while "old Coptic" likely was not, any more than Young thought Chinese to be. Grounded in meaningful monsyllables, such languages inevitably lacked syntactic precision. Translations from mind to sound to hand worked poorly here, inescapably leading to native scripts that had to be loose figurations of what words and phrases represented. Well-constructed alphabetic scripts, which required well-formed languages, embodied in contrast the very connections that these others lacked. Someone who thought this way about Egyptian language and writing was hardly likely to go much beyond establishing correspondences between native signs and meaning. Any elements of phoneticism in the ancient scripts were limited to the needs of invading Greeks and Romans to express their words in signs that the subjugated populace might read. After all, Young insisted, "the genuine language [of Coptic] bears very evident marks of great antiquity; its construction is simple and often awkward; and a great number of its words are monosyllables."[45] Ancient, awkward, simple, and monosyllabic: these were for Young the marks of a language whose speakers would not have developed a phonetic script without foreign influence.

CHAPTER 16

LETTERS FROM PARIS

Immersed in his work on the Rosetta scripts, Young could still be mistaken for a physician with curious extracurricular interests; de Sacy, however, was a star on the rise. He was installed as professor of Persian at the Collège de France on April 4, 1806, and in the course of that same year he dedicated his *Chrestomathie arabe* (1806) to Napoleon, who found him congenial despite his Royalist and religious sympathies. In 1808, de Sacy became the legislative representative for the department of the Seine; six years later, in March 1814, Napoleon granted him a baronetcy.[1] After Napoleon's abdication and during the conflict that followed, de Sacy's renown remained so great that Wilhelm von Humboldt arranged special protection of his property by the entering armies. On October 24, Louis XVIII named de Sacy royal censor, and the following February he became rector of the University of Paris.

Some months before, on August 20, 1814, Young took up his pen and wrote a letter to de Sacy in excellent French.[2] This letter was the first in an exchange that shows the main lines of thinking about the Egyptian scripts at this time apart from Champollion, who was in only intermittent contact with both men—though, as we will see, these communiqués were of no small significance. Young's letter arrived just as the restored monarchy coalesced, during a summer in which English tourists flooded Paris while the French public stood by, growing increasingly bitter and resentful about their situation.[3] Although nothing in this correspondence directly reflected these upheavals, the absence of any reference to them may well amount to a response of its own, a refusal on the parts of both men to sacrifice a shared scholarly pursuit on the altar of national pride. At any rate, in his letter Young inquired whether Åkerblad was still working on the Rosetta inscriptions, and whether de Sacy would himself be interested in receiving "some attempts of my own, which have enabled me to obtain a literal translation of the greater parts of the words, but without concerning myself with the value of the characters of which they consist; this mode of entering upon the investigation appearing to be by far the least liable to error."

Replying on September 23, de Sacy invited Young to send the work along. De Sacy also shared his dim view of Åkerblad's alphabet, clearly still angry that Åkerblad had not shared more of his work years before. "He never wanted to defer to my wishes," de Sacy complained. He disagreed with Young's opinion that Coptic was not particularly significant for the decipherment of the Egyptian scripts, and he advised Young that the middle script "alone can lead to the discovery of the value of the hieroglyphic characters." Along the way de Sacy warned of a certain Champollion who, having just published two volumes on the ancient geography of Egypt, and who also knew a great deal of Coptic, "claims to have read that [Egyptian] inscription." De Sacy would much like to see what Young had accomplished.[4]

On August 21, the day after writing de Sacy for the first time, Young sent a second letter, now apparently lost, to Åkerblad. Young apparently had heard that Åkerblad had returned to the Egyptian scripts and was studying "the Indian languages, with a view of facilitating that of the Inscription of Rosetta."[5] From Åkerblad's replies (which we will soon discuss in detail), it seems that with this communication, Young was probing to see whether Åkerblad had progressed with the Egyptian script over the years. Åkerblad's response was delayed by several months, no doubt due to Young's having sent the letter to Paris, not knowing that Åkerblad had been in Rome for the past decade. He learned of his mistake a month later, when de Sacy explained the situation in his letter.

Young labored over his next missive to de Sacy, making many deletions and insertions as he sought to convey just the right tone. This careful drafting suggests that Young wished to dissuade de Sacy from thinking that he had closely studied the letter by Åkerblad to de Sacy, even though it had been available in print for some time. Young defensively noted that he had only "hastily" looked at Åkerblad's letter "last winter" and that he had not then been "disposed to place much confidence in the little that I recollected" and so had not been "materially influenced" by it.[6] Young was, of course, sensitive to any claim that he had relied excessively on someone else's work, for reasons we've discussed; this defensive pattern would continue as he deepened his involvement with the scripts.

Young sifted through Åkerblad's work, finding that Åkerblad agreed with his own results in the "sense of the words which the author has examined," even though the two methods of arriving at such meaning were completely different. But Young disagreed with Åkerblad's contention that the Greek was the original and the Egyptian its translation. He thought the reverse was true chiefly because some Greek phrases abbreviated or missed words that Young identified conjecturally in the Egyptian. "It appears improbable that a translator should amplify in this manner the terms of his original," Young wrote to de Sacy, "although it is very natural to abridge them by the omission of superfluous repetitions." Their principal disagreement concerned the character of the Egyptian script. Åkerblad may well be correct, Young averred, in respect to his provision of several proper names, but this did not entail that the script must be intrinsically alphabetical. Conceding that "the value of the individual letters still requires much laborious investigation," Young noted that the inscription contained at least a hundred different characters; such a large number could not possibly be accommodated to Åkerblad's alphabet.[7] With this letter Young apparently included a handwritten "conjectural translation" of the Egyptian inscription. This was perhaps not the English translation that was available thereafter in the *Archaeologia* but, more likely, a Latin version that was never printed, the Latin guaranteeing that de Sacy could read the "conjectural translation" with understanding. (Young did generate a Latin "translation" at the time; see Young 1814b, 18. Cf. fig. 16.2.)

De Sacy would not reply until the following July. On October 21, in a third letter, Young notified de Sacy that although he had "succeeded in ascertaining the sense of the greater part of the words with scarcely any remaining doubt," he had discovered only a few "similar words in Coptic" that could capture a related meaning.[8] This may be, Young suggested, because the language had changed markedly in the half-millennium between the Rosetta inscription and that of "the oldest Coptic books extant." Young did not doubt the existence of at least a filial connection between ancient Egyptian and Coptic, but he set aside the possibility of a deeper connection that could be used to read scripts intended to represent the original language.

Young did furnish de Sacy with a list of eighty-six words that might correspond to the Egyptian script, accompanied by a list of their (partially) corresponding Greek-Coptic cognates. Once again, he arrived at his sense of the Egyptian by means of spatial correspondence with the Greek and by inferences drawn from his conviction that the latter was based on the former rather than the reverse. Despite his doubts about its validity for the pre-Alexandrian period, Young also provided de Sacy with a thirteen-sign addendum to Åkerblad's Coptic-Egyptian alphabet. Twelve of these signs were based on Young's scan of proper names from the Greek to the Egyptian. The thirteenth was "merely the mark of the termination of a proper name," a sign in the Egyptian for a proper name that did not have alphabetic significance but rather indicated that a sequence of signs had to be read in a certain way.

According to Young, the appearance of phonology for Greek proper names indicated nothing more than the insertion into a purely logographic script of foreign elements. The insertions were effected by mapping Egyptian to Greek sounds insofar as possible. In his letter to de Sacy, Young did not indicate how this was accomplished, that is, whether something like a rebus principle had been used, in which part or all of the sound of a word was detached from its sign. He instead merely listed both his and Åkerblad's Coptic-Egyptian alphabetic cognates, preferring as usual to stick closely to what he took to be the evidentiary basis. Later, in his *Britannica* article, Young relied upon a rebus-like connection to extract sounds from a limited set of signs, though he also simply redeployed Åkerblad's alphabet, albeit with modifications, for Greco-Roman names. Åkerblad himself had not used anything like such a principle, for he presumed that the demotic was, in both language and script, quite directly conformable to late Coptic. Less convinced that the language had remained so stable, Young was unprepared to allow more than limited use of the rebus principle in refashioning a sign.

And so matters rested until the end of November, when Joseph Banks handed Young a letter, recently arrived at the Royal Society, from a Frenchman who had some questions about the Rosetta inscriptions that could only be answered by someone with direct access to the artifact. With his request the correspondent had enclosed a copy of his latest publication, and a sheet of paper, since lost, listing the specific parts of the Rosetta inscription for which he required clarifications. The writer, of course, was Champollion, and tucked in the packet were the two volumes of *L'Égypte sous les Pharaons*.

Young, who had already received a copy of the *Pharaons* from Gurney on his way back from Paris, was primed to find little of value in the work. De Sacy had further softened the ground with his derogations of Champollion and the *Pharaons*. But Young had another reason to be disappointed, as he expected to find a translation of the Rosetta inscription in its pages. Young had told de Sacy that "a friend," that is, Gurney, had informed him that Champollion's work included "his interpretation of the Inscription." Of course, it did not—as de Sacy noted. "I am sorry to learn, from your account," Young continued, "that I shall be disappointed in the expectation of finding, in this work, the details which would have given me so much pleasure." To the extent that Champollion expected a sympathetic reception of his materials, he was badly mistaken.

After what was probably a monthslong delay, Young replied on Banks's behalf to Champollion's letter, asking first whether he should forward the *Pharaons* volumes to Antiquaries.[9] Young mentioned that he had sent de Sacy a copy of his "conjectural translation with explanation of the final lines of the hieroglyphic characters," having previously sent his "translation of the Egyptian inscription at the beginning of the previous October."[10] He continued that he had "succeeded" with the "interpretation of the hieroglyphics" at the end of that month. Young thereby established the dates, with epistolary backing, of his priority in Egyptian "translation" and hieroglyphic "conjecture," and put himself on record as understanding that Champollion had discovered essentially only what had appeared in his *Pharaons*.

Young turned next to Champollion's queries. He "had much pleasure and interest," he wrote, in comparing "the two copies of the inscription" that Champollion had said he possessed with the stone itself, "[i]nsofar as I've been able to distinguish the features on a day that was not very favorable." He remarked reassuringly that "those who wish to give themselves the trouble to study this inscription will find both copies sufficiently exact to be assured of the sense of most of the words." The *Vetusta* plate seemed to Young to be "nearly perfect" for the lines in question, though "sometimes," he conceded, "the French copy is the more exact." Still, there were sufficient obscurities on the original that "only by comparing the different parts of the stone can one be assured of the true lesson."[11]

Figure 16.1 reproduces the signs Champollion asked about, along with the copies Young made directly from the Rosetta Stone and their representations in print.[12] As Young noted to Champollion, there are subtle differences among these versions. Young had resolved them by comparing the signs in question with similar ones elsewhere in the inscription, all without any recurrence to Coptic. Champollion, in contrast, had already linked the signs to Coptic in work begun two years earlier on the language's "Theban dialect." That work, as we've discussed, involved the production of a dictionary consisting of what he termed "Egyptian roots."[13] In his letter, Champollion gave no hint as to the specific analysis, made in his unpublished Coptic notebooks, of this sequence from the Rosetta inscription.

Although Young had not yet studied Champollion's "interesting volumes," he had read enough to understand Champollion's convictions that Coptic was an essentially unaltered descendant of the Egyptian language and that the original Egyptian script had also been alphabetical.[14] The "Coptic language," Young could read in the *Pharaons*, "is really the Egyptian language written with Greek characters." He could also find Champollion explaining that "[t]he Egyptian alphabet, properly speaking, was composed of 25 signs. We know that the Egyptians used it until the period when they adopted the Greek alphabet. Of the 24 elements that compose the latter, 18 corresponded exactly to the value of so many Egyptian letters; the six others were foreign to [the Greek] language." Champollion remarked in a note that he had already labored over the Rosetta intermediary inscription and on "the alphabet which we have adopted" for it, but that he reserved "this important subject" for a subsequent work.[15] Nevertheless, he asked the reader's indulgence since he intended to

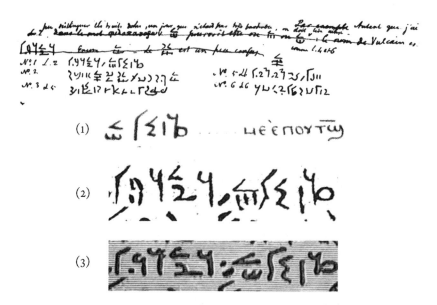

FIGURE 16.1. (*Top*) Detail from Young's reply to Champollion's request (© The British Library Board ADD 27281 f5). One of the requests concerns graphemes from the second line of the Rosetta Egyptian. The latter signs are reproduced from (1) Champollion's 1812 notes (Bibliothèque nationale de France), (2) the *Vetusta Monumenta* plate, and (3) the *Description de l'Égypte* plate made from the cast by Alire Raffeneau-Delile used by Champollion.

deploy some of his conclusions concerning the script, which he did by providing a considerable number of Coptic identities, and perhaps it was news of this list that had excited Gurney. Young was, of course, hardly interested in Coptic connections, focused as he now was on establishing the sense conveyed by the Rosetta intermediary through comparison with the Greek.[16]

In his reply of May 9, Champollion encouraged Young to keep the *Pharaons* volumes at the Royal Society, commended his "scholarly works on the inscription of Rosetta [and] the results you have obtained," and thanked him for his "extreme helpfulness." Which was all well and good—but Champollion had not as yet seen anything that Young had written. Although Young had mentioned his "translation" of the Egyptian, done "nearly altogether through a very laborious comparison of its different parts with one another," de Sacy had not yet given Champollion the pamphlet containing these materials. This delay is unsurprising: Champollion was in Grenoble and de Sacy was not well disposed toward him in any case. In closing, Champollion reiterated his wish for a cast of the Rosetta and instructed Young to send the missing pamphlet to his brother who, having rallied to Napoleon, was working in Paris as secretary to a delegation from Grenoble's electoral college.[17]

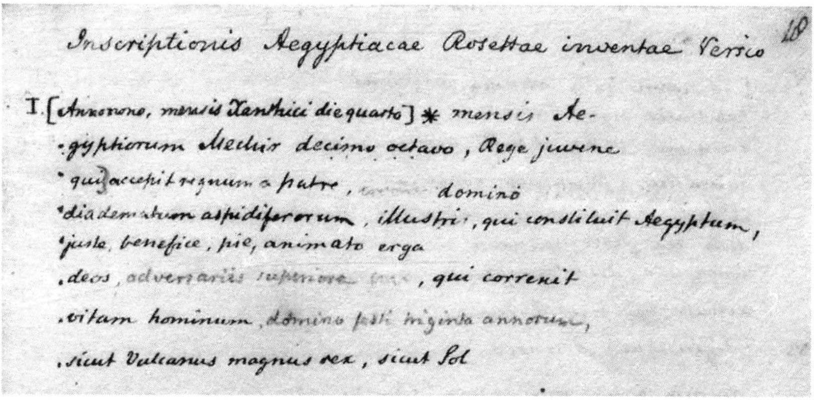
Inscriptionis Aegyptiacae Rosettae inventae Versio 18

I. [Anno nono, mensis Xanthici die quarto] * mensis Aegyptiorum Mechir decimo octavo, Rege juvene
qui accepit regnum a patre, domino
diadematum aspidiferorum, illustri, qui constituit Aegyptum,
juste, benefice, pie, animato erga
deos, adversariis superiore, qui correxit
vitam hominum, domino post triginta annorum,
sicut Vulcanus magnus rex, sicut Sol

FIGURE 16.2. Young's Latin "conjectural translation" of the section of the first Egyptian line that he matched to the Greek (de Sacy quoted the material in brackets). © The British Library Board ADD 27281 f18.

Two months later, on July 20, de Sacy wrote to Young after a silence of ten months. Much that could naturally be expected to delay correspondence had transpired in the meantime. Napoleon had escaped from Elba on February 26, and during the subsequent tumult, de Sacy prudently retired to his residence.[18] It was not until just after Louis XVIII had been restored to the throne for the second time that de Sacy returned to his neglected correspondence. In this letter, de Sacy referred to a "Latin translation" that Young had sent of the "Egyptian inscription."[19] No such Latin version of Young's "conjectural translation" appeared in print, but he must have sent a transcript in his letter of October 3 since de Sacy quoted from its first line (fig. 16.2).[20] De Sacy reported that he had also received "another English translation that I do not have at the moment, having lent it to M. Champollion at the request of his brother according to a letter that he said he had received from you." This was the missive delivered by Boughton. In closing de Sacy reported that he had just received it back from Champollion.[21]

Documentary housekeeping aside, de Sacy had serious questions about Young's working method.[22] He counted five major elements: first, a spatial comparison of the Egyptian with the Greek lines using a "compass" (de Sacy's word) to find corresponding loci; second, a search for frequent repetitions of a given sign sequence; third, work to determine thereby the "value of various character series" and to recognize their "correspondence with such and such a word or such and such a series of words in the Greek inscription"; fourth, from that and "with the aid of proper names, you fixed the value of a more or less large number of letters"; and finally, given these letters Young then had the "means to find other words corresponding to the Coptic language."

Despite de Sacy's close reading of Young's materials, their central import was lost on him, as he mistakenly imagined Young had sought Coptic analogs to the sign sequences he uncovered. Quite the contrary: As we have seen, Young had first produced his "conjectural translation" by way of a search for meaning given approximately corresponding loci between the Greek and the Egyptian. Only then did he search for an appropriate word in his Coptic dictionary. The resulting Coptic analogs were moreover highly limited, a point he made explicitly to de Sacy in his letter of October 21.[23] Although he kept his opinion from de Sacy, Young remained skeptical about the generality of the alphabetic letters that he himself, de Sacy or Åkerblad had divined except in special cases, especially the writing of Greco-Roman proper names. In his summer *Memorandums* he had conceded that much: "[o]n comparing the proper names with each other Mr. Åkerblad's mode of reading them appears to be in great measure justified." Nevertheless he wished to subject Åkerblad's alphabet to various tests, such as making "a comparison of all the forms and connexions of letters." He also wanted to "discover the names of the months" in order to see if it were possible "to apply his [Åkerblad's] alphabet to a single Egyptian word with any ~~probability~~ certainty of accuracy," which he doubted.[24] In his 1819 *Britannica* article Young explained that the writing of a particular sign or sign sequence with something like Åkerblad's alphabetic characters could only have occurred in the post-Alexandrian era for the explicit purpose of conveying words and names foreign to the Egyptian language.[25]

De Sacy was particularly mystified by an insertion that Young had made in the Egyptian text, as he explained in a letter written eight months later (July 20, 1815).[26] Young had inserted two dates into his Latin version of the *first line* of the Egyptian, where the stone is badly damaged. The first of these two, which is set in brackets (fig. 16.2), is "Anno nono, mensis Xenthici die quarto" (In the ninth year, on the fourth day of Xanthicus); the second, this time not bracketed, is "mensis Aegyptiorum Mechir decimo octavo" (the eighteenth of the Egyptian month of Mechir). The bracketing of the first added date indicates that Young did not think it was present in the Egyptian. The absence of brackets for the second addition suggests that he thought it was nevertheless readable as such on the stone. But the Gough-Porson translation includes the following dates in the *sixth line* of the Greek: "on the 4th day of the month Xandicus, *and* of the Egyptian Mechir the 18th decree."[27] In this instance, then, Young had repurposed the Greek by moving most of its sixth line into the Egyptian's first.

"What I cannot conceive," wrote the puzzled de Sacy of the unbracketed addition, is that "you could, by means of simple conjecture, without reading the Egyptian text, and without explaining it with the assistance of the Coptic language, recognize in the Egyptian inscription things that are not present in the Greek"—not present, that is, in the *first line* of the Greek, for de Sacy knew at least from Ameilhon that the date is present in its *sixth line*.[28] He reasoned that the only way Young could have arrived at the insertion was through an alphabet for the Egyptian script that was related to Coptic: "If," de Sacy wrote, "you in effect read the words ⲙⲉⲕⲓⲣⲭ ⲁⲃⲟⲧⲭ ⲛⲧⲉ ⲡⲓⲣⲉⲙⲡⲕⲏⲙⲓ," that is, the Coptic translation of Young's second Latin insertion, "I see the reason for your determination, but then you would have possessed the alphabet and would only have had to read to translate."

De Sacy had good reason for puzzlement. To him, Young seemed to behave contradictorily, at once embracing and rejecting the possibility that the script was alphabetic. In his second letter, of October 3, Young was explicitly skeptical about the possibility of assigning alphabetic significance to the majority of signs. Yet, despite these doubts and his professed lack of progress in this direction, Young noted that he had "found at least a dozen words which may be recognised" in the Egyptian signs. This unspecified dozen were certainly included in the much longer list of eighty-six words that he sent de Sacy eighteen days later. Young did not specify which words could and which could not be so interpreted alphabetically. In the absence of explanation, de Sacy became sure that Young had used Coptic alphabetic signs to move the Greek dates from their position in the sixth line to the beginning of his Latin translation of the Egyptian's first. He pointed out that although Young expressed little confidence in Åkerblad's alphabet, nevertheless "it would seem that you have made another that brought you the reading of a large number of words because you say that the language greatly resembles the Coptic or Thebaic."[29]

De Sacy may have understandably remained skeptical concerning Young's use of "Coptic or Thebaic"—and by extension of the possibility of an antique Egyptian alphabet—because in his October 21 letter Young did in fact provide many Coptic equivalents, together with Greek ones, for all the Egyptian sign sequences whose meaning he thought he could identify using his method. Most of these equivalents could not, however, readily be constructed from Åkerblad's alphabet even with the extra signs that Young had added to it. Particularly confusing for de Sacy would have been precisely the Greek month Μεχειρ (*Mekeir*) insamuch as it appears in Young's insertion for which he had given the Coptic as ⲙϩⲏⲣ and either of two possible Egyptian sequences, namely, ⲁⳋ, ⲁⳋ. Looking at the list from Åkerblad that Young had enclosed with his letter of October 21, de Sacy would have noted the inclusion of signs that (with a stretch) might be linked to those for Μεχειρ. He quite reasonably wondered just how Young could be so confident about his insertions—unless Young had produced a considerably expanded alphabet that he was not handing over. After all, despite Young's skepticism toward the applicability of any alphabet at all and his doubts concerning Åkerblad's version, Young had not repeated these reservations when he provided his list of words. This silence may have suggested, at least to de Sacy, that Young had progressed precisely along such lines. Young's customary politesse may have added to de Sacy's confusion. Knowing de Sacy's views on the matter, in his October 3 letter Young had tactfully left open the possibility (which he forthrightly doubted elsewhere) that future alphabetic success might be achieved through "long and laborious study."[30]

Despite his perplexity over just how Young had produced his Latin version of the Egyptian script, the following July de Sacy wrote that he would not press for his "secret," though he did hope to "have a foretaste of [the] discovery."[31] Setting this question aside, he admired Young's translation overall, remarking that it seemed probable because Young had sought to convey the repetitions of ideas and not to produce a set of repetitious phrases. If the translation proved to be correct, then Young might well be right that the Greek inscription was produced on the basis of the Egyptian. Altogether Young seemed to be much further along than Champollion, about whom de Sacy now issued his warning: "If I have one counsel to give you, it is not to communicate too much of your discoveries to Champollion. It could happen that he would then claim priority." In *L'Égypte sous les Pharaons*, de Sacy reported, Champollion "seeks to make one believe that he discovered many words in the Rosetta Egyptian inscription. I am truly fearful that this is only quackery; I add that I have many reasons to so think."[32]

Champollion was not the only one claiming the ability to read the inscription. According to de Sacy in this same July letter, an unnamed person in Holland also claimed to have discovered "the alphabet," while Quatremère in Paris "boasts that he can read much of the inscription." Backtracking somewhat from the position expressed in his 1802 letter to Chaptal, de Sacy found this latter possibility hard to credit. "I take it for certain that Coptic is not close to ancient Egyptian," he wrote, perhaps in deference to Young's explicit skepticism on the matter. Intriguingly Young included this comment in the *Criticum*, but omitted what followed. "I cannot persuade myself," he finished, "that if M. Åkerblad, Quatremère or Champollion had made real progress in reading the Egyptian text that they would not have quickly made it public. It would be a truly rare modesty, of which none of them seems capable to me."

As we have said, de Sacy's letter was very delayed: more than six months had elapsed before he answered. Distant for more than a decade by this time from the Rosetta issues, de Sacy would reasonably have decided not to reply until he could revisit his notes and earlier thoughts about the subject. De Sacy had distanced himself as well from his erstwhile client Quatremère, and he had widened his critique of Champollion's claim to have read the scripts in order to impugn his character. He also seems to have abandoned Coptic as a potential intermediary for translating the Rosetta Egyptian. These maneuvers were complex, reflecting de Sacy's awareness of a number of potential allies and adversaries, including Young himself. But the wider French context seems also to have played a role. During the Hundred Days, de Sacy had retreated to his country house, where he seems to have put aside the whole matter of the Egyptian scripts. After the Bourbons were restored to power, both Champollions meanwhile became targets of a backlash, as we will see. These changes may have further chilled relations with the Champollion brothers. Still, some trust remained: de Sacy had lent his copy of Young's *Archaeologia* piece to Champollion via his brother on the basis of the latter's conveyed request from Young, and the borrowed work had been returned just as de Sacy finished his July letter.[33]

Young replied quickly, on August 3.[34] He noted that the body of his enclosed materials had been completed "more than nine months ago" in continuation of his October 21 letter, but he had withheld it upon further news from de Sacy. In the meantime, he had been "very fully engaged in pursuits of a very different nature," an oblique reference to his major preoccupation of the intervening months during which he had been occupied with the completion of his massive *Treatise on Consumptive Diseases*.[35] By October 1815 he was assembling the material related to the Rosetta scripts that would appear in the May 1816 issue of *Criticum*, which he would in time send to both de Sacy and Champollion.[36] In his letter of August 3, Young finally gave a detailed account of how he had worked with the Egyptian and hieroglyphic scripts during this productive summer. To de Sacy he explained:

> You are at a loss to imagine how it was possible for me to recognise the words ⲙⲉⲕⲓⲣ and ⲁⲃⲟⲧ at the beginning of the inscription, without being in possession of an Egyptian alphabet. I answer, that the word "Month" is found several times very distinctly marked, in the 28th and 29th lines, and that having observed the same characters in the first line, with the epithet Egyptian, before the characters which answer to the word "Reigning," at the beginning of the Greek inscription corresponding to the the passage of the Greek which contains it, I thought myself fully authorized to conclude, that the Egyptian inscription began with the date; and this opinion was afterwards confirmed by the discovery of a similar group in the latter part of the inscription, where the date is repeated.[37]

Åkerblad had made a largely similar claim concerning the putatively missing date at the beginning of the Egyptian script. In a letter to Young sent the previous January 31, Åkerblad wrote, "I believe that the beginning of our inscription differed from that of the Greek, in which the date appears in the 6th line, while in the Egyptian the date seems to have been placed at the beginning, *as would be the case in all oriental writings of this nature* [emphasis added]" (fig. 16.3, *top*).[38] A careful student of Coptic, and unaware of Young's similar additions, Åkerblad had not inserted the dates on the basis of anything beyond a conviction that this would be the usual way to begin work in, to use his words, "oriental writings." His conviction rested on several linked assumptions. First, he was certain that Coptic—an instance of such an "oriental writing"—had to be very close to the Egyptian language. Its eventual script, based on the Greek script, even adapted extant Egyptian signs to express sounds not present in Greek. These convictions, coupled to the norm for date placement, meant that the dates had to have been there in the Egyptian.

Young's reasoning, which was altogether different, had nothing to do with Coptic per se or with any claim concerning date placement in "oriental writings." We can detect elements of his reasoning in his procedures for handling the scripts. For instance, because Young's Latin version of the Rosetta inscription immediately followed pages in the *Memorandums* that are altogether devoid of Coptic, we may infer that Young had inserted the dates into the first Egyptian line before he had even considered the script's relationship to Coptic. Though he did not explain the details of his cut-up procedure beyond his remarks to de Sacy, it's reasonably certain that he proceeded in the following way. First of all, when he first divided the Greek text to correspond to the Egyptian, Young decided to break the Greek corresponding to the Egyptian first line after its initial mention of "sun" (HEΛIOΣ) due to the solar deity's importance. This occurs at the end of the second Greek line.[39] Using the Gough-Porson translation, we find that the first Egyptian line should correspond to "In the reign of the young prince . . . restorer of the life of man . . . like the great Vulcan king, even as the Sun," with the second Egyptian line continuing "the great king of the upper and lower districts, descended from the Gods Philopatores."

The next step involved word counting in the Greek. Young identified Egyptian signs for the first word in the Greek text, βασιλειας (reign), from their multiple appearances in the Greek and the Egyptian. Such a sequence also occurred not far from the beginning of the first Egyptian line. Those signs were preceded by others. Among these was the sequence that appeared, as Young explained to de Sacy, "several times very distinctly marked, in the 28th and 29th lines [of the Egyptian]."[40] By word count Young had identified these signs as rendering the Greek for "month." That same sequence also occured in the extant portion of the Egyptian near the beginning of its first line. Hence Young thought it reasonable to move the second date in the Greek sixth line, which includes the word "month," to the head of the Egyptian first, putting the sixth

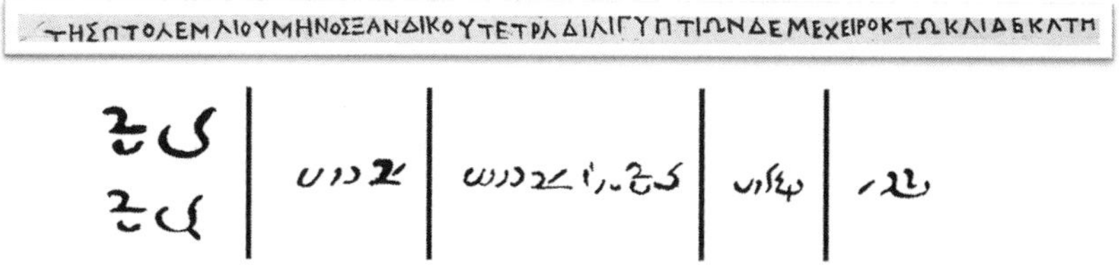

FIGURE 16.3. (*Top*) The dates from the sixth Greek line that Young included at the beginning of the first Egyptian (from the *Vetusta* plate). (*Bottom first*) The sequences for Egyptian "month" (both occur in Egyptian line 28); (*second*) "Egypt"; (*third*) "month of Egypt" from Egyptian line 1; (*fourth*) "reign" or "reigning"; (*fifth*) "Mekeir" (Μεχειρ) (all bottom images © The British Library Board ADD 27281 f97, f57, f82).

line's first Greek date ahead of it in brackets to signal that it was now missing but would have been positioned in the gap created by the broken-off piece at the upper right (see fig. 7.4, *bottom*).[41]

None of this experimentation required any knowledge of what Åkerblad called "oriental writings." Contra Åkerblad, Young remained unconvinced that extant Coptic was closely similar to its ancestor and that Coptic letters were more than superficially linked to the ancient Egyptian scripts. Indeed, Young would shortly write, "It seems probable that the introduction of the Coptic character was only coëval with that of Christianity."[42] Though Young *had* sent de Sacy additions to Åkerblad's alphabet, these were, in his view, limited to Greco-Roman words or their close analogs in Egyptian and were not, he thought, used for general Egyptian writing. Ever skeptical that anything like a full-fledged alphabet could ever characterize the Egyptian script, by the spring of 1815 his acute ability to recognize patterns—the foundation of his work—had also led Young to a striking result concerning the relationship between hieroglyphic and "enchorial" signs. He wrote to de Sacy on July 20:

The difficulty of the analysis [of the hieroglyphs], you will easily imagine, was not trifling; and I should not have been able to overcome it, but for the advantage of the intimate connexion between the hieroglyphic and Egyptian inscription, which, as you observe, might naturally be expected. . . .

But to return to the alphabet: after having completed this analysis of the hieroglyphic inscription, I observed that the epistolographic characters of the Egyptian inscription, which expressed the words God, Immortal, Vulcan, Priests, Diadem, Thirty and some others, *had a striking resemblance to the corresponding hieroglyphics; and since none of these characters* could *be reconciled, without inconceivable violence, to the forms of any imaginable alphabet, I could scarcely doubt, that they were imitations of the hieroglyphics, adopted as monograms or verbal characters, and mixed with the letters of the alphabet.*[43]

Here Young made three notable points. First, there was no way to force even the Egyptian characters for a significant set of words into an alphabetic mold. Second, many of the signs for these words seemed to bear "a striking resemblance to the corresponding hieroglyphs," though the differences between the two scripts were sufficiently striking that they "might be supposed to belong to different languages, for they do not seem to agree even in their manner of forming compound from simple terms." Third, the hieroglyphs were neither phonograms nor broadly interpretable signs of ideas but logograms, signs for specific words or phrases. At the time every other investigator of the Egyptian scripts was also convinced that the hieroglyphics were not phonographic. If Young's implications were correct—if the Egyptian characters were "imitations" of the hieroglyphs, however different the scripts might otherwise be in respect to the underlying language—then it seemed all but certain that they could not be alphabetic except when used to express Greco-Roman names or words. About this he now had little doubt. He was "not surprised," Young wrote in the same letter, "that, when you consider the general appearance of the inscription, you are inclined to despair of the possibility of discovering an alphabet capable of enabling us to decipher it; and if you wish to know my 'secret,' it is simply this, *that no such alphabet ever existed.*"[44]

The correspondence with de Sacy had been interrupted but at least it had resumed. Not so with respect to Åkerblad, from whom Young continued to await a reply to a letter he had sent on August 21, 1814.[45] Åkerblad, who had not worked on the scripts for a decade, may not have been eager to return to them. De Sacy's criticism had discouraged him; his diplomatic activities had proved absorbing; and he had, in any case, heard that many "literary persons in France, England, and Germany" had turned their attention to the Rosetta inscription. Åkerblad finally replied on December 15 when he sent a short note accompanied by a "Dissertation," which included a remark that he had identified the same sign sequence as Young for a particular Greek word, indicating that in his August letter Young had included at least a few of his own identifications. Six weeks later Åkerblad posted a second, longer reply.[46]

Åkerblad's certainty concerning Coptic and the alphabetic character of the Egyptian script had not altered in the decade since he had last addressed the question. He remained convinced that it was "only from the Coptic language that we can expect any assistance," with the principal difficulty being the finding of "words and expressions, hitherto unknown" in Coptic that might then be linked to the Egyptian inscription. The associations would be difficult to make for many reasons, not least the likelihood that the Egyptian language changed between the production of the inscription and "the earliest Coptic works which we possess." Young, of course, thought that Coptic had likely altered so much as to make such connections problematic. Moreover, Åkerblad persisted in believing that the Egyptian inscription was alphabetic. In adapting the Greek alphabet, he wrote Young, the Egyptians "did indeed add to it some of the old letters," but "their orthography remained vague and undetermined, as their [later Coptic] books demonstrate," making the process of identifying an appropriate set of letters for the original script that much harder. Moreover, "alphabetical characters, which without doubt are of very great antiquity in Egypt, must have been in common use for centuries." They mutated over time "as has happened in all other countries," Åkerblad ventured, eventually constituting "a kind of running hand."

Åkerblad's long letter of early 1815 contained his French and Coptic versions of the first five Egyptian lines of the Rosetta inscription, followed by a line-by-line discussion of his Coptic identifications together with some corresponding Egyptian sign sequences. Åkerblad's extraordinary performance in this letter reveals the manner in which he deployed Coptic together with shrewd conjectures concerning place-names and locales to generate the meanings of the lines. He did not, however, rely extensively on his alphabet. The alphabet—derived from the word identifications—remained sufficiently incomplete to provide comparatively little assistance in identifying further words. Still, the alphabet did lead Åkerblad to make several corrections to his previous identifications. In later extracting the letter for inclusion in the *Criticum* together with those from and to de Sacy, Young omitted many of Åkerblad's excursions and also added references back to the word list that he, Young, had sent to de Sacy the previous October (1814).[47]

Young did not reply to Åkerblad until August, nearly seven months later. In the meantime, he completed his *Treatise on Consumptive Diseases*. Although work on that project alone may account for the delay, Young may have been reluctant to court controversy by engaging Åkerblad further. Despite Åkerblad's persistence with Coptic, and despite Young's agreement that a few signs would have been repurposed for Greco-Roman words, he remained unconvinced by Åkerblad's admittedly ingenious performance. "I must now confess," he wrote, "that all the learning and ingenuity, which you have displayed . . . only serve still more to convince me of the extreme hopelessness of the attempt to read the Inscription of Rosetta, by means of any imaginable alphabet, into tolerable Coptic, and of the necessity of adhering strictly, in the first instance of the plan, which I have adopted, of comparing the inscription with itself and with the Greek only."[48] Young agreed, overall, with Åkerblad's renditions of the first five Egyptian lines, but they had only cemented his view that "it is evident, from your mode of treating the subject, that you have been very little, if at all, indebted to the Coptic, for the sense which you attribute to any of the words." Anyone reading this statement when it appeared later in the *Criticum* would think that Åkerblad must have worked much as Young himself had, with Coptic introduced at the end, and the "alphabet" an after-the-fact addition tied essentially to the Coptic construal. The implication was clear: the alphabet was valid if and only if the latter were correct—which was unlikely, in Young's view, since he did not think Coptic was a reliable guide to the spoken language of ancient Egypt.

Young listed twelve differences with Åkerblad's readings of these five initial lines of the Egyptian script. The first is notable simply for its minute detail, showing just how certain Young was that word order and meaning were the surest guides to understanding. According to Young's English rendition, Åkerblad had translated the date in the first line of the Egyptian script discussed above as "Month of Egypt 18 Mechir." Young objected, finding the separation between the epithet "Month of Egypt" and the month's name "Mechir" by the day number "improbable," because in Egyptian line 27, the character that stands for the number 30 is immediately next to the month called "Mesore," while in line 28 the number is also next to the month in question (in the latter case, "mensem secundum celebrari"). These reasons, he continued, are "stronger than any connexion you can discover between the characters and the sounds." Do not rely, Young insisted, on links between sign and alphabet— for these, being spurious, will inevitably lead to the wrong result. Figure 16.4 shows Young's manuscript signs, with Latin beneath, for the relevant parts of the two lines in question.[49]

Young made additional critiques based on the occurrence elsewhere of signs that he interpreted differently from Åkerblad, including, in one instance, his identification of a sign signifying the plural. All pointed to Young's deep skepticism about the usefulness of Coptic and of any generally applicable alphabet. It is worth quoting his remarks, for they are among his clearest objections to working in any way other than by location, Greek, sign repetition, and divination of such things as plurals and pronouns:

> With respect to your illustrations of the inscription from the more modern Coptic, I shall only observe in general, that as you have seldom expressed any great degree of confidence in your own conjectures, you cannot be surprised if I have still less disposition to be satisfied with them. The nature of my objections, in many particular instances, will occur to you from the inspection of the readings which I have attempted in my letter to Mr. de

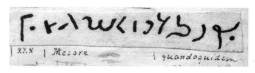

Line 27

Line 28

FIGURE 16.4. Young's rendering of the signs in Egyptian line 27 for "XXX Mesore" and, in line 28, for "mensem secundum celebrari VIII X XXX diem." © The British Library Board ADD 27281 f 96, fo7.

Sacy: among these, however, you will observe several words which have also occurred to yourself; and such a coincidence, as far as it extends, cannot but be satisfactory to us both: *but I apprehend that if you had simply made a complete alphabetical enumeration of all the forms, which you have been obliged to attribute to the respective letters, even in the first five lines, you would yourself have been alarmed at the inextricable confusion of heterogeneous elements which you have unavoidably introduced.*[50]

What did Young mean by an "inextricable confusion of heterogeneous elements . . . unavoidably introduced"? Simply this: Åkerblad's alphabet, presumably even with Young's additions to it, scattered the putative signs higgledy-piggledy among other signs that could not be so interpreted. Nearly every sign sequence, identified for meaning in Young's fashion by location and repetition, that contained such "alphabetic" characters also contained others that could not be matched in Åkerblad's or any other substantially phonologic list.

In addition to his piece for the *Museum Criticum* with its letters to and from both de Sacy and Åkerblad, during this period Young completed a piece for the *Quarterly Review* on two recent books, one concerning language by John Jamieson, the other on Moses by Joseph Townsend.[51] Jamieson, trained in theology, was a pastor in one of the Scottish Secession churches who was instrumental in effecting the 1820 reunion that became the United Secession Church. Educated in Edinburgh as a physician, Townsend was also vicar of Pewsey and a defender of biblical inerrancy. The second volume of Townsend's *Character of Moses* attempted to confute such expanded chronologies as would be required by James Hutton's uniformitarian geology on the basis of putative linguistic affiliations. Though Young congratulated Townsend on his "pains" in collecting such "a large mass of very interesting materials," he was less impressed with the newfound evidence on behalf of Moses. "We are not aware," Young continued, "that he has, by any original researches, contributed much to confirm the 'probability,' which we are by no means disposed to call in question, that in the period subsequent to the deluge, and prior to the dispersion of mankind, the whole earth was of any one language." As ever, Young pointedly insisted on probative evidence for claims concerning the origins and affiliations of languages and scripts, and Townsend's spate of comparisons did not come close to meeting the demand. Jamieson was hardly better, though he had "confined himself to a much more limited department" in an effort to demonstrate that the Greeks, Romans, and Goths were all descendants of Scythians. His arguments were both historical and etymological. Young found neither persuasive. Although the historical was "much less unexceptionable" than the etymological, even the latter seemed "almost entirely imaginary."[52] Imaginary as well, of course, was the full-scale alphabetic structure with its Coptic affiliations that Åkerblad had found in the Rosetta Egyptian. Given his own strictures on probative evidence, to the end of 1815 Young felt that he lacked sufficient grounds to probe the connection he suspected between the hieroglyphic and Egyptian scripts. He thought that the Egyptian must be an "imitation" in some sense of the figurative signs because at least some of the Egyptian characters seemed similar to the corresponding hieroglyphs. But how far did this "imitation" reach? As yet he did not know.

CHAPTER 17

THE PAPYRI OF THE *DESCRIPTION DE L'ÉGYPTE*

The tumult of 1815 reverberated across the Channel, unsettling the status quo in many realms, including the study of ancient Egypt. Earlier that year, Young had heard from several Paris specialists—Åkerblad, Champollion, and de Sacy—with the results we have discussed. On April 21, less than a month after Napoleon had retaken the reins of government, Jomard, the editor of the *Description de l'Égypte*, visited London where his "interest and curiosity" had been aroused by Young's "success," as Jomard perceived it, in the "interpretation" of the Rosetta Stone.[1] On returning to Paris, he wrote up his thoughts on the matter and conveyed them to Young, whom he had not met in London.[2] Jomard's material arrived at the end of April, touching off a period of intensive work by Young on papyri printed in the French collection. Of particular interest to Jomard, though not to Young, were what the Egyptian scripts revealed about the culture of ancient Egypt.

Hardly a newcomer to the hieroglyphic problem, years before Jomard had published a short piece on Egyptian writing in the *Description* in which he conveyed the prevailing opinion in France concerning the ancient scripts.[3] He first of all rejected any claim that this writing encompassed three distinct types.[4] Instead Jomard argued that, since the word "hieratic" meant sacred, it must refer to characters that were essentially the same as the hieroglyphs. These, he insisted, citing Porphyry, were allegorical and symbolic, while the "popular" or "profane" script, visible on the Rosetta, was a different form of writing entirely. As such, it had nothing directly to do with the hieroglyphs, though he conceded the possibility of some loose graphical connection between them.[5] Unlike the hieroglyphs, which Jomard considered purely symbolic, the "profane" script was altogether alphabetic, indeed *the* original alphabet: "The invention of the alphabet is commonly attributed to the Phoenicians, on the testimony of several authors; but the Egyptians can claim their right to the glory of such a beautiful discovery."

To explain why Clement and Porphyry referred to "two types" of sacred characters, namely, hieroglyphic and hieratic, Jomard asserted that the latter differed from the hieroglyphs sculpted on monuments in precisely the same manner that cursive handwriting always differs from carved letters. "At base," he concluded, "the signs were the same," which made the hieratic script also completely nonalphabetic.[6] This conclusion had immediate practical import. To decipher the hieroglyphs scholars should therefore turn to the large papyrus written in these cursives, which Jomard would include in the *Description*.[7] Jomard recommended matching characters, considered as symbols, with what could reasonably be conjectured about Egyptian religious beliefs (e.g., the scarab might signify regeneration).

Had Young read this essay—which is doubtful—he would not have been impressed. He did not think the hieroglyphs were altogether symbolic in the manner Jomard described; he believed many were also nonphonographic logograms for specific words or phrases. He did, however, read the draft of Jomard's piece for the *Description* on the numerical and other signs of the Egyptian, in which Jomard provided what he called an "exact catalog," a "complete series of all the forms" of the script, classified in a "methodical order." A tremendous believer in the revelatory power of exact measurement, Jomard freely projected his faith in exact measurement onto the culture of ancient Egypt. "He was convinced," writes one historian, "that the Egyptians had been innately geometrical and that a rational system lay behind all the ruins of Egypt. He further believed that this system was decipherable and would give considerable insight into ancient Egypt and the influence of ancient Egypt on the modern world, if one but knew how to study it."[8] Generalizing that attitude by considering hieroglyphs and their allied cursive as elements in an organized system, Jomard set out to produce a "rational" categorization of them.

Seeking "a line that could direct him through this maze" of signs, Jomard scrutinized the inscriptions available to him, paying special attention to the monuments on which they were inscribed. He classified the signs according to three principal categories: first, by their general composition and distribution, such as the directions in which the signs point, whether they are disposed vertically or horizontally, and so on; second, according to "the nature of the signs," for example, their frequency, whether simple or complex, and what objects are figured; and third, according to possible meaning as suggested by the proximity of signs on tablets to depicted scenes. This analysis resulted in two elaborate tables eventually printed in the 1822 delivery of the *Description*. Jomard divided the graphemes into eleven classes, ranging from "human figures" through "complex signs or combined figures" to "legends or phrases enclosed in a line and usually called scarabs." These classes comprised three major categories: "works of nature," "works of men," and combinations of the two. The resulting tables, Jomard noted, had already been substantially completed while he was in Egypt busily measuring monuments. A particular discussion of numbers, always Jomard's special interest, followed. If all the numerical signs could be found, why "would we not have a hope of then making entirely new discoveries in ancient astronomy and in history"?[9]

Jomard sent at least the substance of what would be printed in the *Description* to Young, though not his "methodical table" of hieroglyphs. In return, Young promised to forward remarks he was "going to have printed on the Rosetta inscription, without however entering into the interpretation of the hieroglyphic part." This was the *Criticum* piece that would go to press the following fall (1815). Young notified Jomard that he intended to continue his work once he'd received "copies of other fragments known of trilingual stones."[10] Salt, now the consul-general in Egypt, had promised to help. Young added that he had learned from Clarke's travel account of one more such stone stashed "in the house of the Institute in Cairo." Though the stone was badly damaged, Young was eager to see it. Having dispensed with these niceties, Young turned to his main purpose: to request copies of Jomard's plates with their ordered sequences of hieroglyphs from which Young hoped to make an "Index Hieroglyphicus."[11]

Young handed his reply to Boughton, who delivered it to Jomard in Paris where the English were arriving in droves following the second Restoration, much to the consternation of Parisians whose feelings of grievance needed little encouragement. John Scott, an Englishman, noted the intense reactions provoked by the removal from the Louvre, "a palace of their capital from which they were excluded," of artifacts acquired as spoils of Napoleon's military triumphs. "Every Frenchman looked a walking volcano, ready to spit forth fire. Groups of the common people collected in the space before the Louvre, and a spokesman was generally seen, exercising the most violent gesticulations, sufficiently indicative of rage, and listened to by the others with lively signs of sympathy with his passion."[12]

Still busily finalizing his table of hieroglyphs in advance of its engraving, Jomard in his reply to Young on August 17 made no mention of the disruptions in Paris and focused instead on Egyptian matters. In response to Young's request, he described a "trilingual monument" that he had seen at Menouf in Lower Egypt. Like the Rosetta, this was a large piece of black stone bearing three inscriptions. Of these, one was clearly Greek; the other was written "in letters like those on the Rosetta stone." Like the damaged stone Young noted in Clarke's account, this one was badly preserved.[13] "Unfortunately its characters are all poorly visible, and one can distinguish almost nothing even in the case of the Greek," Jomard explained. Even so, some observations could be made, and some inferences hazarded. "Since the three inscriptions fill the space in a regular way, one must conclude that they were carved at the same time, and in consequence that at that epoch the hieroglyphs were not altogether lost." Jomard further suggested that "given the analogy of this monument with that of Rosetta," the three inscriptions carried the same sense.[14]

Mildly discouraged by the prospect, growing ever more remote, of finding another readable example of the Rosetta inscription, Young informed Jomard that he had been copying every hieroglyph that he could find. He had not yet heard anything from Salt, charged by the government "at my urging" to spare "no expense" in finding and bringing back any useful fragments. Although the French government had sent Salt a copy of the extant volumes of the *Description* "to aid his research in the antiquities of Egypt," as yet Young did not have access to a set. Plaintively remarking that "the more I pursue this work the less I find myself confident at ever arriving at a perfect interpretation of all the characters," he nevertheless looked forward to the printing of Jomard's table. It is perhaps indicative of Young's isolation from the Egyptological ferment of Paris that, despite having exchanged letters with Champollion and others on the subject, he claimed, strangely, that such a table "would interest me no doubt more than any other person."[15] On the other hand, since his correspondence with Champollion had concentrated on the Egyptian inscription and not the hieroglyphs, perhaps Young's insistent interest in the table merely reflected the extent to which his focus had narrowed. Young included a copy of his *Archaeologia* paper with the letter.

Jomard answered on February 25 (1816), thanking Young for his "ingenious remarks on the Egyptian languages." Though Jomard had not yet carefully read the paper, he had absorbed enough to see that Young agreed with "many of our scholars" concerning Åkerblad's efforts. Jomard flattered Young, telling him he was the only one to have "gone so far in the discovery, or at least in the publication of such interesting results." No doubt Young would eventually solve the entire puzzle, Jomard seemed to suggest, if only further Rosettas could be found. For this, Jomard continued to count on the good efforts of Salt. Having laid his faith to the production of new sources, Jomard next thoroughly hedged his bets with the Englishman, agreeing that Coptic was insufficient to "read Egyptian" and promising to send the catalog of hieroglyphs on which he was working once it had been printed.[16]

In addition to exchanging letters with Jomard, Young had engaged de Sacy in a drawn-out correspondence on the nature of Egyptian hieroglyphic and Chinese scripts. The previous summer Young had sent de Sacy the long piece in which he had, as usual, downplayed the usefulness of Coptic. After a six-month delay, de Sacy finally replied on January 20, pleading that he had been saddled with "a surfeit of distasteful occupations" that had kept him from his correpondence.[17] Turning to the matter at hand, de Sacy agreed with Young concerning the putative structure of the ancient Egyptian language. Observing that its nouns lacked declension and its verbs conjugation—two points on which they agreed—de Sacy averred that all "primitive languages" are so constituted "so long as a nation retains the almost exclusive use of hieroglyphics." As Chinese had "never escaped the yoke of hieroglyphs," it remained "primitive" by de Sacy's lights. Egyptian never truly emerged from this same state, though at some point alphabetic characters began to be used "concurrently" with the figurative signs, but even so the development of grammatical form had "scarcely begun." On these points, de Sacy thought, he and Young were in accord.[18]

Beyond this excursion into Chinese, de Sacy remained unconvinced by Young's method of reading and unpersuaded by his identification of additional letters, though he still seemed to think that the Egyptian text was phonographic, such signs having been introduced at some point well after the nonalphabetic hieroglyphs. But he phrased his letter carefully, couching his skepticism in elliptical and respectful language: "It would be reckless of me to censure the work of people who have devoted so much time to the examination and analysis of this monument." It does seem that de Sacy either did not recognize, or preferred to overlook, Young's conviction that the script in its pre-Alexandrian form could not have been alphabetic at all, and that the letters that Young had added to Åkerblad's list were produced in antiquity not to write the Egyptian language per se but principally to capture the sounds of Greco-Roman names and words lacking Egyptian equivalents.

To be fair, anyone reading Young's *Archaeologia* piece might have been confused by just what Young did think about the Egyptian script. For though the *Archaeologia* demurred explicitly on the question of Coptic, it had not addressed the general question of letters, and Young's inclusion of the additional signs in his October 21 missive to de Sacy suggests at least a tentative pro-alphabetic stance. Yet Young had overtly rejected this stance in a letter to Åkerblad the year before. This letter—included in the first *Criticum* article, which de Sacy had not yet seen—asserted the "extreme hopelessness of that attempt to read the Inscription of Rosetta, by means of any imaginable alphabet." Yet even here he had added "into tolerable Coptic," leaving open the possibility that some other rendering into letters might work.[19] Then, on May 4, Young received copies of his *Criticum* piece, one of which he forwarded to de Sacy the next day with an utterly explicit note: "The *enchorial* inscription is assuredly not *alphabetic* except in a very limited sense."[20]

In mid-May, de Sacy reported that Åkerblad now expressed "doubts about his Egyptian alphabet" and worried that he had been a bit "too positive and too assured" in the letter that he had permitted Young to include in the *Criticum*. Conveyed "in confidence," this communication should not, de Sacy warned, be passed back to Åkerblad. Perhaps to secure Young's silence, de Sacy larded his letter with fulsome compliments. These encomia contrasted markedly with de Sacy's criticisms of Åkerblad, whom he found insufficiently forthcoming, and of Champollion, whom he found pretentious and politically unsavory. To be sure, de Sacy had good reason to ingratiate himself with Young. The foreign secretary of the Royal Society of London with the admirable reputation as a natural philosopher might be just the one to crack the problem that had defeated him, and a prominent friend in England was surely useful insurance against further political disruptions in France.[21]

To his credit, Young evinced no sign of encouraging de Sacy's sycophancy. More significant for Young was an engraving of a manuscript "in Egyptian characters" that de Sacy included with his letter. Young in short reply thanked him and announced that he had compared the characters on the manuscript with "the hieroglyphs of the great papyrus in the *Description de l'Égypte*," which, we shall see, he had obtained not long before. About de Sacy's engraving he wrote "that lines four, five and six agree nearly perfectly with" those in the *Description*. Furthermore, a grapheme at the end of de Sacy's line five and the beginning of line six corresponded in the Rosetta hieroglyphs "to the epithet Epiphanes." This, he continued, had an equivalent in the "cursive script" (on mummy wrappings) of a certain sequence, and a somewhat different but graphically quite similar series of signs on the Rosetta enchorial itself. (Young allowed significant pattern variation among the enchorial sequences provided that at least several sign elements appeared in every version and that there seemed to be enough strokes to effect a reasonable match.) Applying this method, Young found "Epiphanes" six times in the Rosetta Greek and identified by count and position the corresponding loci in both the enchorial and the hieroglyphics (fig. 17.1).[22]

With this letter Young intended more than merely to announce the identification of "Epiphanes" in the enchorial—of which de Sacy would hardly have been surprised to learn, given its multiple appearances in the Rosetta Greek. Rather, he intended to establish the existence of a graphically similar sign sequence in the cursive script. What Young did not explain was just why this close similarity between the cursive of mummy wrappings and the enchorial was so important. In the early months of 1816, he had obtained a copy of volume 2 of the *Description* that contained engravings of three papyri, one in hieroglyphs, and two in the "cursive hand." The new material had striking implications for

Enchorial Word List

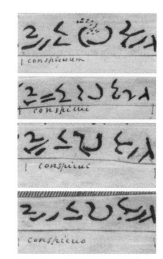

From the Enchorial+Hieroglyphic Strips

FIGURE 17.1. Young's enchorial and hieroglyphic signs for "Epiphanes." © The British Library Board ADD 27281 f 57, f100, f82, f91, f92, f98.

the relationship between all three scripts, and consequently for the question of whether or not the enchorial could be fundamentally alphabetic. Young's remarks to de Sacy hinted at what he had found in those plates.

Jomard, we noted above, had visited London in early 1815 to promote the *Description* volumes (and to take a cast of the stone).[23] As the *Description*'s de facto publicist, he had his work cut out for him. In London the distributor was to be the bookseller Thomas Payne, assisted by a Frenchman by the name of Béfort who lived near Payne's store. Béfort proved ineffective, as Jomard discovered. To help sales along, he disseminated an advertising prospectus unhelpfully printed only in French.[24] Sales were further impeded by both the great expense of the volumes and the extreme complications of their production.[25] The *Description* material had arrived in two distinct "deliveries," that is, two separate printings spaced several years apart. The first group, dated 1809, was printed in 1810; the second was dated 1812 and printed in 1813. Three full sets of antiquities plates were eventually available for purchase, comprising volume 1 from the first delivery, and volumes 2 and 3 from the second.[26] Each plate was placed in an envelope, while the text volumes were delivered in folio gatherings. A total of 843 plates were planned, of which 800 had been printed by 1814, ahead of Jomard's trip.[27]

It remains unclear how many of these sets Payne managed to sell, though he did have one of the costly vellum sets available by April 1815, and according to the terms of the sales agreement, he was to be sent either nine or ten additional copies (presumably of the cheaper version). Documents concerning the *Description*'s distribution preserved in Paris indicate that the duke of Wellington received a set in April 1816, while the British Museum received another at about the same time.[28]

So, although the *Description* plates and text had been available for purchase in England since around 1810, Young was unlikely to have more than glimpsed them for quite some time. Early in 1816 he obtained a copy of the second volume of the plates from William Richard Hamilton, who had been undersecretary for Foreign Affairs since 1809 and who had ensured the safe transport of the Rosetta Stone to England after being sent to Egypt at the behest of Thomas Bruce, Lord Elgin (whose acquisition of the Parthenon frieze he later assisted).[29] For the first time Young now had access to a large number of hieroglyphic and "cursive" signs. The volume lent by Hamilton included four color plates of a papyrus with five hundred columns of hieroglyphs; many of the columns were headed by figures of what appeared to be priests and gods. A second papyrus occupied six further plates, this time with columns in cursive also headed by various figures. A third filled six plates with columns, also in the cursive, headed by poorly preserved figures.

Comparing the beautifully colored plates with the other papyri, Young noticed that the columnar headings on all three papyri bore signs that were quite similar to one another. These similarities led him to suspect that the accompanying texts might also be similar, in which case he might be able to identify corresponding sign sequences between the hieroglyphs and the cursive. Young made a careful, sign-by-sign copy of the papyrus's five hundred columns, numbering each and writing notes on the copies as he did so, using French for information gleaned from the *Description* and Latin for his own comments.

Turning to the cursive papyri, Young sought to identify columns that, in them, were headed by figures similar to those that headed columns in the hieroglyphic one. He found several in the cursives that seemed to match reasonably well with plates 74 and 75 of the hieroglyphs. He then set to work copying the

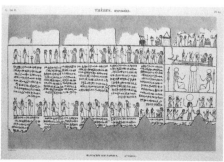

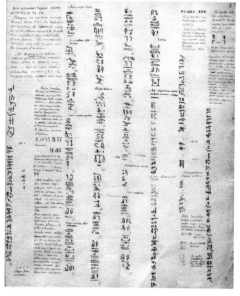

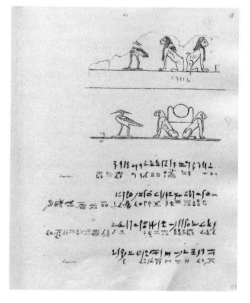

FIGURE 17.2. (*Top left*) One plate each from the color hieroglyph and the two "cursive" papyri; (*top right*) first page of Young's copy of, and notes on, the *Description de l'Égypte* color papyrus (© The British Library Board ADD 27283 f2); (*bottom*)Young's comparisons of hieroglyphic plate 75 with "cursive" plate 71 (© The British Library Board ADD 27283 f35, f38).

cursive texts, producing dozens of pages on which he placed hieroglyphs directly beneath what he decided were the corresponding cursive signs, with the comparison becoming more precise and sure-handed as he went along (fig. 17.2).[30]

By means of these comparisons, Young elaborated the link between the cursive script and the hieroglyphs. Here was the substance of the matter at which he had hinted in his letter to de Sacy. Because the cursive script had to be closely related to the hieroglyphs, and because it seemed to match the enchorial, it followed that the enchorial had to be strongly connected to the hieroglyphs as well. Young had for over a year suspected as much, but at the time he could only "scarcely doubt" the connection.[31] Now he had probative evidence. Agreeing with his French correspondents that hieroglyphs were nonphonographic logograms, he declared that neither the cursive nor the enchorial scripts were alphabetic. Any alphabetic elements that either possessed must have been introduced during Greco-Roman times. Young had at last found conclusive evidence for what he had long suspected, evidence made possible by the printing in the French *Description* of three sets of papyri.

In early August, not long after finishing this research, Young described the work in detail to Archduke John of Austria. Formerly a field marshal who had risen through the ranks of Napoleon's continental army, the archduke was for many reasons an unlikely confidant, but he'd met with Young while visiting England in July 1816, and he was interested in the Rosetta question. "I have now fully demonstrated the hieroglyphical origin of the running hand [the enchorial]," Young informed the archduke. Having "made copies of the respective passage in contiguous lines [of the *Description* hieroglyphic and cursive papyri] . . . I found that the characters agreed throughout with each other, in such a manner as completely to put an end to the idea of the alphabetical nature of any of them." That being so, Young continued, it might be possible "to translate back the whole of the running hand of the Stone of Rosetta into distinct hieroglyphics, and thus to compare it with a multiplicity of other monuments." But imperfection of the "imitation" of hieroglyphs by the running hands (i.e., of both the mummy scripts and the enchorial) foreclosed this result. The enchorial equivalents, for instance, required the addition of a sign.[32] Furthermore, he continued, "a loose imitation of the hieroglyphical characters may even be traced, by means of the intermediate steps, in the enchorial name of Ptolemy." That Young could perceive such a "loose imitation" testifies to his skill at tracing graphic patterns (fig. 17.3, *top and center*).

By the late summer of 1816, Young had decided that the cursive script was directly linked to the hieroglyphs, that the enchorial was a degraded form of the cursive (i.e., the hieratic), and that none of the three could have had phonographic signs until after the Alexandrian conquest, and then solely for writing Greek and Roman proper names and the occasional foreign word.³³ He conceded that "something like a syllabic alphabet may be discovered in all the proper names" from Greco-Roman times once other examples were obtained. If only the missing fragments of the stone's hieroglyphs could be found, it might be possible to solve "this mystery which appears to involve the very interesting point of the direct transition from pure hieroglyphics to alphabetical characters." But the general purport of the hieroglyphs involved meanings, not sounds. In the absence of a native reader of the signs, how could one hope ever to uncover their significance? The question troubled Young. After all, "in China it is reckoned sufficient for the labour of half a life to learn a single hieroglyphical language, with all the aid of oral and lexicographical instruction . . . Equally absurd would it be to pretend to decipher, as if by inspiration, by means of any general principles, an unknown Egyptian inscription, in the absence of all personal and almost all traditional instruction."³⁴

When, eighteen months later, Young wrote his long article "Egypt" for the *Britannica*, he discussed how phonography might work in the hieroglyph and enchorial signs for "Ptolemy." To do so he used the identifications that Åkerblad in 1802 had fashioned from proper names for the enchorial script, though he compacted and modified them slightly (fig. 17.4, "supposed enchorial alphabet").³⁵ He explained that he had effected a linkage between hieroglyphic and enchorial signs for the name Ptolemy by assuming that a limited set of hieroglyphs had been repurposed via the rebus principle to write Greek proper names and that the surrounding cartouche functioned to signify that the enclosed characters had been rendered phonetically.³⁶ Because this would later form one of the central elements in the controversy between Young's and Champollion's several partisans, the passage concerning Ptolemy is worth quoting in full (with reference to fig. 17.3, *top*):

> The beginning and end are obviously parts of the ring, which in the sacred [hieroglyphic] character, surrounds every proper name, except those of the deities. The square block and the semicircle answer invariably in all the manuscripts to characters resembling the P and T of Åkerblad, which are found at the beginning of the enchorial name. The next character, which seems to be a kind of knot, is not essentially necessary, being often omitted in the sacred characters, and always in the enchorial. The lion corresponds to the LO of Åkerblad; a lion being always expressed by a similar character in the manuscripts; an oblique line crossed standing for the body, and an erect line for the tail: this was probably read not LO but OLT; although, in more modern Coptic, OILI is translated a ram; we have also EIUL, a stag; and the figure of the stag becomes, in the running hand [enchorial]

something like this of the lion. The next character is known to have some reference to "place." In Coptic MA; and it seems to have been read either MA, or simply M; and this character is always expressed in the running hand by the M of Åkerblad's alphabet. The two feathers, whatever their natural meaning may have been, answer to the three parallel lines of the enchorial text, and they seem in more than one instance to have been read I or E; the bent line probably signified great, and was read OSH or OS; for the Coptic SHEI seems to have been nearly equivalent to the Greek SIGMA. Putting all these elements together we have precisely PTOLEMAIOS, the Greek name; or perhaps PTOLEMEOS, as it would more naturally be called in Coptic. The slight variations of the word in different parts of the enchorial text may be considered as expressing something like aspirations or accentuations.[37]

According to Young the final seven signs were not phonetic. Rather, they were adjectival qualifiers: the snake and ankh meant "immortal"or "ever-living," while the twisted rope meant "loving" or "beloved." So altogether Young found, on this interpretation, P-T-OLT-EIUL-M-IorE-OSH or OS. Significantly, he was not at all concerned with whether a character had to visually correspond to the object or event signified. So, for example, the sign that "is known to have some reference to 'place' " in the hieroglyphs does not have an obvious visual connection to the notion, and so to the original word connected to "place," being a sort of horizontal loop. However, the lion sign, used phonetically for LO or OLT, did at least correspond in Greek at any rate to λιονταρι—though in modern Coptic, Young wrote, "OILI is translated a ram; we have also EIUL, a stag; and the figure of the stag becomes, in the running hand [enchorial] something like this of the lion." Which, as he saw it, simply indicated that the ancient language could not have been altogether close to extant Coptic even in the later Greco-Roman era.

Of the six identifiable cartouches on the Rosetta, only four are complete. A fifth is nearly so, and a sixth is only partially evident. But all involve at least the sequence that Young identified as Ptolemy. He examined similar cartouches found elsewhere, including an inscription "on a ceiling at Karnak" that he read as "Ptolemy and . . . Berenice the savior gods." "Berenice" also appears in the Rosetta Greek, and Young had early decided on the matching enchorial sequence of signs, but the corresponding part of the Rosetta hieroglyphs is missing.[38] Two cartouches on the Karnak ceiling also contain the sequence that he identified as "Berenice." One, the leftmost, reads from right to left, the other from left to right, the order being determined by the directions in which the figures face. The leftmost of the two is preceded by a cartouche containing the sequence for Ptolemy (fig. 17.3, *bottom*).[39]

Notably, Young's identification of the name Berenice, the wife of Ptolemy Soter, did run through Coptic, just as it did in the case of Ptolemy. Although he thought the language changed considerably over time, nevertheless Young felt justified in recurring to Coptic when reading Greek proper names. Again, his analysis is worth quoting at length since the interpretation of the sequence would become a point of contention following Champollion's public reading of the same signs in his *Lettre à M. Dacier*, in which he controversially failed to acknowledge Young's earlier effort. Young wrote:

> The first character of the hieroglyphic name is precisely of the same form with a basket represented at Byban El Molouk, and called, in the description, "panierà anses"; and a basket, in Coptic, is BIR. The oval, which resembles an eye without the pupil, means elsewhere "to," which in Coptic is E; the waved line is "of," and must be rendered N; the feathers I; the little footstool seems to be superfluous; the goose is KE or KEN; Kircher gives us KENESOÜ for a goose; but the ESOÜ means gregarious, probably in contradistinction to the Egyptian Sheldrake, and the simple etymon approaches to the name of a goose in many other languages. We have, therefore, literally BIRENICE; or, if the N must be inserted, the accusative BI-RENICEN, which may easily have been confounded by the Egyptians with the nominative. The final characters are merely the feminine termination.[40]

Note the phonological lability: for Young, a phonetic sign might represent a syllable, or it might signify a letter, though he sometimes preferred a syllabic reading in order to keep close to Coptic pronunciation, presumably on the grounds that, being a very late application of the sign system for the sole purpose of rendering foreign names and words, the Egyptians would have kept as close as possible to the sounds of their language—and Young did not doubt that Coptic provided reasonable access to such late Egyptian, though not to earlier forms of the language.

Young identified an additional fifty-eight signs as denoting a characteristic (termed, in his list, "attributes or actions"), such as those for respectability or illustriousness. These were, he thought, of a piece with the ideographic nature of the hieroglyphs and suggested nothing beyond that. This lacuna, and the subtle rigidity in his point of view that it suggests, should not surprise. Young expressly disdained writing that merged the syllabic with the alphabetic, seeing the mixture as similar to a childish confusion of words with things:

> In this name [Berenice] we appear to have another specimen of syllabic and alphabetic writing combined, in a manner not extremely unlike the ludicrous mixtures of words and things with which children are sometimes

FIGURE 17.3. (*Top and center*) The Rosetta hieroglyphs (© The British Library Board ADD 27283 f92) and enchorial signs (© The British Library Board ADD 27283 f8) that Young linked to "Ptolemy" and (*bottom*) the cartouches for the names Berenice and Ptolemy from the *Description de l'Égypte*.

amused; for however Warburton's indignation might be excited by such a comparison, it is perfectly true that, occasionally, "the sublime differs from the ridiculous by a single step only."[41]

Which was not to say that all shifts in written forms were useless or meaningless. Some shifts were practical, reflecting adaptations to emergent needs for clarity and speed. In the *Britannica* article of 1819, Young repeated his longstanding view, expressed first in the 1816 letter to the archduke, that hieroglyphic writing, "the original representation, had passed, in its degradation from the *sacred* character, through the *hieratic*, into the *epistolographic*, or common running hand of the country." That result, "having long been used in rapid writing, and for the ordinary purposes of life, appears to have become so indistinct in its forms, that it was often necessary to add to it some epithet or synonym, serving to mark the object more distinctly." He compared this addition to spoken Chinese, averring that "when the words are translated from written characters into a more limited number of sounds, it is often necessary . . . to add a generic word, in order to determine the signification, and to read, for example, *a goose bird*, when *a goose* only is written, in order to distinguish it from some other idea implied by a similar sound." A similar situation also obtained "even in English" when "we might sometimes be obliged to say *a yew tree*, in order to distinguish it from *a ewe sheep*, or *you* yourself, or *the letter* u." Nevertheless "the enchorial character . . . though drawn from the same source, can scarcely, in this form, be called the same language with the sacred hieroglyphics, which had probably remained unaltered from the earliest ages." Similarly, "the running hand" that "admitted all the variations of the popular dialects, bore but a faint resemblance to its original prototype."[42]

By this point, it should be clear that although Young happily conceded that orthographic shifts could and did occur as writers labored to effect accurate transcriptions, he maintained that Egyptian hieroglyphs had never changed from the originals, neither in shape nor in meaning, despite their limited re-purposing for the conveyance of Greek proper names. They otherwise carried a syntactically limited sense, albeit one that was essentially unambiguous. By contrast, their convenient mutations, first into hieratic and finally into encho-rial, had inevitably degraded this pristine, if narrow, clarity—but not through the introduction of fully developed phonography. Rather, the signs themselves took on various meanings, especially in the enchorial, as the latter had to con-vey the changeable nuances of popular vocabulary. Indeed, the enchorial script had mutated so far that the graphical connections between its signs and those of its hieroglyphic progenitor had become difficult to trace.

At the end of his article Young produced three plates containing, in addi-tion to the "supposed enchorial alphabet" for foreign names, 204 interpreta-tions of hieroglyphic signs together with what he took to be the enchorial sequences with the same meanings. These included twenty-seven cartouches with the names of "kings" and one cartouche for a "private person." He speci-fied hieroglyphic sequences for Arsinoe and Cleopatrides, and for Arsinoe he included the corresponding Rosetta enchorial. For many other sequences Young included the Coptic word that he took to have a similar connotation—but without any claim, quite the contrary, that the Egyptian articulation of the sign sequence would be precisely, or even approximately, the same as the Coptic. In fact, none of Young's identifications involved either alphabetic or syllabic significations, apart from those for the Greek proper names Berenice and Ptolemy, for which he added fourteen hieroglyphs whose sounds he had traced through the relevant cartouches. Of this Young remarked that the "*sup-posed enchorial alphabet*, which is subjoined [fig. 17.4], is applicable to most of the proper names in the inscription of Rosetta, and probably also to some other symbols which have been the prototypes of the characters: it is taken

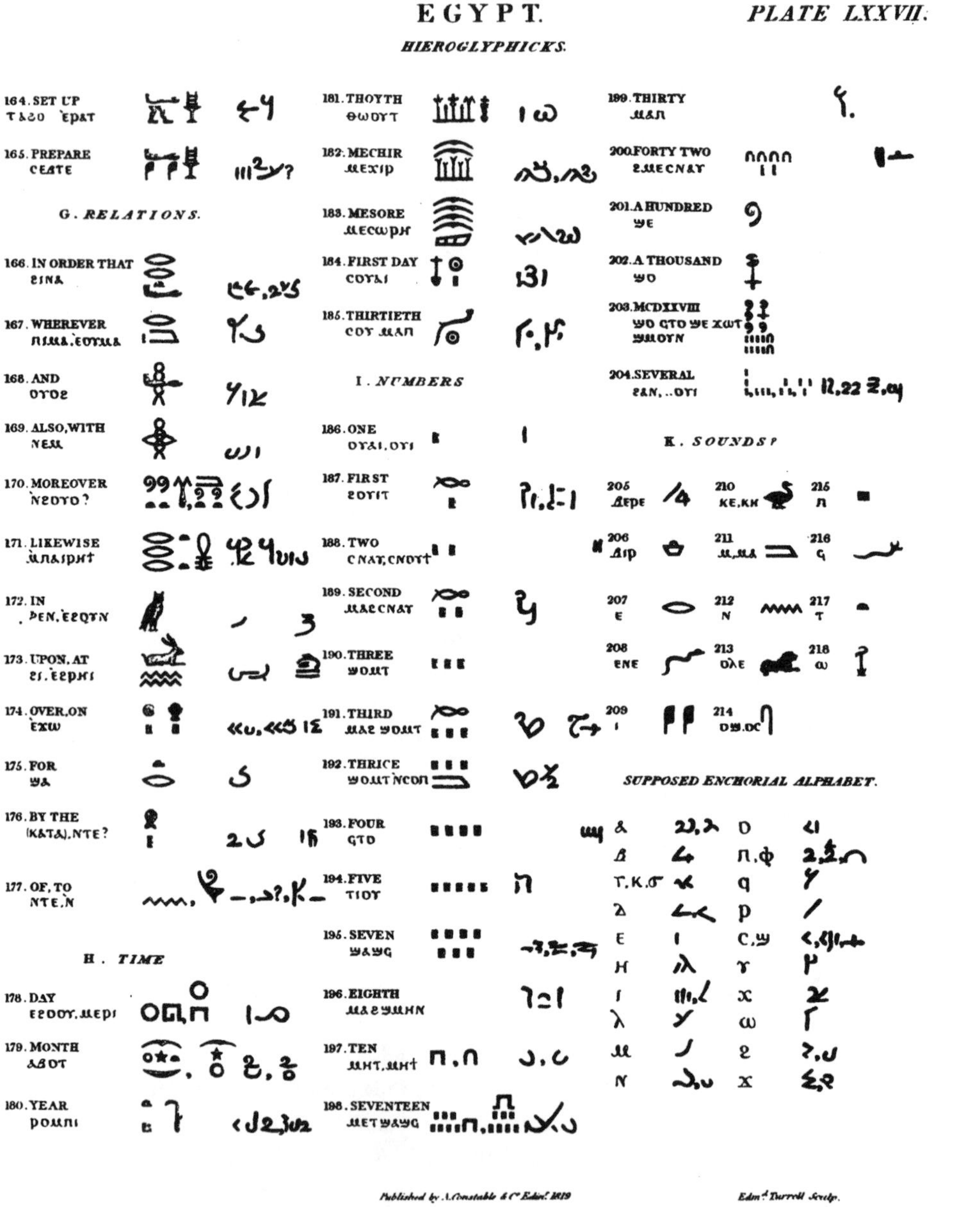

164. SET UP
ΤΑϨΟ ⲈⲢⲠⲀⲦ

165. PREPARE
ⲤⲈⲂΤⲈ

G. RELATIONS.

166. IN ORDER THAT
ϢⲓⲚⲀ

167. WHEREVER
ⲚⲓⲘⲀ, ⲈⲞⲨⲘⲀ

168. AND
ⲞⲨⲞϨ

169. ALSO, WITH
ⲚⲈⲘⲀ

170. MOREOVER
ⲚⲈⲞⲨⲞ ?

171. LIKEWISE
ⲘⲠⲀⲓⲢⲎⲦ

172. IN
ϨⲈⲚ, ⲈⲈϦⲞⲨⲚ

173. UPON, AT
ⲈϨ. ⲈⲈⲢⲎⲒ

174. OVER, ON
ⲈⲬⲰ

175. FOR
ϢⲀ

176. BY THE
(ΚΑΤΑ), ⲚⲦⲈ ?

177. OF, TO
ⲚⲦⲈ, Ⲛ

H. TIME

178. DAY
ⲈϨⲞⲞⲨ, ⲘⲈⲢⲒ

179. MONTH
ⲀⲂⲞⲦ

180. YEAR
ⲢⲞⲘⲠⲒ

181. THOYTH
ⲐⲰⲞⲨⲦ

182. MECHIR
ⲘⲈⲬⲒⲢ

183. MESORE
ⲘⲈⲤⲰⲢⲎ

184. FIRST DAY
ⲤⲞⲨⲀⲒ

185. THIRTIETH
ⲤⲞⲨ ⲘⲀⲠ

I. NUMBERS

186. ONE
ⲞⲨⲀⲒ, ⲞⲨⲒ

187. FIRST
ϢⲞⲨⲒⲦ

188. TWO
ⲤⲚⲀⲨ, ⲤⲚⲞⲨⲦ

189. SECOND
ⲘⲀⲈⲤⲚⲀⲨ

190. THREE
ϢⲞⲘⲦ

191. THIRD
ⲘⲀⲈ ϢⲞⲘⲦ

192. THRICE
ϢⲞⲘⲦ ⲚϢⲞⲠ

193. FOUR
ϤⲦⲞ

194. FIVE
ⲦⲒⲞⲨ

195. SEVEN
ϢⲀϢϤ

196. EIGHTH
ⲘⲀϢⲘⲎⲚ

197. TEN
ⲘⲎⲦ, ⲘⲎⲦ

198. SEVENTEEN
ⲘⲈⲦϢⲀϢϤ

199. THIRTY
ⲘⲀⲂ

200. FORTY TWO
ⲂⲘⲈ ⲤⲚⲀⲨ

201. A HUNDRED
ϢⲈ

202. A THOUSAND
ϢⲞ

203. MCDXXVIII
ϢⲞ ϤⲦⲞ ϢⲈ ⲬⲰⲦ
ϨⲘⲞⲨⲚ

204. SEVERAL
ϨⲀⲚ, .. ⲞⲨⲒ

K. SOUNDS?

205 ⲆⲈⲢⲈ	210 ΚΕ, ΚⲎ	215 ⲡ
206 ⲆⲒⲠ	211 Ⲙ, ⲘⲀ	216 ϭ
207 Ⲉ	212 Ⲛ	217 Ⲧ
208 ⲈⲚⲈ	213 ⲞⲨⲀⲈ	218 Ⲱ
209	214 ⲞϢ, ⲞⲤ	

SUPPOSED ENCHORIAL ALPHABET.

Ⲁ	Ⲟ	
Ⲃ	ⲡ, ⲫ	
Ⲧ, Κ, Ⲥ	ϭ	
Ⲇ	ⲣ	
Ⲉ	Ⲥ, Ϣ	
Ⲏ	ⲩ	
Ⲓ	Ⲭ	
Ⲗ	ⲱ	
Ⲙ	Ϧ	
Ⲛ	Ⲭ	

Published by A. Constable & C.º Edin.ʳ 1819

Edmᵈ Turrell Sculp.

FIGURE 17.4. The last of Young's three plates containing what he took to be the meanings of hieroglyphic and "enchorial" sign sequences, together with a limited list of possible hieroglyphic sounds (note the question mark) divined from the Ptolemy cartouche and the "supposed enchorial alphabet."

from the alphabet of Åkerblad, but considerably modified by the conjectures which have been published in the *Museum Criticum*."[43]

For Young, Egyptian hieroglyphs had ever remained unambiguous logograms for words and phrases, with the sign bearing a figurative resemblance to the object or actions represented. In his view such a system was intrinsically primitive because it lacked syntax and the ability to convey tone and emphasis. Hieratic shared this problem. Enchorial was even worse, for two reasons. First, the enchorial signs had for the most part lost their figurative aspect, so that the system's basic elements had become hard to disambiguate, leading to the loss of primitive simplicity. To Young's way of thinking, hieroglyphs and their progenitors in hieratic and enchorial could never have evolved into a phonological system without having altogether abandoned the simplicity of their representational structure, with its comparative lack of ambiguity. Only the necessity of representing foreign names and words had allowed the entry into Egyptian writing of phonography, and this was strictly limited to a small number of signs.

Young was nevertheless willing to admit that some element of phoneticism had entered in the hieroglyphs even for the names of native pharaohs through the repurposing of entire Egyptian words. This did not extend to the use, for that purpose, of the "supposed enchorial alphabet" that he had identified for foreign names. In a brief, rather obscurely phrased section of the article, he discussed a name on "the obelisc at *Heliopolis*" that "may also be observed in several other inscriptions, but with the substititions of two other names for that of the father." That name, he asserted, referring to a remark by Pliny, is "Ramesses, for we have RE, 'the sun,' MES, 'a birth,' and SHESH, 'a pair.' "[44] Here Young creatively applied material from the 1775 Coptic-Latin dictionary of Scholtz and La Croze. For the Coptic ⲢⲎ, ⲰϨⲰ, and ⲘⲈⲤ the dictionary gives the Latin "sol," "nasci," and "aequalis," respectively. Young interpreted the second ("grow") as signifying "birth" and the third ("equal") to signify "pair." Young wrote nothing about which characters on the obelisk in question corresponded to the Coptic words, though correspondence for "sol," namely, the circle, was widely accepted. Nor did he say which signs he associated with ⲰϨⲰ and ⲘⲈⲤ. This is the only example concerning a pre-Alexandrian name in Young's article. Although it is brief and mysterious, it does indicate that he had considered the possibility that ancient Egyptian monosyllables might have remained sufficiently stable to permit the use of late Coptic to render the sounds of pharaonic names by means of the rebus principle. As was usual with him, Young neither elaborated nor did he explore further possibilities. Still, anyone who, like Champollion, carefully read the article would have noticed the possible extension.

In his letter to the archduke Young also revealed just what he thought about ancient Egyptians and their society. Even if every sign could be read, and every letter in every name identified, which he doubted, the value of all this effort remained uncertain. He did not expect to discover much worthwhile in the written remnants of Egyptian antiquity—no Pliny, no Plutarch, certainly no Euclid. "[A] few historical details are the utmost that we could reasonably expect to obtain," he wrote. Even though the ancient Egyptians were known as sophisticated observers of astronomical phenomena, Young expected little from those reports either. Since "the great mass of Egyptian monuments of all kinds relates exclusively to the religious and superstitious rites observed towards the ridiculous deities and the idolized heroes of the country," he continued, "we can entertain but slight hopes of finding any very accurate records of astronomical phenomena, among the monuments of *so foolish and so frivolous a nation.*"[45]

If idolatry and superstition marked the Egyptians as foolish and frivolous, for Young it could scarcely have been otherwise. He may have shed the outer trappings of Quakerism, but subtle elements lingered in his suspicion of Egyptian religious practices, such as the worship of theriomorphic deities.[46] The multicolored hieroglyphs may have seemed frivolous as well, in excess of what was strictly necessary to convey an idea. Unfortunately, it seemed certain that the hieroglyphic signs could carry considerably different meanings when grouped in various ways—a quality sure to irritate Young, who valued simplicity in language. He detested language in which the constituent words taken together meant something different from the same words taken individually, and he despised idiom, which to him only repackaged meanings better expressed in standard forms. In 1824 he told an acquaintance that he "*hated*" idiom. "In fact," he continued, "every idiom seems to me in the nature of a proverb, and to abound in idiomatical phrases seems to me a deformity of the same kind as to interlard every speech with proverbs: there is something in it like the affectation of being very fashionable."[47]

Young's attitude toward ancient Egypt was commonplace in England, where understanding rarely extended beyond a narrow biblical framework and where Egypt as the scourge of the Hebrews elicited special distaste. "From evangelical preachers to their High Church critics," remarks one historian of the period, "countless sermons made Egypt the target of prophetic vitriol and an exemplar of the destructive potency of holy wrath."[48] An 1802 painting by J.M.W. Turner of the tenth Mosaic plague exemplified the association of ancient Egypt with biblical horror of idolatry.[49] Such connections were pervasive. In an 1821 review of an Egyptian tomb reconstruction exhibited in London by Giovanni Battista Belzoni, a circus strongman turned excavator of

Egyptian antiquities, the *Literary Gazette* made the link explicit. "At what time and in whose reign," asked the reviewer, "were the Jews bondmen in Egypt? Will an enquiry into this throw any light upon sacred history, and afford another testimony to the truth of the old Testament?" For these commentators, it was as if ancient Egypt existed for no purpose other than to shed light on contemporary spiritual conditons of the English.[50]

Young's disdain for ancient Egyptian religious beliefs was not uniquely English. Similar views also circulated in France. Abbé Nicholas Halma, for instance, was incensed by arguments that used "pagan" carvings to contradict the truth of the Scriptures, and French savants who attempted to use the zodiacs from Dendera and Esneh to confute biblical chronology predictably became targets of his ire.[51] According to Halma, "superstition and immorality" characterized these monuments, "much more than the state of the sky at the epochs of their construction." As a "carnal, somber and fanatical people," the Egyptians produced only dangerous nonsense and "veiled the confusion of their morals with the cloak of their religion," which was a bit rich, coming from a dogmatist like Halma. Halma urged comparison of Egyptian works with those of the finest French artists, as the former displayed only "the most shameful indecency united with all the horror of human sacrifice."[52] A translator of Ptolemy's *Almagest*, Halma pointed out that even Ptolemy had not cited a single observation from the Egyptians, and this silence, he averred, testified to their intellectual and spiritual bankruptcy.

Halma in France and Young in England may have been certain that the worshippers of zoomorphic idols could never have produced a true philosophy, much less have been the ancestral source of such a thing. But among French philologists and savants gripped by the artifacts, papyri, and inscriptions obtained by Napoleon's invading expedition, matters were rather otherwise. Spurred by these objects, Jomard, de Sacy, and many others had renewed the notion, which dated to Plato, of ancient Egypt as a fount of wisdom.[53] But the hieroglyphs could not have encapsulated that wisdom, for such recondite matters could only have been inscribed with clarity in alphabetic texts. De Sacy, for example, opined: "One must not imagine that the hieroglyphic inscriptions incised on the monuments were the only written works of Egyptians' history, and the unique depositories of all the sciences that are the fruit of meditation or experience, in a word, of divine and human knowledge." As attested by ancient sources, the Egyptians also "possessed a great number of books written in their language," and these attestations, coupled with "extant" fragments, "permit us to conclude that the writing employed in these books was not at all the hieroglyphic writing reserved for the monuments, but an alphabetic writing."[54]

If Young were correct in supposing that none of the Egyptian scripts was alphabetic, then the purported wisdom of that ancient land would vanish into the abyss. The specific question of whether the "Egyptian" script—Young's "enchorial," Champollion's "Demotic"—was alphabetic resonated beyond philology to interact with a long-standing idealization of Egypt as a source of esoteric knowledge. When Young identified that writing with hieroglyphics, he undermined a great deal more than the question of a script's nature. In Grenoble, Champollion would scrutinize Young's position on these matters, only to arrive at an even more extreme conclusion. Unlike Young, however, Champollion's developing views only deepened his appreciation of ancient Egypt, which he shared with other French savants of his time.

CHAPTER 18

SEEKING UXELLODUNUM

As Young, in London, was writing up his research on "enchorial," the mummy scripts, and his projected phonography of Greco-Roman Rosetta cartouches, the Champollions' fortunes had sharply worsened in Grenoble. Champollion-Figeac had worked himself into a curious conflict with the prefect, Casimir de Montlivault, whose spies were already watching the brothers closely, even as the greater danger seemed to emanate from the prefecture. In February 1816, it was rumored that an employee of Montlivault's had repurposed, as wrapping paper, the pages of a valuable work belonging to the Grenoble library. Naturally, Champollion-Figeac demanded an inquiry, from which it emerged that the thief was none other than the prefect himself, who had in fact sold the precious sheets.[1] As punishment for their anti-Royalist activities, and perhaps as payback for exposing this shameful behavior, Montlivault forced the Champollions into internal exile to their hometown of Figeac.

Distant and isolated, Figeac was perhaps the worst destination imaginable for two so enmeshed in the dynamic political, intellectual, and social universes of Paris and Grenoble. Just getting to Figeac was an ordeal. Located in the Célé valley, where the Midi-Pyrenees meets the Auvergne, Figeac is among the remotest parts of France, "the prototype of the tiny *village de la France profonde.*"[2] Accompanied by Champollion-Figeac's young son, Ali, the brothers journeyed for two weeks into the interior, on a five-hundred-kilometer odyssey around the tableland plateaus and dormant volcanoes of the Massif Central.[3] To foil spies, they adopted a speaking style fashionable among the Ultras. To keep faith with political allies still under threat, Champollion took every opportunity to send back to his good friend Michel Augustin Thévenet information about the political situations in the half dozen departments and various towns they visited while making their way toward the final destination, the family home on the rue de Boudousquarie.[4]

Despite being under police surveillance in Grenoble, which had only tightened as the year turned, the brothers had continued their anti-Bourbon activities. Grenoble's minister of police noted that "they have discussions, and the motives and purpose of such meetings are contrary to the interests of the Government."[5] Rather than allowing them to reorganize in Grenoble, Montlivault exiled them to the remote countryside where they had fewer friends and so, he hoped, could do less damage. "They have not lived in this place for a long time," he noted, and for that reason, "they are not known."[6] Despite effecting these deprivations, Montlivault remained uneasy. "Their intelligence and high social rank," he said, "render the example of their conduct and the expression of their opinions all the more dangerous."[7]

Montlivault's selection of Figeac as the site of exile was astute. In Figeac the brothers enjoyed few of the routine pleasures of Grenoble—frequent visitors, close ties to the metropolis, a lively intellectual milieu. Toward the end of the eighteenth century, when the French central government appealed to the nation's small towns and parishes for reports of weak or missing infrastructure, thousands of grievance reports (*cahiers de doléances*) poured into Paris. A correspondent outside Figeac described the area as "the most atrocious and abominable corner of the world. Its only possessions—if they can be called that—are rugged rocks and mountains that are almost inaccessible."[8] The writer Prosper Mérimée, then working as inspector general of historical monuments, was assigned to the Midi-Pyrenees during the summer of 1837 as part of a three-month tour of the Auvergne. In a letter to a friend, Albert Stapfer, written shortly after concluding his travels, Mérimée described a landscape of "beautiful natural horrors" (*belles horreurs naturelles*), especially the volcanoes, which "appeared to have gone extinct just the day before" so that if the locals "had any taste for the picturesque," which to Mérimée's disappointment they did not, "they would burn a Bengal fire at night or shoot Roman candles out of the craters." Against this sublime and menacing backdrop, regional villages like Figeac were expectably provincial. "One of the saltiest provinces of France" was how Mérimée epitomized the place for Stapfer, full of country people "always putting their hair in the soup."[9] The Champollions arrived on April 3. On April 15, Champollion mailed a *cahier de doléance* of his own to Thévenet. "Figeac," he lamented. "May the Devil take it!"[10]

Fleeing Grenoble, Champollion had stashed his Egyptian materials in a locked cabinet where they would remain until the summer of 1817, when he persuaded Thévenet to ship materials relating to the Rosetta inscriptions to Figeac. This shipment would not arrive until late August, delaying his work by more than a year, though he continued with his Egyptian studies as best he could in the meantime.[11] To distract themselves, the brothers dreamed up schemes and projects. Champollion penned anti-Bourbon plays, verse, and screeds. One titled, unpromisingly, "Men as They Are" faulted the government and especially its hangers-on for their vanity. So pressing was their need to see themselves in a flattering light, Champollion observed, that such people gave up their integrity, leaving them knowing "neither who they are nor what they want." Their weakness made them "[d]espicable playthings of political and religious tricksters who flatter and nourish man's worst passions in order to stun and impose their yoke." His conclusion was broadly pessimistic: "Most of them are bestial, in my opinion, [but] that's man."[12]

Isolated and ignored, Figeac offered relatively safe opportunities for Champollion to tweak proprieties in print and on the stage. But the lure of the antique had not abandoned him—nor, perhaps, had good sense. Figeac was removed from a great deal, including the worst of the violence that swept France after the second Restoration. The reprisals included the notorious Didier Affair, when the lawyer and fédéré Jean-Paul Didier and some two dozen comrades in arms were captured, tried by military tribunal, and summarily executed over the course of several bloody days in Grenoble.[13] As a founding member of Grenoble's association of fédérés, Champollion could easily have been among this group. In Figeac, at least, he could lay low.

After Paris and Grenoble, what was it like for the Champollions to find themselves at loose ends in such a backwater? Though it was published a generation after the Uxellodunum adventure, Mérimée's short story, "The Venus of Ille," suggests one answer. Written in 1835 and published in 1837, the story, set in the Catalan-speaking eastern Pyrenees, features a familiar comic type: a pretentious antiquarian whose learning reassured him of his superiority over his more ignorant neighbors. Although the antiquarian of Rousillon is intolerant of those nearest, he eagerly and admiringly cultivates Mérimée's narrator, a visitor introduced with great excitement "as a famous archaeologist who was going to raise the province of Roussillon from the obscurity in which it had been left by the neglect of the learned."[14] The visitor, a stand-in for Mérimée's educated reader, evinces an equally sneering contempt for the antiquarian, as we will see.

The visitor is taken to meet the antiquarian, described as "a little old man" with "powdered hair, a red nose, and a jovial, bantering manner" who was "vivacity itself" and "never still for two minutes at a time." Securing the visitor's agreement to stay for a meal, the antiquarian puffed himself into a minor mania: he "talked and ate, got up, ran to his library, brought me books, showed me engravings, and poured out drinks."[15] As the meal unfolded, it became clear that the antiquarian was interested in putting the archaeologist's expertise in the service of local glory; specifically, he wanted the visitor's opinion on the meaning of an inscription on an ancient copper statue, the eponymous Venus, found on his property and widely believed to be cursed. If the inscription could be deciphered, the relic might be significant enough to put his small town on the map.

The following day, the antiquarian showed the statue to his guest. Conversation soon turned to the Latin inscription on the pedestal. The antiquarian was convinced that the inscription ("Tvrbvl") was a corruption of the name of a local village ("Boulternère"). "Boulternère, Monsieur, was a Roman town," the antiquarian assured the narrator. "This Venus was the local goddess of the city of Boulternère; and the word Boulternère, which I have just shown to be of ancient origin, proves a still more curious thing, namely, that Boulternère, before becoming a Roman town, was a Phoenician one!"[16] The antiquarian paused to observe his listener's response, while the archaeologist, acutely aware of this attentive observation and exhibiting significant restraint in the face of it, "had to repress a strong inclination to laugh." Here, the antiquarian's lecture took an etymological turn:

> "'Tvrbvlnera' is, in fact, pure Phoenician," [the antiquarian] continued. "'Tvr,' pronounced 'tour'—'Tour' and 'Sour' are the same word, are they not? 'Sour' is the Phoenician name of Tyre; I do not need to recall the meaning to you. 'Bvl' is Baal; Bâl, Bel, Bul are slight differences of pronunciation. As to 'Nera,' that gives me some trouble. I am tempted to think, for want of a Phoenician word, that it comes from the Greek υηρός [ynros], damp, marshy. That would make it a hybrid word. To justify υηρός I will show you at Boulternère how the mountain streams there form foul pools. On the other hand, the ending 'Nera' could have been added much later, in honor of Nera Pivesuvia, the wife of Tetricus, who may have rendered some service to the city of Turbul. But on account of the pools, I prefer the derivation from υηρός."

The antiquarian paused again to take a pinch of snuff "with a satisfied air" before winding up his speeech: "But let us leave the Phoenicians and return to the inscription. I translate, then: To the Venus of Boulternère Myron dedicates at her command this statue, this work of his hand." On this point, Mérimée's narrator politely kept his thoughts to himself vis-à-vis the antiquarian while favoring the reader with his private response: "I took good care not to criticise his etymology."[17]

Had he been more forthright, he might have drained the power of the statue from the minds of the villagers and perhaps have averted the crisis that followed. The misadventure unfolded in a provincial picaresque ornamented with moments of high Gothic menace. The antiquarian's dissipated son was to be married to a more virtuous, and wealthy, woman from the next town. The archaeologist accepted an invitation to the festivities, which, unbeknownst to him, had been ill-starred from the outset. The bridegroom had unwisely bested a stranger in an impromptu tennis match, during which he carelessly placed his engagement ring on the copper finger of the Venus and promptly forgot all about it. During the wedding ceremony, he gave his bride another ring, a last-minute replacement of dubious provenance. Meanwhile, the statue's fingers had mysteriously curled around the first ring. The wedding and subsequent party were marred by falsity and vulgarity, and someone was later heard stomping upstairs to the nuptial chamber, from which ominous noises emanated. The following day, the groom was found dead with the missing ring on the floor beside him, and the bride suffered a nervous breakdown. The tennis-playing stranger, known only as "the muleteer from Aragon," was briefly suspected as the murderer, but he was released on his own recognizance, and the statue blamed instead. The antiquarian's wife then had the statue melted down and recast as a church bell, but the statue continued to haunt the village and was rumored to have blighted the surrounding vineyards for years.

Obviously we cannot shade the antiquarians whom the Champollions encountered in and around Figeac too strongly in the colors of Mérimée's satire. But his portrait of the antiquarian of Roussilon, obsessed with a quixotic project as a matter of local pride and identity, may not be so far from the truth either.[18] Champollion-Figeac was careful to distance himself from those who, like Mérimée's antiquarian, sought to exploit the possibility of phonetic shifts to link modern toponyms to ancient names. He had this to say about etymologists who sought to find in "Capdenac"—the town that he did think to be the site of the Roman settlement—some root of "Uxellodunum": "It is said in Capdenac that Caesar made them cut off their noses; and it may be remarked that torture was not unknown in the wars of antiquity, and morevoer that the same letters comprise the words *MANVS* [hand] and *NASVM* [nose]. But no manuscript says this." He noted the self-servingness of the local story: "The amputation of the nose instead of the hands seems to have been supposed by those who wanted to find the etymology of the name: Capdenac in Cap-dé-naz, *point de nez*, which is [to say] without a nose."[19]

Unlike Mérimée, Champollion enjoyed his encounters with local antiquarians, whose distance from the power plays of Paris refreshed and even amused him. On August 27, 1817, having been permitted to leave Figeac, Champollion diffidently asked his brother for advice about his next professional move. "If you have some projects for me," he wrote, "I give you carte blanche for the choice of the theater on which I must play out the rest of my role, from Lisbon to Petersburg inclusively." Well and good: so far, this was consistent with the terms of their relationship as they'd been established long before in Paris. But, Champollion noted, "I'm in no hurry. . . . I should not fear staying in Figeac for a long time yet. I would take advantage of the advantages of the little town, without adopting its vices or faults. As for the ridiculousness, it is a good thing: it amuses."[20]

He relayed a story about a visit from one Raulhac, "a doctor from Aurillac [who] came through Figeac to meet me. I could not refuse to accept a dinner from one of my admirers, as he gallantly put it. My Amphitryon," as he called him, referring to the Greek general whose name, since Molière, had evoked hospitality "is an intrepid etymologist. He has found the key to all the myths of antiquity, [and] the history of Hercules in particular, unknown to this day, will be exhibited by him in his true light." Dr. Raulhac really could have come straight out of "The Venus of Ille," as Champollion went on to tease him in absentia for the provincial fixity of his belief that every major deity in the ancient world was a personification of fire, "so that one can say of Mr. Raulhac that in the history and worship of the peoples of Antiquity this perspicacious scholar sees only fire. He particularly esteems my work," Champollion continued, "because in it he read that the word *Berber* means burning, which has given it the etymology Barbor, the name of an extinct volcano of Auvergne, Monte Barbara in Italy, Barbarism in Africa, a volcano of the island of St. Barbe and the powder magazine [Sainte Barbe] of our warships." Knowing full well the dangers of such monomania, an amused Champollion nevertheless signed off tenderly, using a local and stubbornly untranslatable expression: "Ce que ç'est de nous!"[21] *What more is there to say!*

Antiquity was everywhere in Figeac, at least for those who knew how to look. In 1838 Mérimée noted the area's significant ancient settlements.[22] Shortly after the Champollion brothers arrived, Baron Jean-César-Marie-Alexandre Chaudruc de Crazannes, a lawyer and the subprefect of Figeac, commissioned Champollion-Figeac to investigate sites of potential archaeological interest in and around the town.[23] The commission brought the Champollions into contact with amateur archaeologists and antiquarians with whom they discussed the work over the course of long evenings at the homes of like-minded neighbors.[24] Champollion hiked with his brother and various companions between different sites near Cahors: Luzech, Puy d'Issolu, and especially Capdenac, which bordered Figeac. They also visited Reilhac, where they inspected tombs that were either reflective of a Gaul settlement or presumed to be so.[25] For Champollion, the work had a morbid charm. In a letter, he joked darkly to his friend Dufour, a medical student, that he would soon send along notes to his discoveries of "old carcasses," a jest that was perhaps in reference to their excavation, outside Reilhac, of the remains of a dozen Gallic warriors.[26]

Of all the amateur antiquarians who became involved with the Uxellodunum project, Chaudruc de Crazannes was the most closely associated with antiquarians and archaeologists in Paris.[27] He authorized Champollion-Figeac to judge whether ongoing excavations at nearby Capdenac provided evidence sufficient to show that Capdenac was the site of the ancient Roman settlement of Uxellodunum, the famous hilltop town that provided the setting for the final showdown between Caesar's army and the legendary Gaul resistance led by Vercingetorix. Accompanied by his brother and local antiquarians, Champollion-Figeac set about testing sites and archaeological findings against the account given in the eighth commentary of Caesar's *Gallic Wars*.[28] No mere distraction, this was a chance to bring provincial research up-to-date along a path that Champollion-Figeac had already outlined in his Grenoble antiquarian and philological studies. Moreover, their work reprised the methods that Champollion had used in his *Pharaons*, and in particular to his connections between topographical features and place-names, cementing his focus on linguistic linkages that promised to enhance his access to antiquity.

Like Heinrich Schliemann scouring the plains of Asia Minor for the site of ancient Troy using only the volume of Homer he held in his hand, the brothers began their search for Uxellodunum with only a limited textual account to guide them. This account, written not by Caesar but by one of his officers, Aulus Hirtius, tells the story of the seige of Uxellodunum, a Gaul hilltop town that fiercely resisted Caesar's efforts to subjugate the area.[29] The Gauls were fighting for their freedom in the most elemental sense: once conquered by the Romans, they could expect to be sold, as other vanquished Gauls had been, into slavery. In 51 BCE, at the time of the siege, the area was inhabited by the Cadurci, a tribe that had been allied with Vercingetorix against Caesar in his fury to subdue this part of Gaul. Defended by steep cliffs on all sides, and nearly surrounded by a river, Uxellodunum was a natural fortress, and it was then under the control of Lucterius, a member of the Cadurci tribe. It was to Uxellodunum that he and his ally Drappes, of the Senones, retreated when their forces came under attack from two of Caesar's legions, led by Caninius. The site had a special feature—access to an underground spring—which ensured that Lucterius's troops would not die of thirst even if their access to the surrounding river was cut off. As Caninius's forces increased their attacks, the resistance forces stubbornly dug in, until several thousand Gallic soldiers occupied the settlement whose name Hirtius recorded as Uxellodunum.

Lucterius's occupation of the hilltop brought Caesar to the area in due course. His forces circumvallated the hill and constructed a ten-story tower over what they believed, as it turned out correctly, to be an access point to the source of the underground spring. Lucterius's forces counterattacked with missiles, including casks filled with tallow and kindling, which were set alight and rolled, flaming, down the slope. Meanwhile Roman sappers, protected by the tower, excavated the cliff until they found and destroyed the source of the spring. Lucterius's forces soon surrendered. In a move for which the siege is principally remembered, Caesar amputated the hands of all those who fought him there, a sadistic parting shot intended to deter similar exhibitions of resistance. Unsurprisingly, the punishment also ensured that the battle would persist in bitter memory.[30]

In France, the seige of Uxellodunum has been told and retold as a story of heroic underdogs resisting a ruthless imperial invader. It remains an important anchor of French identity, linking local pride with national honor. This was true even as Champollion hiked the countryside with his brother in search of the original site. "The description of antique Gaul has been like a debt owed to the country," Champollion-Figeac wrote. Finding Uxellodunum was important as "the last refuge of the liberty of the Gauls," the site where Caesar, "having already vanquished Vercingetorix in Alésia, was forced to fight the same battle all over again with Lucter at Uxellodunum." The question of the original location was disputed as passionately "as that of the birthplace of Homer: many towns compete stubbornly for the honor."[31]

Neither the place-name nor any credible derivation of it could be found on later maps of the area. But distinctive topographical features could be gleaned from Hirtius's account. The site had to be high enough to command good views, strongly defended on all sides by steep precipices, and nearly surrounded by a river. Such hilltop outposts were not uncommon in this part of France. Many if not all of these towns had been Roman settlements and, as such, could be expected to contain ruins and other traces of Roman occupation. But the underground spring was one distinctive feature that could perhaps be located, as were the large boulders Hirtius reported at the foot of the cliffs. Hirtius had further noted the presence of an important road running along the top of the plateau from the main settlement at Uxellodunum to fields that could be cultivated for grain in case food ran short.

Hoping to map these clues decisively to the landscape, the brothers undertook some fieldwork. In September the brothers headed to Capdenac to join an excavation under Causse's direction, where they sought evidence to support a conjecture that a particular depression in a nearby rock was the last remaining trace of "the famous Gallic spring" that had provided water to the Gauls during the siege. The excavation uncovered remnants of the original walls of a fortification as well as pottery fragments and other objects, which they sketched.

Champollion-Figeac published a detailed account of this unusual period of work in 1820.[32] After summarizing the general problem and recapitulating the main features of Hirtius's account, Champollion-Figeac surveyed three hundred years' worth of arguments to prove this or that French town was the original site of Uxellodunum. These arguments derived mainly from faulty analyses of toponyms gleaned from old maps. Weaving place-names and their literal meanings into convoluted arguments required contortions that Champollion-Figeac found impossible. For example, he described the mind-bending effort of the early modern scholar Franciscus Junius to run *lou Puech d'Ussolu*, an obscure toponym wrapped in a local idiom, backward in time, to arrive at Uxellodunum.[33]

Topographical description had, of course, formed the basis for Champollion's method in *Pharaons*. Now Champollion-Figeac used the method to mine Hirtius's text for clues that could be mapped to the landscape. The site's nine principal features, as described by Hirtius, could be compared with topographic analyses of each of the four candidate sites; these comparisons were, in turn, supplemented with discussion of typonyms on extant maps from the seventeenth century onward.[34]

Once he established rough correspondences between Hirtius's textual description of the landscape and a particular spot in Capdenac, Champollion-Figeac next deployed archaeological evidence that he and his brother had gathered during excavations with their local antiquarian and archaeological allies in September 1816.[35] There the excavators discovered a circular opening that was "practically vertical" within the cliff face—perhaps the remains of the very passage that gave Caesar's forces access to the secret spring. Venturing through, they found more Roman pottery, cylinders of red clay, and, finally, an underground water source similar to the one described by Hirtius. Champollion-Figeac even claimed to have found evidence of the means by which the Romans interrupted the flow of water.[36]

This mostly circumstantial evidence was not sufficient, however, to secure the Capdenac site as Uxellodunum. Hirtius's account was less than detailed, and any number of ancient hilltop forts were dotted about the area, all of which typically contained potsherds and coins, if not human remains. As it turned out, many of these sites were circumscribed by rivers, and some were even fed by underground springs. But Capdenac was the most promising, Champollion-Figeac insisted, because there Hirtius's text closely matched the terrain: "As the natural state of the one presents all the topographic circumstances of the history of the other, Capdenac will therefore remain the natural type which the Latin description has reproduced in a copy."[37] Shifting between a terrain and its topographic description was of course precisely what Champollion-Figeac's brother had done in the *Pharaons*. Still, if the evidence was suggestive, it was hardly definitive. Something more was required—a bit of evidence that could put the argument over the top.

At this fraught point Champollion-Figeac presented his capstone exhibit: an inscription, "recently discovered and until now unpublished," which not only reflected "the great influence that Lucter[ius] wielded among the Cadurci," an influence that was preserved and transmitted through generations from the Gauls, but also promised to show that he had exercised his influence at just the right time and in just the right place to make Capdenac the most likely site of Uxellodunum. In the village of Pern, two leagues from Cahors, an inscription had been discovered bearing the name Lucterio and a description of this character as "a priest sent by the village of Cahors to the temple erected in honor of Augustus at Lyon." On October 8, 1816, Champollion-Figeac traveled to Pern to investigate, accompanied by the antiquarian Guillaume Lacoste, "who shared my eagerness." At Pern, they located an altar stone that did indeed bear a Latin inscription, but its style disappointingly suggested an origin later than Roman Gaul. Looking closer, however, they noticed that the stone "was caught under a second wooden step which half-covered it." Champollion-Figeac and Lacoste freed the stone, "especially curious to know if its other side might not provide traces of an older inscription, perhaps even that of Marcus Lucterius."[38]

Their hypothesis proved correct. On the stone's reverse side they discovered an inscription that, though somewhat fragmented, could be reconstituted as an official statement linking the monument to Augustus (fig. 18.1). Champollion-Figeac conceded that many inscriptions of this type had previously been found in Gaul but pointed out that this was the only one to mention the city of Cadurci, filling a long-standing historical lacuna. Moreover, it gave the "true orthography" of the name of Lucterius, different from what had appeared in the various editions and translations of Caesar's commentaries, an additional point in favor of the inscription's great age and authenticity. Champollion-Figeac presented the discovery as the final piece of evidence that he needed to clinch the site's location.[39]

The Champollions' investigation of Uxellodunum fit within an existing scholarship on the early history of France that was itself concerned with the meaning of a particular word. The word, or rather suffix, in question was "dunum," which had, or was projected to have, a number of meanings descriptive of the place to which it was attached. Semantics in connection to topography had for more than a decade gripped Jean-François, but these interests also had a context: they participated in a long-standing tradition. Debate about "dunum" had opened three generations earlier with an argument at the Académie between Camille Falconet (and his ally, Fenel) and Nicolas Fréret.[40] A bibliophile, book collector, consulting physician to Louis XV, and doyen of the medical faculty at the University of Paris, Falconet entered the fray in 1753 with his *Remarques sur la signification du mot DUNUM*. His opponent, Fréret, had written a contrasting report for the Institut on the word in 1745.[41] Though the dispute had resonances beyond etymology, the crux of the disagreement concerned whether the "dunum" referred to a high place or to an inhabited one. The argument found an echo in Champollion-Figeac's study of Uxellodunum, as he unpacked Fréret's suggestion that both "uxell" and "dunum" may have originally been Celtic words. Following Fréret, Champollion-Figeac suggested that the etymology of "dunum" in Uxellodunum was best understood in terms of persistent Celtic traces in the language. But if that were true, then the Uxellodunum study fit within the conversation about non-Roman and pre-Roman heritage that was being used to consolidate French national identity in a manner consistent with the interests and activities of groups like the Académie celtique.

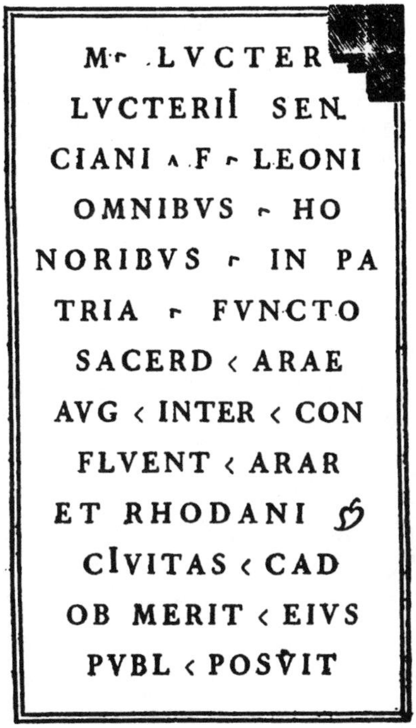

Champollion-Figeac's attention to "dunum" aligned with more recent work on the issue, work that focused less on finding a single correct mean-ing than on how meaning might have

M͟ ͙.LVCTER
LVCTERIĬ SEN͟
CIANI ᴧ .F ᴦ LEONI
OMNIBVS ᴦ HO
NORIBVS ᴦ IN PA
TRIA ᴦ FVNC͟TO
SACERD ‹ ARAE
AVG ‹ INTER ‹ CON
FLVENT ‹ ARAR
ET RHODANI ⟆
CĬVITAS ‹ CAD
OB MERIT ‹ EIVS
PVBL ‹ POSV̌IT

FIGURE 18.1. The "Uxellodunum" inscription.

varied over time. In 1765 Charles de Brosses had explored the "prodigious effect of metonymy on the derivation" of words such as "dunum."[42] Unlike both Falconet and Fréret, de Brosses sought a solution that could take multiple semantic possibilities into account, convinced that words may change accord-ing to rules interior to language itself. Consonants may invert themselves, so that the syllables of "dunum" might, when sounded out backward, sound like "mount." They might also change through rules created by imaginative lan-guage users, who might shift a word's spelling and even meaning according to metonymy. "All these derivations . . . [consist in] deriving the name of one thing from the name of another relative to it; as when we say to drink a bottle, which is to say, to drink the wine that is in it."[43]

Commenting on the "dunum" dispute, de Brosses envisioned "a current of metonymy" as a sort of pressure acting within the stream of a develop-ing language, by means of which a word's meaning might shift over time in recoverable ways. One might thereby spy, he hoped, evidence of movement from nature to human skill or art. "When we find a word, such as 'dunum,' that constantly has the same form and two meanings, such as *Dun* for both *mons* [mountain] and *oppidum* [town]," he wrote, "we cannot fail to sense that of these two meanings, one is necessarily primordial, the other secondarily adopted through metonymy." The insertion of time into language opened a question of priority: "Nothing shows better which of the two is primordial," he continued, "than when one means a thing of nature, and the other a thing of art," for "[t]he expression of a material, natural thing, in which art has no part, is obviously primitive."[44] Consequently, *mons* must have come first. Champollion-Figeac, however, distanced himself from such arguments, reminding readers that local idiomatic expressions for "town, dwelling, habitation" were nothing other than Latinized words for "Dun, Dyn, and Dynas" in Gaulois or Breton. In so doing, he asserted a linguistic continuity and staying power similar to the one his brother had posited for the ancient Egyptians.[45]

The origin and meaning of "dunum" had long preoccupied Champollion as well. During his student days in Paris, Champollion discussed with his brother the "dunum" conflict between Fréret and Falconet as glossed by de Brosses, and connected the argument to the open question of Uxellodunum's location. Falconet, he wrote, "maintains that Dunum means 'a high place,' which suits Capdenac pretty well as Exellodunum [*sic*], since Capdenac as you know is situated on a height," whereas Fréret "says that [the word] signifies an inhabited place." Dissatisfied, Champollion turned to his dictionaries. "I have searched among the Hebrew roots and I have found that *Dôme* means town; witness *Médine* [which means] city in Arabic; and *Dinas* in Bas-Breton signifies town. In Greek, I have found that *Dinè* signifies a gulf." But Falconet, he said, wanted to claim that "dunum" had another significance, of mountain. To check this hypothesis, Champollion turned to Chinese, where he found the syllable "CHAN-I, HAN-I, TA-I, which has absolutely no analogy to *Dunum*" as signifying mountain. Signing off, he asked, "What do you think? Who is wrong?" Regardless, he added, it would be advisable to "consult all your dictionaries of etymology, above all the Celtic folio on Uxellodunum."[46]

We have already seen how word roots had for some years been central to Champollion's work on ancient Egyptian and would also appear in his subsequent proposals for an Egyptian grammar and dictionary. Preoccupation with the meaning of topographical words like "dunum" echoed central facets of Champollion's *Pharaons* work as well. More broadly, the Uxellodunum adventure suggests the ways in which Champollion's preoccupations with word roots and grammatical order were converging with new approaches to the study of the past. But this continuity may have been more apparent than real; significant ruptures were coming, and a detail from Champollion's reading suggests the changes already afoot. Seven months after the Uxellodunum adventure ended, Champollion asked his brother for a copy of Abbé Sicard's *Traité des signes pour l'instruction des sourd-muets* (1808). While we do not know exactly why he requested this work, the pedagogical focus suggests he may have sought guidance in preparation for a larger undertaking, spearheaded by Jomard, to improve French primary education.[47] Sicard, who headed a school for the deaf in Bordeaux, intended his program for students who were unable to benefit from language instruction that depended upon making assocations between letters and sounds. Sicard's sign language developed instead from a lexicon of gestures, each indicating a word's root meaning (what Sicard called the "seul primitif"). Sicard systematically coupled these gestures with others indicating the root word's grammatical function in context.[48] The work naturally appealed to Champollion who had long been preoccupied with the identification of root words and with their systematic variation to express grammatical categories such as number and gender. Significantly, Sicard's method required neither an alphabet nor a syllabary, and his minimalism soon found an echo in Champollion's understanding of the Egyptian scripts, which was about to undergo an extraordinary shift. For now, though, Champollion was venturing few novel claims. Of Sicard's theory of signs, Champollion only wrote blandly, "He seems to have grasped some ideas that went through my head."[49]

CHAPTER 19

THE MASTER
OF CONDITIONS

The Champollions' archaeological interlude ended with the summer of 1816. In November, Champollion-Figeac secured permission to leave Figeac but delayed departure until the following April when he returned to Paris to become an assistant to Bon-Joseph Dacier. It was just at this moment that their unstable father—Jacques Champollion, the old bookseller of Figeac—threw the family into turmoil.[1] If Champollion-Figeac's presence through the winter had kept the lid on his father's deepening alcoholism and accumulating debts, the pot boiled over upon the elder son's departure. In nearly daily letters, Champollion regaled his brother with news of the family crisis interspersed with reflections on his struggles to keep up with the latest research on ancient Egypt and to secure funding for his pedagogical projects.

Though Champollion's letters are dense with incident, reflective of the chaos engulfing him, they evince a striking arc. Champollion initially kept vigilant watch over his Parisian foes, especially de Sacy. But these paper conflicts with professional rivals were replaced by a struggle with a far more psychologically powerful figure—his father, whom he dubbed the "patron," whose selfish, underhanded, and vain behaviors the brothers alternately rued and condemned. It was only after Champollion had secured the family estate from the predations of his father that his resentments toward other authority figures, such as de Sacy, matured into an equally fierce but more impersonal position linked directly to his independent research interests, particularly regarding the authority of biblical sources regarding ancient Egypt.[2]

Having a bookseller for a father ensured that books and learning pervaded Champollion's early life. As a boy, Champollion had enjoyed the run of his father's bookstore where he commandeered an out-of-the-way corner in which to read and eventually copy whatever pleased him.[3] Although we don't have his father's inventory, he probably stocked the bestsellers of the day, which ranged from pornography to anticlericalist and anti-Royalist polemics to such unclassifiables as Louis-Sébastien Mercier's *L'An 2440* (1771), a utopian fantasy of a future society organized according to the ideals of the philosophes; it had for years been a bestseller in France.[4] The suppression of secular books under Robespierre was a tremendous blow from which the family's business never recovered. By the time Champollion returned to Figeac, the patron had run up such staggering debts—in sum, at least 8,000 francs—that when his debtors demanded repayment in the spring of 1817, there were no funds to draw upon.[5] Worse, the father had already sold a vineyard belonging to the family without consulting his children, despite the fact that, after the death of their mother, a portion of the estate was theirs by maternal rights, according to the terms of the original marriage contract.[6] He spent the money resulting from this questionable liquidation, Champollion reported with disgust to his brother, "to eat with his whore or on debauchery."[7] Meanwhile their two older sisters, still living at home, were threatened with economic ruin, and their own shares of the estate were endangered as well.

Although Champollion was, by this time, in his midtwenties, he was nevertheless accustomed to relying on his brother's help with virtually all of his difficulties. Now he had to act quickly and decisively without Champollion-Figeac's usual support. The initial result was poor: Champollion was over his head. One of his father's creditors threatened to send the bailiff to enforce repayment by whatever means necessary, stirring up the specter of the poorhouse for his two sisters. Champollion wrote to his brother, "We are perpetually overwhelmed. Judge how painful it would be for me to witness this scene. I am almost tempted to be angry with you for leaving in the midst of this mess. There is no way out of this problem except to get through it, one way or another. Write to me as soon as possible telling me how to escape this labyrinth of trouble."[8]

To no avail: On May 13, the bailiff appeared on the doorstep. Dismayed, Champollion reported that the catastrophe he'd predicted had finally come to pass: "The thing happened, by the fault of the one whom the sky has unfortunately given us for a father." He complained that the difficulty was "a little bit your fault as well, since you left here without forcing our big baby of a father to make arrangements with his creditors," and he lamented that, as a result of this negligence, resolution of the problem "was left to me, who doesn't have the beginning of a notion for affairs." To keep the wolf from the door, or at least the father's creditors and their intimidating legal representatives, Champollion tendered 300 francs of his own—a filial gesture that would soon go badly awry. Meanwhile his father made scenes, complaining about suppers served too late for his schedule of carousing. Champollion's mounting frustration became more difficult to conceal. "I am anxiously looking forward to a plan," he wrote his brother. "I am astonished that you neglected to write to me immediately on receipt of my last letter, which sounded the tocsin on this whole affair. So

I wait and burn with impatience. I have never spent days more dreadful than the two that have just passed. I attribute a portion of my suffering to you for having left without settling all the affairs, of which the most urgent . . . ought to have been to put some limits on the *patron* who has behaved atrociously towards me through it all."[9]

Perhaps his brother was overwhelmed with troubles of his own. Or perhaps Champollion-Figeac had decided that it was time for Champollion to learn to deal with problems rather than leaving their solution to him, as had been his habit in Paris and Grenoble. In any event, Champollion rose to the occasion. His letters through the spring and summer of 1817 are filled with intrigue as he relayed, with surprising relish, every twist in the conflict.

His first priority was to deal with the most pressing debts. Fortunately his friend Thévenet was willing to loan him a sum sufficient to cover the shortfall. "I have finally found," he wrote to his brother, "the means of getting the money to avoid the distressing scenes that the *patron* was preparing for us. Independent of him, these plans are complete." This loan, he continued confidently, "will not be a problem," because it came "from one of my friends on whom I count most." He had, moreover, acted completely autonomously: "The *patron* does not know what arrangements I have made." Champollion had learned to keep not only his own counsel but also his own accounts: "It is from my own money that I have paid a portion of my father's debt, and the receipt is in my wallet." Yet he had not finished needling his brother for dereliction of filial duty. After prodding his father for payment, he lodged one final complaint with Champollion-Figeac: "I asked [our father] to see if he could earn some money. This has been fruitless. Provided he eats, he does not mind anything. He often infuriates me. I don't know what I might do. A man of this caliber is less than a brute. That's why you should have seized control of the situation before you left."[10]

Over the following weeks, Champollion arranged to sell a portion of the family property, a garden, over which he and his siblings still had some right, being originally the property of his mother; he planned to use the proceeds to dispatch his father's remaining debts.[11] His father, naturally, opposed the plan and devised a stratagem to undermine it. On June 16, 1817, he sent a letter to Champollion-Figeac in which he asked for money, claiming to have reached an "understanding" with one d'Ay on some matter related to the outstanding debts, with the sale of the garden as a means to settle them. Aware of the mischief, Champollion enclosed his father's note with a letter of his own, in which he asked his brother to "do me the pleasure of reprimanding our good father, who, not content with ruining me, wants to exhaust your finances as well. Be a little firm with him in your reply." Having fully taken the reins of the family crisis, the newly self-assured Champollion told his brother exactly what to do: "Do not believe a word of his understanding with d'Ay. It's a tale. He wants to keep his garden. Instead, offer to accept the garden as payment for what he owes you and me. You could then sell it according to your fancy and get our money back." In closing, Champollion for once did not complain about his brother's absence. He only exclaimed, "This is all so dull!"[12]

Never one to quit any field that could be used to stage a tantrum, their father had yet one more disappointment in store for his children. On July 23, Champollion wrote his brother that the sale of the garden also offered a chance to take "control [of] the affairs of the household once and for all." Their father had already overstepped once by selling the vineyard without their permission. Now he threatened to repeat the performance by selling the garden behind their backs. "I wrote to Bord [the sales agent] telling him the garden belonged to us [and that] consequently Bord should conclude nothing with my father without my consent, yours, and that of our sisters." Champollion insisted that the garden be sold at a price sufficient to repay the debts, and the estate arranged so their father no longer had any opportunity to ruin it. "By overseeing and consenting to such sale only on condition that we shall regulate the use of the funds which it will draw from it," he told his brother, "we will prevent him from spending those funds without rhyme or reason, as usual." In addition, the estate would be arranged "so that he can no longer touch it and so that when he has eaten all his possessions, which he may have already done, he will renounce the tyrannical puffs of control that send mustard right up my nose."[13]

Having presented his plan, he asked Champollion-Figeac for formal consent to proceed. The matter was urgent: "The patron may appear to resign himself but he always seeks to dodge the yoke. I am holding firm. . . . If this opportunity of binding the prodigal father escapes us through our fault or our weakness, it is all over with the family. In less than a year he and our sisters will be in the poorhouse." If they could act quickly, he wrote, this undesirable outcome could yet be avoided. Seeking his brother's cooperation, Champollion acted characteristically: he anticipated as many possibilities as he could and prepared himself to act. It is just here we see the real-world correlate to his ability—expressed until now only in his Egyptian scholarship, especially the Coptic notebooks—to hold various possibilities in mind, developing a set of potential outcomes differentiated according to variables he had identified but not specified, leaving their values indeterminate. "I await only your answer," he wrote to Champollion-Figeac, "to finish all this according to my plan, if it seems good to you, or according to yours, if you have a better one. *I am the master of conditions.*"[14]

The affair was soon concluded. The last outstanding debt was the one the father owed Champollion for his initial advance of 300 francs to ward off the debtor, Delclaux, whose demand for repayment had touched off the whole sorry business. The father left this final debt unpaid. "The *patron* has neither heart nor soul," Champollion wrote, stung by this defiance. No wonder: Champollion never had much money and was more than once reduced to relying on the generosity of a small group of loyal friends like Thévenet. Having redeemed his father's debt by incurring an equal one of his own, Champollion was mired in a terrible predicament that nevertheless only provoked a sarcastic response from his father. "When I asked him for my 300 francs which I used to ward off the first alarm," Champollion reported ruefully, "he laughed in my face."[15]

To respond to such faithlessness would be pointless, as Champollion knew all too well. But his father's betrayals nevertheless did find an indirect rejoinder in Champollion's renewed devotion to his study of ancient Egypt. Champollion's return to his scholarship was greatly abetted by Thévenet's shipment of "my famous case," as Champollion called it, which arrived safely in Figeac in August despite sustaining heavy damage to the locks, evidence of ongoing surveillance.[16] By this time, he also possessed or had at least reviewed a copy of Young's *Archaeologia* piece, on loan from de Sacy through his brother, in addition to the materials in the "famous case." Reunited with his materials, Champollion resubmerged himself in ancient Egyptian writing systems.

In particular, he sought more complex links between ancient Coptic inscriptions and his concept of the linguistic root. The coin record offered an opportunity to unify philological concerns with archaeological evidence to explore how each could shed light on the other. In reference to a coin presumably discussed in a letter, and a passage in Diodorus on an Egyptian law against counterfeiters, he remarked to his brother, "I doubt very much that Egypt had ever had metal money like Greece and Rome." While he conceded that "Egyptian texts of the Middle Ages" attested to the presence of coin specie "imprinted with a sign that fixed its value," he found this evidence more broadly unconvincing. "Nothing really indicates that this was a domestic metal currency. It could well be pieces of round porcelain, as I have suspected, bearing on one side the imprint of the symbolic eye and on the other the head of Typhon." Such roundels, he noted, "are found in very large quantities in Egyptian ruins and in tombs." He knew of such pieces, all "bearing an *imprint* or *stamp* from which the Egyptian name *La Brigue* is derived from [ⲧⲱⲱⲃⲉ] (th[eban]) [ⲧⲱⲃⲟ] (memphitic] and [ⲧⲱⲃⲉ] as well as the same root [ⲧⲱⲃⲉ] or [ⲧⲱⲃⲟ], *Sigillare*, to impress with a stamp." This was familiar ground: the same Coptic root had figured prominently in Champollion's early notebooks. Such stamps were complexly indicative: the stamp pointed both to itself, expressing its function as a signifying element, and to its meaning, which could not be understood except insofar as it pointed to an external historical reality. "This [stamp] refers obviously to the beautiful Egyptian bricks of Thebes bearing the imprint of a great seal in hieroglyphics as well as the bricks of Babylon which are loaded with cuneiform characters. Moreover, Roman bricks, like terracotta lamps, also bear a maker's mark."[17]

Even while in exile, Champollion carefully followed recent work linking Coptic and Egyptian. His correspondence from this period included discussion of Wolf Frederik Engelbreth, who in 1811 published a volume amending Zoëga's translations of Bashmuric fragments in the Borgia collection. In early summer of 1817, Champollion was reading Engelbreth in preparation for a review he would soon write for Millin. "I have examined, pen in hand, the Engelbreth volume," comparing it to Zoëga's prior publication "based on the same originals." Before advancing more than a few pages into Engelbreth's volume, he had uncovered "sixty variants," artifacts of "the way Zoëga and Engelbreth have read the original manuscripts. Sometimes one is right, sometimes the other." From this close examination of these Coptic variants, Champollion increased his own store of results. "The belly of my dictionary, [of] so-called Egyptian, is gradually filling."[18] He reported to Thévenet that the work now ran to a thousand manuscript pages.[19]

The scholars of Paris, particularly de Sacy, continued to preoccupy him, though with less of the painful intensity that had marred his student years and subsequent period in Grenoble. This conflict extended well beyond the argument with Quatremère. De Sacy had been attempting to thwart Champollion's investigation of demotic since at least March 1810, when he wrote to Champollion-Figeac in order to convince him to discourage his brother from this line of inquiry. At that point, Champollion had been at work on the demotic for nearly two years, despite being routinely stymied by lack of access to clear, accurate, and complete copies of the Rosetta inscription. When it was clear that Champollion had continued this research, de Sacy took his monitoring of Champollion's activities a step further by reporting them to Young, as we have seen, caustically writing on July 20, 1815, about Champollion's claim to have "discovered many words in the Rosetta Egyptian inscription" and warning Young not to let Champollion know too much of what he, Young, was doing.[20] On January 20, 1816, de Sacy wrote Young that he had heard "no more of Champollion. His political conduct during the three-month reign of Ahriman [i.e., Napoleon] did him little honor, and he has no doubt not dared to write me. He will have elsewhere seen, in a report that he himself provoked and with which I was charged, that I was no dupe of his charlatanism. He is about to play the role of a jay adorned with the feathers of a peacock."[21] The effects of such remarks were not lost on Champollion, even if he did not know precisely what was being said to whom in Paris and London. During the exile, when Champollion-Figeac proposed to ask the help of, among others, de Sacy, Champollion unequivocally rejected the idea: "As to the Rabbi [his usual epithet for the solemn de Sacy], don't bother yourself to see him. This is a man who knows our predicament and, if he had a little soul, he would sacrifice all his special little notions and be less cruel." Although Champollion acknowledged de Sacy's "talent, learning," and "superior knowledge," he expressed deep reservations about his character, calling him "a Tartuffe whom I would hate if he were worth it," who "has no soul" and "is no longer a man but an ambulatory book."[22]

In June, as the family crisis worsened, Champollion shifted his attention from de Sacy to a less imposing but equally irritating personification of Parisian intellectual vacuity, Louis Ripault. Ripault had traveled with Napoleon's scientific entourage to Egypt, where he served as librarian of the Institut d'Égypte before publishing a description of the principal monuments of Upper Egypt in 1800. On June 16 Champollion wrote to his brother: "I beg you to continue to give me your details of the Grand Oeuvre of Ripault." These, he mocked, "are too pleasurable for me to give up." Ripault's error, "the idea that each hieroglyph has a different meaning according to the matter being treated," was, Champollion judged, "a dream."[23] Two weeks later, irritated by Ripault's election to the Institut, Champollion again wrote Champollion-Figeac: "The Institut is going from bad to worse." With the addition of the "Sphynxinet Ripault," as Champollion dubbed him, its members were only adding to "the already rich collection of *savants* who know nothing and great doers who do nothing." In the next breath, he generalized this objection to include de Sacy. Unlike his other complaints, this time he attended as much to the specifics of de Sacy's contributions as to his untrustworthiness: "I should like to know how M. de Sacy, with his immense erudition, has caused *historical* [our emphasis] knowledge to be made? He translated, translated, translated, and that's all! The only really useful things he has produced are his Arabic Chrestomathy and grammar."[24] Although the criticism contained ad hominem elements, for once it was not wholly personal. Rather, it reflected the sort of work Champollion had come to value. This focus on historicity would soon develop into a thoroughgoing critique of source material for knowledge about ancient Egypt.

It was still only midsummer; the excruciating affair of the *jardin* would not conclude for another two months. Champollion turned his attention to another competitor on the Rosetta inscription, Samuel Friedrich Günther Wahl, professor of oriental languages at Halle, an Arabic specialist and translator of the Koran. His 1784 *General History of Eastern Languages and Literature* included more than a dozen tables comparing examples of scripts, including Coptic, Arabic, and hieroglyphs. Wahl's work attempted to locate correspondences between Coptic and Hebrew letters, ancient versions of these alphabets, and demotic characters found on the Rosetta Stone. On the basis of this analysis, he made a conjectural translation of about a dozen lines from the Rosetta demotic script into Coptic.[25] On examining Wahl's paper, Champollion immediately noted his weak grasp of Coptic. "He gives an atrocious mélange of the Theban and Memphitic dialects," Champollion opined, and his transcriptions were clogged with "grammatical debris or barbarisms to which I do not believe the Egyptian language is susceptible." Wahl's faulty Coptic blinded him to elements of linguistic structure in the Rosetta inscription that Champollion claimed to perceive on the basis of his knowledge of Coptic word order, inflections, and so on. Wahl had offered "mostly isolated words without grammatical inflections. This is not how the Rosetta inscription is arranged, wherein all the grammatical forms are as strictly observed as I find them in the Coptic manuscripts."[26]

Some weeks later, having more fully reviewed Wahl's work, Champollion concluded that Wahl's results were so faulty as to be merely a "teutonic dream." Neither his proposed Egyptian alphabet nor his reading of the inscription "exhibit common sense."[27] Meanwhile Ripault, unaware that Champollion would not return to Grenoble until October, had offered to meet in Lyon "in order to make an offensive and defensive alliance" and "promising to give me the explanation of the last two lines of the hieroglyphic inscription" of the Rosetta Stone. Although Ripault "propose[d] to unite our efforts," Champollion was not holding out much hope, noting Ripault had "only very slight notions of the Coptic language (he thinks I have enough for us both)." Nevertheless, Ripault had promised to share his "sublime notions about the composition of the letters and the arrangement of the lines." If this was mockery—"sublime notions" does seem to twist a knife—there was also a glimmer of maturity. "Such beautiful things are neither to be despised nor allowed to increase my spleen," he reflected in late August, "but an interview is impossible."[28] The trip from Figeac was simply too far to go. He asked, considerately, for Ripault's Paris address, for the polite purpose of sending a reply. His sharper assessment came later and was not shared with Ripault. The following week, in a further letter to his brother, Champollion gave his final assessment of both Wahl and Ripault: "The German Günther Wahl had no more discovered Egyptian alphabetic writing than Ripault had the hieroglyphics. All their results are foolishness."[29]

The tenor of Champollion's score-settling now shifted, becoming less personal and more intellectual as he turned his attention to source criticism, condemning Egyptian studies based solely on the Bible. An early indication of the change came in mid-June, as his antagonists, including de Sacy, were massing their support around Young in England. Advising his brother as he prepared a response to Engelbreth for publication, Champollion condemned the work of "all the savants who learn Egyptian from the Bible."[30] This source-focused criticism grew even more pointed on receipt of a hostile review of the *Pharaons* volume. Anonymously authored, as usual for such reviews, it had appeared in the *Monthly Review* for the first quarter of the year.[31] The review began with a dismissive remark, namely, that it would "perhaps . . . be well for Europe if the French were suffered to acquire [Egypt]; since the destructive character of the climate would render it an efficacious drain for the superfluous young men of France, who otherwise become troublesome neighbours." The viewer continued by speculating that French influence in Egypt could be socially beneficial as "the sympathetic licentiousness of French and Aegyptian manners would facilitate an amalgamation of the people, that might be favourable to the re-civilization of an important corner of Africa."[32] "Re-civilization" by amalgamation of French with Egyptian society reflected the common opinion in both France and Britain that contemporary Egypt could bear no comparison to its exalted Pharaonic past.[33] The reviewer clearly subscribed to this view, or at least the English version of it, but went a step further to suggest that the present state of affairs could be remedied if only the French and the Egyptians were willing to recognize and take advantage of their supposed mutual degeneracy.

The initial offenses against Champollion as a young and perhaps "superfluous" Frenchman were compounded by a more specific error, as the review's author wrongly assumed Champollion to be a disciple of de Sacy. Well into the review the author remarked positively (if incorrectly) that under "the protection of a master [de Sacy] so advantageously known by the comprehensive curiosity and judicious penetration of his researches, it is natural to expect a disciple of no ordinary endowments; and M. Champollion has not rendered a slight or transient service to the classical antiquary."[34] Champollion did not bridle at this slight, perhaps because to be bracketed with de Sacy would have furthered Champollion's Egyptological credentials, at least in England. The close conjunction of admiration of de Sacy with praise of his presumed student might also serve to irritate de Sacy, another benefit not to be too quickly discounted. Although the review was filled with fighting words, Champollion's response was relatively temperate. Champollion wrote to his brother that the review was "much as one might wish to my glory. De Sacy is presented as placing me ahead, [as seeing] me as his pupil, the one who does him much honor. That must enrage him. Good."[35]

The critical substance of the review remained to be addressed. There wasn't much: "The Englishman reproaches me only with respect to the synonymic table of geographical names," Champollion observed—that is, the *tableau ana-lytique* with which he closed the introduction.[36] This was the same table over which he had come to such grief with respect to Quatremère, whose 1808 work on Egyptian toponyms the reviewer had singled out for, if not praise, then priority, saying that the work "may be considered as having shewn the radical identity of the extant Coptic with the primaeval language of the anti-ent Aegyptians."[37] A further doubt, however, had been raised, and it was this that Champollion addressed in his letter to Champollion-Figeac. The reviewer "adds that I could have given him more material of interest if I had not ne-glected to use the documents provided by the Sacred Books (the true sources of Egyptian history according to him)."[38]

Champollion pointed out the many ways in which a misplaced faith in biblical sources led the reviewer into risible errors of historical fact: "[The reviewer] says with admirable confidence that Sesostris is nothing but Joshua, that Menes lived very shortly before Moses, that Egypt has no other reliable and authentic historical document than the story of Joseph in Genesis which was written . . . by Joseph himself, on documents he had from Jacob, Isaac, and Abraham" and "a thousand other foolishnesses of this [type]." He even included a disparaging reference to Ossian, sensing in the reviewer's remarks an agreement with the Ossianists that "the gods of Homer spoke Welsh or Bas-Breton." Lumping the hostile English reviewer with the inept German Wahl, Champollion concluded that his counterparts in England and Germany "had not yet emerged from the nursery."[39] This was a dismissive finale, not a defensive one. He was slow to perceive a threat in the reviewer's stance. Rather, it was unthreatening childishness that Champollion projected, not without reason, on his critic, and he was free enough, still, to be amused, even when the review veered into sensitive territory. Although the reviewer had made a point of saying that "De Sacy would have deemed it meritorious to have studied more attentively the true sources of Aegyptian history," Champollion was moved to nothing but amusement by this appeal to the (here conflated) authorities of his detested former teacher on the one hand and of the equally unreliable Bible on the other.[40] Neither could say anything about *history*. To his brother, he wrote, "[the reviewer] adds that my opinions on the Books seem rather more expressive of the prejudices of an infidel than the just impartiality of a philosopher. I am only sorry for one thing," he joked, "that in this respect I shall die in final impenitence."[41]

That the Bible should hold all of the "true sources of Egyptian history" re-mained a sticking point. On August 27, as the affair of the *jardin* drew to a close, Champollion made a fuller statement of his antipathy to biblically based work. In a further letter to his brother, he insisted that reading Egyptian history out of Genesis entailed missing the nuances of the different epochs. Considering, for example, the second volume of Claude-Emmanuel de Pastoret's *History of Legislation*, a multivolume study of legal history, Champollion focused on its slapdash treatment of ancient Egyptian government. "I was, so to speak, piqued at the noble peer's contention that the Egyptians were subjected to a despotic government like the other nations of the East." Champollion located the source of this blind spot in Pastoret's reliance on the Bible. "All the proofs he brings of the despotism of the pharaohs are taken from Genesis." As a result, "[h]e denies the existence of the moderate government which many historians have recognized in the scattered remnants of Egyptian history." Pastoret's reliance on the Bible had led him astray in another direction as well, as he'd confused the norms and structures of pharaonic Egypt with those of later arrivals—specifically the Hyksos, foreign rulers of Egypt during the Second Intermediate Period (1700 to 1550 BCE) and known, in Champollion's time, as the "Shepherd Kings." Here and elsewhere, Champollion spoke of them as "Arabs," "Bedouins," or "Bedouin Arabs," emphasizing the group's foreign- ness relative to the ancient Egyptians. He complained that Pastoret "did not consider that the facts he cites in such numbers prove nothing, since they appear at the unhappy time of the Reign of Shepherds [Hyksos] foreign to Egypt. Arabs seated on the throne of the Pharaohs could only establish Arab despotism, [according to which] the victors treated the vanquished as allies sometimes treat their allies."[42]

Champollion was concerned to preserve the distinctiveness of ancient Egyptian culture, to keep it separate from the histories of conquerors who arrived later. This concern was consistent with his argument, implicit and explicit, about the continuity of Coptic and ancient Egyptian, for that was based on his conviction that Egyptian culture, and by extension, language and writing, were impervious to change, especially change brought about by foreign influences. His desire to set Egyptian culture apart put Champollion at odds with de Sacy and others who insisted on assimilating Egyptian to bibli- cal history, making Egypt a mere bit player in Christianity's grand narrative. But this, Champollion insisted, was impossible. Attention, for example, to the history of Egypt as recorded in sources apart from the Bible showed that Joseph's trajectory, from the miserable pit into which his brothers threw him to his exalted position as the pharaoh's advisor, could only have happened in a context where foreigners were unusually well tolerated—which is to say, when foreigners were themselves in positions of power. "Joseph could make a fortune only at the court of a Bedouin Arab," Champollion wrote. "Under the kings of the Egyptian race, Jacob's son would have remained in the lowest ranks of the people." Instead he rose to prominence under a government that "could only have been adopted by a ferocious Bedouin and supported only by a people crushed under the saber of the foreigner."[43]

As the preceding material suggests, this newfound personal authority, expressed in terms of the insufficiency of biblical sources for reliable knowledge about Egypt, had a dark side. During the summer of 1817, Champollion had endured a descent into painful identification with every single oppressive authority figure in his life. After a hard fight, he'd scored a decisive victory against his father, one that extended, symbolically, to the hated authorities of his student days, especially de Sacy. But just when he could freely take aim at the sources and biases of these authorities, Champollion slipped into the very same mire that had often foiled his adversaries. In an effort to keep Egypt a world apart, he again veered into racist caricature. Writing to his brother about the *Monthly Review* article, Champollion cast the reviewer's pro-Bible stance in starkly anti-Semitic terms: "Besides, the Biblical opinions of the author of the article, which employs four pages to the exposition of ideas of his own, would suffice to disgust me with all possible Hebraicisms [*tout les juiveries possibles*]." History suffered when seen through such "Hebraic eyeglasses [*lunettes hebraïques*]."[44] On the following day, August 2, at the height of the miserable struggle over the *jardin*, Champollion made an even less temperate statement, condensing his anger at his father with the Arabs of de Sacy's interest, the Bedouin invaders brought to mind by Pastoret, and fantasies of oriental despotism: "If the proprietor [his father] had not attracted to the house by his extravagances and follies a cloud of Arabs, Bedouins, and Jews who demanded the tithe, I should not fear staying in Figeac for a long time yet."[45] This obnoxious mood did, however, prove transient, even as his characteristic defiance deepened; at some moments during this difficult summer, he seemed quite like his hated father. Perhaps he had some inkling of the conflict blowing in from across the Channel. The hard vicissitudes of Figeac had at least prepared him for the fight to come.

CHAPTER 20

ABANDONING
THE ALPHABET AT
GRENOBLE

Despite his extensive work on the Egyptian dictionary and grammar, Champollion seems to have done nothing of substance about the Egyptian (demotic) text of the Rosetta inscription from 1808 until the autumn of 1814 when he had sent the Royal Society a copy of his *Pharaons*.[1] Some time thereafter Champollion began working directly with the demotic. In a letter to Thévenet written on July 18, 1816, while still in Figeac, Champollion asked him to send, from Grenoble, a set of papers left behind in the exodus of the preceding year. These papers comprised the engraving he had of the Rosetta "Egyptian and Greek inscription," as well as a poor copy of the *Vetusta* hieroglyphs, together with his notes on demotic.[2] When he returned to Grenoble in the autumn of 1816, Champollion still lacked a copy of the superior *Description* engraving, which had only recently been printed, based on the cast taken from the stone in London by Jomard in 1815.[3] If only he had a copy of the *Description*'s plate, Champollion lamented in a letter to his brother, he would be able "to place beneath each hieroglyph the corresponding French word and even the Egyptian cursive." This work, he claimed, was "three-fourths finished" despite the lack of a clear and complete print. By June 1818, Champollion finally possessed two good copies of the Rosetta inscriptions: one that Jollois, managing editor of the *Description*, possessed and one of the *Description* plates.[4]

With this material at last available, Champollion began an extensive division of the hieroglyphs into semantic equivalents, bracketed with what he took to be the corresponding demotic, together with a list of putative alphabetic signs for demotic that was clearly based on Åkerblad's work.[5] It is likely that he undertook the division in part because he had read Young's "interpretation" of the hieroglyphic text in the *Archaeologia* and was certain he could do better. As he perfected an alphabet for the demotic via links to Coptic equivalents, he produced a list of the meanings of specific words (fig. 20.1).[6]

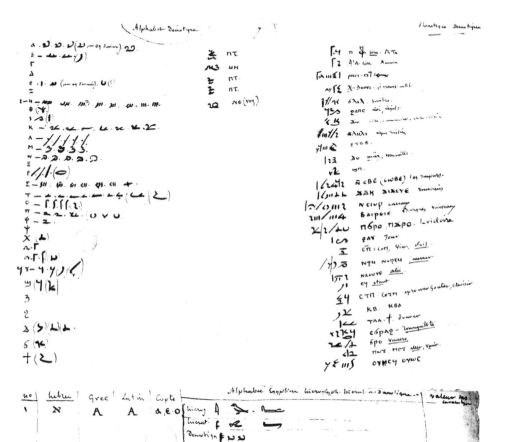

FIGURE 20.1. (*Top*) Pages from Champollion's attempt at demotic alphabetic signs and words via Coptic before leaving Grenoble for Figeac and (*bottom*) his earliest effort with hieroglyph parallels. Bibliothèque nationale de France.

Champollion thought groups of hieroglyphic signs represented specific *words* or *word phrases* and not difficult-to-interpret conceptual sequences; that is, he considered the grouped hieroglyphs to be principally logograms but with many signs functioning as qualifiers.[7] Given the same text, two native speakers of ancient Egyptian who could read hieroglyphs would consequently vocalize largely the same sequence. But someone who did not know the language would not be able to read, or for that matter generally to understand, such a text even if some of its component signs depicted a specific object or action.

On May 21, 1814, several months before the publication of his *Pharaons*, Champollion had made a cryptic remark concerning the manner in which the hieroglyphic sign groups that form words might have been constructed. In a letter to his brother, he wrote that "a single hieroglyph, that is, an isolated one, has no meaning, but they are arranged in groups that I can easily distinguish . . . You see, this purely material result can only encourage me to insist on my opinion that the hieroglyphic system, like that of the Egyptian language, is entirely syllabic."[8] To grasp what Champollion may have had in mind here—this is the only such remark from the period—recall that he was convinced that the ancient language had to be essentially Coptic, which builds words by combining minimally significant elements, or morphemes. Each word, that is, consists of a number of semantic subunits that are essentially monosyllabic.[9] As shown by his subsequent proposal to Napoleon for a dictionary, Champollion viewed hieroglyphic writing as syllabic in the sense that sign *groups* that did stand for words were built from subunits, each of which represented a semantic monosyllable. Only a literate speaker of Egyptian would have been able to understand, and to vocalize, such a hieroglyphic passage because of the large number of morpheme units involved. If this hypothesis was correct, it would explain an odd feature of the Egyptian scripts that Champollion had long noticed: there appeared to be many more hieroglyphs than demotic signs. To the extent that demotic was alphabetic, then there could not have been a close connection between it and hieroglyphs beyond their common link to the language. While a few hieroglyphs might, with considerable ambiguity and incompleteness, have been repurposed to represent the (Greek) letters in foreign names, their overall essence lay in logograms and semantic qualifiers (e.g., to disambiguate gender), whereas most demotic signs seemed to represent consonants, vowels, and diphthongs.

Champollion's efforts to divide hieroglyphs into semantic groups owed something to cognate investigations undertaken by Young that, as we have seen, were published in two sets of materials in the *Museum Criticum* starting in 1815. The first of the two *Criticum* publications was included in the journal's edition for 1815 and was available early in 1816. Though the second was not published as a *Criticum* issue until 1821, it was printed in 1816, and Young later wrote that "several copies were immediately sent to Paris, and to other parts of the Continent." Ignoring de Sacy's 1815 warning against communicating in detail to Champollion about his discoveries, Young sent offprints of both *Criticum* pieces (fig. 20.2, *left*).[10] Champollion definitely had *Criticum 1* before leaving Isère since he critiqued Young's Coptic identifications in a manuscript written while there, and he likely had *Criticum 2* as well.[11] He had seen the *Archaeologia* in 1815, which was reprinted in *Criticum 1* (albeit without its "conjectural translation of the Egyptian inscription"), so nothing was new there, but *Criticum 1* added four letters to de Sacy, two from Åkerblad, and one from Young to Åkerblad; *Criticum 2* published two further letters, one to Archduke John, the other to Åkerblad. The substance of these letters, which were written and sent between August 1814 and August 1816, recorded Young's developing efforts to extend Åkerblad's attempt to decipher the Rosetta Egyptian inscription using Coptic alphabetic equivalents.[12]

Young's results were, at best, equivocal. On August 3, 1815, Young, we saw above, told de Sacy that the Egyptian signs are "imitations of hieroglyphics, adopted as monographs or verbal characters, and mixed with the letters of the alphabet." Again, by "monographs or verbal characters" Young meant that the signs that "imitate" hieroglyphs are, like the latter, representations for words, word phrases, or concepts that could be articulated in various ways. Beyond that Young could not go. "None of these characters could be reconciled, without inconceivable violence to the forms of any imaginable alphabet," he wrote. A year later, in the letter to Åkerblad discussed above, Young distanced himself even further from Åkerblad's position. Noting that while Quatremère and Champollion had tacitly and substantially adopted it, he nevertheless remained "persuaded that if they will take the trouble of making the comparisons which I shall point out in this letter, they will be fully convinced that we have both been attempting an impossibility," that is, "to read the inscription of Rosetta into Coptic."

Three central points in the *Criticum* letters must have surprised, if not shocked, Champollion. Young claimed, first of all, that the enchorial (demotic) script was essentially derivative in every respect, including signification, of the hieroglyphs—a point at which he hinted in *Criticum 1* and asserted directly in *Criticum 2*. In both pieces, he also claimed that any attempt to produce a generally applicable alphabet for the enchorial script could not work beyond Greek transliterations. Finally, he concluded that Coptic could provide no substantial help either, since the enchorial was not phonographic, a conclusion he expressed in *Criticum 2*.[13] Each of these three claims ran directly counter to Champollion's convictions at the time. Although Champollion's hieroglyphs were signs for words, word qualifiers, or epithets, they had nothing beyond a pictorial link to demotic, while Coptic provided the essential clue to the language and to the signs of the alphabetical script itself.

Little wonder that Champollion's work began to change. At some point during his first postexile year, he made a lengthy set of "general observations" concerning the Rosetta hieroglyphs (fig. 20.2, *bottom*). He presented at least some elements of this work to the Grenoble Academy on July 24, 1818.[14] One is tempted, Champollion wrote in a manuscript entitled "General Observations on the Hieroglyphs of the Rosetta Inscription," to think that the five cartouches on the Rosetta hieroglyphic inscription contained the name "Ptolemy" together with his titles. A positional comparison of these signs with those of the demotic text suggested as much. And, indeed, he agreed that the cartouches did contain signs signifiying titles. But none among them in fact signified "Ptolemy": "the hieroglyphs of the small cartouche of the Rosetta inscription do not at all represent this name."[15] Why not? Because the same sequence occurs on monuments at Thebes together with others that Champollion identified as meaning "eternal" and "beloved of Ptah." Since "everything concurs in irresistibly proving that the temples of Thebes existed or were constructed at epochs infinitely prior to the reign of the Lagides in Egypt," it followed that the signs in question simply could not be read as "Ptolemaios." This observation marked a substantial shift. Not long before, Champollion had essayed a brief link between some of these signs and the Greek letter A (see fig. 20.1, *bottom*), apparently on the tacit assumption that Greek proper names might have been transliterated by appropriating signs, but he limited the attempt to this single letter.[16] He had now rejected even that much: since the same set of signs had been inscribed epochs before at pre-Alexandrian Thebes, they simply could not be interpreted as sounding "Ptolemaios."

The signs were, however, in the proper positions to represent the *person* of Ptolemy, if not the name itself. If they did not correspond to the sounds of the accompanying Greek and of the demotic as Champollion read it, what then did they signify? Champollion devised a clever solution: the cartouched sequence at Thebes would have represented some ancient "king or hero." Then, "not being able to render the Greek word ΠΤΟΛΕΜΑΙΟΣ in hieroglyphic writing," the Egyptians simply appropriated the signs for this ancient luminary to represent Ptolemy. The sign sequence gestured toward a ruler while failing to specify one, and this despite the reproduction of Ptolemy's name. Champollion wrote his brother on April 19, 1818, that he had now returned to Clement of Alexandria's view that Egyptian hieroglyphics possessed considerable semantic lability in that Clement's "symbolic" hieroglyphs could bear a variety of relations to their referents, for they might well be imitative, figurative, or

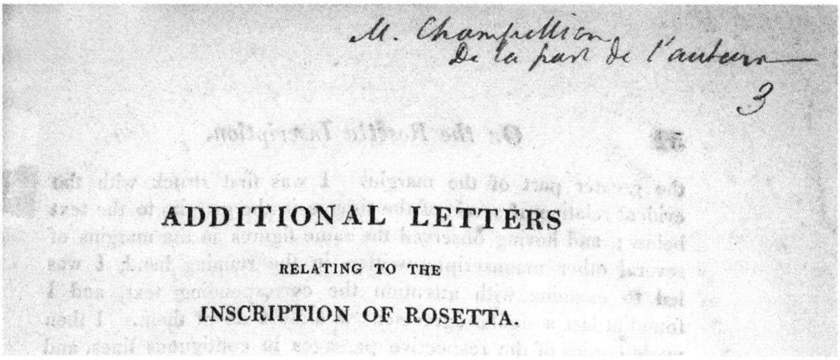

M. Champollion
De la part de l'auteur
3

ADDITIONAL LETTERS
RELATING TO THE
INSCRIPTION OF ROSETTA.

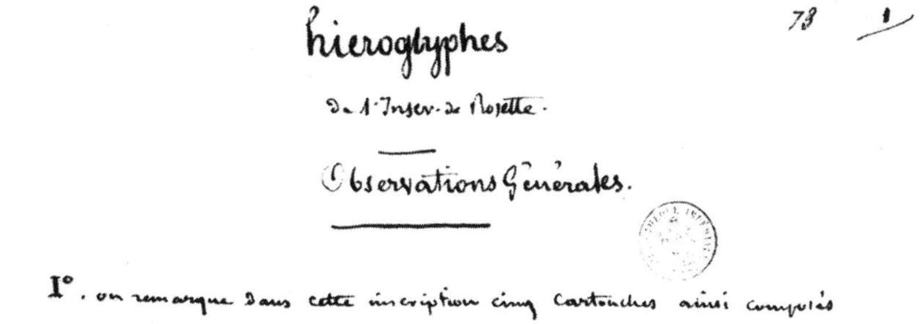

hieroglyphes
de l'Inscr. de Rosette.
Observations Générales.

1°. on remarque dans cette inscription cinq cartouches ainsi composés

FIGURE 20.2. (*Top*) Young's inscription to Champollion on the copy of the *Criticum* that he sent and (*bottom*) part of the first page from Champollion's "general observations" on the Rosetta hieroglyphs. Bibliothèque nationale de France.

even allegorical (cf. book 5, chap. 4 of the *Stromata*). "I see that it's necessary to come back to Clement of Alexandria," he wrote, carefully, "except for some small modifications."[17] Champollion had jettisoned his earlier notion that the hieroglyphs necessarily and directly represented the words of Egyptian as spoken; many among them might instead map to ideas or concepts. These certainly could be read as specific words or word phrases, but they could also be articulated in other ways (as, e.g., signs that signify "a priest at prayer" could be spoken in that way or, with equivalent meaning, as "a praying priest"). Such ambiguous signs gestured toward a set of potential equivalent meanings, a set that was bounded, semantically, by some hitherto unrecognized element of the script. As a direct result, attempts to read an isolated hieroglyphic passage would easily founder on the shoals of the system's semantic variability.

By offering a basis for comparison, the Rosetta's demotic inscription promised to prove distinctly, if not uniquely, valuable, as the inscription could be mined for confirmation of any projected hieroglyphic meanings. As he progressed, Champollion detected correspondences between sequences in the demotic text and those in the hieroglyphic passage that were so close that the meaning of "the hieroglyphic text is nearly word-for-word equivalent to the Egyptian cursive," that is, demotic. Even if hieroglyphic signs were not always representative of the words of Egyptian per se, in the Rosetta inscription they nevertheless had been arranged to carry the same meanings, in the same order, as demotic alphabetic sequences, no doubt due to the need to make the connotations of all three texts as similar as possible. But the hieroglyphs also contained something else: a unique set of qualifying signs. Indeed, on the Rosetta, Champollion now averred, "the sole difference between [the hieroglyphic and demotic inscriptions] nearly resides in the formula or protocol that precedes or follows the name of the king in whose favor the decree of the priests was carried."[18]

Qualifying signs now became important to Champollion's explorations of both scripts. Young had noticed that some signs seemed to indicate various "attributes and actions," but he thought of these as having fixed significations (positing, for example, different qualifying signs for "king" and "kingdom"). Champollion, in contrast, entertained a broader notion of how such qualifiers might work. Assisted by a consideration of inscriptions besides those on the Rosetta, Champollion identified a sign group on the first line of the Rosetta hieroglyphic inscription as "a qualification always preceding the cartouches that seem to enclose proper names on the obelisks and other great monuments." The group could signify that what followed pertained to variant properties of kingship, including a king's realm, or it might designate the status of kingship proper—kingliness, as it were. Which of these might be the case depended on neighboring signs.[19]

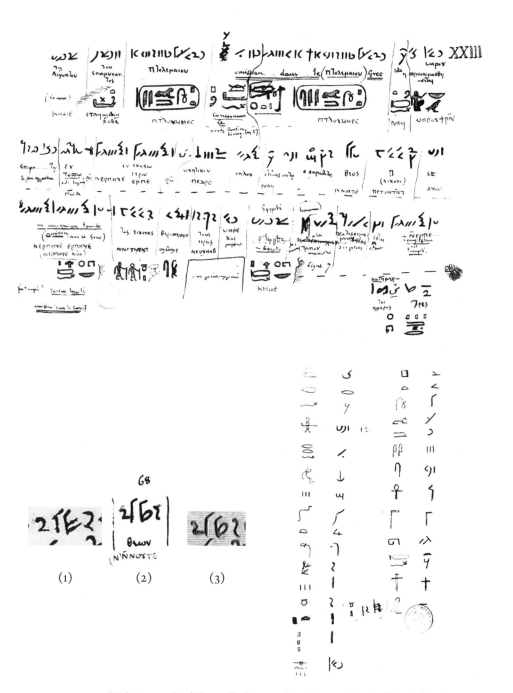

FIGURE 20.3. (*Top*) An example of Champollion's concordance between hieroglyphic and demotic texts; (*bottom left*) a comparison of (1) Champollion's drawing of graphemes from the third demotic line (2) with those same signs on the *Vetusta Monumenta* and (3) on the *Description de l'Egypte* plates; (*bottom right*) Champollion's positional paralleling of some hieroglyphic with demotic signs. Bibliothèque nationale de France.

Immediately following his discussion of hieroglyphs Champollion developed a line-by-line, semantic concordance between the Rosetta's hieroglyphic and demotic texts, taking the material as far as he could given the incomplete state of the Rosetta hieroglyphic inscription. Where the hieroglyphic inscription could not be interpreted, Champollion provided what he took to be word-by-word Greek—and from that, Coptic—correlates of the demotic sign sequences only, as they were all he had in any case. Convinced at this point that many, if not most, hieroglyphs were not directly representative of words, Champollion attempted to isolate hieroglyphic sequences in the Rosetta that paralleled, in sense, the words that he continued to spy in the demotic. In the course of doing so, he generated an uncommented list of parallels between hieroglyphic and demotic signs, drawn from his division of the Rosetta scripts (fig. 20.3).[20] Whereas Champollion's initial results derived from his perceptions of graphical similarities between the signs, he soon found a good deal more.

At this point a notable change occurred in Champollion's understanding of the relationship between the two kinds of signs. Immediately following his list of parallels, he produced five pages of equivalences between *sign groups* (fig. 20.4, *top left*).[21] This material had potentially radical implications since hieroglyphs, in his sense of the term, were syllabic. If groups of demotic signs mapped directly to hieroglyphic groups, then it followed that the demotic script could not be thoroughly alphabetic. Champollion generated lists of words and word phrases for demotic signs, just as Young had done in his letter of October 21, 1814, which Champollion read in *Criticum 1* (fig. 20.4, *top right*).

Champollion's conception of demotic evolved quickly. Investigating the sequence corresponding to "Alexandria," Champollion suggested that the word is constituted from both "phonetic" (alphabetic) and "ideographic" elements (fig. 20.4, *center*). He identified the phonetic component as "the name Alexandria." The "ideographic" element qualified "the name Alexandria" by conveying the manner in which it should be understood—in this case, as a toponym. Consequently a demotic sequence could contain both phonetic and nonphonetic elements. The nuance Champollion discerned here in the composition of a place-name becomes even more distinct, and even less phonetically dependent, in succeeding notes. These begin with nineteen loci in the demotic where he identified the sign sequence for another toponym, perhaps the most important of all: "Egypt." Young had expressed doubts about Champollion's identification of such a sign sequence on the Denon papyrus in a letter to de Sacy included in *Criticum 1*, which Champollion had now seen. Following the last of his identifications, Champollion added a critique of Åkerblad's demotic sign identification and Young's "proposal" to read the Coptic analog of the sequence as ⲏⲩⲉϩ (fig. 20.4, *bottom left*).[22] And here we find a decided, indeed

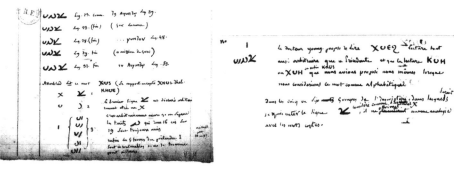

FIGURE 20.4. (*Top left*) Equivalences between hieroglyphic and demotic sign groups; (*top right*) a page of demotic meanings; (*center*) the demotic for "Alexandria"; (*bottom left*) the demotic for "Egypt"; (*bottom right*) the first complete rejection of the script's alphabetic character. Bibliothèque nationale de France.

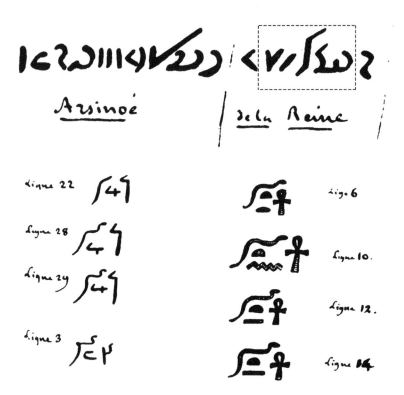

FIGURE 20.5. (*Top*) Champollion's sign sequence to demonstrate the nonalphabetic character of demotic and (*bottom*) his illustration of a close parallel between demotic and hieroglyphic signs. Bibliothèque nationale de France.

a momentous, shift: *Champollion rejects altogether the alphabetic character of demotic itself.* Even his own earlier Coptic analog for the demotic sequence, which he thought signified "Egypt," was "arbitrary," having been proposed "when we considered this word to be alphabetic" (fig. 20.4, *bottom right*).

Champollion had now leaped through Young's assertion that the demotic was not alphabetic to make it his own by doing what Young had not: offering direct and extensive support for the claim. Ironically, in light of subsequent events, it was precisely Champollion's intimate knowledge of Coptic that provided him with what he took to be conclusive evidence that the script could not be alphabetic, now not even for Greek names or words. At one location in the Greek Rosetta, for example, a form of the word for queen, βασιλισσα (*basilissa*), appeared, the Greek in question being βασιλισσης Αργινοης. Turning to the "corresponding part" of the cursive (here the demotic), Champollion found that the sequence that signified aspects of kingship occurred in the middle of the group that he identified as meaning "of the Queen" (fig. 20.5, *top*). This proved "incontestably" that the sign group could not possibly be alphabetic. How so? His path wound, as ever, through Coptic, this time using Åkerblad's letter associations. The final sign in the group that was "supposed to represent the letters мп or мϕ should not be found in the word expressing Queen. In effect мпоүро signifies of the King in the Coptic language but the words of the Queen can only be expressed by нтоүро [or by a variant of the second letter in this last] according to the dialects, words in which the п article of the masculine gender can never appear." So in Champollion's understanding, Coptic itself precluded the alphabetic identification of a cursive sequence for "queen." The argument depended upon his consistently maintained belief that, if the cursive were indeed alphabetic, then it had to correspond closely to the extant Coptic script since such an alphabetic script would have expressed the sounds of—in his view, substantially unchanged—ancient Egyptian.[23]

Champollion buttressed the claim by offering examples based on the unlikelihood of Coptic equivalences, especially those provided by Åkerblad. More crucially, he identified further linguistic qualifiers, qualifying signs that he called "determinatives." The more of these he found, the less room remained for alphabetic assignation. In this way, the demotic became thoroughly equivalent to "ideographic" hieroglyphs. "Let's examine," he continued, "if these determinative signs are not in effect found placed before or after the groups of signs which appear in the Egyptian cursive text where the Greek bears, King, Queen, Reigning, Royalty or Royal Power, and Royal rights."[24] To that end, he first deployed the "Arsinoe" sequence. Here the angle-like sign in the figure to the left of the basic sequence (fig. 20.5, *top*) must, he argued, have transformed the semantics from meaning "king" into "queen."

Generalizing, Champollion inferred that the sign mutated a masculine significance into its feminine correlate. The final sign in the figured group—to the right, at the beginning of the basic sequence—must signify the genitive ("of the queen"). With breathtaking confidence, Champollion advanced the Arsinoe evidence as decisive. "Let's therefore conclude," he wrote, "with new assurance that all [previous] reading of this group is impossible and that it belongs to a writing system other than alphabetic." To dispel lingering doubts, he provided further evidence.[25] Particularly illuminating is his use of graphical similarities between demotic and hieroglyphic signs to suggest the nonalphabetic nature of the scripts (fig. 20.5, *bottom*).[26] "[It] is difficult not to be struck by the very marked analogy between the Demotic and hieroglyphic groups," he wrote. From this observation, "one will also be forced to conclude that the signs of the one or the other system are of the same kind, that is to say veritable signs of objects."[27]

Champollion provided 102 exemplars, many of which separated the demotic from any possible alphabetic interpretation in one or more of three ways: by arguing that the sequence in question could not reasonably be assimilated to a Coptic equivalent, by identifying numerous instances of "determinatives," and occasionally by finding graphical parallels between a sequence in the demotic and another in the hieroglyphic inscription. Along the way he further asserted that the hieratic and demotic scripts were themselves graphically equivalent, differing principally in the material facility with which each could be inscribed: the "constitutive signs of the two groups only differ in general by this, that the ones, those of the hieratic text, were traced rapidly by a reed on the smooth surface of a papyrus, and that the others, those of the Demotic text, were possibly engraved by a chisel on difficult-to-work granite." Moreover, demotic had borrowed its signs from hieratic.[28]

Champollion derived his newfound conviction that all three scripts—hieroglyphic, demotic, and hieratic—were fundamentally equivalent from his identification of numerous instances of qualifying signs or determinatives. His point was not that an alphabetic script cannot have determinatives. Obviously such a script could, and did. A dollar sign "$" and a percentage sign "%" are determinatives of a kind. The question is the extent to which such things appear. Just before he rejected alphabetic structure altogether, Champollion began to think that qualifying signs occurred in the demotic more frequently than would ordinarily be expected in alphabetic writing.[29]

His notebooks document the development of this line of inquiry. Toward the end of his time in Grenoble, Champollion began an extensive essay concerning demotic that also explains the turn his work took at about this time to hieratic.[30] The roughly written manuscript is replete with insertions and deletions likely reflecting its conception as a draft to be turned into a formal article—which, we shall see, Champollion did produce in Paris. In the rough draft, he described how his efforts had led him to a completely new understanding:

> We recognized, after finding many pages of hieratic papyri traced line by line in [*ellipsis in which Champollion intended to write a number*] columns of the [*illegible*] hieroglyphic, the intimate connection that existed in Egypt between the three different systems of writing, the *hieroglyphic*, the *hieratic*, and the *vulgar* or that of the intermediate Rosetta text; these relations that we believe we can prove can only exist between systems of writing that differ, no doubt, in the forms or combinations of signs, but not at all in the nature of these same signs.
>
> Such is the lengthy research that requires one to surrender, to abandon the general idea that the intermediate inscription was like an alphabetic

script, and to reach not only the conviction but also the demonstration that the signs which compose them pertain to a script formed purely of signs of objects or ideas and that none are signs of sounds or articulation.[31]

The signs of all three scripts were neither alphabetic nor generally representative of specific words or even word phrases. Rather, they represented concepts, and as such they could be spoken in various ways. Nevertheless, Champollion remained certain that the sequence of signs followed the syntactic order of the spoken language.[32]

With his novel conceptualization in hand, Champollion turned to hieratic. The availability of what he identified as hieratic texts from papyri and mummy wrappings, particularly the plates of papyri in volume 2 of the *Description*, provided a much greater supply of material with which to work out the implications of his latest ideas. He drew up a brief article entitled "On the Hieratic Writing of the Ancient Egyptians, Explanation of Plates," that was printed at Grenoble in 1821.[33] Though he later disavowed it, the short article gives important insight into Champollion's views of the relationship between two of the scripts.

The article, probably an abridged version of the talk that Champollion gave to the Grenoble Academy, comprised just seven pages of text followed by a series of plates, most of which reproduced inscriptions taken from the *Description*. Champollion began with an elementary description of the two scripts he intended to discuss. One of these, the hieroglyphs, consisted of "more or less exact images of natural objects." The other appeared on canvas or papyrus and was made with either a reed or a brush. This script consisted of "various features, entwined among one another, of bizarre appearance, and formed of straight or curved lines." Scholars differed as to whether these latter were the "hieratic" or the "epistolographic" characters mentioned by the Greeks, but all agreed that they were alphabetic, "that is to say, composed of signs destined to recall the sounds of the spoken language." But, he continued, "a long study and above all a careful comparison of the hieroglyphic texts with those of the second kind considered alphabetic has led us to the opposite conclusion." These manuscripts were not at all alphabetic, for the hieratic script in which they were written was "a simple modification of the hieroglyphic system"—a sort of "tachygraphy," as he put it, or shorthand variant of the hieroglyphs. The characters were accordingly "signs of things and not of sounds."[34]

Champollion presented proof of this latter claim in a "general table of hieratic signs," where he paired these signs with what he took to be the corresponding hieroglyphs according to visual similarities between characters written in the two scripts (fig. 20.6).[35] But Champollion did not note just any visual similarity. Rather, he systematically juxtaposed complex hieroglyphic signs with what appear to be their visual *simplifications*. Although he did not

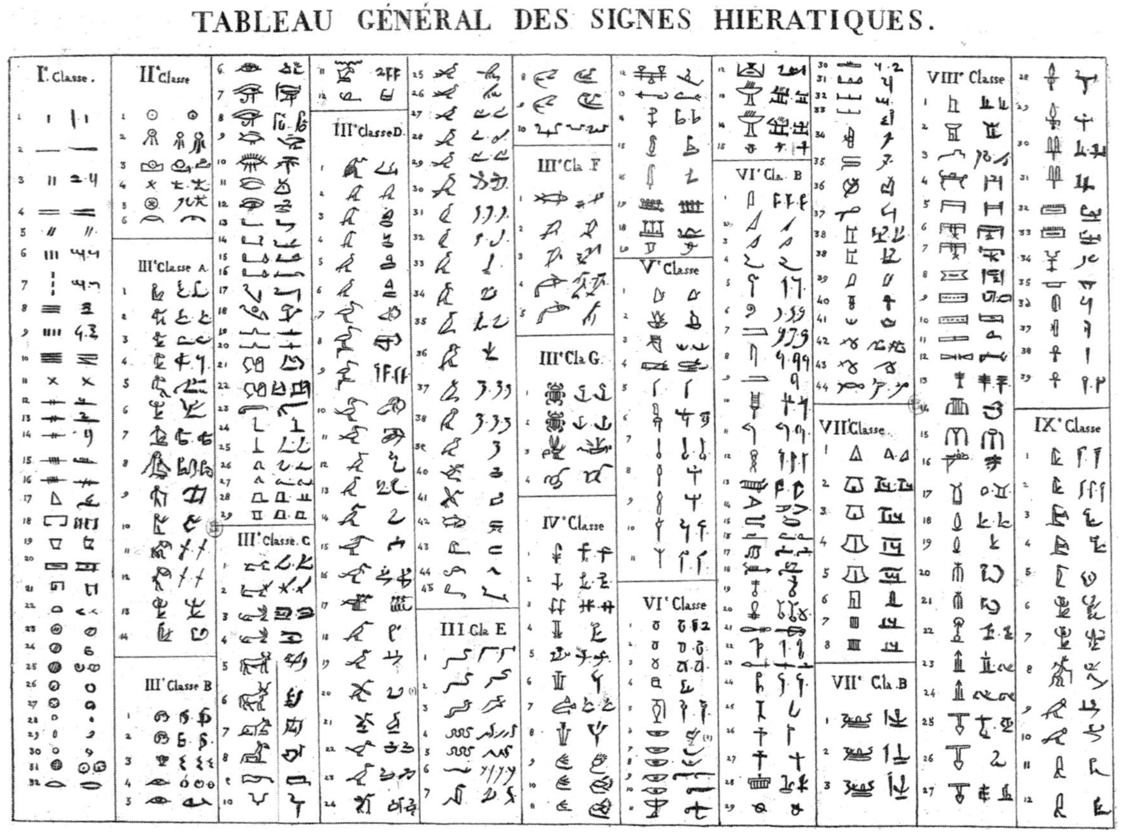

FIGURE 20.6. Champollion's table of hieratic signs paired with hieroglyphs, printed at Grenoble in 1821.

explain his reasoning, apparently this was what he meant by a hieroglyphic "tachygraphy." Readers were referred to the table, where they could inspect for themselves the similarities he perceived.[36]

In this chapter, we have discussed two remarkable documents, one buried in Champollion's archives and another that Champollion explicitly disavowed. Both presented novel but incompletely developed ideas. No full version of the reasoning behind the claims made in the short piece on hieratic was ever printed; neither was the early scribbled draft on demotic, though we shall see that Champollion eventually presented more developed versions of each. Half-baked and tentative, these early efforts complicate the tidy story of Champollion's achievement as it has been told for the past century. Nevertheless, they exemplify a process that will be familiar to anyone who teaches, and that has been frequently observed by historians of ideas: knowledge advances fitfully, "in stages rather than on a continuum; through false starts, corrections, forgotten facts, rediscoveries; and thanks to filters and schemas that simultaneously blind and open our eyes."[37] Until this point, Champollion had agreed with the general view that hieroglyphic writing represented concepts or words and phrases while the Rosetta intermediary script was alphabetic. According to his new conception, hieroglyphs did indeed represent such things, ordered according to the rules of the language, but they differed only in form and not in meaning from the other Egyptian scripts. Moreover, hieratic could not be alphabetic because it was a graphical shorthand of the nonphonetic hieroglyphs. Around this time, Champollion produced a third unlikely document that was also nearly lost. With this document, which we will turn to next, he continued to experiment with the gestural and abbreviated evocativeness he spied in the Egyptian scripts, but at greater personal risk in a different arena entirely, reporting on a rebellion in Grenoble.

CHAPTER 21

DEMONSTRATIONS

Much had transpired during Champollion's last years in Grenoble. He was promoted to professor of history at the lycée, and he succeeded in marrying Rosine Blanc over the objections of both his brother and her father, who was skeptical of Champollion's ability to provide for a family. Still, the settled comforts of bourgeois domesticity seems to have held little appeal for Champollion, and his political activities continued much as they had during the Hundred Days. As late as 1820, he continued taking risk after risk. Although no documentation explicitly links Champollion's extraordinary turnabout in respect to the Egyptian script's alphabetic nature to his ongoing anti-Bourbon agitation, de Sacy's denigrations of Champollion during this period of upheaval suggest a connection.

In his letter to Young written on July 15, 1815, only a week after the second restoration of the monarchy, de Sacy had accused Champollion of "charlatanism." Six months later, in a further letter to Young, he found fault with Champollion's "conduct during the three-month reign of Ahriman," de Sacy's derogatory name for Napoleon.[1] Even though Champollion would not have seen these remarks, he had by then long viewed de Sacy as an antagonist. He was aware of de Sacy's loathing for Napoleon and might easily have inferred a corresponding coldness to himself. We suspect de Sacy's hostililty derived from a perception that Champollion intended to launch two paired revolts: a nationalist one that sided with the Liberal opposition to the restored monarchy and one against the opinion, memorably upheld by de Sacy years before in his letter to Chaptal about the inscription, that the intermediate Rosetta script was alphabetic. By rejecting the orthodox view of the script's alphabetic structure, Champollion had rebelled against all that was sacred under the terms of the restored monarchy—or so it might have seemed to de Sacy, whose career if not identity was tied to its fortunes.

De Sacy was not completely wrong. Although his hostility is evident, what's missing is proof that Champollion saw his own work in terms that went beyond philology. As we've seen, Champollion premised his rejection of alphabetic structure on his conviction of the unusual semantic lability of the signs and on his identification of nonalphabetic elements. To have come even this far, he first had to perceive the scripts as potentially either fully alphabetic or not alphabetic at all, holding the ambiguity unresolved long enough to explore both possibilities systematically. Surely he knew that de Sacy would not appreciate a wholesale rejection of the alphabet, and no doubt he enjoyed the prospect of tweaking de Sacy on this point. But Champollion's linguistic views were linked to his political agitation in ways that went beyond personal antagonisms to reach the deeper sources of his ways with language.

Among Champollion's published writings from this period was a political pamphlet, *Attention*, which hints at the interplay between Champollion's Egyptological work and his political activities. Composed in response to an anti-Royalist disturbance that erupted in Grenoble in the spring of 1820 and published in Paris that same year, *Attention* was soon banned and most copies were destroyed.[2] A surviving copy is preserved in the municipal library of Grenoble. Although the document provides a vivid record of the conflict, it is relevant here for a different reason. The formal principles of its composition not only repeat Champollion's tweakings of Egyptological orthodoxy but also comment on his unorthodox insights, revealing facets of his general view of writing. As Champollion's political views are also explicit in *Attention*, we have an opportunity to suggest and provide evidence for links between his politics and his scholarship.

On May 4, 1820, Louis Antoine, the duke of Angoulême, arrived in Grenoble to a chilly welcome. A crowd of unhappy Grenoblois followed the royal cortege from Mont-Saint-Martin to the prefecture some eight kilometers distant, sarcastically shouting, "Vive le roi! Vive la charte!" all the way.[3] The protest, which reprised similar ones elsewhere, displeased the duke's entourage, who might well have agreed with the expressed good wishes for the monarch but certainly preferred to leave the Constitutional Charter of 1814 out of it.[4] As for the duke's opinion of the paired slogans, no one could be sure. A tepid supporter of the constitutional project, the duke had proved too passive and reserved to be a reliable ally to the opposing side and was by this time a cipher, distrusted by all. Meanwhile, the very existence of the constitution threatened the stability of the monarchy.[5]

The following day, the duke convened a military exercise in a field near the Porte de France. The protesters gathered once more, and a member of the royal retinue ordered four gendarmes into the crowd with swords drawn. The shouts of the previous day rose up again, this time with such renewed force that even the "threatening presence" of the gendarmes "could not stifle the sacred cries that had been heard since the beginning of the revue," Champollion wrote.[6] When the exercise concluded, the restive crowd chased the departing entourage. As violence finally threatened to erupt, the meaning of their slogan shifted, becoming something more ominous. The constitution came to suggest not an addition to royal power but an existential threat to it. As the cortege returned to the prefecture, a gendarme ordered the arrest of a demonstrator. A doctor defending the threatened man asked the gendarme, "Général, the sacred cry of *vive la charte*! Is it so forbidden?" The gendarme replied, "You know exactly what you mean by *la charte*. It is a rallying cry." The doctor's response was not recorded. Perhaps he had, in fact, said nothing. In any event, the gendarme lost his temper. He called the cry "seditious" and issued an order: "Slay these partisans."[7]

The protest concluded peacefully, but the power of the *charte* had encroached too far upon the privileges of the *roi*. A crackdown ensued. Within twenty-four hours, demonstrators were identified collectively as "a seditious movement," rights of assembly were restricted so that "no more than two people could walk together" in public, and the sons of a local notable were arrested, even though "[t]heir sole crime was to have changed their cry of 'vive le roi' to 'vive la charte.' " The "sacred cry" underwent yet another mutation. It was now uttered only in secret, with hushed sarcasm: "Vive le roi!" went the new slogan. "Mer[de] . . . pour le rest."[8] *Long live the king! Lives of sh—for the rest of us.*

Attention was thoughtfully composed. Champollion told this story as an eyewitness report, giving a firsthand account of the struggle. Providing a steady but mobile point of view, Champollion described a series of scenes through which the transformations of *Vive la charte!* run like a red thread. These formal decisions—to use the eyewitness narrator with an attentive but flexible point of view—were not arbitrary. Champollion, who moonlighted as a playwright and polemicist, was an accomplished writer in several genres. He understood that different arrangements of the same information could create more or less dramatic effects. Important writers of his day, including Stendhal and Balzac, shared this understanding of composition. We should not be surprised, therefore, to find Champollion using up-to-the-minute storytelling devices in *Attention*. By setting this inflammatory essay squarely within emerging conventions of modern narrative, Champollion reprised his preoccupations with meanings that were ambiguous because incompletely spelled out, arising from sequential composition, abbreviation, and evocative gestures.

Politics had been injected with fresh force into dramatic writing after Napoleon's invasion of Egypt required the effective public relations services of the theater.[9] As theaters of war gave way to those with plush seats, writers sought ways to present potentially confusing battlefield scenes to audiences accustomed to lighter fare. One way to compel an audience was to set a hero at the center of the story, a decision that encouraged the audience's identification with the central character's point of view.[10] This use of a character as the audience's proxy witness was also commonly used to dramatize unfamiliar experiences for audiences beyond the theater; indeed, by midcentury it had become a defining feature of realist writing.[11] In *Attention*, the steady gaze of the witness lends coherence to rapid and potentially confusing shifts of scene, and the identification compelled by the presence of a central figure makes the reader a witness to events as well. These qualities appeared in contemporary war writing and the dramatization of military events on paper and on the stage. Comparing nineteenth-century accounts of the Battle of Waterloo, literary historian Harry Levin notes that while Thackeray's *Vanity Fair* (1847) makes Waterloo "an off-stage noise and a stock-exchange report," in Hugo's *Les Miserables* (1862), it is "an apocalyptic vision and a tour of the battlefield." But before Hugo and Thackeray, and at least as early as Stendhal's *Charterhouse of Parma* (1839), the fateful battle is something else entirely. In Stendhal's hands, the battle of Waterloo is "a personal experience," just as it is in Champollion's *Attention*.[12]

In 1840 Balzac published an essay in the inaugural issue of the *Revue Parisienne*, a short-lived literary magazine that, thanks to his editorial involvement, also provided him with a platform for his aesthetic program. This first issue included a long exposition, unsigned but attributed to Balzac, of the principles of effective realist writing. According to Balzac, the most pressing problem facing the modern writer was the problem of scope, in particular the effective presentation of large-scale events—battles, frontier-opening exploits, ambitious colonial projects, and so on. He lodged especially bitter complaints against novelists who gave exhaustively detailed accounts with no regard for the audience's tolerance for information delivered via firehose. Balzac insisted that confusing, crowded situations—in cities, on battlefields—required the writer to make choices. Bad art was overinclusive, he wrote, bogged down by excessive detail.[13] "A book should amuse," he wrote, "or it should instruct," and since all authors inevitably presumed upon the reader's scarce attention, the accomplishment of either amusement or instruction could be effected only by means of careful selection of events representative of the most important facets of the otherwise bewildering whole.[14]

To demonstrate the need for authors to distill large-scale events into representative examples, Balzac anatomized a handful of contemporary popular novels, which he diagnosed as suffering from a lack of such distillation, including two sprawling historical novels in which battle featured prominently: James Fenimore Cooper's *The Pathfinder* and Eugène Sue's *Jean Cavalier*, both published in 1840. Weighing the broad scope of each book against its formal success as a novel, Balzac faulted both writers for overwhelming readers with irrelevant details. Finding the result boringly digressive, Balzac was mystified as to why the author felt free to send the reader on a wild goose chase. "I find it impossible when the author does not marry events and men to the accidents of nature, and fails to explain one by the other," he griped. While conceding that it was, in theory, "possible to paint the movement of the encampments and the great *tohubohu* of a battle by putting the lens of the general to the eye of the reader," Balzac opined that success would require much "typographical space" and "the rarest efforts of talent," resources ordinary writers could not expect to command.[15]

Having criticized the strategies of his contemporaries, Balzac next offered some praiseworthy examples both old and new. This breadth of reference reflected his long engagement with the narrative problems posed by war, in particular his extended and ultimately doomed effort to write *La Bataille*, a novel of a single battle, a project to which he devoted himself from the late 1820s to 1833.[16] Turning back to the early decades of the century, Balzac found much to admire in Walter Scott's Waverley novels (1814–21), which included *Ivanhoe* and *The Antiquary*, two works of special salience for the story we have told so far; and he praised *The Charterhouse of Parma* (1839), a novel written in two feverish months by none other than Stendhal, who based his account on his much earlier personal experiences of the Battle of Waterloo. Consistent with the qualities of modern writing he had just singled out for praise, Balzac admired Stendhal's selection of scenes and his use of a stand-in, the benighted Fabrice, as a witness to the events those scenes described. As had Champollion with his account of the disturbance at Grenoble, Stendhal used brief scenes to convey a larger experience.[17] Stendhal's Fabrice experiences Waterloo as a mélange of disjointed moments, an experience so bewildering that Fabrice only understands that he has been in a battle after he reads about it in the newspaper. Balzac attributed Stendhal's success to his ability to resist the temptation to "paint the complete picture of the battle at Waterloo." Instead, the author merely "followed the rearguard of the army," and "offered two or three episodes" sufficient to show the defeat. For Balzac, the effective writer's vision was tachygraphic: "So powerful are these brushstrokes that the mind sees beyond them: the eye embraces the whole battlefield and the grand disaster."[18]

So far, Champollion had done nothing more than prove himself au courant, if not ahead of his time, in the composition of *Attention*. To identify connections to his concurrent work on the Egyptian script, we turn to Champollion's handling of the "sacred cry," *Vive la charte!* Though the slogan remained virtually the same throughout the essay (with one important exception, to be discussed), differences in local conditions qualified its meaning. As the slogan appeared in different contexts, its meaning changed from a tentative affirmation of equality between the monarchy and the rights secured by the

constitution to a cry of sedition and worse. Yet no explicit lineage of these semantic transformations is given. The narrator never says, "This happened, and
as a result, the slogan's meaning changed, too." Rather, Champollion conveyed
the semantic transformations of *Vive la charte!* narratively, showing rather
than telling the reader what occurred. Even shifts in tone—from shouting at
the outset, to frantic cries during the most serious confrontation, to digusted
muttering at the end—were, for the most part, only suggested. He left much
to the reader's imagination. Nevertheless, as the slogan's meaning shifted, the
slogan itself remained materially the same—composed of the same words, the
same letters, the same punctuation. Just as nonalphabetic elements shaped
meanings in the Egyptian scripts, the formal properties of *Attention*—that is,
the juxtaposed micronarratives that contained and shaped the meaning of the
"sacred cry"—altered the semantic content of the (nearly) invariant text string,
Vive la charte! Champollion's handling of the slogan as a compound object of
both imagined sound and real text suggests much about his developing views
of the Egyptian scripts and perhaps of the general nature of writing as well.
For him, writing was a material trace that might point toward or indicate a
meaning but did not, on its own, determine it. Meaning remained for the
reader to construct from a particular sequence of rhetorical gestures, much as
Champollion's sequence of signs in Egyptian writing mapped to the language's
syntax but reserved its sense for the reader to manufacture.

By emphasizing the constitution over the king, *Attention* pointed to wider
concerns about the appropriate structure of the French polity. Compared to
a symbolic royal figurehead, a mere *document* seemed a slim guarantor of the
rights of individuals. The question was pressing since, without such a guarantor, the identity of the individual as a citizen remained fragile and easily devalued. The final scene's transformation of *Vive la charte!* is especially important
in this regard. By this point in the demonstrations, *charte* and *roi* were fully
opposed; the triumph of one ensured the defeat of the other. If the monarchy
were to prevail (*Vive le roi!*), the lives of the king's subjects would become
worthless (*Mer[de] . . . pour le rest!*). Each possibility depended on whether the
state's authority was coextensive with the exercise of royal power or whether it
could be discovered elsewhere. But this was no real choice. The fragility of the
charte must be embraced; the alternative was unacceptable. When the duke's
gendarme gave his order—"Slay these partisans"—the choice for those on the
receiving end was stark: solidarity or oblivion. Significantly, when the cry itself
changed—when *Vive la charte!* became *Mer[de] . . . pour le reste*—the scatogi-
cal statement was never quite made. The expletive was not quite deleted. The
slogan's full meaning did not appear on the page but was only apprehended
in the mind of the reader. The effect was, in a word, tachygraphic. It was also

vulgar, profane, rude. Demotic. The word becomes unavoidable. Champollion winks across centuries.[19]

Champollion had abandoned the alphabet at Grenoble—abandoned, that is, the general certainty that one or more of the Egyptian scripts must be alphabetic. This certainty, shared by de Sacy among others in France, no longer held sway over his own beliefs about the nature of the scripts. In place of the jettisoned alphabet, Champollion had posited such uncertain and questionable stopgaps as qualifying signs and the use of concrete proper names, like Ptolemy, to refer to abstract concepts like *the person of the king*. But then so much that seemed certain and unquestionable had vanished—the promise of Napoleon's early victories, the rights of the younger Champollions to their family's estate. Even de Sacy's scholarly interests had only masked a deeper hypocrisy, at least as Champollion saw it, by allowing de Sacy to put a respectable face on his petty intrigues. Having overcome disappointments at the hands of his father and his teachers, and swept up in political events following a similar disappointment with Napoleon, Champollion, with *Attention*, tested his freedom in a characteristic way: by exploring it. Just as he explored what freedom might mean politically, he explored the Egyptian scripts in the freedom secured by his separation from the commitments of de Sacy and Young. With respect to the scripts, this took the form of a toleration for ambiguity. On the streets of Grenoble, Champollion located the ambiguity in the slogan itself, that is, its potential for revision. The "sacred cry" might become anything—a taunt, or a joke, or even political form, the basis for a polity shaped by a constitution. *Something* could change. To call it a "sacred cry" was an intriguing choice for a man who found little sacred. Perhaps he was only repeating what others were saying. Or perhaps this potential for change was the sacred element.

In the early part of 1820, Champollion was still protected by the liberal Isère prefect, François René Choppin d'Arnouville. He was replaced in July by Charles Lemercier de Longpré, Baron d'Haussez, an Ultra with a decided antipathy to Champollion. These changes occurred as a result of the right-wing Ultra domination of the Chamber of Deputies following the assassination of the heir to the throne, the Duc de Berri, on February 13 at the Paris Opéra. What happened next is well known: Revanchists who had long thought Louis XVIII too tolerant of liberalism now had a pretext for furious reaction. They forced him to accept the resignation of his relatively moderate prime minister, Elie Decazes, while a bill went before the Chamber that increased the power of the upper tier and instituted the press censorship that so drastically limited distribution of *Attention*. The new minister, Armand-Emmanuel Richelieu, left office at the end of 1821, to be succeeded by Jean-Baptiste de Villèle, an Ultra whose authoritarian program found rapid support in the legislature, freshly stacked with right-wing peers. After Berri's assassin, Jean-Pierre Louvel, was guillotined on June 7 at the Place de Grève, the streets of Paris thronged with demonstrating students. Although little came of these events in Paris—the economy was in good shape, unemployment was low, and the Garde Royale remained loyal and restrained—back in Grenoble Baron d'Haussez moved quickly against the opposition, taking steps that would nearly cost Champollion his life.[20]

The following March, Champollion's friend Thévenet started a rumor that Louis XVIII had abdicated. In response, several hundred Grenoblois converged on the prefecture waving the *tricolore*, and d'Haussez struggled to persuade his troops to venture out against them. "D'Haussez and Pamphile then issued blood-curdling proclamations," writes Robert Alexander, "put Grenoble in a state of siege, and ordered arrests."[21] Champollion was already under suspicion. On March 3, the Royal Council had suspended his position at the lycée, and during the revolt, he had been seen among the rebels. D'Haussez attacked with renewed ferocity, attempting to have Champollion charged with treason, an offense punishable by death. Champollion escaped this fate thanks to the arrival in Grenoble of Claude-Victor Perrin, Duc de Bellune, who had been sent by the king to prevent destabilizing repercussions.[22]

Tried in civil court rather than for sedition, Champollion was acquitted, but his position at the lycée was a lost cause. As this setback only confirmed his father-in-law's low opinion of him, sheltering with Rosine's well-to-do family was also out of the question. Nothing was left but to abandon Isère, leaving Rosine with her family, and to regroup under his brother's protection in Paris.

CHAPTER 22

ICONOCLASM AT THE ACADÉMIE DES INSCRIPTIONS ET BELLES-LETTRES

Champollion returned to Paris in mid-July, joining his brother in an apartment at 28, rue Mazarine in the sixth arrondissement, steps from the Institut d'Égypte and just across the river from the editorial seat of the Commission de l'Égypte.[1] The neighborhood was familiar: as a student years before, Champollion had often stopped here to visit a family friend, Faujas.[2] Located on the second floor of an old building centered on a courtyard, the apartment included a large room fronted by a bay of windows. Once used as a studio by the painter Vernet, the bright room was now lined with Champollion-Figeac's books. For the next two years, Champollion—unemployed, without commitments apart from Rosine, who would join him only later—worked here, in the abundant light and quiet.[3]

For some years Champollion-Figeac had been reestablishing connections with Parisian colleagues. Jomard, now leading the nearby Commission de l'Égypte, reportedly greeted his 1817 return to Paris "with the greatest joy."[4] Champollion-Figeac was soon immersed in the most up-to-date Parisian scholarship, helping to establish new scholarly societies, including the Société de géographie, the Société des méthodes, and the Société asiatique de Paris. All the while he continued to write. After his 1818 essay on the chronology of Egypt's Greek kings won a prize competition set by the Académie des inscriptions et belles-lettres, he drew close to Dacier, the society's seventy-six-year-old secretary, beginning a collaboration that would last more than a decade.[5] Through Dacier he secured an office at the Institut and employment as the older man's "right hand, if not [his] alter ego," a prelude to his eventual appointment as conservator of manuscripts at the Bibliothèque royale.[6] Despite enjoying the support of Dacier and Jomard, the Champollion brothers nevertheless continued to face opposition from the usual suspects, the Throne and Altar conservatives who were now in the ascendant. Champollion-Figeac found himself stymied by the Académie despite his standing, in the end, three times for membership; before he could make a fourth attempt, the organizers of the block engineered changes to the electoral prerequisites to ensure that neither he nor his brother would be elected.[7]

In late 1820 or early 1821 Champollion had begun drafting an article on hieratic script, of which the brief publication with plates ("De l'écriture hiératique des anciens Égyptiens") was a précis, on which he continued work in Paris. Thanks to Champollion-Figeac's close relationship with Dacier, Champollion was invited to present his recent findings at the Académie. The first of his two lectures occurred on August 23, just a month after his arrival; the second followed in September. Neither talk was ever published, but Champollion's manuscripts remain, including revisions that were done well after he gave the talks at the Académie (fig. 22.1).[8] The audience was surely startled by what they heard, for Champollion announced in public that his work had led him to conclude that none of the Egyptian scripts were alphabetic at all. De Sacy above all must have been unpleasantly astonished. The manuscripts for the lectures convey a striking dynamism as Champollion indefatigibly tracks his claims against a changing body of evidence; if at times he seems volatile, staking claims only to alter them or strike them out, he devotes just as much energy to methodically and deliberately working out their implications.

The formal manuscript of the talks on hieratic runs to eighty-nine pages, though only the first thirty-five are written with great care, as in figure 22.1. Likely intended for publication, the neatly produced manuscript was preceded by a draft in rough hand.[9] Champollion divided the work into eight parts followed by a table giving a "concordance of the hieroglyphic manuscript with the hieratic manuscripts."[10] He began by canvassing and critiquing earlier views of the Egyptian scripts, paying particular attention to how earlier scholars had distinguished between them. Warburton, Champollion asserted, had drawn a false distinction between the hieroglyphs and other kinds of Egyptian writing, the latter of which he, like many others of the time, thought to be alphabetic.[11] The error derived from his failure to pay substantive attention to monuments, "the study of which alone leads to real knowledge of the theory of the Egyptian graphical system, because only the monuments offer innumerable applications [of it]."[12] Weighing theories such as Warburton's against the breadth of available evidence, Champollion emphasized the evidentiary shortcomings of his predecessors. "In such cases," Champollion opined, "one must proceed only with the assistance of facts, and the monuments are the only certain facts that have

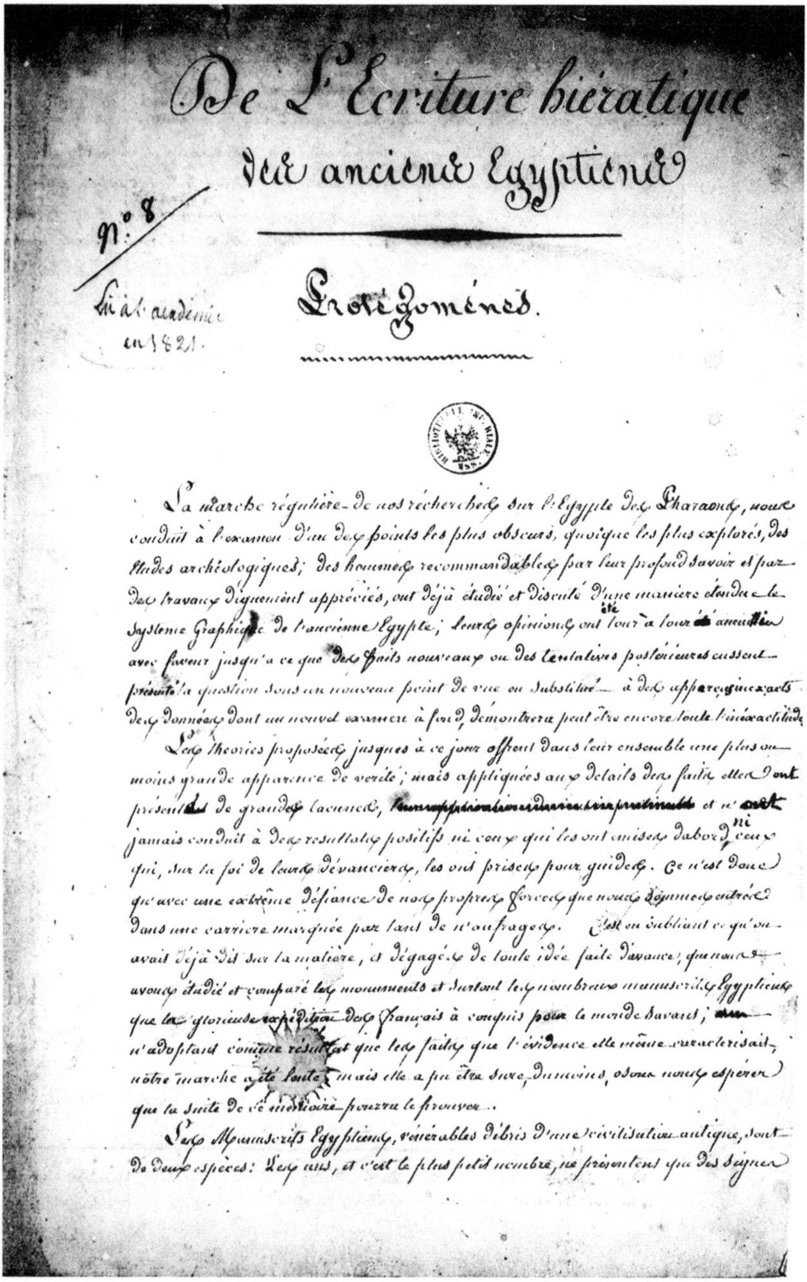

FIGURE 22.1. Prolegomenon to Champollion's talk on hieratic at the Académie des inscriptions et belles-lettres, August 23 and September 7, 1821. Bibliothèque nationale de France.

come down to us without alteration and above all without being denatured . . . by ignorance of the spirit of the age or by a half-knowledge that is even more corrupting."[13]

Turning to Rigord and Montfaucon, Champollion faulted both for being "no less wrong than Warburton." The trouble with Montfaucon derived principally from his poor choice of monuments and the infidelity of the engravings that accompanied them, vulnerabilities that made his *Antiquité expliquée* a markedly "weak help for Egyptian studies."[14] Caylus, that "illustrious academic," fared better, because he at least thought that hieratic and demotic, though alphabetic, derived from the hieroglyphs. Barthélemy was criticized for drawing affinities between the nonhieroglyphic scripts and Phoenician characters. Zoëga surpassed his predecessors because he had examined "a quite large number of Egyptian manuscript fragments or inscriptions" in a "writing other than hieroglyphic." Nevertheless, he too thought the nonhieroglyphic scripts to be alphabetic. Humboldt had fallen prey to a similar mistake, believing the script of the papyri to be "covered in part with alphabetic signs."[15] The situation had changed, Champollion wrote, with the French expedition to Egypt, which resulted in the discovery of so much new source material.[16]

Champollion noted that the opinion presented in the *Description* allowed for only two types of script: "the one vulgar and used by the people, the other secret and used by priests." But the nature of these scripts was at issue, too. The second sort of writing was "composed of isolated signs, arranged in columns and representing animals, plants, etc.," while "the other, vulgar script was formed of characters analogous to those of alphabetic scripts arranged in horizontal bands."[17] This last remark, written in a looser and more rapid hand, reflects a change from the draft version, which had attributed that opinion directly to Jomard "who in this manner reduces the types of Egyptian writing to two, and not to five like Warburton or two or three like Zoëga." Jomard also "recognized the numerous manuscripts that Zoëga termed Hieratic," that is, the papyri, "as belonging to an alphabetic script, a system known by the people and vulgarly used, and so he always called these volumes alphabetic manuscripts."[18] In other words, Jomard had lumped the script of the papyri with demotic and had set both apart as alphabetic from hieroglyphic. This division of the scripts distinguished his views from those of Zoëga who had instead separated the "hieratic" script of the papyri from the "epistolographic" but had nevertheless considered both to be alphabetic. In the formal manuscript intended for publication, Champollion did not specify who was responsible for this idea, an omission that seems intended to avoid antagonizing Jomard, and was perhaps made at the suggestion of Champollion-Figeac.

The "unanimity of opinion" on these counts had, Champollion continued, at first convinced him of the alphabetic character of the papyri. Indeed, he had claimed in his 1814 *Pharaons* that he had thereby been able to "read the Egyptian text of the Rosetta inscription" and that the results he had obtained "must apply equally to the reading of the alphabetic manuscripts," that is, the papyri.[19] For this reading, which he had not published at the time, he used Åkerblad's alphabet for the intermediary script via Coptic analogs, which latter Champollion always maintained was "the ancient language of the Egyptians." Finding that "diverse attempts on the intermediate text of the Rosetta monument" required examination "of the canvas rolls and of the papyrus [from the *Description*], he had brought "together all the signs whose forms evidently differ," thinking that this "should provide an ensemble of the alphabet of the manuscripts on canvas or on papyrus." More than three hundred "mutually quite distinct signs" resulted.[20]

This efflorescence of "alphabetic" signs did not initially faze him. Certainly, Champollion recalled thinking, there would be more signs for letters than were normally to be expected in an alphabetic system since "the number of hieratic signs could have grown considerably through change of form" over time. Yet failure stalked his effort at massaging the group into anything like an alphabetic or even syllabic form. About that Champollion did not equivocate. "[A] purely <u>alphabetic</u> script composed of many hundreds of signs frightens the imagination [and] little satisfies the mind," he wrote. Even if so many signs could be "well fixed and determined," the result would nevertheless "approach the richness of <u>hieroglyphic</u> scripts." This situation, in which the immense number of hieroglyphic signs were to be matched by a huge quantity of alphabetical ones, would constitute "an absurd luxury" because "[t]he constant and unique purpose" of "an <u>alphabetic</u> script" was "to express many sounds with few signs." The principal goal of any alphabetic script was, after all, "to reduce the art of painting the sounds of a spoken language to elements that are as limited as possible in number, creating only those that are indispensable."[21]

To buttress this point, Champollion compared the Egyptian scripts to other ancient ones. He knew that the "absurd" multiplicity of signs might simply reflect the situation, common in other alphabets, in which the same letter takes different forms depending on the context. But, he observed, in those cases the forms tended to link letters to one another, for example, by means of ligatures. Such systems "give different forms to the same letter" depending on its position—first, last, or somewhere in the middle—within a sequence, so groups of letters can be written easily, in a single pen-stroke. But "if we cast our eyes on one of the Egyptian manuscripts said to be <u>alphabetic</u> we remark on the contrary that all the signs are formed in isolation from one another, that there is no link among them, no connection—no adherence." Moreover, in Arabic, Kufi, and Naskh, concatenations of letters form *groups*. But in the Egyptian scripts, grouped letters "do not touch one another at all and cannot have been traced with a single stroke of the pen without lifting it, as are most of the words that compose a page of Arabic, of Persian or of Syriac." [22]

Champollion considered a similar objection made on the same basis but with syllables: "Might the signs of the Egyptian manuscripts belong to a type of syllabic alphabet" composed of many elements "such as that of the Hindus or the Ethiopians?" He thought not. "In the Deva-nagary alphabet, that in truth reaches 444 signs, the 34 basic syllabic signs always retain the same general form although successively affected by sixteen different vowels." In contrast, "the signs of the Egyptian manuscripts, which rise perhaps to more than double the number of the Hindu characters, all have rigorously connected forms and do not seem to retain among themselves any analogy of formation." The same absurd proliferation that threatened the argument for an alphabetic system also endangered one based on syllables, for the Egyptian syllabary would have to be composed "of many hundreds of diverse signs, each expressing a different syllable." How then to expect "any analogy of form" between syllables expressing similar sounds, for example, "between the sign expressing the syllable BA and the signs that should express the syllables BÉ, BÈ, DI, BÔ, B, BOU; a resemblance that analogy, ease of expression, convenience of practice and the purpose of every alphabet imperiously command." [23]

Such a comparison of forms, he continued, had initially suggested that the papyri consisted of "alphabetic or syllabic texts intermixed with hieroglyphics signs." But neither would that possibility work. The number of signs "in the so-called alphabetic texts" that appeared similar to those in the hieroglyphs amounted to "at most a sixth of the totality of the signs." In which case the remaining number of alphabetic or syllabic signs would be too vast. Champollion consequently separated hieratic altogether from phonography, and he did so in the overtly public setting of the Académie des inscriptions. "So striking a consideration, this immense number of collected signs, seemed conclusive to us, and thereby authorised us to think that the writing of the papyrus could not in any way belong to a purely alphabetic system, that is, to result from an alphabet." As a result, Champollion was forced "to relinquish the hope of finding either an alphabet or a syllabary." [24] Since the *Description* considered the papyri script to be essentially the same as the demotic of the Rosetta intermediary, it followed that neither could demotic be alphabetic, in which case de Sacy, Åkerblad, and others had been quite wrong two decades earlier. Abandoning phoneticism altogether, Champollion drew the only possible conclusion:

[T]he signs of these papyri, though offering little analogy with those of the hieroglyphic texts, nevertheless correspond one by one to them, [so] that one and the other are nothing but different signs of one and the same idea or part of an idea; that these signs are of the same kind, with this sole difference, that the linear signs of the so-called <u>alphabetic</u> papyri do not at all present, like the signs of the hieroglyphic text, the image of physical objects, and seem to be for the most part arbitrary.[25]

Champollion piled example upon example to drive his point home. He produced pages of sign correspondences between hieroglyphs and hieratic, claiming semantic equivalences while pointing out differences between the ways in which the two sets of signs were arranged and simplified for orthographic convenience. For instance, Champollion rendered hieratic equivalences for the hieroglyph of a kneeling man, remarking that "this hieroglyph is so complicated and so difficult to execute rapidly that it is reduced in the so-called alphabetic text to a much simpler form" (e.g., fig. 22.2).[26] Champollion chased these parallels, providing what he termed a "theory of the forms of hieratic writing," which he divided into several parts accounting for the disposition, order, nature, and classes of signs as well as "hieratic anomalies."[27] All of this was based on the grand papyrus of the *Description*, the very one that Young had so carefully scrutinized in reaching effectively the same conclusion.[28]

The work was slow, difficult, and certain to raise hackles. Champollion was fighting strong convictions in favor of the view that the nonhieroglyphic scripts were alphabetic. Undoing those convictions would be a challenge. In his own defense, Champollion pointed out that despite enormous energy devoted to the study of Egyptian writing, little had come of it. "As a whole, the theories thus far proposed offer more or less the appearance of truth; but applied to the details of the facts they present large gaps, and have never led to positive results," he wrote. Such views were difficult to relinquish, particularly in the absense of a clear and convincing alternative. "It is therefore in extreme defiance of our own strengths," he continued, "that we have entered upon a career marked by so many shipwrecks." It was just here that the question of evidence became crucial. "We have studied and compared the monuments and, above all, the numerous Egyptian manuscripts that the glorious French expedition conquered for the scholarly world; adopting nothing but the facts presented by the evidence proper, our path was slow but safe."[29]

In a subsequent revision Champollion changed his mind once more. Having demolished the concept of alphabetical Egyptian scripts, Champollion ran full-speed into a new problem: If the scripts were not alphabetical, into what sorts of groups could they be arranged to reveal at least some systematic

FIGURE 22.2. Champollion's hieratic sign parallels for the hieroglyphs that represent a kneeling man. Bibliothèque nationale de France.

organization that reflected the underlying language? Champollion examined the "number of hieratic characters" with a view to forging a "comparison of the grammatical sequence of the signs of the two systems." Then, in the midst of this section, the handwriting abruptly loosens, and Champollion signals his intention to produce an additional four sections encompassing classification, number signs, other general considerations, and a conclusion. Only the first of the four was written. A note added in 1840, no doubt by Champollion-Figeac, remarked that the missing three sections remain altogether "unknown." In the midst of the single extant section, on sign classification, Champollion struck the following passage:

> If we consider therefore that the hieroglyphs cannot express the sounds of the spoken language, then from all we have shown will follow with no less probability the existence and use of two types of Synoptic Tables of hieroglyphic signs among the ancient Egyptians:
> 1. the <u>curiologic table</u> of these signs classed according to their figurative value, in a natural order;
> 2. a table containing the entire series of these same signs divided according to the true meaning of each of them and in classes methodically determined by this meaning; we will call this collection the <u>tropical table</u>.[30]

Canceling the remainder of the sheet and the top of the following one, he also eliminated material concerned with the classification of signs according to their "tropical" (semantic) value in distinction from their purely "curiologic" structure (their shape).[31] Why the redaction?

The talks on hieratic took place on August 23 and September 7, 1821. Three months later, on December 23, Champollion's brief piece on the script, with plates, was published at Grenoble. In January an object arrived in Paris after a journey from Marseilles that, since its description during the Napoleonic invasion of Egypt, had occasioned a great deal of animated discussion and would now do so again: the stone "zodiac" carving blown and sawn out of the temple of Dendera. Champollion soon became involved in controversy over what it represented. His involvement in these debates, together with new hieroglyphs from an obelisk taken from the isle of Philae, signaled the beginning of another significant change in his views. For his intrusion in the affair led Champollion to suspect that his scheme for what he termed a "tropical" classification of Egyptian signs based principally on meaning rested on weaker ground than he had assumed. The arrival of this fresh evidence challenged Champollion's convictions about the nature of the ancient scripts—and may explain the long redaction we have just discussed. On August 22, 1822, almost exactly a year after he had given his hieratic talks at the Académie, Champollion would give another in which he developed related claims for demotic, but with a telling change, as we will see.[32]

PART 4

READING THE PAST

CHAPTER 23

THE OBELISK
FROM PHILAE

Although William Bankes is now primarily remembered as a sometime Tory politician dogged by scandal, as a young man he cut an impressive figure as a wealthy adventurer and collector with a special interest in Egypt.[1] The second son of Henry Bankes, William attended Trinity College, Cambridge; following the death by shipwreck of his older brother, he became heir to the family's estate. Economic security freed him to travel, and his trips in turn provided many opportunities to expand his collection of antiquities. On a first trip to Egypt in 1815, he copied hieroglyphs and inscriptions while appropriating antiquities; what he could not take, he earmarked for removal on a subsequent visit. On that second trip, which occurred in 1818, Bankes carried out a special project assigned to him by Young. Via Bankes's father, Henry, Young issued his request, prefaced by diplomatic phrases: "I doubt not that so enlightened and enterprising a traveller will be as willing as [William] is able to assist in promoting the investigation of the hieroglyphical antiquities of that singular country; and I trust that a few hints of what has already been done will enable him to effect this purpose, with considerably less labour than might otherwise have been bestowed on it."[2]

Young described the hieroglyphs he particularly wanted Bankes to document. These represented "the names of the kings, whom they commemorate, and those of the deities to whom they are dedicated." The deities were "generally distinguished by a hatchet or a sitting figure which follows them," while the kings' names were enclosed "by an oval." Young included hand-drawn specimens, as he called them, of hieroglyphs representing gods or conceptions, together with the cartouche-enclosed signs of four names, including those he had identified as "Ptolemy" and "Berenice" (fig. 23.1). Young also wanted "the last fragments of the Rosetta Stone." Clarke had apparently seen a duplicate at the "Institute at Cairo, but in imperfect preservation."[3] Via Salt, who was then the consul-general, Henry Bankes transmitted Young's wish list to his son William, who was by then in Egypt.[4]

On Bankes's first trip to Egypt, an obelisk on the island of Philae, near Aswan, had captured his attention. It is not hard to understand why. By this

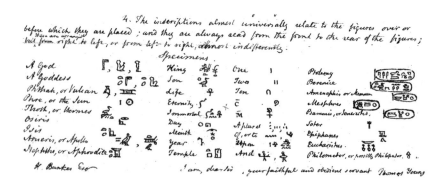

FIGURE 23.1. Hieroglyph signs for which Young asked William Bankes to seek out examples. © The Trustees of the British Museum.

time, these striking objects were rarely found in situ. They had excited so much greed, and for so long, that only a few remained in Egypt. Musing on some ruins while attached to Napoleon's army in Alexandria, the artist Denon had reflected that an obelisk "at once becomes a trophy of conquest, and a very representative one at that," because "every one of them is itself a monument."[5] But the obelisk at Philae had another irresistible quality: it was found close to an object that seemed to be its base, from which it had become separated over the centuries. The presumptive base bore a Greek inscription including royal names.

While making his way back to Philae on his second trip, Bankes stopped north of Luxor to visit the ruins of two temples at Hu, known under the Ptolemies as Diospolis Parva.[6] The propylaeum on one of these temples bore a Greek inscription in which the name "Cleopatra" was followed by "Ptolemy," an inversion of the usual sequence. It was believed that the inversion reflected Cleopatra VII's regency for her son by Caesar. Among the hieroglyphs on the same building, Bankes discovered another curious pairing. One cartouche contained Young's sequence for "Ptolemy" and was located near a male figure, while the other nearby cartouche was located near a female figure.[7] Bankes conjectured that the latter represented Cleopatra.

Moving on to Philae, Bankes next uncovered "a little temple" that, like the plinth found near the Philae obelisk that he had seen in 1815, also bore an inscription in Greek for "the same sovereigns," namely, Ptolemy and, in two instances, Cleopatra. Turning to the obelisk, Bankes searched for the signs Young sent.

Like every obelisk, the Philae has four faces; each one is covered in hieroglyphs. These inscriptions contain six cartouches: one on each of three sides and three on the fourth. Two of these cartouches contain the signs that Young had identified as representing Ptolemy. Of the four remaining cartouches, one occurs twice, but neither of those appears on the two faces containing the name Ptolemy. Labeling the sides of the obelisk A, B, C, and D, and using Young's identifications, side A bore one non-Ptolemy cartouche, side B bore a Ptolemy followed by two other different cartouches, side C had the same non-Ptolemy cartouche as side A, and side D had the second Ptolemy (fig. 23.2).

Bankes at once recognized Young's signs for Ptolemy. Then, just below the one on side B, he saw signs that were identical to those on the Diospolis Parva temple that he had just linked to Cleopatra. He did not even need the obelisk's base to effect the identification, which in any case posed no problem because the obelisk's inscription clearly could not be a direct version of the one on the plinth. Whereas the Greek on the base contained the name Cleopatra twice near the name of Ptolemy, the Cleopatra cartouche that Bankes identified was not the same as the only non-Ptolemy cartouche on the obelisk that occurs twice. Bankes's Cleopatra did occur near a neighboring Ptolemy, but only once. Bankes relayed his observations to Salt and Young.[8]

The Philae obelisk had already become a flash point for disputes between the French and the British regarding the ownership and removal of antiquities. By the time Bankes arrived in 1818, the dust had settled sufficiently to permit Giovanni Belzoni to extract the obelisk, together with the base, at Bankes's behest. According to Belzoni's later recollection, the extraction was a complicated affair, suffused with greed and beset by misfortune.[9] Both obelisk and plinth reached England in 1821. When the monument arrived at the Royal Navy dockyard at Deptford outside London, Bankes commissioned the lithographer Charles Hullmandel to make prints of all four faces and the Greek-inscribed plinth (fig. 23.2).[10] Both obelisk and plinth were shipped to Kingston Hall (now Kingston Lacy), the Dorset estate of the Bankes family, where they were installed on the lawn, upon a pedestal of bricks. The monument remains to this day, though the inscriptions have suffered greatly from erosion over the decades.[11]

As in the children's game, paper triumphed over rock: although the granite obelisk suffered under the onslaught of Dorset's weather, its images proved both more durable and easier to circulate. The obelisk had first been sketched by Bankes in 1815.[12] The base, with its Greek inscription, was copied by him and by the mineralogist Frédéric Cailliaud, whose drawing arrived in France in November 1818.[13] Cailliaud's rendition eventually came to the attention of the historical geographer Jean-Antoine Letronne who, in late 1821, became involved in Egyptian matters while preparing a lengthy work on Egypt under the Greeks and Romans.[14] After obtaining Cailliaud's copy of the plinth's Greek

FIGURE 23.2. (*From left*) First, the trio of cartouches from side B of the William Bankes obelisk, the topmost one being identical to Young's "Ptolemy"; (*second*) the Ptolemy cartouche from side D; (*third*) Young's signs for "Ptolemy" in his letter to Bankes (© The Trustees of the British Museum); (*fourth*) the second cartouche of the trio on the obelisk's side B that Bankes identified as "Cleopatra."

from Jomard, Letronne published a transcription and French translation.[15] According to his version, the Greek began: "To the king Ptolemy, to the queen Cleopatra his sister, to the queen Cleopatra his wife, [to the] gods Euergetes, greetings." Because the dedication mentioned two queens named Cleopatra—one a sister, the other a wife—Letronne claimed that the dedication on the plinth must concern Euergetes II (ca. 182–116 BCE). He opined that Euergetes II had at first married his sister, wife of Philometor, his deceased brother, but had then "repudiated" her to marry her daughter instead, who was also his niece and also named Cleopatra—hence the dedication.

Letronne went on to discuss how all this might have come about in the light of dynastic intrigue. The point for our purposes, however, is that the name Cleopatra appeared twice just after Ptolemy on the plinth. Neither Letronne nor anyone else in France had as yet seen any drawings of the obelisk, but he nevertheless conjectured, correctly as it turned out, that the obelisk's hieroglyphs were not an essentially similar rendition of the Greek inscription.[16]

Then, in November 1821, Bankes published Hullmandel's lithographs.[17] He sent copies to several recipients in Paris, including Denon and Jomard at the Académie des inscriptions.[18] Jomard handed them to Letronne, who in turn gave copies to Champollion and to Champollion's old friend, Saint-Martin.

Years before, Saint-Martin had been de Sacy's student in Arabic, Turkish, Persian, and Armenian. He was also, we have seen, a friend of the young Champollion during the latter's two years of study in Paris, and he had proved supportive in Champollion's fight with Quatremère. Enough had changed in the intervening decade to make that delicate balance of loyalties unsustainable. In 1820, a year before Champollion returned to Paris, Saint-Martin was elected to the Académie with de Sacy's support and had become a fervent devotee of the Restoration.[19] In an early recognition of the widening incompatibility of their political commitments, Saint-Martin and Champollion had forged a tentative peace pact three years before, with Champollion's stipulation that they never discuss politics.[20] It seems that Saint-Martin did not hold up his end of this bargain. Champollion complained of Saint-Martin's "hypocritical Jesuitism," to which Champollion-Figeac responded drily that Saint-Martin "expected mountains of benefits from Sacy, but I don't know whether in this world or the next."[21]

On February 8, 1822, Saint-Martin gave a talk at the Académie about an Egyptian antiquity recently arrived in France, the zodiac that had been extracted from the temple of Dendera in Egypt, drawings of which had occasioned much debate years before concerning its age.[22] He had not by then actually seen the object, but he worked from a print engraved by Louis-Jean Allais from the drawing made by the engineers Jollois and Devilliers during the Napoleonic expedition for the *Description* (fig. 23.3).[23]

In his talk, Saint-Martin tried to integrate problems of Egyptian history raised by the zodiac with his commitments to Restoration politics, all the while presenting himself, and his argument, as a model of evenhandedness. To start, he presented the major claims concerning the object that had appeared since 1800. These were mainly concerned with whether the inscriptions suggested that it predated the origin of the world according to Mosaic chronology. Such claims had years before been based on a drawing of the object by Denon and later on the *Description* plates. Although Saint-Martin's tone was restrained in comparison with the vituperativeness that typically infused discourse about these matters, he staked out a provocative position, agreeing with none of those arguments and finding points to criticize in all. He temporized that the Dendera zodiac was neither so old as savants, such as the mathematician Fourier had calculated—with dire implications for biblical chronology—nor so young as others had claimed. He doubted that much could be done by way

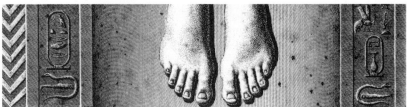

FIGURE 23.3. (*Top*) The Dendera zodiac as depicted in the *Description de l'Égypte*'s engraving based on the drawing by the engineers Jean-Baptiste Jollois and René Édouard Devilliers; (*bottom*) the two cartouches engraved at the bottom mentioned by Antoine-Jean Saint-Martin.

of interpreting the object in astronomical terms (as had been done by such savants as the mathetician Fourier), but he was also certain that "the Dendera planisphere is a production of the art and science of the Egyptians," even if it had perchance been produced during Greek or Roman times (which, however, he also doubted).[24] In support of this latter point, Saint-Martin discussed two cartouches appearing beneath the female figure on the side of the zodiac to either side of the figure's feet (fig. 23.3). These, Saint-Martin asserted, must like others contain

> the name of the king under whom the ceiling containing the piece was made, and the name of his father. The name of the king is composed of two characters; the first is a rectangle open below, the second an elongated vase. The same name is found at Thebes in the ruins of Medinet-Abou. As to how these two characters should be pronounced, and in consequence which king they're about, we know nothing; understanding of hieroglyphic writing is yet too little advanced for us to be able to do so with certainty; what's certain, is that this prince is not one of the Ptolemies, their name being too well known from the Rosetta inscription for us to be mistaken.[25]

He spied the same two signs in the same configuration in a cartouche in plate 7, volume 2 of the *Description*, a fact that will be of considerable import.

Saint-Martin turned next to the Philae obelisk. In a letter dated February 10, Letronne had reported to Saint-Martin that Champollion "agreed" with him concerning the presence on the obelisk of cartouches containing the signs for "Ptolemy" (of which more below).[26] Five days later, Saint-Martin spoke about the object at the Académie, where he attempted to identify the rulers represented in its cartouches. He was certain that all of them concerned the Lagides, that is, the regents of the Ptolemaic era. Like Champollion, Saint-Martin had read Young's *Britannica* article.[27] He noted the signs in the two cartouches for Ptolemy, though he did not mention the Englishman or Champollion.

The Ptolemy cartouche on side B of the obelisk, he explained, was followed by "another cartouche, at a small distance, containing the name of the queen, his wife." That the hieroglyphs in the cartouche referred to a female personage was clear because the feminine signs (again identified as such by Young in the *Britannica*) preceded the others. And that the name in question was Cleopatra followed because Saint-Martin decided, echoing Letronne, that the Ptolemy in question had to be Euergetes II. (Letronne had, of course, already so identified the Ptolemy named on the plinth.) Saint-Martin's assertion effected for the first time a claim that both plinth and obelisk at least referred to the same rulers. He based this conclusion on the other signs, drawing his arguments from complex conventions for expressing Egyptian royal names that, if not perfectly understood, could lead to confusion.[28] Saint-Martin explained that signs for "proper names" of Egyptian rulers were made by associating their phonetic values with ideograms—and he did not limit the assertion to the Greco-Romans of the post-Alexandrian era, though he thought it more common during that period. Saint-Martin's argument closely reiterated the one that, as we saw above, Young had briefly advanced in his *Britannica* article for the name Ramesses:

> To express proper names the Egyptians reduced their symbolic characters to nothing but simple signs of sounds. This was done in two ways: first, they considered the sense of the name in the spoken language; then they made the sound and the meaning; so for Ramesses they took the character *sun*, pronounced *ra*, and a scarab that signifies *generation* and so is pronounced *misi*, and in one way or the other they had the equivalent of *son of the sun*, in Egyptian *ramisi*. In the second place, among the phonetic hieroglyphs and homophones, they chose those that offered the most flattering and happy hints: in this way it was possible to write in many ways one and the same name. This method was perhaps more used for foreign names that had no significance in the common language.[29]

Champollion would respond in short order precisely because Saint-Martin claimed that phonetic characters had been used before the Greco-Roman period whereas, by this point, Champollion rejected any such usage at all, whether for foreign or native rulers.

Champollion's first response appeared in April in a brief piece entitled "On the Egyptian obelisk of the isle of Philae." Although his analysis was based on Bankes's prints, which he already had in February from Letronne, Champollion was notably unimpressed.[30] He disparaged "this English engraving, of very small proportion, executed by artists poorly accustomed to the style of Egyptian monuments" as being "most inferior in all respects to the beautiful drawings of the obelisks given by the Egypt commission." Nevertheless he judged the engraving at least sufficient to "resolve the following question: Was the obelisk of Philae erected by a king of the Egyptian race, or does it belong to the time of the Ptolemies or the Lagide kings, as [stated on] the Greek inscription on the plinth?"[31] He did not address the most pressing question, namely, whether the obelisk's cartouches could be linked in some fashion to Cleopatra. Instead, he aimed to separate the inscription on the plinth from the hieroglyphs on the obelisk, despite the contrary view that he claimed was "generally believed in England" and despite what Letronne had told Saint-Martin in his February letter. The Greek inscription, Champollion insisted, might not refer in any way at all to the obelisk. In that case, one could not reason from the Greek Cleopatra on the plinth to any cartouche-enclosed name on the obelisk. Although he did not explicitly make this claim, it was his central point.

Champollion's principal evidence depended on the Greek inscription's referring to a "stele": "what the Greeks called a *stele* cannot in any way at all be confounded with an *obelisk*." For, he continued, a stele is a rock slab such as the Rosetta Stone, and among the Rosetta's hieroglyphs is a sign that resembles such a slab. In contrast, the hieroglyphs on the Philae include no signs for such a stele or slab; instead they have signs for two obelisks. Champollion surmised that there was probably another obelisk "still existing among the ruins." Such a doubling would not have aroused suspicion, for obelisks were known to appear in pairs.[32] The Greek inscription by the Philae priests promised to memorialize their recognition of the "gods Euergetes" on a monument, in which case they would have done so on a stele like the Rosetta, Champollion concluded, "and not on one or two obelisks."[33]

Although Champollion's response amounted to little more than a development of Letronne's 1821 conjecture that the obelisk hieroglyphs did not constitute essentially the same passage as the Greek inscription, it also advanced a subtle critique of Saint-Martin's argument.[34] Beyond the absence of any sign on the obelisk resembling a slab, Champollion looked to cartouches that could contain royal names. One of these, he noted, did refer to Ptolemy. Observing that the name Ptolemy could be found on hieroglyphic inscriptions on many Egyptian monuments, he cataloged a half-dozen such instances that could be found in the *Description*'s plates.[35] Although Champollion did not admit that he had first known hieroglyphs for the name from Young's *Britannica* piece, he pointed out that "an English savant inserted, in the *Encyclopedia Britannica*, the hieroglyphic names and surnames of nearly all the Lagides" and expressed skepticism about this result: "we also think that this work is at least premature."[36] This was somewhat disingenuous. Just before asserting Young's list to have been "premature," Champollion had himself claimed to have found "only three different hieroglyphic legends representing the three names of the Lagide queens, and it is very remarkable that all the queens of this dynasty bore in effect only three names, those of *Berenice, Arsinoe, and Cleopatra*."

In his article on Philae, Champollion did not specify just how he had identified signs for the names Berenice, Arsinoe, and Cleopatra; nor did he explicitly assert that the Cleopatra signs were present on the Philae obelisk. However, the first two had been identified as such by Young in the *Britannica*, and though we have no direct report on this matter from Bankes himself, Bankes was claimed to have identified the name Cleopatra on the obelisk based on his earlier conjecture for such signs in the temple at Hu. Moreover, on at least one of the lithograph copies of the Philae that Bankes distributed in France, he had indicated in the margin the cartouche representing Cleopatra.[37] Champollion's omission would soon generate considerable enmity, as Salt reported that Bankes later became convinced that Champollion had identified the Cleopatra cartouche from Bankes's marginal annotation.[38] Still, given Champollion's

pointed skepticism concerning Young's "premature" identification of signs for
"nearly all the Lagides," he would hardly have taken Bankes's marginal notation at face value, assuming he had seen it in the first place, which we can be
reasonably sure he had. Bankes had merely pointed to the Cleopatra cartouche
without further analysis or explanation.[39] As a basis for a priority claim, that
would certainly not have satisfied Champollion.

We do not know whether Champollion had attended the meeting at the
Académie at which Saint-Martin spoke on the Philae obelisk, but he certainly
had heard enough about it to form his Philae article in partial opposition. Not,
significantly, to Saint-Martin's claims concerning phonetic structure, which
he avoided altogether, but rather to his explicit identification of the Ptolemy
in question as being Euergetes II: "one can only conjecture uselessly as to
[which] Ptolemy [is the] author of the Philae obelisk," he wrote, contradicting what, after all, had been Saint-Martin's central claim. In April, following
his own publication, Champollion undoubtedly had access to the article that
Saint-Martin wrote following his talk at the Académie, including its claim
concerning phoneticism.

In May or June Champollion turned for the first time to the Dendera zodiac.
Given his political difficulties, Champollion surely would have preferred to
avoid the subject. The object's arrival in France had not only revived public
interest in Egyptian antiquity but had also raised considerable ire among the
Ultras, who thought the zodiac an "infernal" pagan relic useful only to attack religion. By early summer, Champollion could no longer remain aloof
from what was quickly becoming a furor. Saint-Martin, of course, had already
weighed in. In July, the physicist Jean-Baptiste Biot heightened the drama by
claiming to use astronomy to produce an unequivocal date for the zodiac's
construction. We now turn to the zodiac's remarkable effect on Champollion's
subsequent career.

CHAPTER 24

A SINGULAR
AND PUZZLING
ARTIFACT

Having been exploded and sawn out of a temple ceiling, transported down the Nile and across the Mediterranean, and quarantined for a month at Marseille, the Dendera zodiac finally arrived in Paris in early 1822.[1] The "infernal stone," as it was known to religious conservatives who feared its challenge to biblical chronology, sparked widespread enthusiasm for all things Egyptian. Champollion remained aloof from the excitement, seeing little to celebrate in an object ripped from its surroundings and so denuded of its material connection to ancient Egypt. In an anonymous note printed in late 1821 in the *Revue Encyclopédique*, Champollion invited readers to imagine themselves at Dendera, looking up and through the hole where the zodiac had been: "Today, solely as a result of the zodiac's removal, the astronomical room is uncovered, and the rest of the ceiling is menaced with complete destruction; it's as though the allies had removed a part of the ceiling in the great gallery of Versailles to take some paintings; what will become of the rest of the roof and of the gallery itself?"[2] Relocated to Paris, the zodiac stood alone, a singular and puzzling artifact, symbolizing an atomized conception of history according to which the past could only be reconstructed from disconnected, unrelated objects. "Who knows," he went on (presciently, as we shall see), "whether we won't argue about the place it occupied in the great monument from which it has been torn, about its orientation, about the sculptures that surrounded it, etc.?"[3]

Champollion's protest availed nothing. The zodiac was immediately absorbed into the French scientific milieu, as it gave frank expression to the paradox at its heart: Of what use was the past, except as fuel for a technocratic future? The retreat of the Marquis de Laplace, the doyen of French science, at Arcueil outside Paris followed this pattern precisely: it was a shrine to the future as far as it was scientific, yet every detail of architecture and landscape gestured toward antiquity. Six months after the zodiac's arrival, Benjamin

Dockray, an English visitor, was entranced by the play of past and present at Arcueil: "On alighting, we were conducted through a suite of rooms, where, in succession, dinner, dessert, and coffee tables were set out," in anticipation of the gathering, which included some of the most influential and creative scientists of the day: the atomist John Dalton, the chemist Claude-Louis Berthollet, the physicist Jean-Baptiste Biot, the mathematician Joseph Fourier, and, of course, Laplace himself. Dockray was led "onwards through a large hall, upon a terrace, commanding an extent of gardens and pleasure-grounds. There was a sheet of water in front, and a broad spreading current pouring into it from some rocks, where was seen a sculptured figure, an antique found in the locality, representing the genius of the place." This was no mere decoration, as Dockray discovered: "It is in these grounds that are still remaining the principal Roman works near Paris, the vestiges of Julian's residence, as governor of Gaul." Stunned by the view, Dockray was ambushed by his host and the other guests, figures of such grace and elegance that Dockray, observing their approach, was moved to wonder, "Was it possible not to think of the groves of the Academy, and the borders of the Illyssus?"[4]

Given the many exciting and contentious developments in science that had taken place in recent years, the conversation over dinner that night might easily have turned to the workings of matter and force, or the nature of light. It didn't. Table talk orbited instead about the things of antiquity: "the zodiac of Denderah and Egypt, Berthollet and Fourier having been in Egypt with Napoleon; the different eras of Egyptian sculpture; the fact that so little at Rome—of public buildings—is earlier than Augustus, &c." No doubt Fourier had tales to tell about his trek up the Nile, while both he and Berthollet could regale the dinner party with romantic stories about Cairo and the vicissitudes of the expedition. Dockray didn't record any remarks by Biot, but a week later, on July 15, he spoke on the subject at the Académie. There Biot made an elaborate attempt to turn the Dendera zodiac into a triumph of astronomical representation and measurement.

Convinced that the object was a planisphere drawn by the ancient Egyptians to represent the positions of the stars at the time of its production as seen from Dendera, Biot (who was granted direct access to the zodiac by the king) developed a complex procedure to measure signs on it that he took to be stars, from which he expected to calculate the zodiac's date of creation. Many of his exceedingly tiresome arguments—a single paragraph could go on for pages—accordingly aimed to explain why *this* symbol had to be interpreted as *that* star. Biot insisted, for example, that a recumbent cow sign, which had long been taken to symbolize Sirius, did not represent the star's true position, even though the cow has a distinct star over its head. Rather, Sirius the star was represented by the symbol to the cow's right, a hawk perched on a lotus leaf. Why? Because a line drawn through that symbol and the center of the zodiac would then be parallel to the Dendera temple's long axis, and Biot took this to symbolize the importance of Sirius itself, expressed in the very architecture of the temple. The star-topped cow was placed nearby as a sort of pointer to the neighboring position of that important luminary. In the end Biot produced an astonishingly precise date: the Dendera zodiac represented the heavens as they stood in the year 716 BCE. The zodiac therefore dated neither to remote Egyptian antiquity, as had been suggested, nor to the Greco-Roman period, and the ever-industrious Biot had proved it to be so by an irrefutable astronomical calculus. Or so he argued.

The engineers Jollois, Devilliers, and Charles Hippolyte de Paravey contested Biot's effort on astronomical grounds. Others took exception as well, using different arguments informed by various interests. Some strove to save France from the dangers of impiety, while others were perturbed by this latest invasion of astronomy into history. Abbé Halma, for instance, wrote in connection with the Dendera zodiac that of "all the errors of which one has endeavored to convict the sacred books, none would be more grave than the epoch" that the impious astronomers have "assigned to the creation. . . . They obtain warrant for so claiming by means of some ancient monuments that were furnished by paganism and interpreted by *philosophie*."[5]

Around this time, someone took a sufficient interest in the zodiac debates to collect, annotate, and bind a series of reprints on the subject. All of the articles were printed between 1822 and 1824, during the high point of the affair's revival in France. Impressed on the spine of the resulting book was the word *Zodiaque*. A pamphlet in this collection is dedicated "by the author to M. le Comte de Chabrol, councilor of state, Prefect of the Seine, as a testimony of his old friendship." The pamphlet's author, Honoré Dalmas, had been in Egypt with Napoleon's expedition, where he had acted as administrator of one of the provinces and was by this time a civil engineer in Castelnaudary. He likely had known Chabrol—who was, like Dalmas, an engineer in Ponts et Chaussées—in Egypt. Whoever annotated the *Zodiaque* pamphlets was also able to calculate, knew some astronomy, and was familiar with Egypt. As we shall see in a moment, this person also had access to the plates in the antiquities volumes of the *Description*. Though we cannot be certain, it seems probable that the collector and annotator was Chabrol himself.[6] Saint-Martin's piece, included in the volume, was of special interest because the controversial question of the zodiac's age might be solved, Saint-Martin was suggesting, if only the names in the cartouches next to the feet of the figure that borders the zodiac could be read. Chabrol, if he was the annotator of

FIGURE 24.1. Gaspard de Chabrol's cartouches from the Dendera zodiac, based on the *Description de l'Égypte*'s rendering (cf. fig. 23.3, *bottom*).

the pamphlet, thought the point sufficiently important to pencil drawings of them in the margins (fig. 24.1).

Between the cartouched hieroglyphs noted by Saint-Martin and Chabrol, and the mysterious star-shaped signs scattered on the zodiac itself, the zodiac suggested several methods by which it might be dated. Responding to Biot's claims in a signed letter to the *Revue*, Champollion cautioned against attempts to do so by construing the zodiac as an exact planisphere, as Biot had. Since Champollion was friendly with Biot, whom he'd met years before in Grenoble, he may have given warning prior to publication, perhaps around mid-July when Biot had first spoken at the Académie on the subject.[7] In any event, Champollion's formal response to Biot's astronomical dating appeared toward the end of July, just a week after Biot read a second paper on the zodiac before the Académie.

"Here," Champollion wrote, we have "a new opinion on the presumed epoch of an astronomical tablet [*tableau astronomique*] that, for twenty years, has been successively the occasion for and the subject of a crowd of theories [*foule de systèmes*], all contradictory." Though commentators were not inhibited by their ignorance, the fact remained that few had training adequate to the task: "Those who wanted to explain it, and who claimed to draw from it rigorous consequences, were more or less well prepared to attempt this difficult enterprise depending on the type and direction of their studies." Champollion insisted that it wasn't enough "to possess the learned theory of modern astronomy." One needed "also to have exact knowledge of that science as the Egyptians themselves conceived it, with all of its errors and in all its simplicity." This was necessary because "Egyptian astronomy was in its essence mixed with religion." Without such an understanding "the courageous explorer of the Dendera monument will find himself on dangerous terrain; he chances mistaking a religious [*culte*] object for an astronomical sign, and considering a purely symbolic representation as the image of a real object." But the major barrier to any attempt at understanding this Egyptian monument had to do with writing itself, because of the difficulty of distinguishing, on these ancient tablets, images that truly represent, either literally or figuratively, the bodies or the *celestial signs*, from images that pertain solely to the system of *Egyptian writing*, and which appear on the zodiacs solely as simple signs of *ideas*, with which their forms have no sort of relation whatsoever. This distinction requires a long acquaintance with the Egyptian monuments; and one can say that up to now few archaeologists have perceived the extreme importance of doing so.[8]

Sidestepping Biot's calculations, Champollion addressed himself to the core problem as he understood it: Biot's identification of four principal stars on the monument. Using as his source the engraving of the zodiac that had been sold together with Sebastien Saulnier's brochure (fig. 24.2), Champollion urged his reader to look at the object as a whole.[9] Many images of men or animals appear on it, nearly all of which are accompanied by a small group of hieroglyphs either above or next to the image. Nearly every such group includes a starlike symbol. There are thirty-eight groups altogether, thirty-three of which sit on the circle held up by the kneeling figures. According to Biot, four among these represent the locations of the principal stars that he had used to ground his calculation. But if that were so, Champollion countered, then it ought to be the case that the remaining twenty-nine should represent stars as well. Why? Well, why not? Why would the Egyptian designers have drawn so many similarly placed figures only to have just four among them represent physical objects? The "analogy of position," Champollion averred, argued for "an analogy of expression." But this could not possibly work because all of the images were symmetrically arrayed on the circle, and nothing astronomical could correspond to the remaining twenty-nine in the array. Champollion didn't need to know much about astronomy to reach this conclusion, because in Biot's interpretation the circle in question was centered on the monument's center, calculated by Biot to be the North Pole. Consequently the circle itself must represent a parallel of celestial latitude, and even a quick glance at a celestial globe would show that nothing of the kind occurs: there simply is no latitude with such visibly noteworthy and symmetrically disposed stars arrayed along it.

If the objects that looked like stars in all but one of Biot's groups did not represent stellar locations, then what were they for? "We recognized," Champollion explained, "that every hieroglyphic group, placed on the head or by the side of the image of a god, a man, an animal, etc., expresses its proper name, or at least a particular qualification devoted to it." Moreover, each such group must be read from the direction toward which the heads of the figures face. The star-figured symbols were therefore the *last* in each group naming the figure.

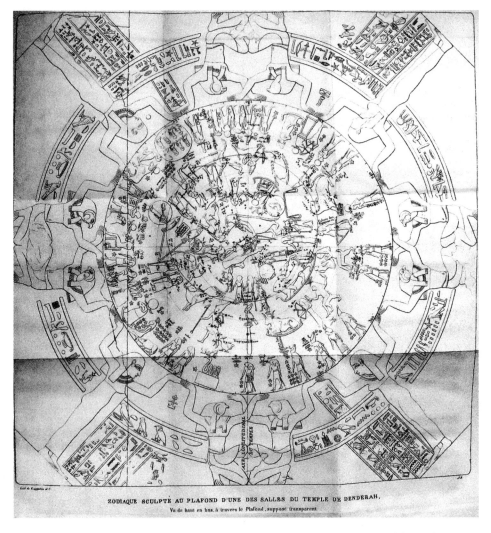

FIGURE 24.2. Jean-Baptiste Biot's marked star signs on an engraving by François Gau of the Dendera zodiac.

Their regularly occurring terminal positions implied that they were particular kinds of elements in the hieroglyphic script. "The positions of real stars that one would like to deduce from these figured stars are therefore completely without foundation," he continued, "because the figured stars are all parts of the *proper names* related to representations of characters." Taken in isolation, these figured stars "could hold the true place of the constellations, if we were to attach an elevated importance to the relative place that these stars occupy and consider that as fixing their absolute positions," but this was something "which I do not think we should do. These stars, we repeat, are nothing but hieroglyphic signs that contribute to forming a proper name."[10] As such, these signs should be considered "as a sort of letter, and not as the imitation of an object."

"A sort of letter": the Dendera tablet is a text, not an image. It does not *depict*; it *names*. It does so, moreover, by means of "simple signs of ideas," for the zodiac's stars are precisely the type of "logical signs"—determinatives that qualify other signs—about which Champollion would speak at the Académie the following month. He had first discussed signs of this type in his lecture at Grenoble where he had emphasized their importance for specifying the meaning of certain sequences. As instances of this type of sign, the zodiac's stars were not, he would soon insist, merely addenda to scripts otherwise comprised of signs for "ideas." Not at all: with his critique of Biot, Champollion had begun to consider that such elements, if not truly central to the scripts, were nevertheless frequently deployed in a spectrum of contexts. Precisely because he remained certain that the Egyptian scripts had been designed to represent the course of Egyptian speech—of ancestral Coptic—through signs linked to words but not to sounds, Champollion fastened on these logical signs as elements that helped ensure the viability of such a scheme. Without qualifiers to indicate what a passage was about or to designate gender or number, signs for words might proliferate essentially without limit, and any connection to spoken language would be lost. The logical signs seemed to him helpmates that refined basic meanings represented through signs for words. As such, they were a further indication that Egyptian writing was not natively phonographic. But this was not the only stimulus to the evolution of Champollion's understanding in the summer of 1822. Other developments would soon lead him in a considerably different direction.

CHAPTER 25

A MOMENTOUS CHANGE

On August 16, just weeks after Champollion sent his letter about Biot's work to the *Revue*, Saint-Martin spoke again at the Académie des inscriptions. Published in the September issue of the *Journal des Savans*, Saint-Martin's talk preferred tantalizing specimens of Egyptian writing that had recently surfaced in Paris. Known as the Casati papyri for the otherwise obscure traveler who brought them to France, this clutch of papyrus rolls bore inscriptions in hieroglyphs, hieratic, and Greek. One of the Greek rolls also contained text in a script like that of the Rosetta Egyptian. Saint-Martin was allowed to handle this last roll and two others written in Greek, but only for an hour, under Casati's watchful eye.[1] Despite its brevity, the encounter gave rise to a fresh set of conjectures.

Casati's papyrus joined the Rosetta Stone, Saint-Martin wrote, as "the only considerable monuments that have been found" containing both Greek and demotic for the same passages. On the roll were thirty-seven lines in demotic and six incompletely legible ones in Greek. What he could read of the Greek convinced Saint-Martin that the roll was a contract for a sepulchre written on the ninth of Choiak (December) in regnal year 36 at Diospolis the Great (Thebes). He identified several names as well: Lysimachus, Ptolemy, Léon son of Ronnophris, and son of Orus, the last of whom would do the work for 603 pieces of copper.

Spurred by motivations similar to those prompting his analysis of the Philae plinth's Greek inscription, Saint-Martin was again principally interested in determining under whose reign the contract had been concluded. He settled on Ptolemy Soter II, an apparently odd result since that sovereign had not reigned continuously for thirty-six years. He justified the anomaly by explaining that having been "named king by his mother Cleopatra, [Ptolemy Soter II] reigned six years with her. This ambitious woman then chased him away and gave the crown to Alexander, another son, while Soter was obliged to content himself with the isle of Cyprus. Only after the deaths of the one and the other could Soter again seize the scepter, which he held for eight years until his death." Saint-Martin asserted that the regnal length would have been counted from its initiation with Soter II's mother.[2]

FIGURE 25.1. (*Left*) "Ptolemy" reproduced from the Casati papyrus in 1823; (*right*) Champollion's own versions of "Cleopatra" and "Ptolemy" from the Casati in August–September 1822 (Bibliothèque nationale de France).

Hoping to corroborate his dating, Saint-Martin searched for the names of Ptolemy and Cleopatra in the accompanying, more completely legible Egyptian writing. Paradoxically, it was a very slight difference between the Rosetta sequence for Ptolemy and the corresponding signs on the Casati papyrus that secured Saint-Martin's identification on the latter of signs he identified as Cleopatra (fig. 25.1). In making this identification, Saint-Martin followed Åkerblad and Young in claiming that the angle bracket signified "T." The Rosetta sequence for Ptolemy does not contain what Saint-Martin referred to as a "trait" or "diacritic" mark, which appears in an otherwise similar set of signs on the Casati. (In the version of the Casati that Young later published, one such diacritic appears to the right of the angle bracket, while Champollion, in an unpublished manuscript based on a direct examination of the papyrus, placed the diacritic beneath it.)[3] Saint-Martin convinced himself that the name Cleopatra was present on the Casati because the angle bracket and the singular accompanying diacritic appeared in a sequence just to the right of signs that also included such a mark; except for that, these signs were identical to those for Ptolemy on the Rosetta. The extra mark thus seemed to firmly cement the nearby presence of a sign for "T," in which case the name of Cleopatra, he decided, must also be present phonetically.[4]

Champollion learned of Saint-Martin's claims in mid-August, either directly during the talk at the Académie or from a secondhand account. Except for Young himself, no one since de Sacy or Åkerblad had worked to any comparable extent on the demotic script. A great deal was at stake: Champollion's politically unsavory reputation made his scholarship hard to accept under the best of circumstances. Now it seemed that the more palatable Saint-Martin might overtake him. Champollion-Figeac pressed his brother to recapture the field by giving another lecture at the Académie. Champollion at first demurred on the grounds that only his hieratic work was as yet "in a state of perfection," and he had already gone public with that. Nevertheless Champollion-Figeac prevailed—and so, six days after Saint-Martin's August 16 talk, Champollion gave a lecture of his own at the Académie, this time on demotic.

In that lecture, never published but preserved in manuscript, Champollion developed a series of significant claims: first, that demotic, like hieratic, shared its fundamental nature with hieroglyphs; second, that the scripts were never intended to be alphabetic; third, that they incorporated syntactic structure; fourth, that neither hieratic nor demotic was a "degraded" form of hieroglyphs though related to them; and, finally, that the two scripts differed from one another only in the shapes of their respective signs.[5] Although Champollion was lecturing in response to Saint-Martin's recent presentation of new work, his true interlocutor was Young. A major aim of his talk, but not his earlier one on hieratic, was to argue that Young's views, as expressed most recently in his *Britannica* piece, were wrongheaded, with the further implication that Saint-Martin's contributions would appear to be comparatively trivial since they seemed to be at best a codicil to Young's. For, Champollion argued, neither of them, nor even de Sacy or Åkerblad, had understood the relationship of the ancient scripts to the native language.[6]

After acknowledging that de Sacy had "presented the first true result" by identifying which demotic sign sequences corresponded to the names Alexander, Alexandria, Ptolemy, Arsinoe, and Epiphanous, Champollion turned to Åkerblad, who had expanded the set of sequences to include the names Berenice, Pyrrha, Philinus, Irene, Areia, and Diogenes, as well as signs for Egypt, king, Greek, temples, and others. Champollion observed that Åkerblad's "Egyptian alphabet" for the demotic required multiple signs for one and the same putative vowel or consonant. Such a scheme could not possibly work, he argued, because "a system of alphabetic writing formed of elements susceptible of taking such varied forms, and above all of signs representing at once three or four different vowels, in addition to certain consonants, cannot exist among any people however barbarous it once had been." Previous decipherments of Palmyrian, Persian, and Sassanid had, he insisted, involved sign-letter consistency.[7]

Champollion rejected Young's expansion of Åkerblad's system on similar grounds, claiming that it was "even more variable than" the Swede's. The main problem was that "in the entire text of the [Rosetta] inscription, this savant [Young] was unable to find a single Egyptian word similar, not only to those contained in the oldest Coptic texts or in the most modern ones, but also to alleged Coptic words." If the expanded alphabet were correct, then, based as it was on Coptic analogs, there should be sign sequences in the demotic that were phonographically similar to Coptic words. Both Young and Akerbald had been led into error by their inadequate knowledge of Coptic and their belief that the language was vastly degraded from ancient Egyptian. "The more these two scholars were imbued with a strongly established prejudice tending to the belief that the <u>Egyptian language was transmitted to us in a state of complete corruption by the books of the Egyptian Christians or Coptes</u> [*sic*]," the further they were led astray. The result was "inextricable confusion of forms or values" within their presentations of the demotic, a confusion that was "particularly striking in their alphabets and above all in the signs representing vowels. We think, on the contrary, that this idea rests on no solid grounds, that certain Coptic texts, in Thebaic or Memphitic dialect, for example, preserve very purely the language of the ancient inhabitants of lower Egypt, or the Thebaid, or of Nubia." These texts, he continued, "offer us only pure Egyptian words, if we abstract from the Egyptian language the easily recognized Greek words that were introduced through a long and inevitable mixture of the two peoples."[8]

Neither Young nor Åkerblad—nor, by implication, Saint-Martin—had realized that the Egyptian writing system could fully represent the spoken language without conveying its actual sounds. Champollion instead maintained that the scripts could be read by identifying Egyptian signs with Coptic words—*not letters*—and the order of signs with Coptic grammar. Young and Åkerblad had, in his opinion, constructed alphabetic fantasylands that pointed to imaginary soundscapes that traduced the structure of the underlying language. Champollion's certainty about this ran so deep that he drew up a version of the Rosetta demotic entirely in Coptic, titling the manuscript "Rosetta Inscription. Oral Language" (fig. 25.2, *top*).[9] His unique insight lay precisely here:

> So my work on this [demotic] text took a sure path; it was in truth ever slow, but it led to results founded on a well-established principle. Ceasing altogether to seek alphabetic analogies in the groups [of characters] of the inscription, and penetrating to rules that necessarily preside over the combination of signs of ideas, I reached the point of placing under the major part of these groups, with no effort, without supposition, without changing anything, finally without omitting any sign of the Egyptian text, the words of the Greek text that consistently correspond to them. This work is so complete, its parts justify and prove one another. . . .
>
> This aperçu loses nothing of importance, even though the intermediary Rosetta text does not at all express the sounds of the Egyptian language: it's entirely evident that in using a script composed of signs of ideas, the Egyptians could not proceed to make combinations of many such ideas, except in the same order that was already adopted for expressing them by means of the sounds of the spoken language. The thoughts, the judgments, in a word the generation of ideas, was essentially tied to the state of the language spoken.[10]

As Champollion saw it, any reading of a written passage had to consider the exact sequence of the signs used to construct it because the sequence represented speech. This was precisely the procedure he had followed when

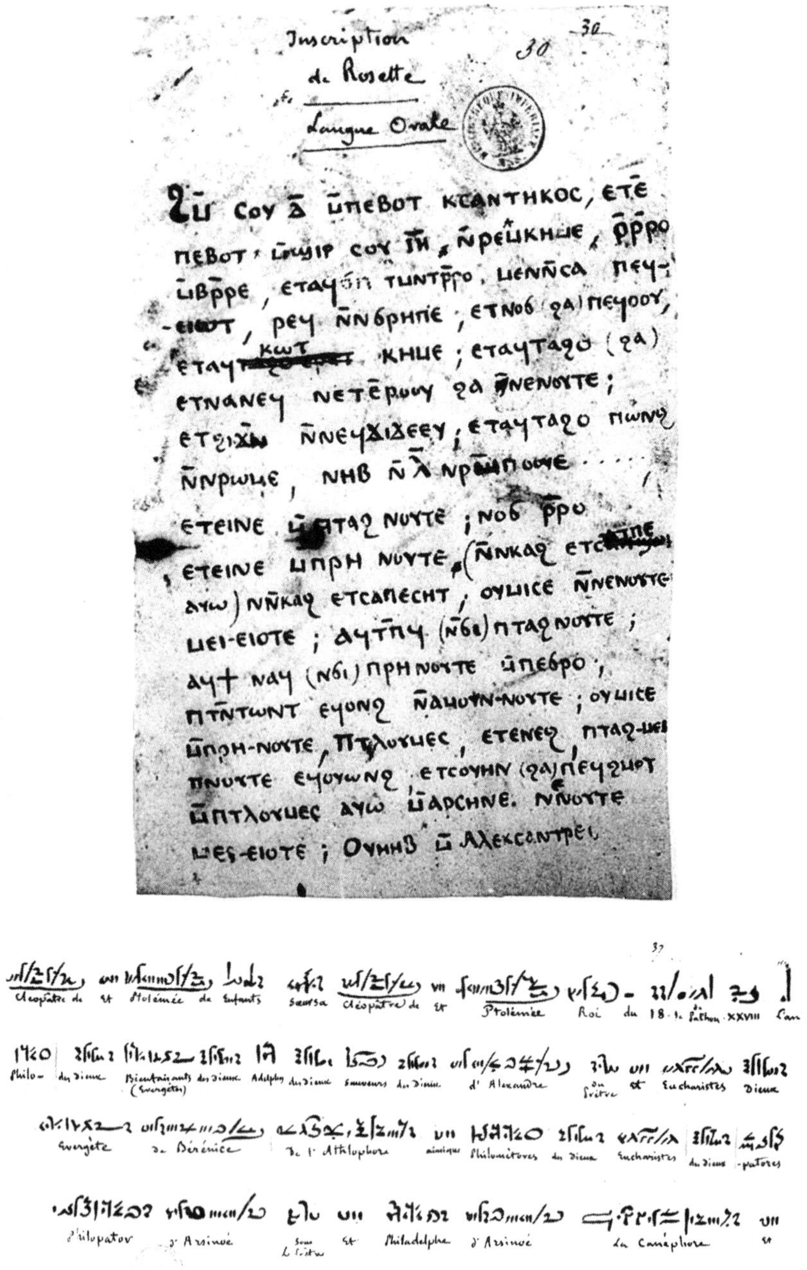

FIGURE 25.2. (*Top*) Champollion's rendition of part of the Rosetta demotic into the Coptic language and (*bottom*) his rendering of a demotic line into French. Bibliothèque nationale de France.

constructing his Coptic analog of the Rosetta demotic. Such a close relationship between spoken language and script had a direct implication. Unless the
language itself had radically changed at the syntactic level, all systems of inscription used to express that language must be essentially similar to one another, though allowances could be made for differences relating to a particular
subject or audience. Young was consequently mistaken in considering demotic
a "degraded" form of hieroglyphs for precisely the same reason he was wrong
to consider Coptic a degraded form of ancient Egyptian. Herein lay the core
of Champollion's own originality: "the only point in which our work accords
with [Young's] is the <u>non-alphabetic nature</u> of the intermediary inscription of
the Rosetta; all that he affirms concerning the origin and composition of this
writing system is contradicted by the results we will exhibit."[11]

In the many pages that followed, Champollion developed a basic set of comparative sequences for demotic signs intended to capture the sense of Coptic
words or phrases. For each word or phrase, he also provided the corresponding
word from the Greek text, together with the hieratic and hieroglyphic signs
that he identified as having the same connotations (fig. 25.3).[12] Although this
catalog exceeded what he had produced back in Grenoble, both in developing
parallels among the scripts and in establishing their differences, he had arrived
at the scheme's essence in Isère. Young's *Britannica* article had served only to
sharpen and clarify his own essentially unchanged understanding.

Champollion rejected not only an alphabetic reading of the Egyptian scripts
but also any other sort of phonetic reading for the native language. The reason
lay in what he insisted was the language's foundation in semantic monosyllables. One might think that such a language could be easily represented phonetically, though not alphabetically, simply by assigning signs to its monosyllables. Not at all: a language built out of such monosyllables, he insisted, "is
less susceptible than any other language to being linked to a <u>syllabic</u> system of
writing" precisely because each of its syllables "ever recalls <u>one and the same
idea</u>." Therefore "each sign corresponding to an <u>Egyptian syllable thereby
becomes a true sign of an idea and not a syllabic</u> character." And so "every
syllable in the Egyptian language is a word, that is to say it has a fixed value,
<u>a value it conserves</u> whether employed in isolation, or whether this syllable
combines with others."[13]

This feature of the language militated against the creation of a phonographic
script constructed on a syllabic basis. "In supposing that the Egyptians ever
possessed a truly <u>syllabic</u> script," Champollion explained, "it would be incontestable that each syllable of their language was represented by a sign particular
to that syllable [and] having no relation to the form of the sign representing
a very close syllable." The inevitable consequence would be that "in thinking

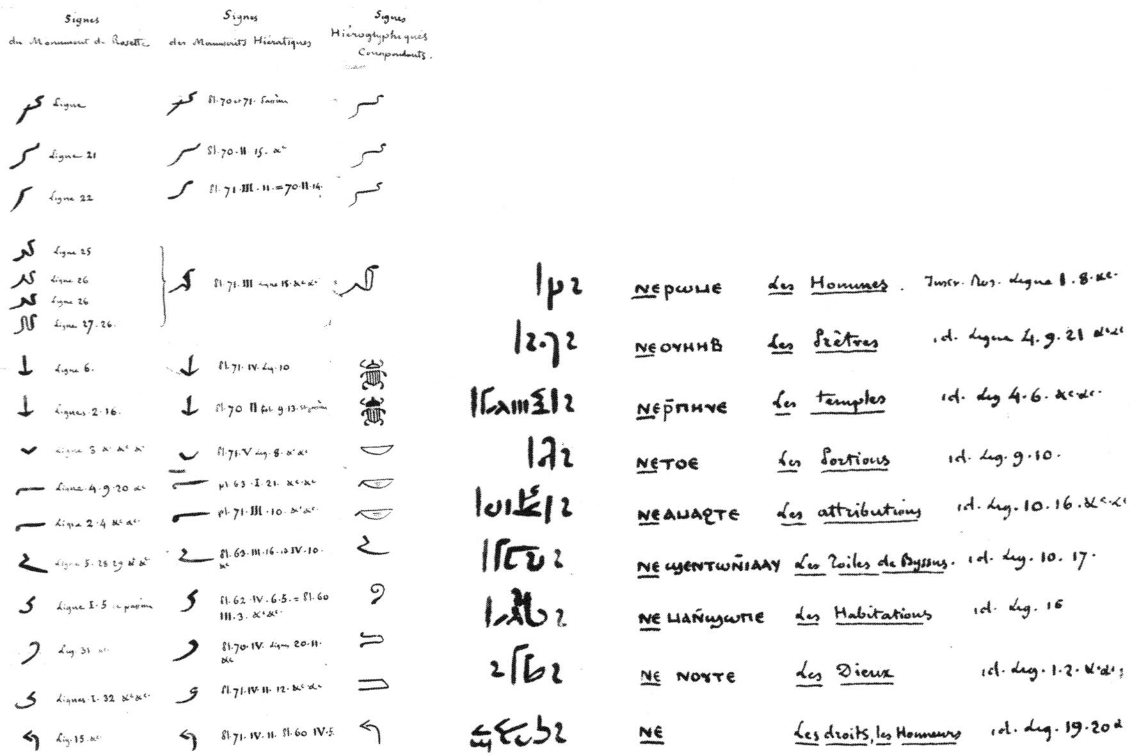

FIGURE 25.3. (*Left*) A page from Champollion's sign parallels for demotic, hieratic, and hieroglyphs, and (*right*) a set of Champollion's demotic "logical signs." Bibliothèque nationale de France.

to have created a <u>system of syllabic writing</u>, the Egyptians would have missed their aim because their language is so constructed that in proceeding thusly they would have invented, without wishing to do so, a true <u>system of ideo-graphic writing</u>," that is, a system of signs for words.[14] Moreover, any attempt to reduce signs to stripped syllables—to something closer to a true alphabet—was also improbable. In that case, the signs would have lost their original semantic values, an unlikely outcome given the conservatism of Egyptian culture.[15]

Champollion also observed that, despite all this continuity and sameness, the scripts differed from one another in important ways. Although all three were tied to language insofar as the sign sequences represented words or word phrases, the scripts nevertheless differed graphically because all three had been designed to embody a descending religious and social hierarchy. The "pure," inherently sacred hieroglyphs on monuments captured the speech of Egyptian deities; hieratic reflected the formal religious speech of the priestly caste; and demotic was designed for popular use. Because it was written on papyrus and not carved into monuments, hieratic diverged from hieroglyphs principally in-sofar as its signs had substantially lost any "figurative" resemblance to objects. Demotic differed from the other two because of its use in daily life; in order to convey practical affairs in an efficient manner, the sign system demanded compression: "It was important, in effect, not to overload the memory of the people with too great a multiplicity of purely arbitrary signs not offering in their outlines, [and] deprived of all imitative intention, any other way of being grasped and held with ease." This need for efficiency overwhelmed other con-siderations, such as the aesthetic value and nearness to divinity encapsulated in the more figurative scripts, outweighing "all the advantages of the art of the script, however great and desirable it would otherwise have been."[16]

Champollion insisted on a significant further result reflective of the specific purpose for which demotic had been created. A script intended for popular use could not presuppose an audience of sophisticated readers (e.g., priests) able to recognize and resolve semantic ambiguities in the absence of syntactic cues. Therefore that script must have incorporated an ample supply of qualifying or "logical" signs to capture both semantics and syntax, spelling out meanings for the less literate. These signs "determine[d] the <u>logical</u> nature of either a sign or of a group" by "warning the eye to consider this sign or group as corresponding to a noun, an adjective, a verb of the spoken language." In Grenoble, Champol-lion had decided that many demotic signs must act as "determinatives," that is, as graphemes indicating what the sentence in which they appear is about. He now extended this position by linking it to a social framework according to which such signs, created for practical purposes, were for that reason less prevalent in the hieroglyphs themselves. "The use of <u>logical signs</u> is constant and habitual in an Egyptian text written for the people," he argued, whereas "the hieroglyphic system, used among the educated classes of the nation and practised by men whose ideas were more developed than those of the individuals of the lower castes, could more easily dispense with these signs before each character of a hieroglyphic group expressing an idea." [17]

These logical signs also clarified by suggesting where semantic units began and ended. In this way, "the well-known forms of the spoken Egyptian language are in themselves able to lead us to the positive distinction of the <u>logical signs</u> which, in the popular script, effect the varied combinations of the ideographic [i.e., word-signifying] groups, coordinating them, linking them, and so to say communicating life and movement to the entire series of groups otherwise formed of arbitrary signs that are altogether mute."[18] Indeed, it was "the discovery of these logical signs that led us to determine with certainty the limits of the groups of the intermediate text, representing the idea expressed by each word of the Greek. Without this preliminary knowledge, all work undertaken with the aim of distinctly separating the groups from one another will yield results that are nearly always erroneous." Young had noticed such things, but his conviction that demotic signs were just degraded hieroglyphs led him to misidentify many groups and pay them little heed. Champollion, on the other hand, conceived that the facets of the writing system that challenged the modern expert eased daily life for the ancient Egyptians by means of a script that, having been "created for the people, altogether fulfilled its aim, and was thus perfectly appropriate to its purpose."[19] To identify these logical signs and describe their role in the representation of language was the major aim of Champollion's lecture, intended to secure his claim to a result that was both significant and novel.

Saint-Martin had claimed to identify in the Casati papyrus evidence for a phonetically rendered Greco-Roman name, "Cleopatra," based on much the same kind of identifications that Young and Åkerblad had previously deployed. Of course, the purpose of Saint-Martin's article had not been to demonstrate the existence of phonography, but rather to use the identifications of names for chronology. After all, he thought the alphabetic character of demotic for foreign words to be a foregone conclusion since de Sacy, Åkerblad, and even Young had so asserted. Champollion had to that point spoken in public only about the nonalphabetic character of hieratic, not demotic. Saint-Martin's identifications accordingly supported contentions that Champollion had rejected in Grenoble and that he continued to sustain in the talk on demotic to be given at the Académie.

Although Champollion had seen Young's *Britannica* article shortly after arriving in Paris the previous summer, there is no indication as late as the month before the talk that he had changed his mind about the scripts' nonalphabetic character, whether for the native language or for foreign words and names. He had, after all, insisted on precisely that in his hieratic talk at the Académie the previous August and September. Egyptian signs for Greco-Roman names simply did not represent sounds. Rather, they were repurposed sequences used in the remote Egyptian past to designate rulers by means of synecdoche or metonymy.[20]

Now, however, a new and especially competitive scholar had encroached on the field of demotic. Saint-Martin had not only presumed demotic phonography but he had used it *to date Greco-Roman events*. This startling move caught Champollion in a dilemma. If Saint-Martin were correct about the presence in demotic of *phonetic* Greco-Roman names, then either the three scripts were not all identical, with both the hieroglyphs and hieratic being nonalphabetic even for such purposes, or, if they were identical, then every one of them had to be alphabetic for foreign words. Both possibilites violated the conclusions Champollion had so painstakingly constructed in Grenoble.

Worse, Champollion could not summarily dismiss Saint-Martin's work. Whereas Young's ambition to understand the nature of the ancient writing made him a robust contestant on a field of inquiry, Saint-Martin appeared to be merely a disinterested scholar who had used phonography solely to sustain claims in chronology. To reject his work would appear ungenerous if not self-justifying, and Champollion's reputation was already shaky due to his politics—on which grounds Saint-Martin, of course, *was* a robust opponent and one with powerful allies as well. Champollion had little choice but to admit Greco-Roman phonography into his scheme, but only for a specific purpose that was, in effect, imposed upon Egypt by its foreign conquerors. In the late years of pharaonic Egypt, demotic (and so by his own reasoning hieroglyphs and hieratic as well) had undergone a limited expansion "when it was a question of introducing, in texts composed of signs of ideas, words or proper nouns foreign to the Egyptian language."

Even Åkerblad's alphabet, Champollion now admitted, could not be altogether discarded. Of its sixty-eight characters at least twelve had been fixed "in a sufficiently positive way."[21] The scripts, however, remained nonalphabetic for every set of signs "that does not express a proper Greek name."[22] Where such a name was expressed, a phonographic reading may work, but only in a "semi-alphabetic" form, one that would inevitably be "limited and incomplete" in comparison to the "ideographic scripts" in their pristine state, for these were "vast as a whole and rich in their details."[23] To this limited end Champollion allowed only fourteen demotic signs, three for vowels and eleven for consonants "or syllables" (fig. 25.4).[24] Åkerblad instead had given thirty-seven signs, while Young had offered nearly thirty in his *Criticum* piece. Moreover, Champollion insisted disparagingly, this ancient phonetic effort amounted at best to "a very imperfect transcription of Greek proper names."[25] When used for the native

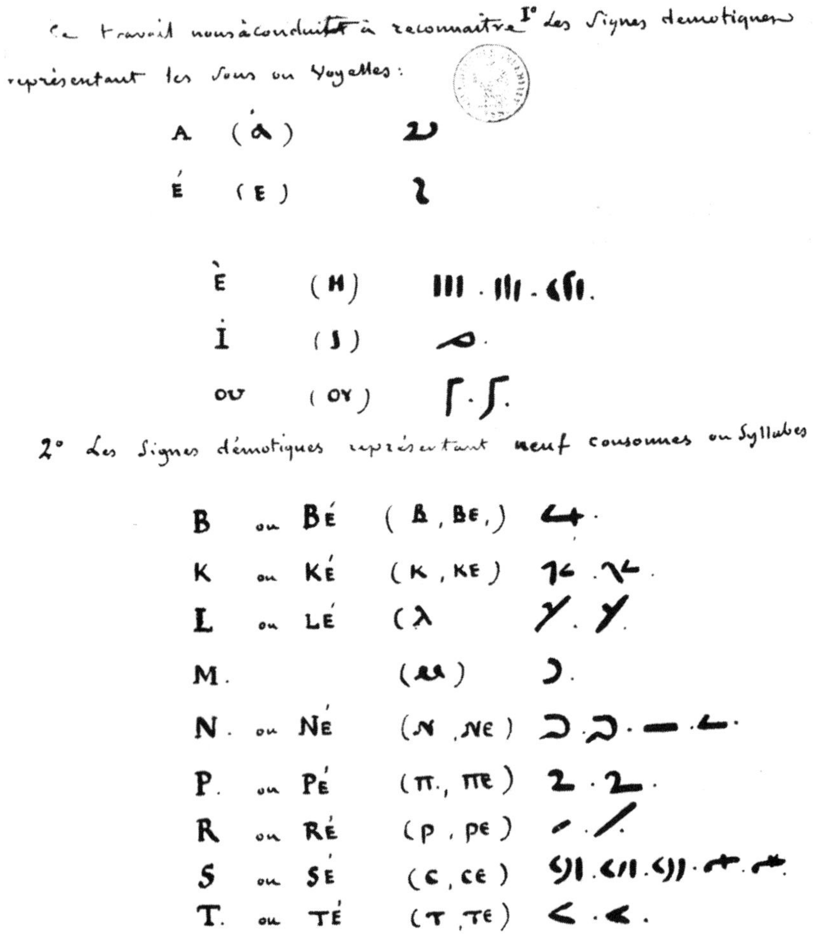

Ce travail nous à conduit à reconnaître 1° les signes démotiques représentant les sons ou voyelles :

A (a)

É (E)

È (H)

i ( ))

ou (OY)

2° Les signes démotiques représentant neuf consonnes ou syllabes

B	ou Bé	(B , BE,)	
K	ou Ké	(K , KE)	
L	ou Lé	(λ	
M.		(ɯ)	
N.	ou Né	(N , NE)	
P.	ou Pé	(π., πε)	
R	ou Ré	(ρ , ρε)	
S	ou Sé	(ϲ , ϲε)	
T.	ou Té	(τ , τε)	

FIGURE 25.4. Champollion's fourteen phonetic signs from his demotic lecture. Though there are none for hieroglyphs, Champollion insisted that the two scripts were fundamentally similar. Bibliothèque nationale de France.

language, the Egyptian scripts remained precise in their nonphonographic representations of words and syntactic structure. Phonography was unavoidable: it was a foreign insertion produced entirely by Greco-Roman domination.

And yet, this late proliferation of scripts from three to four suggests that Champollion had begun to recognize the problem with his scheme. Once the phonographic barrier had been breached, meanings threatened to multiply without limit. If the native "ideographic" scripts consisted solely of nonphonetic logograms and logical signs, there was no way to be certain that a sign, when used phonographically for a Greco-Roman name or word, would not too easily revert to the sound of the semantic monosyllable that it might have natively represented. And if that were so, inscriptions could become so ambiguous as to render any translation extremely treacherous, if not impossible. Though Champollion did not say anything in his demotic talk about this possibility, the table of signs that he produced (fig. 25.4) attests to his preoccupation with the problem and suggests the expedient he developed to deal with it. Instead of sounding an entire native monosyllable when used to transliterate foreign words, the sign would have been stripped to provide a tightly limited but related vocalization—just a consonantal one, say, or at most a consonant plus vowel. Young had, in contrast, been willing to employ entire monosyllables where he thought it expedient to do so, as in his *Britannica* piece.[26] Such stripping would help to eliminate any potential semantic link. This expedient proved fruitful, as it yielded a table of character values that Champollion loosely called "syllables." This was the initial step in a rapidly developing series during which Champollion effectively reversed his long-standing convictions about Egyptian phonography.

CHAPTER 26

WORDS AND SOUNDS

As late as the third week of August, Champollion had not strayed far from his claim concerning the scripts' essentially ideographic nature. He had reluctantly allowed an "imperfect" phonetic rendering of foreign ruler names in demotic, but because the native script simply did "not belong to a system of alphabetic writing," neither, of course, could hieroglyphs.[1] Champollion accordingly found himself in an unstable situation toward the end of August. Something had to give, and indeed it had by September 14. On that day, according to posthumous accounts, Champollion rushed to his brother in a state of high excitement, handed him a sheaf of papers, gasped "Je tiens mon affaire, vois!," and collapsed. According to the earliest extant report, in the sheaf were notes for Champollion's *Lettre à M. Dacier*, which he would write up as he recovered over the next two days and in which he explained his system for the decipherment of hieroglyphs. In the words of that early commentator, the evidence of phonography that Champollion had long sought was "miraculously revealed" in a flash of insight so powerful he fainted dead away.[2]

Despite the frequency and relish with which the story of this sudden insight has been retailed, no contemporary accounts exist to verify its accuracy. Adolphe Rochas, a lawyer and chronicler of the Dauphiné, brought the story into print for the first time in 1856, twenty-five years after Champollion's death. Rochas's version of the event supplied the basic elements—the brother's presence, the sheaf of papers, the gasped phrase, the collapse—while ascribing its cause to a miracle.[3] Since then the story has been repeated with slight variations but no diminution of the melodrama that makes such stories stick. This in no small measure accounts for what one might term the legend of Champollion, anguished genius, who founded Egyptology after fainting in the rue Mazarine. Rochas's portrait drew on two familiar and often merged midcentury types: the visionary Romantic whose unique perception granted privileged access to quasi-divine revelations, and, to a lesser degree, the Great Man whose superior intuitive abilities led him to truths unavailable to coarser minds. Such typological figurations were endemic to the religious and secular literature of Rochas's day, and the messianism of the theological literature readily transferred to secular contexts.[4] Subsequent versions of the story have repeated these elements. The effect is to reduce the insight to a kind of linguistic puzzle-solving in which Champollion's sudden phonographic reading overshadows and inflects the rest of the *Lettre*. Among other things, such a focus reinforces the belief that the decipherment of hieroglyphics can be best understood in terms of code-breaking.[5]

The shortcomings in Rochas's and most subsequent accounts point to the difficulty of knowing just *what* struck Champollion with such force as he composed the *Lettre*. Because his notes from these critical weeks have not been preserved, we cannot follow, step-by-step, just what he did. Nevertheless, evidence afforded by Champollion's earlier notes and by his recent talk on demotic provides clues to the course of his research. In what follows, we will connect that material with internal evidence derived from the construction of the *Lettre*, together with remarks made by Champollion relating to a succeeding work of 1823–24, the *Précis du système hiéroglyphique des anciens Égyptiens*, to describe just what Champollion had "grasped," or, more prosaically, developed, in the weeks leading up to the *Lettre*.

Champollion began his Dacier account unambiguously and in full consonance with the lectures he had previously given at the Académie. Indeed, pace Rochas, Champollion did not see his *Lettre* as a vehicle of any fresh revelation. On the contrary, he presented it as merely the third essay in a trilogy and a natural continuation of his previous two analyses. Having treated both hieratic and demotic, Champollion intended now "to complete my work on the three types of Egyptian writing to produce my memoir on the pure *hieroglyphs*."[6] He hoped to have "succeeded in demonstrating that these two kinds of writing," the demotic and hieratic, "are, the one and the other, not alphabetic, as has generally been thought, but *ideographic*, like the hieroglyphs themselves." In the next breath, he explained just what he meant by "ideographic," that is, he understood these scripts as "indicating *ideas* and not the *sounds* of a language."[7] In Champollion's developed understanding, these "ideas" could be concepts or specific words or phrases, arranged according to the rules that governed the Coptic language. Yet the bulk of the *Lettre* concerned phonography. We shall see in what follows how Champollion resolved this apparent contradiction. The resolution depended on his long-standing conviction that Egyptian writing paralleled the native language and in so doing functioned as a repository of evidence of a culture's identity and history.

Champollion's *Lettre* advanced three connected claims: first, that all three scripts were at base "ideographic"; second, that all three admitted a "syllabic-alphabetic" interpretation when used for foreign words; and third, given the essential identity of the scripts, that he could thereby obtain phonographic interpretations of *hieroglyphic* signs for foreign words from the demotic. He would now develop his August admission of that limited phoneticism, insisting that he had "deduced the series of Demotic signs that, taking a syllabic-alphabetic value, expressed in the *ideographic* texts the proper names of the foreign persons in Egypt."[8] He nevertheless had something more in mind than the names of "foreign persons," something that had perhaps been confirmed on that legendary morning of September 14 and that balanced unsteadily with his opening remarks. The structure of the *Lettre* betrays this instability as Champollion sought to maintain a place for ideography while pushing ahead with a phonetic set of characters for Greco-Roman names and even words. Written in collaboration with his brother, the *Lettre* is a rhetorical construct intended to introduce a novel conception while retaining some basis for his prior assertions regarding hieratic and demotic. It is a palimpsest of Champollion's thought, the new scribbled as it were hesitantly over the old.

We have first of all seen that, more than Young, Champollion had begun to specify a considerable number of logical signs that did not constitute "ideas" (concepts, words, or word phrases). By indicating sequences that represented names, these signs could identify others that might be interpreted phonetically. That possibility, in turn, depended on Champollion's understanding of ancient Egyptian's linguistic structure, in particular its dependence on semantically rich monosyllables. "You have no doubt noticed, Monsieur," he remarked in the *Lettre*, "in my memoir on the Egyptian Demotic writing, that these foreign names were expressed phonetically by means of signs [which are] *syllabic* rather than *alphabetic*."[9] This had a specific meaning, in that such "syllabic" signs were concocted from consonants and diphthongs of the language's monosyllables. In his *Britannica* article Young had instead repurposed entire Coptic syllables, as well as parts of them, for the limited purpose of phonetically representing Greco-Roman rulers' names. Champollion's skepticism derived from his conviction that more was needed if "syllables" were not to revert to their prior semantic connotations, which he thought would surely be one consequence of Young's scheme.

Young had found the name Berenice on a ceiling at Karnak, in the phrase "Ptolemy . . . Berenice, the savior gods." In a remark that irritated Champollion, Young had dismissively claimed that the signs for the name constituted "another specimen of syllabic and alphabetical writing combined, in a manner not unlike the ludicrous mixture of words and things with which children are sometimes amused."[10] Young's condescension prompted Champollion to pen a lengthy, sarcastic, and revealing footnote designed to establish the fundamental difference between what each intended by "syllabic." Young, he opined, had decided that the sign he took to represent the initial syllable in the Berenice cartouche derived from its graphical similarity "with the representation of a crow." "But," Champollion continued, "this English savant thought that the hieroglyphs which form proper names could express entire syllables, that they were a sort of *rebus*, and that the initial sign of the name Berenice, for example, represented the syllable Βιρ which means *crow*" in Coptic. From this faulty starting point, Young was led to "in great part distort the phonetic analysis he tried on the names of *Ptolemy* and of *Berenice*." Despite correctly recognizing phonetic values of various signs in these two names, "his entire syllabic alphabet, established only from these two names, would be altogether inapplicable to the numerous phonetic proper names inscribed on the monuments of Egypt."[11] Champollion objected to such a full and complete use of syllables with inherent semantic content. What he would now produce, instead, was a set of principally phonemic elements for such names, allowing for the occasional inclusion of a qualifying element.

Champollion began by cataloging the names he could phonetically identify in the Rosetta demotic inscription and on the papyri. Four of these, namely, Ptolemy, Alexander, Berenice, and Arsinoe, Young had also listed. Champollion adduced six additional names and two Greek words (ΣΥΝΤΑΞΙΣ and ΟΥΗΝΝ). Turning to the papyri manuscripts, he noted on the Casati four of the same names on the Rosetta, of which "M. St-Martin, in the *Journal des Savants* for the month of September, has given an interesting notice." There, too, were phonetic signs for the names Eupator, Cleopatra, Apollonius, Antiochus, and Antigone.[12] Recurring to his recent publications on the scripts, he claimed that this group of phonetically rendered names in the demotic and papyri scripts implied the existence of a corresponding set in the hieroglyphs. The signs of demotic were "as I have shown, borrowed from the *hieratic* or sacerdotal script," and "the signs of this *hieratic* script are nothing, as has been shown in my various memoirs, but an abbreviation, a veritable *tachygraphy* of the *hieroglyphs*." Consequently, "this third type of writing, the pure *hieroglyphs*, should also have a certain number of signs endowed with the faculty of expressing sounds."[13] Since Champollion had concluded in the years before his return to Paris that the three scripts were variants of one another, once he had admitted the existence of phonetic renderings of foreign names and words in demotic, he would necessarily seek related signs in the hieroglyphs. For that he did not need Young's identifications.

To extend his scheme, Champollion developed the role played in it by the large number of logical signs or determinatives that he had uncovered. Turning to the cartouched signs in Bankes's lithographs, Champollion pointed out that one of them "necessarily contains the proper name of a woman, a Lagide queen," as was evident from the appearance on the cartouche "of the hieroglyphic signs of the feminine type," that is, by a female-designating determinative. Generalizing, Champollion noted that such signs "also terminate the proper hieroglyphic names of all the goddesses without exception."[14] At Grenoble, before he had seen Young's *Britannica* article, Champollion had decided that an angle bracket sign in demotic also acted as a female signifier. In the notes on hieratic, he had pinpointed the semicircle-plus-oval as such a sign as well, and he had explicitly linked the oval to the angle sign.[15] Although Champollion made no reference to them, Young had also made advances in this area. In the hieroglyphs at Karnak that Young had marked as referring to "Berenice," he had identified one such female designator: a closed semicircle followed by an oval. He had also identified an angle bracket as another female designator.[16]

Although both had produced identifications of logical signs, they meant different things to Champollion and Young, with corresponding implications for their views of the scripts. For Champollion the presence of such signs had provided evidence of the manner in which the scripts could represent the Egyptian language in sense, grammar, and syntax in the absence of phonography. Young, in contrast, never believed that any of the scripts were a strict representation of language per se; for him each of the three comprised a mixture of syllabic and alphabetic signs, along with signs that retained a purely sematological sense and so could be read in ways that would not necessarily have reproduced the Egyptian language, and these, at any rate, all dated from Greco-Roman times.[17]

From logical signs Champollion turned to phonetic ones, signaling that he had begun to perceive in the scripts a more fully phonetic structure than he had admitted in his demotic talk. Pressed by Saint-Martin's competitive overtures, and using the signs in the Rosetta and Philae cartouches for "Ptolemy" as a foil, Champollion reconsidered "Cleopatra." In the cartouche that Saint-Martin and Bankes had both linked to "Cleopatra," there are eleven signs (fig. 26.1).[18] Comparing the Ptolemy cartouches from Bankes's plate and the Rosetta inscription, Champollion noted that "if the similar signs in the two names expressed in one and the other cartouche are *the same sounds*, they must have the same *entirely phonetic* nature." From this supposition he generated twelve "consonant and vowel or diphthong" signs corresponding to the Greek letters A, AI, E, K, Λ, M, O, Π, P, Σ, and T.

In the Cleopatra cartouche Champollion identified two signs as feminine designators, and a third, a type of flower sign, corresponded to O. Champollion further identified a sign that looked like a feather or leaf as representing the short vowel E, corresponding to the diphthong AI represented by the two feathers at the end of the Ptolemy cartouche. To the "parallelogram" in the signs for Ptolemy, Champollion assigned the value Π, and to the resting lion, which also appears in the cartouches for the same name, Champollion assigned the value Λ. The sign that he called a "sparrow" appeared twice in the Cleopatra cartouches but not at all in the Ptolemy where, indeed, it should not appear; to this sign he assigned the value A. His "quarter-circle" assimilated to K, the justification being that, if the cartouche in question did represent the name Cleopatra, then a sign sounding K should not appear in the Ptolemy cartouche, which it did not. Finally, and most importantly, to the sign of the open hand, which does not appear in the Ptolemy cartouche, Champollion gave the value of T. He also assigned that value to a corresponding but differently figured sign in the Ptolemy cartouche. This alternative designator for T, which he described as a "segment of a sphere," *constituted a homonym,* about which he would shortly have much more to say.

We have seen that Champollion had much earlier encountered differing "semes" that conveyed the same sense, in particular signs indicating that neighboring characters referred to a feminine person. But he had not previously considered that such signs might also be interpreted, depending on context, to convey sound. He had accordingly repurposed a sign from one that, in his earlier notes on hieratic, designated the feminine, namely, what he referred to as a "quarter-circle" or "segment of a sphere." This same sign was now to be interpreted as sounding T because it appeared in the appropriate place in the cartouche signifying Ptolemy. This attribution made sense because in Coptic, Champollion explained, the feminine *article* is pronounced *ti* or *te* (ⲧⲉ), thereby transforming an Egyptian sign that he had previously identified solely as an unvoiced qualifier into a phonetic representation by extraction of its initial sound. Coptic also provided the rationale for the open hand, for the associated Coptic word was pronounced *tot* (ⲧⲟⲧ).

With this enlargement of the range of potential significations for the qualifying or "logical" signs, Champollion had radically expanded the rebus principle: Instead of applying it to extract sounds solely from signs for words, he extended it to *unsounded* designators. More precisely, Champollion took a word from the language (Coptic, of course) that was gendered feminine and used it as an unvoiced qualifier to signal that signs for a nearby name referred to a female. He then took the first phoneme of the

FIGURE 26.1. (*Top*) The Bankes cartouche identified as "Cleopatra" and (*bottom*) the Bankes for "Ptolemy."

sign when voiced as the corresponding word (ⲧⲉ) and used it to designate an alphabetic character. This was just the sort of multifaceted repurposing of signs that Young rejected, for two reasons: he certainly did not think extant Coptic to be a reliable guide to its ancestral form, and such a procedure required referentially variable signs rather like algebraic ones, which, as we have suggested, is something Young may have particularly despised. Meanwhile Champollion had, so far without much evidence, extended this mutability to the sign's qualifying and phonographic functions. Consequently no reading of these scripts could be essayed without knowledge of the underlying language.

Champollion grounded his argument in a unique set of assumptions about the Egyptian scripts that reached back to his earliest investigations of Egyptian geography and echoed the historicist preoccupations of his brother and their colleagues in ancient history and philology. "One cannot, in effect, consider the *phonetic* writing of the Egyptians, whether *hieroglyphic* or *demotic*, to be a system as fixed and invariable as our alphabets," he wrote. To drag contemporary ideas about the alphabet into the past was a misstep. Reprising his insistence, in *Pharaons*, on the need to understand ancient Egyptians on their own terms, he urged his audience to consider how representation worked for them.

> The Egyptians were used to directly representing their ideas; the expression of sounds was, in their ideographic writing, just an auxiliary method; and when the occasion presented itself to use it more frequently, they well thought to extend their methods of expressing sounds, but did not at all thereby renounce their ideographic scripts, consecrated by religion and continued use for a great number of centuries.

As usual, the problem was that such phonetic renderings of foreign words risked unrestricted proliferation of meanings due to the semantic richness of the syllables of Egyptian. To sort this out, Champollion turned to Chinese writing, pointing out that wholesale repurposing of syllables à la Young did not occur in that context either.

> So [the Egyptians] proceeded as the Chinese did in absolutely parallel situations; in order to write a word foreign to their language, they simply adopted the ideographic signs whose pronunciation seemed to them to offer the greatest analogy with each syllable or element of the foreign word to be transcribed. One therefore conceives that the Egyptians, wishing to express a vowel, would use a hieroglyphic sign *expressing* or *representing* whatever object whose name, in the spoken language, contained either in its entirety, or in its first part, the sound of the vowel, of the consonant or of the syllable to be written.[19]

One and the same sound might be designated by different signs, each of which visually "expressed or represented" an object whose name began with (or was solely) an appropriate syllable. Since the phonetic repurposing of a sign that might otherwise indicate the Egyptian (monosyllabic) word was used for Greco-Roman purposes, there would be little occasion,

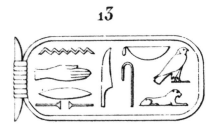

FIGURE 26.2. The Alexander cartouche at Karnak.

in a given text, for confusion as to whether a semantic or logogrammatic meaning was intended. Understood in this way, a hieroglyphic sign used to represent a foreign word would not have evoked an "idea" instead of a sound.

Champollion next exemplified his method by rendering the signs for "Alexander"' from a cartouche drawn from an unspecified monument at Karnak and printed in the *Description* (fig. 26.2).[20] In doing so he relied heavily on the identification of "homonyms" for several signs. Although Saint-Martin had loosely anticipated this notion, he had not elaborated on it.[21] Champollion would now make full use of the idea as a central pillar of his system. He rendered the signs for Alexander phonetically as ΑΛΚΣΕΝΤΡΣ in the following manner. The first two signs (sparrow hawk and lion) he had identified as ΑΛ, and he also had the open hand and mouth as ΤΡ. The solitary feather he had as E "or another short vowel," and the curved sign he had as Σ from the Ptolemy cartouche (read as ΠΤΟΛΜΗΣ). This suggested to him that the cartouche might well contain the name of Alexander, in which case he had to provide readings for the vase-with-ring signs, the wavy sign, and the two "sceptres." These, respectively, had to signify K, N, and Σ, for which Champollion introduced two new graphical "homonyms" for K and Σ.

Champollion still needed to justify these last identifications on grounds other than the apparently arbitrary one of singular necessity. Assimilating the vase-with-ring sign to K, Champollion referred to his previous work on demotic and hieratic. Although he did not provide details, we can reconstruct them from his lecture notes. The initial step ran through the demotic signs Champollion had identified on the Casati papyrus as referring to Cleopatra. One sign in the set, an angle bracket, he had found several times to be equivalent in meaning to an identical sign in hieratic. And that sign had graphical equivalents resembling the vase-with-ring (fig. 26.3).[22] Although all these identifications were nonphonetic, Champollion had already made the angle bracket phonetically equivalent to his "quarter-circle" in sounding K. If, as he insisted, the three scripts were fundamentally equivalent, then the vase-with-ring must be a "graphical homonym" for K.

FIGURE 26.3. (*Left*) The angle bracket in the Casati demotic for "Cleopatra" and (*right*) hieratic equivalents (Bibliothèque nationale de France).

What of the "scepters"? That they were homophones for Σ was "incontestable because these two hieroglyphic signs *are rendered in the hieratic texts by one and the same character*." The argument ran as follows. The Rosetta's Greek inscription had yielded the demotic sign for Σ; the *Description*'s color hieroglyphic and related hieratic papyri provided the corresponding hieratic sign for the hieroglyphic double scepter. That hieratic sign was visually similar to the previously ascertained demotic for Σ. Hence the double scepter had also to be read as Σ.[23]

The coherence of Champollion's argument depended critically on complex movements of this sort between the three scripts, all bound to the reduction of a sign's phonetic rendering to the initial sound of the associated word in Coptic and to the prevalence of nonvocalized qualifiers. Using this method he produced a "table of phonetic signs" (fig. 26.4). Foreign names could be read phonetically in this way; so, too, could some foreign words. One sign that was not present in his table would soon prove especially consequential: the star-shaped one on the *Description*'s plate of the Dendera zodiac. It was absent from Champollion's table precisely because the sign was not phonetic, but neither did it represent, as Biot had assumed, an actual star. It was instead a *qualifier* that indicated the subject of the monument. This solidified Champollion's conviction that "designating" signs, which he soon thereafter consistently referred to as "determinatives," were critical for specifying meanings, thereby undercutting Biot's astronomical interpretation of the zodiac. But Champollion had framed his critique as a contribution to ancient chronology, so that savants might avoid "grave errors concerning the diverse epochs of the history of arts and of the general administration of Egypt." He needed to take one more step by linking his assessment of the Dendera zodiac to his decipherment of Egyptian scripts.

Among the titles that Champollion now read was one taken from cartouched hieroglyphs on the side columns that bracketed the zodiac proper on the *Description* plate (fig. 23.3). There he read several variants of the imperial title Αντοχρατορ, autocrator. He reproduced three of these variants in the *Lettre*

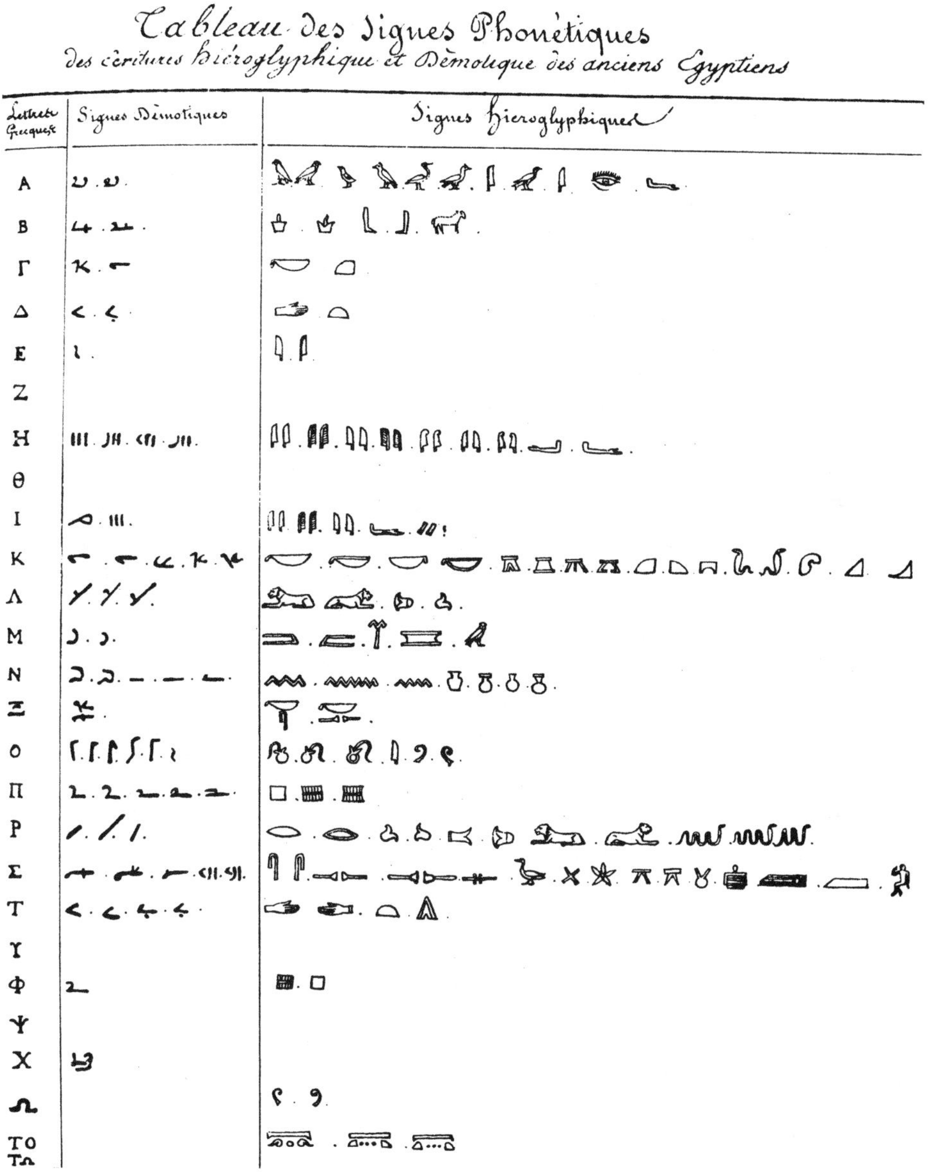

FIGURE 26.4. Champollion's "table of phonetic signs."

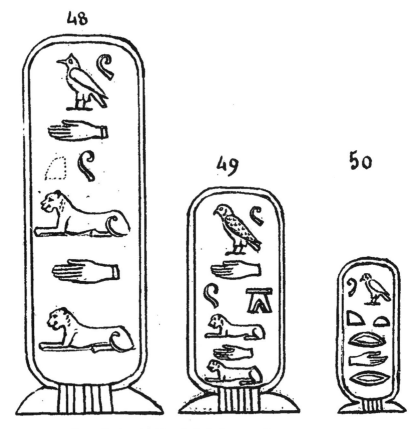

FIGURE 26.5. Champollion's variant hieroglyphs for "autocrator."

(fig. 26.5).[24] The third cartouche in this set was copied directly from the *Description*'s plate of the Dendera zodiac and its immediate surroundings. The representation is, first of all, a mirror image of cartouche 48 (excepting the bird figure, which faces the same direction). Unlike the other two cartouches, which each contain two *rho* (Greek P, Coptic Ρ) signs, there is a semicircle where a second *rho* should appear; and ovals have replaced the lions. Using his method of substituting "homonyms" Champollion accordingly transliterated the cartouche as AOTKRTR.[25]

According to Champollion, the zodiacal cartouche (number 50 in the drawing) was especially significant because it "establishes, in an incontestable way, that the bas-relief and the circular zodiac were sculpted by the hands of Egyptians under the domination of the Romans." The "infernal" stone, bane of the religious right, was after all no older than the Roman Empire. "Our alphabet,"

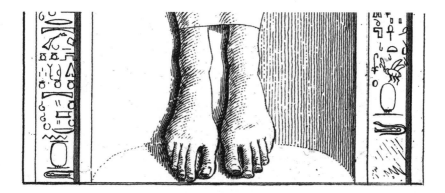

FIGURE 26.6. Dominique-Vivant Denon's drawing of the Dendera cartouches.

Champollion concluded, "acquires by this sole fact high importance, because it greatly simplifies a question that has been been so long debated, and on which most of those who examined it have presented only uncertain and often diametrically opposed opinions." Unfortunately, Champollion's conclusion could only be provisional, as he had only the *Description*'s plate from which to work. He complained in exasperation that the bas-relief containing the cartouche was not in Paris. The zodiac's excavators had, in their ignorance, left that vital element behind: "[T]he stone was cut near that point because the sole aim was to remove the circular zodiac, and so it was isolated from a bas-relief that in all probability referred to it."[26]

Jomard, for one, contested Champollion's dating. That Champollion had found fault with a *Description* plate no doubt contributed to Jomard's pique; the plates had been produced under his supervision, after all. Moreover, the drawing by the engineers Jollois and Devilliers was assumed to be more accurate than any other available rendition because they had been trained in drafting techniques. Of the other available drawings of the object, the best known was Denon's, in which the sky goddess appeared alongside the parallel columns with the important cartouches shown beneath. This drawing differed from the *Description*'s inasmuch as both cartouches were curiously empty (fig. 26.6). But Denon's depictions were thought to be inaccurate, and in any case Jomard, with his personal investments in the success of the *Description*, certainly would not question the existence of the hieroglyphs that appeared in this plate. Jomard continued to insist that the Dendera zodiac was quite old, older indeed than Biot had calculated, to say nothing of Champollion's late date. Jomard challenged Champollion: "Which emperor does this isolated *autocrator* designate?"[27]

Champollion had staked a great deal on his reading of the signs as "auto-crator," yet his evidentiary foundation was weak. Although the cartouche was the sole persuasive evidence taken from the immediate vicinity of the zodiac, the relevant cartouche appeared only in the *Description*'s plates drawn from Jollois and Devillier's original. As Lelorrain had left in situ the bordering panels containing the cartouches, Champollion's reading relied wholly on this essentially secondhand knowledge. Lelorrain's omission had not gone unnoticed. In England the *European Magazine and London Review* hoped that the "remainder of the ceiling might surely be obtained," to be paid for by Louis, who would surely not "fail to possess himself of everything appertaining to this extraordinary specimen of the prosperity of ancient Egypt." [28]

Near the end of the Dacier letter, Champollion ventured a further, apparently surprising expansion of the scheme's applicability, albeit one that Saint-Martin, following Young, had already mooted. "The facts themselves," Champollion declared,

> speak well enough to authorize us to say, with some certainty, that the use of an auxiliary script to represent the sounds and articulations of certain words preceded, in Egypt, the domination of the Greeks and Romans, though it seems natural to attribute the introduction of the semi-alphabetic script to the influence of these two European nations who long used an alphabet, properly speaking.[29]

He offered two considerations in support of this claim, which was all the more startling because it ran completely counter to his previously expressed views. The first reasoned from absence. Suppose per contra that the repurposing of the Egyptian scripts had been based on finding appropriate rebus signs solely for Greco-Roman names and words. Then, he argued, one would expect there to be a clear way of working in order to avoid possible ambiguities. But consistency is precisely what does not obtain because of the omission of vowels and the ubiquitous presence of "homophony." Something more must lie behind the apparent cacophony. Champollion suggested that it might consist in the long-existing presence of phonography, a conviction to which his earlier logic concerning the *absence* of phonography, now led him. The conservative, religious character of Egyptian culture, as Champollion saw it, would never have countenanced a massive manipulation of their writing system, in which case if phonography for royal names were present at any time it must have been there from the outset.

The instigation of this massive shift in Champollion's orientation remains a mystery. In a frequent variation on the legend repeated by Champollion's biographers starting with Hermine Hartleben, the proximate cause of his epiphany on the fateful morning of September 14 was the arrival of a new set of hieroglyphs in drawings from the architect and traveler Jean-Nicolas Huyot, who had made them at the pharaonic monument of Abu Simbel while accompanying Louis-Nicolas Forbin in Egypt.[30] Perhaps Hartleben had access to documents now lost concerning the Huyot material. Neither of the two earliest accounts of the event, by Rochas and by Champollion's nephew, mentions any document from Huyot. In fact, the only references to Huyot that we have come from Champollion himself. The first appeared in a letter Champollion wrote to Young on January 9, 1823, in which he mentioned that he had "drawings from M. Huyot, member of the Institute" that provided "exact, large copies of all the proper royal names carved on the monuments of Egypt; on the basis of these invaluable drawings I made my classification and my *Ideographiconetic* talks."[31] Later that year, eager to differentiate his accomplishment from Young's, Champollion wrote the *Précis*. Huyot's name appeared eight times in this work. One of these references was to Huyot's copy of a dedication to a temple "in Nubia by Ramses the Great."[32] These hints suggest that Champollion did have some document from Huyot, and perhaps Hartleben was correct to attribute his epiphany to that material, but here the trail goes cold.

Two months after his Dacier talk at the Académie, on November 23, Champollion wrote Young a long letter explaining that, while examining the forty-four cartouches "in the inscription of Abydos," he had been able to read the name "Ramessès." It is unclear how Champollion knew of this inscription. Bankes had drawn the cartouches (what Young later called the Abydos "chronological table"), but Champollion only had a copy of the Bankes from Young some time after the Dacier letter had been sent (fig. 26.7).[33] He may consequently have first seen it, as Hartleben suggested, in a now-lost drawing by Huyot.[34] It was this recognition (likely in a Huyot drawing, we agree) that so excited Champollion, but not because it provided him with a new revelation. Rather, what he found was evidence to confirm that phonography for royal names must have extended back to near the origin of the scripts.

How then did Champollion argue for "Ramessès"?[35] His letter to Young in November hints at the details. There Champollion wrote that "[t]he unique point in respect to the Abydos monument, on which I have an arresting idea, is just that this series of kings [on the Abydos] *ends with* Sesotris, Sethos, Sethosis, or Sésoosis, that Manetho tells us also bore the name of *Ramesses*." Champollion continued that he "could not resist the conviction that, so to speak, forces me to recognize in the variations [fig. 26.7, center] of such a frequent cartouche (that you provisionally attributed to *Maenuphtes*) all the elements of the name of *Ramesses*."[36]

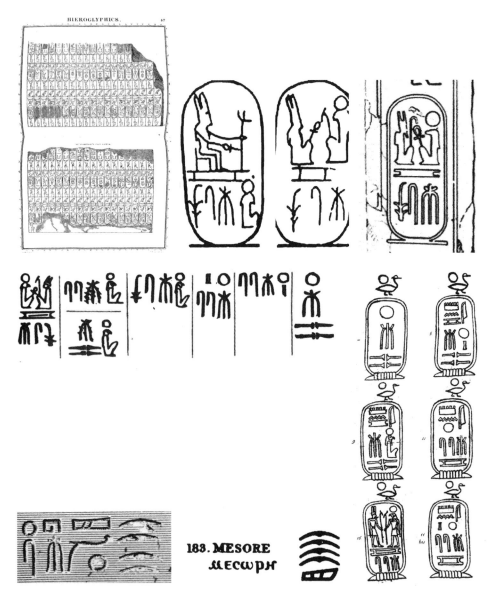

FIGURE 26.7. (*Top left*) Bankes's "chronological table"; (*top center*) the cartouches from the Abydos that Champollion copied; (*top right*) a similar cartouche from the *Description de l'Égypte*; (*center*) the signs from Champollion's letter to Young; (*bottom left*) the Rosetta signs that Champollion used; (*bottom center*) Young's identification of the Rosetta signs for "Mesore"; (*bottom right*) the Ramesses signs from Champollion's *Précis du système hiéroglyphique des anciens Égyptiens*.

Champollion had certainly seen a copy of the Bankes Abydos king list (fig. 26.7, *top left*), but we do not know from whom he obtained it. The list includes two cartouches that Champollion reproduced exactly for Young, and that appeared in the first and third columns of his drawing (fig. 26.7, *top center*). Though Champollion did not mention it, the *Description* contained a similar cartouche that he may also have seen (fig. 26.7, *top right*). The signs in Champollion's second column are partly replicated in others of the Abydos cartouches, but the circle and rectangle do not appear there, while the circle, trident, and double scepters of the sixth column are also absent as a group.[37] Champollion apparently had other sources for his variant sign groups, and of these the most probable is a drawing by Huyot. Although we don't know specifically what he might have had, we can infer a few specifics from what Champollion *did* have on hand—evidence from the Abydos list and the Rosetta inscriptions.

A cartouche from Bankes's table contains a circle over the seated person bearing an ankh (fig. 26.7, *top, third from left*). Champollion read this configuration as signifying the pronunciation of the name of the sun, *Ra* (in Coptic Ⲣⲁ), as had Young before him, with the seated person acting as a nonphonetic qualifier denoting a royal person.[38] The cartouche also contains a three-pronged sign or trident preceded by the signs for a crook or crooks. These signs, which were the crux of the matter, appeared together in a group among the Rosetta hieroglyphs (fig. 26.7, *bottom left*), as Champollion would surely have known.[39] The crook in this Rosetta pair was topped by a circle, while a square-cornered spiral resembling a meander or Greek key crowned the trident. By associating these groups with the Greek inscription, Champollion could have spied a way to read them that connected to the groups on the Abydos and to whatever he may have received by way of a drawing by Huyot.

In his *Britannica* article, Young had identified four curved lines over a rectangle as signifying the month "Mesore," which Young had even transliterated into Coptic as ⲙⲉⲥⲟⲣⲏ. Just to the left of these signs for "Mesore" was the crook-plus-trident, which Young had *not* included in his signs designating the month. Champollion would certainly have noticed these details. But the corresponding Greek line contained a further clue.

The month "Mesore" appears in the forty-sixth line of the Greek inscription together with the numerals XXX and XVII.[40] Along with Young's identification of the signs for "Mesore," it was clear that this sequence appeared in the tenth line of the hieroglyphic inscription. Furthermore, in Young's table of meanings for hieroglyphic signs, the circle-without-a-central-dot accompanied by a rectangular spiral signified "day." The circle-plus-spiral sat atop the two signs that Champollion would have been looking for.[41] Putting this all together, Champollion could easily have decided that the crook-plus-trident, topped by the circle-plus-spiral, must signify something about "day." The corresponding Greek line provides the requisite specification since it contains the word γενεθλια (*genethlia*), which means "birthday." Taken together, all this evidence may well have suggested to Champollion that the four signs together (circle, spiral, crook, and trident) meant something about a birthday, or, more to the point, had phonetic significations to that effect.

Now Champollion had it that in "Thebaic" the word ⲙⲓⲥⲉ also meant something like "to appear" or "to be born," and that corresponded at least broadly to the Greek γενεθλια (*genethlia*). Indeed, Champollion had the Coptic "Thebaic" for day-of-birth (as he would write to Young in November) as ⲅⲟⲩⲙⲓⲥⲉ (in Champollion's Coptic letters). Applying this to the sign sequence in question, Champollion might read it without the vowels ι,ε as ⲅⲟⲩⲙⲥ. The Coptic for "day" is, loosely, ⲉⲅⲟⲟⲩ, which embraces the first three letters of the reduced sequence ⲅⲟⲩⲙⲥ, leaving in the end ⲙⲥ as signifiers for the last part of the word, which accordingly transliterates as something like "ms." And so, together with the circle sign read phonetically as *Ra*, Champollion would have had *Ra-ms*, hence Ramesses. Although he might have produced the entire sequence of linkages entirely de novo, the fact is there were, as we have now seen, substantial hints in Young's *Britannica* article that he would surely have noted and turned to his advantage.

Closing the Dacier letter, Champollion suggested that phonography might well have reached very far back indeed, at least for the names of native rulers. "I am certain," he wrote, that

> the same *hieroglyphic-phonetic* signs employed to represent the sound of proper Greek and and Roman names were also employed in the ideographic texts engraved much before the arrival of the Greeks in Egypt; and that they had already, on certain occasions, the same representative value for sounds in the cartouches engraved under the Greeks and under the Romans. The development of this valuable and decisive fact belongs to my work on the pure hieroglyphic script. . . . I therefore think, monsieur, that the *phonetic* writing existed in Egypt at a very early stage; that it was from the first a necessary part of the ideographic script; and that it was also then employed, as after Cambyses, to transcribe (grossly it is true) in the ideographic texts, the proper names of the peoples, the countries, the towns, the sovereigns, and the foreign individuals whom it was important to remember in the historical texts or monumental inscriptions.[42]

And if Egyptian phonography for such names and places did go far back, it may have been "if not the origin, then the model on which the alphabets of the people of western Asia may have been based." Egypt, often thought during the Renaissance and even later to be the source of ancient wisdom, may have been the source of the writing systems in which Europeans had encoded their own arts and sciences.

Champollion had superseded his own arguments against phoneticism. As he had come to see it by the end of summer 1822, the Egyptian language's meaningful monosyllables could produce phonetic renditions even of pre-Alexandrian names. By stripping away all but the initial sounds, and most vowels, he could reduce to a letter the signification of a sign that might, in other circumstances, indicate a word. Because such a phonetic repurposing did not sound out an entire word, the semantic confusion to which it might otherwise give rise did not and could not occur—provided, of course, that there was a clear indication of when a sign sequence was to be read phonetically and when semantically. Had this complex phonography been present from the outset, it would have changed scarcely at all given the conservative character of Egyptian culture as Champollion saw it. His reported collapse on September 14 was not due to a sudden realization that names and places had long been written phonetically. Rather, it was his discovery of probative evidence for the name "Ramesses," most likely in the Huyot drawing, that clinched the matter.

CHAPTER 27

PARISIAN
REACTIONS

"Just as I finished reading at the Institute a very long memoir, of little importance in itself and mostly composed in Grenoble," Champollion wrote on October 15 to his brother-in-law Andre Blanc, "my good angel led me to one of those literary discoveries that suffice to establish a scholar's glory in perpetuity. By applying the principles established in my earlier works communicated to the Academy, I found the *alphabet of the hieroglyphs* that had long been sought!"[1]

Written less than a month after the Dacier talk, this jubilant letter shows how Champollion saw his own accomplishment in the first blush of its reception. To him, the *Lettre à M. Dacier* was the product of long and steady application, and his discovery the natural development of "the principles established in my earlier works," a description that was correct in essence, but, we have seen, had likely not occurred without Champollion's reading of Young's *Britannica* article. In the weeks immediately following his talk, Champollion believed his auditors most impressed by his decipherments of royal names. Although contemporary accounts of immediate reactions in Paris are scarce, the ones we do have cast Champollion's achievement in a rather different light.[2]

Three months after Champollion's talk, the *Revue Encyclopédique* published an unsigned notice about the Dendera zodiac, the "precious fragment" with which "the zeal of MM. Saulnier and LeLorrain have enriched France."[3] To the remark was appended a lengthy footnote describing the Dacier letter principally in terms of its significance for ancient chronology. The writer contrasted Champollion's approach with the empty and abstract mathematicizing activities of those who, like Biot, invented "many systems" to explain the artifact. These efforts had now been "overturned by the great archaeological discovery of M. Champollion the younger." In a "series of memoirs" the "young scholar pointed out the intimate connection between the *hieroglyphic* and *hieratic* scripts . . . and between the intermediary inscription of the Rosetta stone, an inscription of which M. Champollion presented a complete interpretation to the Institute." In a verbatim repetition of Champollion's line in the *Lettre*, the reviewer contended that the three scripts were "all *ideographic*, painting ideas and not sounds," and reported that, in working out the ramifications of this view, Champollion had incidentally found "a nearly *alphabetic* script" that cast "great light on the age of the monuments of Egypt." Using this method, Champollion had identified Greco-Roman names including, the reviewer claimed, with more enthusiasm than accuracy, the name "Nero" on the Dendera temple.[4] Had Champollion merely presented his method for making phonetic renderings of post-Alexandrian royal names, the reviewer would have remained impressed by its chronological utility, but of equal significance was Champollion's claim regarding the "ideographic" identity of the scripts. This insight yielded hints as to how ancient peoples had "manifested and fixed their thoughts" for posterity. By illuminating the "signs invented by the celebrated peoples of antiquity to manifest and fix their thoughts," Champollion's discoveries reached beyond mere eruditon to the "philosophy of history."[5]

The *Nouvelles Annales* weighed in as well, recapping Champollion's lecture series in a detailed article published just after Champollion's third talk at the Académie. Jointly edited by the poet and expatriot Conrad Malte-Brun and the writer Jean-Baptiste Eyriès, the *Nouvelles* was an expanded reincarnation of the *Annales des Voyages*, founded by Malte-Brun in 1803.[6] Malte-Brun had arrived in France four years earlier, after his prorevolutionary views had drawn the ire of authorities in his native Denmark; in France he pursued his interests in geography and publishing while writing political articles for the *Journal de Débats*. Eyriès, for his part, wrote extensively on geography and was a prolific translator of travelogues and other works.[7] With its focus on place, Champollion's *Pharaons* would have appealed to both editors, and his reputation might have appealed to Malte-Brun as well, given the latter's own political odyssey.[8] In any event, Champollion's lectures found a warm reception in the pages of the *Nouvelles*.

Except for a brief aside noting that hieroglyphic signs could be repurposed "like Chinese writing" to produce "a kind of alphabetico-syllabico script," the *Nouvelles* review focused almost exclusively on Champollion's arguments for the identity of the scripts as principally ideographic.[9] The reviewer lauded Champollion for reversing the long-standing general belief that the hieratic script aimed "to recall the sounds of the spoken language" and for establishing that the script was not alphabetic but intrinsically ideographic. To have recognized, as Champollion did, that hieratic was a simple modification of the hieroglyphs, was something that "no one could have foreseen until M. Champollion the younger." Demotic was, "if possible," even less known than the other script, but the Rosetta inscription enabled Champollion to show that, like hieratic, it too "was ideographic," so that all three scripts consisted of "simple signs to express ideas." Nevertheless, demotic was connected to the spoken language of ancient Egypt in such a manner that each simple sign group ("seme") could represent "a word of the language," indicating that all demotic texts could be "orally pronounced in the same way as Chinese."

In the final two paragraphs, the reviewer turned to ancient chronology, remarking that the *Lettre*'s most significant immediate result was to have "cast considerable light on the chronology of the Egyptian monuments." A proven system for reading Greco-Roman royal names promised to clarify many obscurities, including some that were, at that moment, right before the eyes of Parisians:

> The application of this *alphabet* to the hieroglyphic inscriptions that decorate the Egyptian temples, has read there the name of Alexander the Great, the Ptolemies and their successors, the queens Berenice and Cleopatra . . . and the names and surnames of many Roman emperors, including Augustus, Tiberius . . . on many Egyptian temples, on the two obelisks in Rome, and finally on the celebrated planisphere of Dendera in the Louvre.

Expanding on this point, the reviewer claimed that Champollion's system offered a new and powerful means of taking intellectual possession of an ancient civilization and concluded on a distinctly imperial note: "Here, one may say, is a nearly virgin mine to exploit." If that were not sufficiently redolent of the nineteenth-century travelogue, the reviewer continued by suggesting that ancient Egypt would owe a debt to France for the new perspectives inevitably generated by this fresh route of access into its history, and Europeans, for their part, at long last "will have the means thoroughly to study this old and famous nation" and will "know by what means by [Egypt's] genius manifested thought: therein lies the very origin of its civilization."[10]

Much like the one published in the *Revue Encyclopédique*, this early report saluted Champollion for his demonstration of the identity of the three Egyptian scripts and his production of meanings for the "ideas" represented by their groups of signs. To all these reviewers, the phonetic system presented in the Dacier letter was altogether secondary to the demonstration of the scripts' mutual identities and their fundamentally "ideographic" nature, in Champollion's sense of the term. Despite Champollion's hint, in the Dacier letter, that the phonetics for at least royal names might go far back, and despite the grandiose conclusion drawn by the overheated *Nouvelles* critic, these early reviewers saw the phonetic rendering of Greco-Roman names as little more than an addendum useful mainly for establishing chronology since the Alexandrian conquests.[11]

A third review appeared in Charles-Louis Lesur's *Annuaire historique universel* of 1822.[12] Here, again, the review focused on Champollion's identification of all three scripts as fundamentally "ideographic." The writer speculated that, given this ideographic character, Champollion had wondered how the names of Greco-Roman rulers could have been represented, which led him to compose "an alphabet of phonetic hieroglyphs" dedicated to that purpose. This alphabet had permitted identification of several such names on various monuments "and notably on the famous Zodiac of Dendera, where Champollion recognized and pointed out the imperial titles of Nero," a repetition of the same inaccuracy that had appeared in the *Revue*. The reviewer barreled on, claiming that the identification provided evidence of "the antiquity of this monument [reaching] nearly to the epoch that M. Visconti had fixed, to the time of the Roman domination of Egypt," a point that doubtless reassured those who, like Visconti, had ideological reasons to require a relatively recent vintage for the zodiac. After noting doubts as to whether the alphabet could be applied to Egyptian pharaohs prior to the Greco-Roman period, Lesur remarked—uniquely, at this early date—that further work by Champollion might show this indeed to be the case. If that were so, Lesur supposed that "this hieroglyphic alphabet might be the true key to the entire hieroglyphic system." Reading inscriptions for pharaonic names could thereby establish an exact Egyptian chronology.[13] Lesur by then knew more of Champollion's expansion of his system to the names of ancient pharaohs, about which Champollion had lectured at the Académie near the beginning of 1823, as did the author of a brief remark in the *Journal Asiatique* in 1823.[14]

As Lesur's reference to Visconti suggests, the initial reception of the Dacier letter had one further facet: it reassured those made anxious by the Dendera zodiac's challenge to religious orthodoxy. Champollion reported to his brother-in-law that "[t]his discovery, which I made known to the Institute on 27 September, produced astonishment and forced the applause and approbation of those very people who had long stayed far from me from pure partisan spirit," by which he meant more than just his nearest colleagues, the insular coterie of reactionary Parisian orientalists. "Our good king was informed of my discovery that very evening by the Serene abbot Frayssinous, grand master of the university," Champollion continued. The conspicuously religious Denis-Antoine de Frayssinous likely told Louis of one of Champollion's results that he approved, a point to which we will return. "Two days later the journals proclaimed the matter and, what is no little surprising, the *White Flag*," a conservative publication helmed by the Royalist Alphonse Martainville, "was one of the first to speak of it. If by chance you have read these articles," he boasted, "you will have seen that my discoveries are highly prized."[15]

In the months following the Dacier letter's French debut, Champollion's achievement seemed first of all to have reduced the cacophony of manifold scripts to a single, underlying pattern of signs. Though opinions differed as to what precisely Champollion meant by this, all three scripts were asserted by him to be fundamentally "ideographic," with a limited "semi-alphabetic" repurposing of signs having been developed for proper names and foreign words. This consolidation of the scripts undermined Åkerblad's findings, with which de Sacy, among others, had more or less agreed for the past two decades; perhaps for that reason, other scholars resisted acknowledging the deeper implications at which Champollion had hinted at the end of the Dacier letter, namely, that the "semi-alphabet" might have been developed long before the Greco-Roman domination and used at least on occasion for native rulers. In addition to the insight afforded into the scripts themselves, Champollion's achievement had another, more ironic facet. By reading the word "autocrator" in the Dendera zodiac cartouches, Champollion had landed on the side of biblical chronology and in so doing, remediated his reputation as a young Robespierre.

CHAPTER 28

WORDS ACROSS THE CHANNEL

Upon returning to Paris in September 1822, Young made a point of seeking out Champollion. On Friday, September 27, he attended the Académie des inscriptions et belles-lettres meeting at which the Dacier letter was read, and the following Sunday he received a visit from Champollion himself. If de Sacy's warning of years before had ever carried force with Young, clearly it did so no longer.

Young's initial reaction to Champollion's talk was favorable. Young imagined him rather like another young Frenchman, Augustin-Jean Fresnel who, though he had moved far beyond Young in developing the new wave theory of light, always carefully credited the Englishman as the first to develop the scheme. To Gurney Young enthused that "the author of the book [the *Pharaons*] you brought over has been working still harder upon the Egyptian characters. He devotes his whole time to the pursuit, and he has been wonderfully successful in some of the documents that he has obtained." Two days after the Dacier lecture, Young wrote his friend Hamilton that he had "found here, or rather recovered, Mr. Champollion, junior, who has been living for these ten years on the Inscription of Rosetta, and who has lately been making some steps in Egyptian literature, which really appear to be gigantic. It may be said," Young continued, "that he found the key in England which has opened the gate for him, and it is often observed, that *c'est le premier pas qui coute*; but if he did borrow an English key, the lock was so dreadfully rusty, that no common arm would have had strength to turn it." Were he "ever so much the victim of the bad passions," Young mused, "I should feel nothing but exultation at M. Champollion's success." To Young his very life seemed "lengthened by the accession of a junior coadjutor in my researches, and of a person who is so much more versed in the different dialects of the Egyptian language than myself. I have promised him every assistance in his researches that I can procure him in England, and I hope to obtain from him an early communication of all his future observations."[1]

As Young saw it, Champollion was working from the foundation that he, Young, had erected, merely picking up where Young had left off. But by the time he left Paris, while "waiting for the winds and waves" to subside at Calais, Young's attitude had begun to shift. He wrote to Bankes on October 21 that "Champollion has fully confirmed, and considerably extended, the system of 'phonetic' hieroglyphics, which I had conjecturally proposed from the examination of those of Ptolemy and Berenice, though certainly the extension is so comprehensive, as to require some further collateral evidence."[2] Although Young did not further specify what he saw as Champollion's extensions of his work, he clearly considered himself the first to have asserted the identity of the scripts and to have phonetically rendered hieroglyphs of Greco-Roman names. Like Fresnel, Champollion seemed to Young rather like a talented acolyte who had furthered the master's original insights, and now, with his suggestion that "further collateral evidence" was needed, Young had cast even this derivative expansion into doubt. His initially sanguine assessment, already eroding, was not to last, nor was it ever correct.

The two readers of ancient Egyptian scripts soon exchanged cordial letters. Hoping to bring Young to his side, Champollion sent Young a first missive on November 12 that included two copies, then just printed, of the Dacier letter. Emphasizing his phonetic renderings of hieroglyphic royal names, Champollion told Young of recent successful attempts to confirm "the phonetic values that I attributed to all the hieroglyphic characters." In the drawings from Dendera, in particular, he claimed now to have read "Tiberius Claudius" in a cartouche as well as "Nero." Galvanized by the Dacier letter's positive reception in France, Champollion eagerly awaited a response from Young to "corroborate or rectify" what he, Champollion, had found. In a collaborative overture, Champollion promised to send new inscriptions with "the names of the months" and asked Young for his opinion about whether certain signs might signify "brothers." In reply, Young expressed thanks for the reprints, added a helpful remark concerning the signs in question based on an inscription he had copied from Bankes, and looked forward to receiving Champollion's translation of the "enchorial payprus or at least your remarks about mine."[3]

On November 23, as we have already seen at some length, Champollion wrote, thanking Young for sending copies of his recent letters from Bankes and Hamilton and expressing his hope that the Dacier letter, once printed, would assuage Young's doubts concerning his identifications of the full set of Roman names. Champollion hoped that Young would see fit to test his readings using the materials in the "rich collection of M. Bankes," to which Champollion of course did not have full access. Young had expressed particular doubt about Champollion's reading of the word "autocrator" in the Dendera zodiac, as Young had previously read the signs elsewhere as signifying "Arsinoe." Champollion insisted on his own reading while adding diplomatically that, "assisted by your precious notes and your mnemonic syllabary," he had tried to reconstruct "the diverse cartouches of the Abydos inscription." He also solicited Young's views regarding the crucial phonetic reading of the name of the pharaoh Ramesses that we discussed above.[4] Champollion reported that he was sufficiently certain about this reading to plan a further memoir at the Académie in which he intended to construe phonetically a series of Egyptian pharaonic names. It was a memoir to which "certain academicians appeared to me to accord some importance." Champollion also reported that Belzoni, then making "quite a sensation" in Paris with his displays of Egyptian tombs, had confirmed that other inscriptions at Philae supported his readings.[5]

An ominous silence followed. Three weeks later, on December 16, Champollion wrote again, apologizing for his failure to send a facsimile of the Casati Greek taken directly from the original. In Paris, Champollion had shown his copy to Young and had given him a tracing, which had led Young to provide a quick translation of some parts in his letter to Bankes. To make up the lack, Champollion sent along his "only copy," which he asked Young not to publicize, as the translation of the Casati that Young intended to print was sure to cause trouble with the "conservators" in charge of the document, "even though they're utterly incapable of reading it themselves."[6] It is hard not to be charmed by this petty intriguing, as Champollion attempted to cast Young in the role of coconspirator allied with him against a hostile and plainly stupid bureaucracy.

The charm offensive continued for several months during which Champollion became increasingly convinced that his phonetic readings applied to every pharaonic era. He had meanwhile attracted two surprising champions—of whom we will have more to say shortly—whose influx of support undoubtedly intensified his conviction that he alone was on the right path. On January 9 Champollion wrote to Young again, sending at last a direct copy of the Casati "done by a very able artist," which Young translated himself according to his usual methods and included in a collection of Egyptian inscriptions (fig. 28.1) printed shortly thereafter by private subscription.[7] Champollion claimed to have found "in a hieroglyphic manuscript the name of the king *Osorchon*, son of the king *Schéschonk*," as well as thirty other names of pharaohs, discoveries that justified application of his phonetic scheme to the remote Egyptian past. Huyot had, he now wrote, provided him "large, exact copies of nearly all the proper royal names sculpted on the monuments of Egypt; from these invaluable drawings I made my classification of my *Ideographicophonetic* readings." In none of these letters did Champollion hint that he thought his phonetic scheme might go beyond Greco-Roman nouns and the royal names of both ancient and post-Alexandrian rulers.

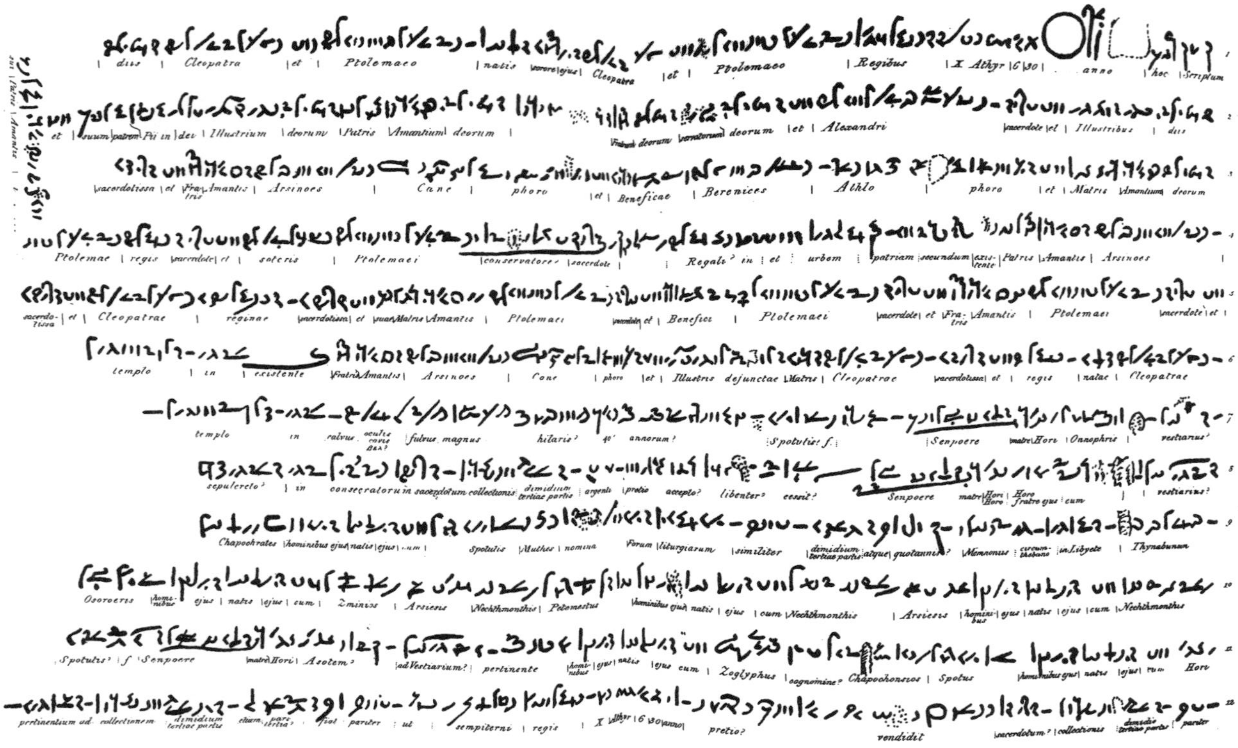

FIGURE 2.8.1. First page of Young's copy and translation of the Casati papyrus.

In his January letter Champollion remarked that he would like to work with an "exact" copy of the inscriptions from the Bankes Abydos tablet, and so he had included a letter directly to Bankes for permission to do so, which he hoped Young would convey. Young had already given Champollion a copy of the inscription, but Champollion told Young that he would not indicate so to Bankes—perhaps sensing, rightly as it turned out, that Bankes was not well disposed toward him. Young did convey the letter, but Bankes refused to reply, for he was furious that Champollion had not admitted seeing Bankes's written insertion of "Cleopatra" on the copy of the obelisk that he had sent to Paris. The following September, he wrote to his friend Sir William Gell that "Bankes would not give [Champollion] a copy [of the Abydos tablet], because he thought him a dirty scoundrel, and would not answer his letter."[8] Bankes's fury was matched by Champollion's when, about five weeks after his January letter to Young, Champollion read a vicious attack against his work in the *Quarterly Review*. The anonymous author of the piece was John Barrow, who years before had with George Staunton written the review of Marsham's *Chinese Languages* that Young had so carefully absorbed. Because Champollion and Young became increasingly suspicious of each other's motives from this point on, Barrow's review is worth exploring.[9]

Barrow began his dismantling with outright mockery. One could scarcely have imagined, he opined, that it might soon be possible to "not only write the name of his present Majesty, but both write and read those of every emperor, king and conqueror of ancient Egypt, from the time of Alexander to Antoninus Pius" in the previously inscrutable Egyptian hieroglyphs. Why, one can even use these characters to "write billet-doux, as we understand the petit-maîtres of Paris are now doing." But what has been gained? "We are not," Barrow carped, "a single iota advanced in understanding the meaning of any one of these sacred characters, unless when so applied in designating the mere names of foreigners."[10]

Barrow continued to think, and he believed Champollion agreed, that the Egyptian hieroglyphs represented specific words of the language. That is, they had been designed to have semantic, but not phonographic, meaning, and this primary semanticity had persisted throughout Egyptian history until it became necessary to represent the names of Greco-Roman rulers. Champollion had therefore accomplished very little, because he had not provided an acceptable way to uncover the characters' proper meanings, that is, the words they originally expressed. Only by doing so would one be able to actually read any of the Egyptian scripts. Although Barrow disingenuously remarked that he had no intention of "at all derogating from the merit of M. Champollion's indefatigable labours," he nevertheless found "nothing" in the Frenchman's work that could "repay him for the persevering siege which he has conducted against the pot-hooks of Egypt, for just so many years as the Greeks sat down before Troy." For there was "nothing, in fact, of originality in his supposed discovery to console him for the laborious investigation he has patiently submitted to." Champollion had accomplished little beyond having worked "merely to complete an invention which had been known to so many of his predecessors." If further pursuit of the matter "had deterred them," that was no doubt because they, unlike Champollion, had known that the result would hardly repay the effort. "Of this sad truth," Barrow went on, Champollion "must surely be aware," given his "ten years' lucubrations to prove that neither the *hieratic*, or sacerdotal, nor the *demotic* or vulgar, writing is alphabetic, (as, he says, was generally thought,) but *ideographic*, like the pure hieroglyphics." But even this was not "a discovery due to M. Champollion, nor are his results quite correct." De Sacy and Åkerblad had preceded him in linking the scripts, claimed Barrow, while "Doctor Young" had developed a method that "successfully extended the catalogue [of Åkerblad's alphabetic system for foreign names], and discovered that the system was equally applicable to the pure hieroglyphs." More followed, equally dismissive. The whole process "was sufficiently obvious," Barrow sniffed.[11]

Barrow did at least find Champollion's Egyptian "homonyms" plausible because foreign words in Chinese were treated similarly. The Egyptians had appropriated "any one *sign*, the name of which either wholly or in part appeared to them to approach nearest to the *sound* of the syllable, or letters required." But the Egyptians, "having no settled system of sounds," would "represent the sound required" on an ad hoc basis, such that "one person might take one character, and another another, to represent the sound required, according as to his ear." Such inconsistent attribution is why "in M. Champollion's alphabet, the same letter of the Greek alphabet" is represented "by eight or ten different hieroglyphics. This is precisely the case with the Chinese." But the Chinese could not have followed the same method as the one that Champollion had attributed to the Egyptians because their language was based on a comparative paucity of semantic monosyllables, with many of their written characters corresponding to the same monosyllabic vocalization. Chinese, he wrote, "being altogether syllabic, and each of their numerous characters a monosyllable, the idea of spelling a word by letters never entered their minds. A proper foreign name, or other foreign word, therefore, written in their language, will always contain only just as many characters as the word has syllables." So, for example, "Pto-le-my" would have been written in phonetic Chinese using three characters, whereas Champollion's Egyptian version deployed six or seven since he had transliterated the Egyptian signs as letters or diphthongs.

Barrow strongly demurred concerning the ease with which phonology had been produced in Egypt because he thought Champollion incorrect in his manner of composing phonetic signs. "We doubt not," he asserted, that Champollion "is mistaken, in supposing that the Egyptians found it easy to form an alphabet out of their hieroglyphics." Why so? Because Barrow considered that, contrary to Champollion's claim, the Egyptian phonography, such as it was, must like the Chinese have been "syllabic rather than alphabetic; and Dr. Young seems to be of the same opinion." And so, given "our estimate of the difficulty attending such a process from the total failure of the Chinese to accomplish it out of their hieroglyphics, we should rather say that the original and unaided attempt of the Egyptians was syllabic rather than alphabetic; and Dr. Young seems to be of the same opinion," contrary to Champollion. There was more, for Barrow minced no words in claiming that, whatever was acceptable in Champollion's work, and there was precious little of that, had already been accomplished by Young. With no subtlety at all, Barrow wrote that "Champollion has little or no claim to originality in what he seems to think a discovery."[12]

These were fighting words. Beyond the direct accusation of having copied Young wherever Champollion's work was at all acceptable, Barrow's critique of his claims concerning phoneticism was especially sure to provoke Champollion. His phonetic reconstrual, entirely different from Young's as he saw it, required stripping the semantic connotations of the languages's monosyllables from their graphological signification when used phonetically. That procedure produced not syllables of the language, properly speaking, but, instead, combinations that, while not necessarily phonemic in the alphabetic sense, nevertheless lacked what Champollion took to be the semantic essence of a syllable for Coptic. Young, in contrast, had simply chosen whatever seemed to work phonetically in very loose connection to Coptic. "The English savant," Champollion had written in the Dacier letter, "thought that the hieroglyphs which form proper names could express entire syllables . . . this point of departure in great measure falsified the phonetic analysis he attempted on the names of Ptolemy and Berenice." Where Champollion perceived a carefully constructed system that avoided the semantic trap, and that accordingly could have been applied even to the names of native Egyptian pharaohs, Barrow saw an inherently defective process that operated altogether differently—one that Young, as we have seen, saw as evidence of a wider cultural infantilism.[13] In a remark seemingly calculated to cause maximal outrage, Barrow closed by undermining Champollion's suggestion, the most radical of all he had ventured in the *Lettre*, that "the natives [i.e., early Egyptians] made use of the same hieroglyphico-phonetic signs, as were subsequently in use." This incredible conclusion, which suggested that "the Egyptians furnished the model on which the people of Western Asia constructed their alphabets," was so insupportable as to seem overly theoretical, that is to say, French: "We consider him to be wandering into the mazes of theory, and venture to pronounce that he will lose himself in the inextricable labyrinth."[14] Accusations by the English that the French were irretrievably addicted to unsupportable theories reach back, of course, to the seventeenth century.

Upon obtaining Barrow's unsigned review, Champollion dashed off an anguished letter to Young.[15] "Everyone who read [the review] reacted strongly against the ignorance and bad faith of the author," he wrote in a fury on March 23, 1823, and "it produced in me the same effect." Everyone knows the "facts concerning my alphabet," he continued, and when and by whom previous attempts had been made. No one understood, "M. de Sacy less than all others," how anyone could have "attributed to [de Sacy] the relations between the three scripts," and neither "does anyone know of an effort by Åkerblad to work on the hieroglyphs." Here Champollion was certainly correct, and Barrow's remarks on this score were misleading to say the least. Among the reviewer's other "absurd" assertions, Champollion noted one of direct interest to Young, as no one would "at all confuse my alphabet with what you published." Against Barrow's suggestion that Champollion's discovery was useless, Champollion reminded Young that the Académie "already knows what to believe about that, and very soon the lettered public will also be convinced that my alphabet is the true key to the system."[16]

The *Quarterly Review* contained much to pique Champollion, and indeed Barrow seemed intent upon winding him up. But Champollion's anger may have reflected something else as well: The critique veered uncomfortably close to convictions he had only recently relinquished. He had, after all, previously argued, even in public at the Académie, that the Egyptian scripts could not have been phonographic, even for foreign names. If, Champollion had long thought, the signs were originally logograms for words, then the language's basically monosyllabic structure would have produced an unavoidable slippage between written sign and meaning. Barrow, of course, held much the same view concerning Chinese. By mid-September 1822, Champollion had changed his mind about the signs' having been used phonographically, but not because he had abandoned his conviction that the signs had originally been produced as logograms. Instead, he now thought that they could have been repurposed even in pre-Alexandrian times by stripping away their initial or final sounds, albeit solely for royal names and epithets. The *Quarterly Review* had attacked Champollion's claims on just that basis by means of an analogy to Chinese, thinking the antiquity of such a process altogether unlikely given the semantic monosyllables of both languages.

Champollion had noticed something else as well. The London edition of the *Quarterly* in question apparently contained a set of advertisements of forthcoming books at its end.[17] Among these was one provocatively titled *An Account of Some Recent Discoveries in Hieroglyphical Literature and Egyptian Antiquities: Including the Author's Original Alphabet, as Extended by Mr. Champollion* by Thomas Young, to be printed in London by John Murray. The title, Champollion fumed to Young, "promises to indicate the first author of an alphabet that I would only have extended." Waxing angrier, he continued, "I will never consent to recognize any other original alphabet than mine when it concerns the hieroglyphic *alphabet* properly speaking, and the unanimous opinion of scholars in this respect will be more and more confirmed by the public examination of every other pretension."[18] Champollion's emphasis on the hieroglyphic alphabet reflects his conviction that only he, and not Young, had understood the fundamental problem: that the scripts could not be syllabic given the nature of the Egyptian language.

In closing Champollion reminded Young of how much he treasured their personal and scientific relations, and ventured to presume that Young surely did not share the contemptuous opinions of the review's author. This obviously wishful thinking reassured no one, least of all Champollion who "impatiently" awaited Young's reply, which he needed in order "to give to the response I'll publish in a spirit suitable to the interests of truth, to yours, Monsieur, and to mine."[19] If Young ever wrote back—and it is not clear that he did, as no record of a reply remains—the response did not assuage Champollion's fears. In any event, their private correspondence ended with this exchange. They were antagonists now, and the battle for priority shifted decisively to the public realm.

PART 5

ANTIQUE LETTERS

CHAPTER 29

GREY'S BOX

Young had been in London for less than a month when, on November 22, 1822, he received a visit from George Francis Grey, an Oxford clergyman and antiquarian. Grey, who had toured Egypt in 1820, handed Young a box of papyrus fragments he had purchased "of an Arab" in Thebes, believing them to contain information related to early Christianity. As it turned out, the papyri, written "chiefly in hieroglyphics," were "of a mythological nature." Grey drew Young's attention to two "in excellent preservation" that also "contained some Greek characters, written apparently in a pretty legible hand."[1] As night fell, Young took a closer look at Grey's gift and noticed an extraordinary coincidence: the Greek writing on Grey's papyrus was a translation of the demotic text on the very same papyrus, the Casati, that Young had recently discussed with Champollion in Paris. This stroke of luck provided Young with an opportunity to attack Champollion's general approach to the hieroglyphs.

The question, as ever, turned on the identification of proper names. On October 21, while in Calais awaiting return passage to England, Young had written his friend Bankes concerning the Casati papyrus, of whose demotic Champollion had provided Young with a tracing while he was in Paris.[2] There, Young later recalled, Champollion "had mentioned to [him] in conversation the names of Apollonius, Antiochus, and Antigonus, as occurring among the witnesses" to the decree recorded on the Casati.[3] Young, however, told Bankes that the Casati contained the names Antimachus and Antigenes, a rather different result. He later explained this inconsistency by claiming that Champollion, in conversation, had misidentified "Antimachus" as "Antiochus," having "omitted the M," and that he had himself subsequently forgotten this misidentification. The arrival of Grey's box in November reminded Young of Champollion's misreading, for it contained a "marvelous accident."[4]

Opening the box, Young withdrew a papyrus covered with Greek, which he "proceeded impatiently to examine." "I could scarcely believe that I was awake," he recalled, "and in my sober senses when I observed, among the names of the witnesses" who signed the document, "ANTIMACHUS, ANTI-GENIS: and a few lines further back, PORTIS APOLLONII."[5] Astonished, Young concluded that the papyrus amounted to "a translation of the enchorial manuscript of Casati." How utterly unexpected and indeed nearly impossible, for Casati and Grey had no connection with each other and had been buying artifacts in complete independence.[6] "A most extraordinary chance," Young marveled, "had brought into my possession a document not very likely, in the first place, ever to have existed, still less to have been preserved uninjured, for my information, through a period of near two thousand years." Even more miraculous was its arrival on his Welbeck doorstep "at the very moment when it was most desirable to me to possess it, as the illustration of an original which I was then studying"; such a coincidence would, "in other times, have been considered as affording ample evidence of my having become an Egyptian sorcerer."[7]

This unlikely document clarified the purpose of the Casati papyrus. Grey's papyrus, Young later explained, concerned the sale of "a portion of the Collections and Offerings made from time to time on account, or for the benefit, of a certain number of MUMMIES, of persons described at length, in very bad Greek, with their children and all their households."[8] At the time of their conversation in Paris, Young said that "Champollion had not thought it worthwhile to give me a transcript of the original Greek endorsement" of the Casati as "he seemed to consider it as not fully agreeing with the Egyptian text, or, at any rate, as not materially assisting in its interpretation," which may have been true. Alternatively he may have been unwilling to give Young, whose Greek was more than usable, a leg up on this score. At any rate, having read altogether sixteen different names on the papyrus from Grey, Young now had excellent grounds for disagreement with Champollion. Since Grey's papyrus seemed to be a Greek copy of the registry given in demotic on the Casati material, Young requested a copy of the latter's Greek in order to check the insight; he received this material from Désiré Raoul-Rochette, who informed Young that Saint-Martin had published the Greek in the September issue of the *Journal des Savans*.[9]

Grey's gift provided Young with a petty victory that he quickly turned to his broader advantage. In a November 23 letter to his friend Gurney, Young bragged that "I have already obtained a little miniature triumph over Champollion in my amusements at Calais. It seems he read the names Antiochus and Antigenes. Grey's MS. has clearly ΑΝΤΙΜΑΞΟΣ ΑΝΤΙΓΕΝΟΥΣ, as I had made them."[10] Writing up his find, Young shifted to the question of the scripts. Grey's papyrus offered evidence "more in favour of the extensive employment of an alphabetical mode of writing" during the Greco-Roman period than the Rosetta inscription had, and some of the words "might be read pretty correctly, by means of the alphabet originally made out by Mr. Åkerblad," a suggestion that left Champollion's work entirely to one side.

It was not Champollion's Dacier talk at the Académie, then, but Grey's papyrus that convinced Young that "an alphabetical mode of writing" had been more extensively deployed than he had thought. The "alphabetic mode" reached, he now admitted, even to "Egyptian names" and perhaps even to some common words of the language during the post-Alexandrian era. However, where a phonetic reading did seem workable, Young insisted that the expression was generally, though not always, syllabic rather than alphabetic—that is, constructed from the entire Egyptian word, not from the stripped forms that Champollion had presented.[11] Overall Young was unpersuaded by Champollion's wider claims concerning the antiquity of phonography for royal names. In his 1823 *Account*, which included his reflections on Grey's papyrus, Young did not mince words to that effect. In response to Champollion's phonetic identification of Ramesses, Young was "sorry to say that I cannot hitherto congratulate Mr. Champollion on the success of his attempts to carry his system of phonetic characters into the very remotest antiquity of Egypt."[12]

In Young's view, Champollion had been led astray by his reliance on Coptic equivalents. In the hieroglyph for Ramesses, for instance, Champollion had identified the initial sign, a circle with a central dot, as having the phonetic value of R. In his *Britannica* article Young had identified the sign as representing the sun deity and had loosely suggested that it might sound RE for the many variants of "Ramesses." Champollion, however, had gone too far in detecting the phonetic presence of "Ramesses" on the Abydos king list. To this Young offered what he thought to be a fatal objection. Namely, "that the circle, which Mr. Champollion considers to be equivalent to the RE or RA of Ramesses, is also the first character of each of the seventeen names immediately preceding it, and indeed of every other" in the Abydos list (see fig. 26.7). The objection was a powerful one: if Young was right, all the names in question were to be pronounced differently.[13]

Having located a fault, Young proceeded to generalize from it, asserting that Champollion had "never been led in any one instance, from the Egyptian name of an object, to infer the phonetic interpretation, that is, the alphabetical power of its symbol." Instead of inference, Champollion had applied the power of lexicon, a relatively weaker force in Young's view. "[T]he letters having been once ascertained," Young speculated, "[Champollion] has ransacked his memory or his dictionary for some name that he thought capable of being applied to the symbol."[14] In this case of "Ramesses," Young surmised that Champollion must have first identified the terminating king name in the Abydos sequence as "Sesostris, Sethos, or Sésoosis" (and hence arrived at "Ramesses") by linking the sequence to the king list of the Greek-era Egyptian priest Manetho.[15] He would then have searched for an appropriate sign, fastening quickly upon the circle with a dot for the sun, hence RA, and so "Ramesses," according to a procedure Young characterized as an unfortunate "ransacking" of links via Coptic for sounds. Couched in palliative and polite verbiage, Young's critique nevertheless amounted to an outright attack on Champollion's effort, if not his entire method. Although Young conceded that Champollion had extended the phonetic list for foreign names, he added that, of course, Åkerblad had really laid the groundwork in this regard. As for anything further, the old Scottish verdict held: not proven.[16]

Young dedicated his 1823 *Account* to Alexander von Humboldt, with whom he had corresponded since 1816.[17] By 1819 their friendship had blossomed, and in a letter of October 26, Humboldt expressed his admiration for Young's *Britannica* articles.[18] Annoyed by Jomard's effort to "arrogate to himself your beautiful discovery of the numerical [hieroglyphic] signs," Humboldt reported that he had "avenged you as I ought" by circulating the *Britannica* essay on Egypt. After reading the *Account*, and no doubt flattered by its dedication to him, Humboldt wrote on June 13 of the "magnificent present and the honor you have deigned to do my name," calling Young "father of all the discoveries made on the mysterious language of Egypt." Despite the evident warmth of these communications, the letter contained much to suggest that Humboldt's embrace of Young was less than wholehearted. No doubt aware of the growing controversy over priority, Humboldt took care not to grant Young too much in this letter. First, Humboldt wrote that he soon hoped to receive the work of "Spohn of Leipzig on the cursive," that is, demotic, inscriptions from his brother (William). Having raised the specter of one competitor, Humboldt immediately produced another, mentioning that Champollion's *Précis* was to appear in several weeks, while de Sacy "who renders, as you know, all justice merited by works crowned with success" would "announce" Young's own *Account* in the *Journal des Savans*.[19] At least in Humboldt's view, Young may have been the "father of all the discoveries," but he was now one player among many in a crowded field. Humboldt was not the only one keeping Young at arm's length. In Paris, Letronne had dedicated his *Recherches* to no fewer than four pioneers in the Egyptian scripts—Young, Huyot, Champollion, and Gau. Which is not to say that distancing oneself from Young implied an equivalent enthusiasm for Champollion. In a letter of March 3, Letronne went so far as to slander him, telling Young, rather cryptically, that the "liberty I took in speaking of a certain charlatan of our country, monopolizer of Egypt, will not please everyone; but about that I don't worry at all."[20]

With the *Account*, Young sought to weaken the significance of Champollion's accomplishments with respect to phonetic signs for foreign names and to denigrate attempts to go further. Underlying the attack was not only

Young's rejection of any concerted reliance on Coptic, but also his disdain for ancient Egypt. "I must acknowledge," Young forthrightly wrote in his *Account*, "that my respect for the good sense and accomplishments of my Egyptian allies by no means became more profound as our acquaintance became more intimate." Aligning himself with the classical tradition of contempt for ancient Egypt, he continued, "on the contrary, all that Juvenal, in a moment, as might have been supposed, of discontent, had held up to ridicule of their superstitions and their depravity, became, as it were, displayed before my eyes, as the details of their mythology became more intelligible."[21] Young's stance is hardly surprising given his Quaker background, in which zoomorphic worship figured as a particularly egregious form of idolatry. His disdain echoed the language that Barrow had deployed years before to describe Chinese popular religion, which he devalued in comparison with a Christianity suitably purged of "Romish" idolatry.[22] Although Barrow and Young could muster little sympathy for polytheistic beliefs, Champollion had long displayed a respect for this element of ancient Egyptian culture, an appreciation motivated not least by Champollion's antagonism to his local apostles of intolerance, the French clerisy.

Champollion apparently did not write Young about the *Account*, or at least no such letter remains. The only known letter from him to Young within months of the book's publication is dated August 21, 1823.[23] After thanking Young for plates from the collection entitled *Hieroglyphics* that Young had produced in 1823, Champollion advanced a claim that surely startled him, and that, we shall see, he had probably arrived at no earlier than the first part of the year: namely, that he was "now certain that a *very great part* of the signs employed in *hieratic* and *hieroglyphic* inscriptions in all ages are nothing other than the signs of *sound*, as is *the greatest part* of all *demotic* or *enchorial* text." So certain of this was Champollion that he informed Young of three lectures he had given at the Académie carrying the unprecedented claim forward. These lectures, he wrote, "will form a volume," which appeared shortly thereafter as the *Précis*. Although we will turn below to the events that led to Champollion's almost complete volte-face in respect to the Egyptian scripts, this was the first Young had heard of it. In the same letter, Champollion added that he wanted to "avoid discussion of priority, which means very little to the public, who demand only illumination without worrying about whence it comes." Such temperance contrasted markedly with Champollion's anguished cry of the previous March. Overall the tone of Champollion's August letter is so respectful and accommodating that we doubt he had yet seen Young's *Account*. But he certainly knew what Young had written by the time the *Précis* reached print in 1824, for Young had offered a full-blown explanation of what

Champollion now claimed, and he insisted on both the novelty and the priority of this work. Still, even here Champollion wrote with unusual care and circumspection. Two events, both serendipitous and highly advantageous to Champollion, had occurred in the meantime. Perhaps these events, to which we now turn, had so improved his confidence that he felt less need than usual to mount a strident self-defense. At any rate, these fortunate accidents smoothed Champollion's path, leaving him freer than ever to pursue the startling claims he had made in his Dacier lecture.

CHAPTER 30

AN OPPORTUNE
ENCOUNTER

Despite the now-legendary status of Champollion's *Lettre à M. Dacier*, its publication initially did little to improve Champollion's immediate prospects. Throughout the autumn of 1822, he remained in dire financial and professional straits. Jomard's enmity continued, and Frayssinous's reactionary influence expanded with his election to the Académie française. In these discouraging developments there was little to indicate the reversal already afoot. But on December 20, 1822, at the opening meeting of the Société asiatique, Champollion caught a hint of the rising esteem in which his work was held, at least in France.

Presiding at the gathering was Louis-Philippe, Orléanist cousin of Louis XVIII, scion of the liberal opposition, and honorary president of the Société, which had been founded under de Sacy's direction the previous April.[1] Louis-Philippe praised Champollion, exclaiming that the "brilliant discovery of the hieroglyphic alphabet honors not only the scholar who made it but the nation!" He asked Champollion to explain "in detail" the "current state of the new science," and inquired as to whether there were in Paris sufficient resources, compared to what the British had, to pursue the subject. Although the meeting had little direct effect on Champollion's material prospects, the attention was surely welcome. Even stronger confirmation of Champollion's improving fortunes would arrive the following month in the form of a cash infusion from the decidedly illiberal Pierre Louis Jean Casimir de Blacas, who had represented the regime at the signing of a Concordat with the Vatican in 1817 and whom Louis had made Duc de Blacas in 1821.[2]

A dedicated antiquarian and a long-standing member of the Académie des inscriptions et belles-lettres, Blacas had helped to excavate the Roman Forum while serving as French ambassador to the Kingdom of the Two Sicilies. In January 1823, he was called to Paris as one of the four "gentlemen of the king's chamber." Several weeks later, he stopped by a salesroom to have a look at various antiquities. Among the items available for purchase were several Egyptian artifacts, one of which Champollion was busily copying.[3] According to Hartleben "a man of a certain age, substantial, and with an unusually commanding air . . . watched in silence" as Champollion worked, admiring his "rapidity and stylish way of copying the hieroglyphs." After they exchanged a few words "about the Egyptian collections in Italy, especially Drovetti's," Champollion guessed the stranger's identity.[4]

It seems that Blacas, for his part, recognized Champollion right away. The notorious Dendera controversy was still fresh, and Blacas had surely heard of the Dacier letter, the presentation of which he may have even attended on September 27. Blacas was likely no less aware of Champollion's views of the monarchy and the church. That day, however, Blacas's antiquarianism prevailed over his monarchism, and Champollion—forfeiting, for once, a chance to irritate a conservative—kept the discussion focused on Egyptian matters. "Abandoning all reserve," Champollion "animatedly explained the advantages that such a historically significant collection would have for the exploration of the ancient Egyptian language" and "complained in strong words that the French government, which had given the enormous sum of a hundred fifty thousand francs for the Dendera zodiac, was robbed by bad luck."[5]

Champollion speedily acquired an influential patron. Blacas brought him to the attention of Louis XVIII, emphasizing how greatly French glory would be enhanced if the young scholar continued to outshine his English counterpart in the matter of reading the ancient scripts.[6] The effort paid off. Louis, who was sensitive to the comparative reputation of France, not least given the widespread view that he was a creature of England and her allies, sent Champollion a "gold box covered in diamonds." The gift, which arrived in February, was inscribed from "Louis XVIII to M. Champollion the younger on the occasion of his discovery of the hieroglyphic alphabet."[7]

The rapidity with which Blacas took up Champollion's cause suggests that his political views were no longer a reason to avoid him. The presentation of the Dacier letter had only enhanced his fame, for he was already widely known for his reading of "autocrator" in the hieroglyphics drawn in the *Description*'s plate of the Dendera zodiac. Ironically, the "autocrator" reading improved Champollion's standing among monarchists and the clergy. His reading, if correct, implied that the zodiac must have dated from Greco-Roman times, eliminating its challenge to religious orthodoxy.[8] Despite the beneficial effect it had on his career, this unsought and accidental reputation as the savior of religious belief deeply annoyed Champollion. As he later wrote to his friend Thévenet, the Ultras, those "vile reptiles," were treating him as "a true defender of Religion and of good chronological doctrines," but they would change their minds as soon as he applied his alphabet to "monuments whose antiquity will frighten them."[9]

At this time Champollion planned to publish a book containing many images of Egyptian hieroglyphs and emblems. On July 24, 1823, he wrote to his

FIGURE 30.1. The goddess Hathor from Champollion's *Panthéon*.

father-in-law, Claude Blanc, about the project, to be titled the *Panthéon égyptien*.
It would be a "deluxe" edition costing three hundred francs, and Champollion
anticipated a lucrative sale because "only works that cost a lot are bought and
sold without difficulty." He had already attracted "enough subscribers to cover
the cost of the first printing, and you may have learned from the *Moniteur* that

the king himself subscribed for three copies . . . I'm certain that at the third or fourth printing I will gain considerable profit."[10] Impressively illustrated, the *Panthéon égyptien* could not fail to secure the attention of a wealthy public enamored of Egyptian objects and expensive editions larded with reproductions of those objects. Perhaps not coincidentally, Champollion's proposed *Panthéon* would undermine the first-mover advantage Young enjoyed as a result of his recently published *Hieroglyphics*.[11]

Blacas's interventions improved Champollion's reputation to such an extent that the king, Champollion continued in his letter to Blanc, had now ordered the publication "without cost" of "another work" that was "already half printed," the *Précis du système hiéroglyphique des anciens Égyptiens*. Louis had also instructed his minister to present Champollion with a full set of the *Description*—an enormously valuable gift, though Champollion regarded it principally as a highly resellable sign of respect. Given his hostile relations with the *Description*'s editor, Jomard, and his suspicion of the plates' accuracy, it's hardly surprising that he would not find the work chiefly valuable for his research. He exclaimed to his father-in-law that this "present is worth seven thousand francs, if I decide to sell it; but the respect and the recognition!"[12]

When Blacas encountered Champollion at the antiquities salesroom, Champollion lamented that the funds spent to acquire the Dendera zodiac had not instead been used to secure the Drovetti collection of Egyptian antiquities, recently arrived in Turin. This collection, comprising hundreds of papyri, statues, and mummies, had been amassed by Bernardino Drovetti during his tenure as French consul in Egypt; since 1816 he had been shopping the collection around Europe.[13] Champollion's lament was especially poignant because he had by then exhausted the Egyptian research materials available in Paris. To completely develop his scheme of meaning and structure for the ancient scripts, he needed the fresh resources now in Turin.

The reasons for the eventual French failure to acquire Drovetti's antiquities remain unclear, though it could hardly have been the price (400,000 lire) per se given the recent purchase by the king of a collection of classical objects for 480,000 francs. One contemporary commenter, François Artaud of the Lyon museum, linked the loss to the "cursed zodiac" of Dendera, fearing that the Drovetti collection might contain objects that would again cast Egyptian antiquity far enough back in time to resurrect the danger to biblical chronology that Champollion had himself just obliterated.[14] Few were eager to reopen that Pandora's box.

Turin beckoned, but Champollion needed money for the trip and even more to print the *Précis*. With Blacas maneuvering in the background to produce the requisite funds, Champollion pressed on with more local and immediate prerogatives. As much as Blacas had improved Champollion's fortunes by moving him closer to the center of royal power, the blessing was not unmixed. Capturing the support of powerful and influential people exposed him to the enmity of those who felt themselves to be less favored. Champollion never fully shed his reputation as a radical either, and his intemperate reactions to criticism, not to mention his lack of generosity in ceding much of anything to Young, created additional resentments. In reply to Barrow's critique, about which Champollion had grumbled on March 23, he gave lectures at the Académie des inscriptions in April, May, and June; this material would appear a year later in the *Précis*. Then, on April 21, Louis-Philippe had applauded Champollion at the Société asiatique in the presence of de Sacy, Humboldt, Rémusat, Dacier, and others. No less than all of France "must swell with pride" now that "a Frenchman has begun to penetrate these mysteries that the ancients unveiled only to well-tested adepts, and to decipher these emblems whose signification every modern person despaired of discovering!"[15] Meanwhile, Blacas continued to work sedulously on Champollion's behalf, offering to finance the longed-for trip to Turin, should royal funds prove unavailable. Meanwhile, currents of opposition and support flowed through the meetings and drawing rooms of Paris.[16] By May 1824, Champollion had finished both *Panthéon* and *Précis*, and Blacas had secured the needed funds. The *Précis* went to press and Champollion to Turin, home of Drovetti's collection.[17]

CHAPTER 31

THE "TRUE KEY" TO EGYPTIAN HIEROGLYPHS

As the only full exposition of his system published during Champollion's lifetime, the *Précis* presented the essence of Champollion's system of decipherment as he wished it to be understood.[1] Prefaced by a fulsome dedication to the king, the enormous work ran to nearly six hundred pages, divided into ten chapters comprising two-thirds of the whole, followed by thirty-two plates, eleven of which listed phonetic hieroglyphs with corresponding signs in hieratic and demotic.[2] Though he provided many elaborations, confirmations, and further sign structures as he grappled with new materials, including inscriptions obtained in Turin and, years later, in Egypt, Champollion did not substantially change his views between the publication of the *Précis* in 1824 and his death in 1832.

Champollion finished the *Précis* rapidly, spurred by the blistering attack on the Dacier letter published anonymously by Barrow in the March 1823 issue of the *Quarterly*. The attack was particularly damaging because Barrow had judged Champollion's results to be useless wherever they were not derivative of Young's. The dismissive critique had two prongs: Even if one granted Champollion's correctness with respect to deciphering the signs, which Barrow doubted, his system for names was simply inapplicable to pre-Alexandrian Egypt.

The *Précis* gave Champollion a chance to reply, which he did, hotly and at length. The review's author, Champollion complained, "all while admitting and repeating my alphabet, which he reprints in abridged form, *puts everyone in position to read the Greek and Roman names inscribed in hieroglyphs on the monuments of Egypt*."[3] Despite now being able not only *"to read and write with the greatest facility"* all the names found on the Egyptian monuments" but also, damagingly, "to write *love notes* with my alphabet as he tells us is already common among the *grand-masters of Paris*," Barrow had boldly declared *"that nevertheless we are not one sole iota advanced in knowledge of the meaning of a single one of the sacred characters.*" Were that so, Champollion continued, then "my alphabet is of absolutely no use, neither for monuments before the Greeks, nor for understanding hieroglyphic texts of any era whatsoever." The criticisms were so misguided, Champollion averred that he would not have responded "had I not seen them reproduced, in major part, in a new work by M. doctor Young," that is, the *Account,* which prompted Champollion to develop his system in ways that could survive any challenge to its novelty.[4]

Out of respect for such an "eminent scholar," Champollion admitted that Young did first spy certain distinctions "concerning the general nature of the scripts," by which Champollion evidently meant their mutual equivalence. He also acknowledged that Young had been the first to mark the existence of several acceptable sound signs for foreign proper names, in particular those for Ptolemy and Berenice. But that was all. For, Champollion insisted, at the time the Dacier letter was read, Young had no general idea concerning hieroglyphic phonography and no certainty concerning the "alphabetic or syllabic or disyllabic" values that he had given to eleven of the thirteen signs in the only two names that he had analyzed. Neither had Young "suspected" the existence of Cleopatra's name on the Philae obelisk, even though "he long had a copy [of it] before me, when I didn't know it except in Paris through the helpfulness of M. Letronne," implying that Young may have withheld this evidence. Turning to the sign equivalences Young had advanced, Champollion declared eight of the eleven purported relations "inexact," which is why "neither Young, nor any other scholar, over thirty years, took the slightest advantage of the presumed small series of phonetic signs, inserted by the English scholar in the *Encyclopedia Britannica* in 1819: only my alphabet gave them to that author and the public." Once Young had seen the *Précis,* he could not possibly "refuse to recognize . . . that his works haven't provided any definite light, whether on the intimate constitution, or on the whole of this system of writing," Champollion declared.[5]

Despite the rapidity of its composition, the *Précis* was a finely tuned polemic in which Champollion defended his accomplishment while accounting for the course of his reasoning. Champollion began by insisting, contra Barrow, on the usefulness of his work for the dating of Egyptian antiquity. This point of departure no doubt reflected his recognition of one important source of his persuasive power. Champollion's work on the Dendera zodiac had recently made him palatable to the regime and the French clerisy. Having finally managed to attract a patron, Champollion needed to secure that support, in part by fortifying himself against a Barrow-style attack. What better way to do that than to make his dating dependent on his breakthrough with the scripts? "My alphabet," he wrote in the preface to the *Précis*, "removes all doubt, dare I say, all uncertainty, and compels us to read an imperial Roman title," that is, autocrator, "on the circular zodiac of Dendera; on the great edifice above which it is placed," as well as names and titles of Roman emperors appearing on another zodiacal monument at Esneh, "one that had previously been thought anterior by *many* centuries to that of Dendera." These and other similar results showed not only that the monuments "belong irrevocably to the time of Roman domination in Egypt," but also that his alphabet, which he had used to identify the names that had permitted him to assign these dates in the first place, was "one of the causes" of his success in doing so, "and one of the *reasons* that have earned me the most flattering encouragements." That Champollion's successful work on the zodiacs relied upon his prior accomplishments with the Egyptian scripts had been insufficiently appreciated by his critics. "One may allow me to display here," he wrote, "the very proofs of my discovery's certainty," which were all the more urgently needed since "I feared contradictions less than pretensions to sharing my discovery."[6]

Champollion's goal for the *Précis* was to "demonstrate, not against the inconsequential opinion of the anonymous [reviewer] of the *Quarterly Review*, but against the much more imposing opinion of M. doctor Young himself" that he, Champollion, had alone correctly grasped the nature of the Egyptian scripts. To do so he would now provide evidence for a staggering new claim: Not only would Champollion demonstrate that his hieroglyphic alphabet applied to royal hieroglyphs throughout Egyptian history, but also that these phonetic hieroglyphs provided the "true key" to a system that was used to inscribe the sounds of the spoken language throughout the history of ancient Egypt. Indeed, these phonetic signs constituted the major part of all hieroglyphic characters, as he intended to show over the course of four hundred pages of argument coupled to twenty-four exemplary texts and tables, all composed in less than a year.[7]

To start, Champollion faulted Young for insufficient knowledge of the range of Egyptian scripts. Although his *Britannica* piece and articles in the *Museum Criticum* did provide a conjectural translation of the two Rosetta texts, Young had recognized only two kinds of Egyptian writing besides hieroglyphs, treating the Rosetta intermediate script as a corrupted form of the papyri scripts, "purely ideographic like the hieroglyphic texts." Greek proper names, Young did admit, could be read phonetically using the alphabet that Åkerblad had long before conjectured "with some extra signs." According to Young, Champollion continued, the Egyptians had worked phonetically only when writing foreign names. When Young published these views, Champollion "without any knowledge of Young's opinions" simultaneously arrived "by means of a sure method to nearly similar results."[8] Despite the resemblance, Champollion would demonstrate that his theory contravened much of Young's, in particular the assertion that the Rosetta intermediary was merely a "corruption" of the papyri script.[9]

Turning to the question of priority, Champollion demonstrated his willingness to cede his claim to unique innovation in respect to the mutual identity of the hieratic and hieroglyphic scripts provided that Young, too, would be affected.[10] He accordingly claimed that Young was not the first to consider hieratic a shorthand for the hieroglyphs; one "Tychsen of Gottingen," Champollion remarked, had asserted as much in 1816. Moving right ahead, he attacked Young's sign identifications in the readings of "Ptolemy" and "Berenice," arguing that Young had been altogether lax in developing his evidence. Champollion aimed, in particular, at Young's "syllabic" rendering of several signs. Were Young correct in his assumption that the signs frequently represented entire syllables of the language, then the very foundation of Champollion's theory concerning the origin of the Egyptian script would collapse. To illustrate this point, he adduced Young's "disyllabic" reading of OLE for the lionlike sign in the Ptolemy cartouche. "For me," Champollion argued, "observing that the *lion*, third sign of the hieroglyphic name of *Ptolemy*, was also the second sign of the hieroglyphic name of *Cleopatra*, I recognized this hieroglyph as being simply the sign of the consonant L."[11] Where Young had read the hieroglyphs as ΠΤΟΛΕΜΑΙΟΣ (*Ptolemaios*), Champollion instead had it as ΠΤΟΛΜΗΣ (*Ptolmns*)—a "skeleton," he wrote, of the full Greek Πτολεμαιος. The choice of "Ptolemy" was astute: besting Young on this centrally important example would certainly undermine any claim on his part to originality.

According to Champollion, Young's approach implied that the Egyptians wrote "*hieroglyphically* the proper names of foreigners by means of characters properly *ideographic* that were used *accidentally* to represent, either a single *letter*, or a *syllable*, or even *two syllables*." Young's system unacceptably deployed "a kind of *ideographico-syllabic mixed alphabet*, nearly like the Chinese when they transcribed foreign words in their language," whereas, according to Champollion, "the Egyptians transcribed proper foreign names by *means of an altogether alphabetic method*, similar to that of the Hebrews, the Phoenicians and the Arabs, their neighbors."[12] All of which countered the dismissive claims of the *Quarterly*'s anonymous critic who, in his rush to "raise a question of priority between M. doctor Young and myself," had failed to notice that there was no basis for a comparison. Champollion insisted that there was no "parity" between Young's work and his own, "between an imperfect, complex system based on an attempted reading of two proper names only, and a simple system, homogeneous throughout, based on a crowd of entwined and mutually supportive applications; between a system, finally, that applies *to nothing*, and a system that applies *to everything*."[13] The main thrust of Champollion's claim to originality lay precisely here, in the assertion that by means of a coordinated use of multiple evidentiary sources, Champollion's system could be used to decipher every scrap of Egyptian hieroglyphics, no matter how ancient.

Champollion may have come to this startling new claim only sometime after his anguished letter to Young of March 23 complaining about the *Quarterly* review. There is no evidence to suggest otherwise. Quite the contrary: even if phoneticism for royal names and epithets extended to native Egyptian rulers, the writing system could have accommodated both the original semantic values of the signs for the language's monosyllabic words and stripped values for the same signs, *provided* there was a way to signal whether signs were to be read semantically or phonographically. And indeed there was, or seemed to be, because royal names were always surrounded by cartouches. All of the evidence that Champollion had tendered in the Dacier letter, as well as in his January 9 remarks to Young concerning the Abydos king list, involved royal names or epithets (such as "autocrator") that appeared within cartouches. But Champollion was now acutely aware that even his Greco-Roman readings were subject to criticism, making any extension of the system to native rulers even more unlikely. He had to find some way to put all doubt to the side. It was likely just around the time of the dismissive *Quarterly* review that Champollion began a concerted search for further evidence. In doing so, he ran squarely into a problem that sent him in an altogether new direction, one that definitively differentiated his system and Young's and thereby foreclosed, or so Champollion hoped, any further challenges to his originality.

To judge from the argument as set out in the *Précis*—the only primary document we have—Champollion sought further evidence for his script by using it to identify common names during the Greco-Roman period. For that to be possible, some sort of phonographic qualifier would have to appear near such names in order to forestall what Champollion still thought would be the usual semantic reading of the characters. Just as cartouches always surrounded royal names and accompanying epithets, this qualifier would indicate that they ought to be read phonetically. As the Rosetta inscription was no help here, Champollion looked instead to the Barberini obelisk, which dated from the time of Hadrian. In the Dacier letter, Champollion had read phonetically, in that obelisk's cartouches, the royal names of Hadrian and his wife, Sabena. Not even Barrow had objected to that. However, elsewhere on the obelisk he identified a sequence of eight telltale characters that were followed by two others that, he wrote, "in all the mansucripts, and in all the funerary steles, on the mummies, etc., always accompany all the proper names of the deceased"

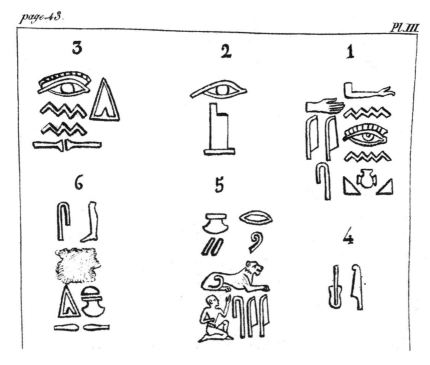

FIGURE 31.1. Champollion's sequences for Antinoüs (1), for designating Osiris (2), and for signifying (4) that preceding or following characters specify the name of a deceased person.

(these are no. 4, fig. 31.1). In consequence, "I did not at all doubt that this group was a proper name, and on the spot I applied my phonetic hieroglyphic alphabet."[14] The result spelled out the name ΑΝΤΝΣ—Antinoüs, Emperor Hadrian's unmistakably nonroyal favorite. Since the two designating signs signaled that the accompanying characters represented a name, here was evidence that Champollion's method could apply to more than just royal names of the time.

This exciting evidence promised to substantiate the validity of Champollion's method, at least for reading Greco-Roman names. But for that to work, and for it to hold also—as he now hoped—for the names of pre-Alexandrian rulers, he needed more examples. And so he canvassed the available evidence. First he turned to an obelisk in the town of Benevento near Naples. On it he found the name ΑΟΥΚΙLΙΟS ΑΟΥΚΙΛΙΟΣ, which he rendered in Latin as "Lucilius." The signs were accompanied by "a hieroglyph figuring *a man squatting and raising an arm*, a character that immediately follows *all the hieroglyphic proper names of individuals*, with the exception of the names of kings, which are sufficiently characterised by the *cartouche*."[15]

Champollion adduced one example after another, all followed by designators for names. But as Champollion accumulated his evidence, he noticed a problem. Although the characters representing names were indeed always accompanied by special signs indicating that they were names, in many instances the characters simply could not be read phonetically. Rather, they had to be read as referring nonphonographically to the individual in question. The accompanying designators in such instances indicated that the characters referred to names, not that the names were to be read phonetically, in which case it seemed impossible to avoid the conclusion that name designators had nothing to do with phonography. "The Egyptians," Champollion wrote, "at the time of the Romans, in transcribing the proper names of foreigners into *phonetic hieroglyphs did not place near these characters any sign that could warn of their phonetic* nature." But if different characters for names could be interpreted some phonetically and others semantically, this could only mean one thing, given the conservativism that Champollion ascribed to Egyptian culture. "With respect to the general system of hieroglyphic writing," he wrote, "we already recognize with certainty that they emplyed two sorts of very different signs: those that expressed *sounds*, and others *ideas*."[16]

Such a system must have existed from the very beginning of Egyptian writing, for otherwise it could never have developed out of a scheme based entirely on signs for semantic monosyllables: the weight of tradition, and the need to avoid ambiguity in a settled script, would have forbidden any such evolution. Phonographic and ideographic signs must have emerged together. The conclusion was unavoidable: this complex script existed from the very beginning of Egyptian writing. But how to prove the long-standing presence of phonography? Although the very same signs appeared in inscriptions preceding the Alexandrian conquest, Champollion had so far discussed only hieroglyphs produced during Greco-Roman times.

Champollion noted a curious fact: In apparently native Egyptian inscriptions, graphical signs are frequently substituted for one another, a pattern that suggested the *ubiquitous presence in such inscriptions of homophony*. To illustrate, he offered three funerary manuscripts with "drawings and legends that undoubtedly resembled one another": the "great manuscript" from the *Description*, which Young had also used; a papyrus obtained from George Annesley, Earl of Mountnorris, containing a late-period Book of the Dead written in hieratic; and Cailliaud's copy of the Abydos king list.[17] Following his procedure for demonstrating the existence of Greco-Roman common names, Champollion sought similar sequences accompanied by signs like the ones he had just used (i.e., signs 4 in fig. 31.1), which indicated that the names of a deceased person preceded or followed them, names that were not Greco-Roman

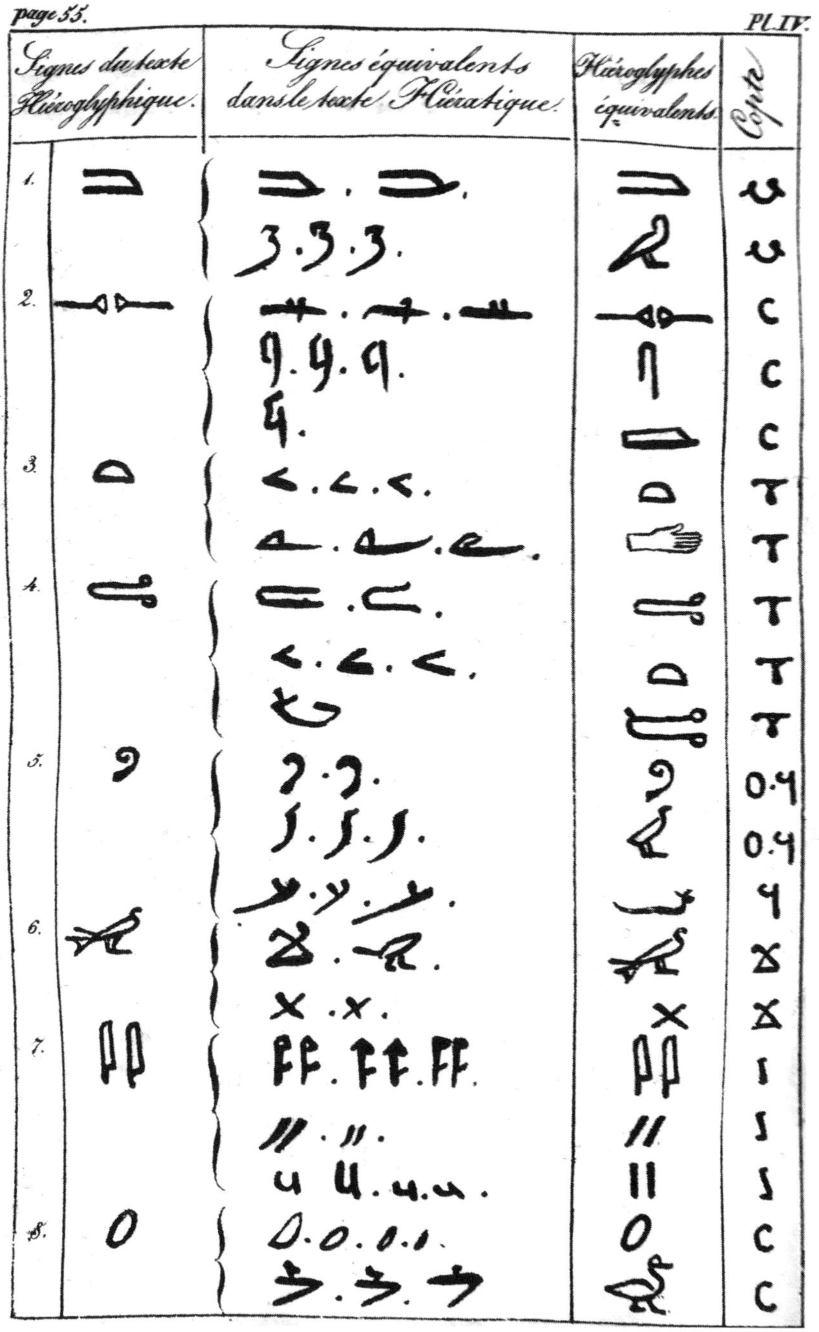

FIGURE 31.2. One of Champollion's tables of homophone signs in the 1824 *Précis*, including hieratic equivalents and Coptic letters.

but natively Egyptian. Collating the characters in these sequences, Champollion produced tables of signs for writing native Egyptian names (fig. 31.2) and observed that these tables repeated the homophonic structure he had found for Greco-Roman names. These repetitions were so closely similar as to seem "*a veritable copy, and so to say a double, of my phonetic alphabet, formed on the basis of Greek and Roman proper names.*" More specifically, "the signs that permute indifferently and incessantly in the hieroglyphic texts are the same ones that the reading of Greek and Roman proper names already led us to recognize as *homophones*, and that permute similarly in these names because they express one and the same consonant or vowel."[18] If the same table of sign equivalents held for native names as it did for Greco-Roman ones, then the former must have been rendered phonetically as well.

This discovery, made in the spring of 1823, was exactly what Champollion needed: a novelty to which no one else could possibly lay claim. With the *Précis*, Champollion's circle of evidence had closed. The Egyptian scripts had *always*, from the remotest of times, been fundamentally alphabetic and ideographic. Champollion devoted the remainder of the *Précis* to applications of the system that uncovered various "hieroglyphic groups and grammatical forms," the "proper hieroglyphic names of Egyptian gods," various "private proper names," sets of "qualifications and royal titles," the "proper names of Pharaohs," and provided as well something of even greater importance, an explication of Champollion's theory in terms of the "first elements of the hieroglyphic system of writing."

CHAPTER 32

THE SEMANTIC
TRAP AVOIDED

Let's pause to recapitulate Champollion's path after his arrival in Paris in July 1821 when he first had access to Young's *Britannica* article. At least through the following December he remained certain that all three Egyptian scripts were, in his words, "ideographic," that they were at base identical to one another, and that none had ever been repurposed phonetically, not even for Greco-Roman names. He publicly maintained that claim in his two lectures on hieratic at the Académie in late August and early September. The conviction derived from his belief that a substantially monosyllabic language like Coptic would never have transformed signs for words into semantically vacuous signs for sounds. As we have seen, this belief in turn rested on the tacit assumption that the signs had been purely semantic from the outset.

At this point, a wave of fresh evidence inundated Paris. In January 1822, Champollion received a copy of a cartouche taken from the Philae obelisk in which Bankes claimed to have read the name of Cleopatra; that copy, from Letronne, may well have included Bankes's written identification. In February, he learned of Saint-Martin's claim, made on the basis of that same cartouche, that "to express proper names the Egyptians reduced their symbolic characters to nothing but simple signs of sounds." Saint-Martin extended the claim to Egyptian antiquity, making particular note of the name Ramesses as evidence for his view. Champollion became embroiled in the zodiac affair, in which he had countered Biot by asserting that the starlike symbols on the Dendera zodiac were not astronomical symbols at all but qualifying signs, that is, determinatives. He had not altered his opinion that all the signs were, and had always been, either logograms or qualifiers for the language's monosyllabic words.

By the time Champollion gave his lecture on demotic at the Académie at the end of August, he had changed his mind, admitting the existence of phoneticism for Greco-Roman names in cartouches. Pressed by his brother, and fearing fresh competitive overtures from Saint-Martin on grounds that he had explicitly denied (i.e., phonography), Champollion reexamined his reasoning and the evidence for it. Developing a point at which he'd hinted in the table of sign sounds that accompanied his demotic lecture, Champollion decided that any sonic repurposing of a character must be separable from the character's basic semantic value.

Signs that retained their semantic character when used for anything but Greco-Roman names or words—most of them did, after all, appear on pre-Alexandrian monuments—posed two problems: they were possibly confusing and they overstepped tradition. These outcomes could be avoided, however, by requiring that a sign used phonetically could no longer have exactly the same sound as the semantic monosyllable it might have otherwise represented. Champollion consequently stripped the basic syllabic sounds to form semantically empty sonic combinations. This, he would insist, marked a critical distinction from what Young had done, since the Englishman had not avoided the use of meaningful signs to produce the sounds for the names of Greco-Roman rulers and some words. At this point, Champollion faced a conundrum. Since Egyptian civilization was highly conservative, it did not seem to him likely that wholesale repurposing of word signs could have appeared so abruptly in the long history of its writing systems. Such stripped, semantically vacuous signs, therefore, must also have been used for royal names from their creation. As this conviction took increasing hold over three crucial weeks, between August 22 and September 14, Champollion scoured new and old sources for supporting evidence.

Shortly before the dramatic mid-September collapse at his brother's, Champollion applied his stripped-syllables scheme to a sequence he identified as "Ramesses" in Huyot's drawings. This recognition led him to compose, under his brother's watchful eye, the Dacier letter, with its final, undeveloped claim that alphabetic phoneticism for royal names reached to the dawn of Egyptian writing. The letter provided novelties principally in countering Young's use of full syllables rather than Champollion's stripped versions. But if, even in their stripped forms, these characters had indeed been used prior to the Alexandrian conquest to write the names of native rulers, then Champollion was confronted by a serious, seemingly unavoidable problem of double-signification—a semantic trap, as it were. For if signs had been originally designed solely to represent the language's monosyllabic words, and if such a repurposing had taken place at some point well after their initial creation, then how could a reader readily have known when to read them semantically and when phonetically unless some sort of purely phonetic markers had been used? Yet the only such signs just seemed to indicate that neighboring characters represented names, and many of those names clearly had to be read semantically since Champollion's phonetic scheme did not fit them.

It is not clear just when Champollion found his way out of the apparent impasse, but the development likely occurred around the time of his anguished March 23 letter to Young concerning the review in the *Quarterly*. Dissatisfied with the lability of a native writing system whose originally semantic signs could have acquired phonographic significance only at a late date, but unwilling to disavow the available evidence, the monosyllabic nature of the Egyptian language, or his enduring belief in the conservative character of its culture, Champollion came to the sole remaining conclusion. The earliest Egyptian writing had to be heterogeneous. Although signs connected directly to the language's semantic monosyllables, the majority of characters were not phonetic and had never been. As he would explain in the *Précis*, these fell into three classes: qualifiers, many "figurative" signs having some visual connection to their referents, and fewer purely "symbolic signs" whose meanings derived from synecdoche or metonymy. In remote antiquity, Champollion suggested late in the *Précis*, the phonetic signs had evolved early out of the rebus transformation of a character representing an object or action into the initial sound of the corresponding word.[1]

In support of this breathtaking claim, Champollion noted the occurrence of homonyms among the characters used for Greco-Roman names. If the Egyptian writing system was essentially unaltered since the origin of the scripts, and if it had always involved phonography, then the character set deployed for purely Egyptian names should be substantially the same as the set used later, during the Greco-Roman period. On this reasoning, identical homonyms must have been deployed in both cases. Using available funerary manuscripts pertaining to native Egyptian names, Champollion produced a table that did indeed comprise essentially the same set of character equivalents as the one that he had generated for Greco-Roman names (see fig. 31.2).

Once the barrier between sign and phonography had been breached for names, questions arose as to whether, and to what extent, phoneticism applied to all of the language's words. After all, in the Dacier letter, Champollion had asserted that phoneticism extended to common Greco-Roman nouns (e.g., autocrator) as well as to names. The *Précis* sustained that radical claim, which necessarily extended to all three scripts, including the hieroglyphs. At this point, two considerations pressed Champollion to include a discussion of Chinese in his argument. Champollion had for a time thought Chinese provided an example that sustained his belief that phonography had never been present in Egypt, or at least not until Greco-Roman times; this possibility assumed the original system had deployed signs only for the language's semantic monosyllables. This could no longer stand in light of Champollion's new claims. Moreover, Barrow had recurred to Chinese in his *Quarterly Review* critique.

At some point in 1822 Champollion read Jean-Pierre Abel-Rémusat's recently published *Eléments de la grammaire chinoise*. From it he uncovered a clue regarding the treatment of monosyllables that could explain the difference between the Chinese and Egyptian writing systems. Although both languages used semantic monosyllables, they did so in different ways. Because the number of Chinese spoken monosyllables was small, the set of signs produced was comparatively limited. Such a set of signs could be modified in a particular manner to produce a compound that conveyed meaning while also representing the compound's pronunciation. In a passage quoted by Champollion in the *Précis*, Abel-Rémusat wrote that the Chinese script incorporated a predominant array of "mixed characters" to represent both sense and pronunciation: one of the signs in such a compound specified meaning, while the other, corresponding to a spoken word, specified its articulation. In this manner the Chinese could generate sign groups for meanings that went beyond the language's monosyllabic words while indicating, using characters for those very monosyllables, how to pronounce the result.[2]

Glossing Abel-Rémusat, Champollion explained that the "nature of the Chinese language led by itself to the invention of a *syllabic* script" because the pronunciation of even such mixed characters was governed by the set of existing monosyllables. Like Chinese, Egyptian consisted "for the most part of monosyllabic word primitives," but with a significant difference. Each of the Chinese monosyllables "begins with an articulation and finishes with vowels or pure diphthongs or nasals," while Egyptian monosyllables did not. Instead, they "usually contain articulations placed before or after their vowel or diphthong," with the result that the "words of their language were infinitely more numerous than the Chinese monosyllables." Such a large number, Champollion reasoned, discouraged the production of a sign for each of the language's words, for that would have amounted "not to representing the sounds of the words, but to creating an excessively imperfect ideographic script."[3] Hence the difference between Chinese and Egyptian writing.

In order to secure his system, which he assumed had been constructed to represent the words, both semantically and phonographically, of essentially extant Coptic, Champollion needed to demonstrate that the occurrence in Egyptian writing of signs that did *not* always or necessarily represent sounds could be interpreted in ways related to Coptic. To that end he turned to his list of determinatives—the qualifying characters that he had long before identified when he thought the scripts to be fully ideographic, some of which Young had also identified. Among these were signs expressing various ideas—of a son, daughter, infant, mother, father, sister, king, location, or place. For instance, the most frequent character for the *concept* of "son" was the figure of a goose combined with a small perpendicular line; Champollion had identified this figure as such using the Rosetta inscriptions in his demotic lecture at the Académie the previous August.[4] Now in Champollion's Egyptian *alphabet* the figure for a goose corresponded to the Coptic ϣ (*schei*), while the figure of a small line could be any of three vowels, including the ε. Turning to the Coptic monosyllable ϣε, Champollion noted that it is found in several words of the language that express a male relationship, including ϣεncon, meaning son of a brother. This yielded a bilevel relationship to Coptic: the figure of a goose treated as a *letter* in the Egyptian script sounds like a Coptic ϣ, while the ϣ combines with ε in Coptic to produce a sequence that, when added to a word, indicates precisely the same relationship as the goose character when the latter works as an Egyptian *qualifier* instead of as a letter, hence the multivalency of the writing system, which would have been present from the outset.[5]

Champollion produced several such examples, mapping sign groups, when used as qualifiers, to Coptic words expressing similar relationships. His intimate knowledge of Coptic dialects proved essential here. Effecting the correspondences required a good deal of interpretation as he linked an array of character homonyms to the segmental roots of relevant words. Champollion claimed, for example, to find Coptic traces of ancient Egyptian forms of gendered terminating pronouns: in Egyptian he located different signs for the past tense by affixes that were different for masculine and feminine subjects, for which he found relics in Thebaic, Memphitic, and Bashmuric words.[6] He systematically proceeded to cover as many possibilities, and possible objections, as he could. Deities' names could be written in several ways— phonetically, "figuratively" (as an image, reflecting their sacred nature), or "symbolically" by metonymy or synecdoche, with qualifying signs (determinatives) comprising the "symbolic" set.

Chronology remained contentious, as experts continued to fret about the ages of Egyptian monuments, not least, of course, the zodiacs. Champollion had little use for these arguments, as they stemmed from premises he believed to be false. "Two contradictory opinions seem still today to divide the scholarly world on the lesser or greater antiquity of the monuments of Egypt," he wrote. "Both are nearly mutually exclusive, and rest in general, one may say, entirely on simple considerations based on partial apperceptions the accuracy of which may be too often disputed." One claim held that all monuments inscribed with hieroglyphs that were constructed "according to rules of architecture that have nothing in common with those of either the Greeks or the Romans"—essentially every temple in Egypt—were built before the conquest by Cambyses. Were that the case, then it would be "natural" to consider the zodiacs at Dendera, Esneh, and in the tombs at Thebes (Luxor and Karnak) to have been built before the sixth century BCE, even, according to "calculations based on a purely gratuitous assumption," at a "prodigiously remote" period.[7] Some experts in Greek and Roman antiquities, on the other hand, considered the Dendera and Esneh temples to have been built just before the reign of Tiberius, from which the conclusion had been drawn that "all the other temples of Upper Egypt could not belong to times much before the common era . . . thus enclosing all the epochs of Egyptian art within the compass of a small number of centuries." This claim, however opposed to the first, itself rested on a diverse set of uncertain assertions. Champollion complained that his own alphabet had been deployed in the service of the latter assertion, with the result that many edifices had been wrongly attributed to the Greco-Roman period.[8]

According to Champollion, the first claim—that no Egyptian temple was constructed after Egypt's conquest by the Persians under Cambyses in 522 BC—should have provoked skepticism in any case. For "people who were so attached to signaling by the most important structures their respect for religion, the fundamental principle of their social organization, and which conserved that religion, their mores, nearly its very liberty, after the Persian domination," were unlikely "to have constructed no public edifice from the time of Alexander the Great to its complete conversion to Christianity." Two new facts had recently shed additional light on the contentious issue. Letronne's research on Greek and Roman inscriptions had demonstrated that some monuments in "Egyptian style" were constructed "in whole or in part" during the Greco-Roman era. Second, Champollion's alphabet confirmed Letronne's dating through the resultant reading of Greek or Roman royal names among the hieroglyphs inscribed on them. For these reasons alone, the claim, Champollion insisted, "must be much modified." The question concerning the antiquity of the monuments accordingly came down to distinguishing pre- from post-Cambyses structures, and the principal way to do that was to deploy Champollion's alphabet to identify "the proper names of Pharaohs anterior" to the Persian conqueror.

Champollion worked this vein in some detail, attempting to prove the applicability of his alphabet to all eras. Success in this regard was crucial, since the *Quarterly* reviewer had asserted that Champollion's efforts had little to recommend them beyond some minor advances that had been anticipated in any case by Young. Barrow had specifically challenged Champollion that if he could "produce the name of Cambyses, as he has done those of Alexander and his successors, though even this would not prove the fact, we should feel inclined to lean to his supposition." Champollion proceeded to find, not Cambyses, who after all had come to Egypt as a conqueror and would hardly have been memorialized in the Egyptians' "sacred writing," but the name instead

FIGURE 32.1. Drawing of an alabaster vase bearing the name of Xerxes in cuneiform and hieroglyphs by Caylus.

of Xerxes, the third successor to the Persian conqueror. This he found, not on a monument, but in a "cartouche engraved on a lovely oriental alabaster vase in the king's collection" (fig. 32.1).[9]

"These facts," Champollion declared, "destroy, it's true, all the systems hitherto advanced on the nature of Egyptian hieroglyphic writing; they reduce to nothing all the explanations of Egyptian texts or monuments hazarded over three centuries: but scholars will easily, in favor of the truth, sacrifice all previous hypotheses that are in conflict with the fundamental principle that we have just recognized." He himself had long thought hieroglyphs exclusively to be "signs of ideas," a belief in which he had persisted "up to the moment when the evidence of facts presented Egyptian hieroglyphic writing to me under a completely unexpected point of view, forcing me, so to say, to recognize the *phonetic* value in a crowd of hieroglyphs comprised in inscriptions that decorate Egyptian monuments of all eras." The statement is, as it stands, entirely accurate, but it does not reflect the sinuous path that Champollion had trod, or the manner in which his continued belief that Coptic was essentially unaltered ancient Egyptian had long governed his understanding. He was certainly correct in asserting that "no one will at least contest my priority in this altogether novel way of considering the hieroglyphic system of the ancient Egyptians of all epochs."[10] Whether all would accept his system was another matter.

Having, he thought, provided sufficient evidence to support his claims, Champollion carefully laid out the "first elements" of his system in the ninth, and by far the lengthiest, chapter of the *Précis*. He worried that, in so doing, the system might be subject to "miscomprehensions, obscurities, illusions," though he had made every effort to avoid the eighteenth-century shibboleth of system. Champollion did not mean that he had avoided generating a scheme of interpretation for Egyptian scripts, but that he had based his work on a comprehensive empirical examination of surviving inscriptions.

Carefully constructed to counter a series of likely objections, this portion of Champollion's *Précis* provided the outline of his system together with applications to Greco-Roman and native Egyptian royal and common names and a list of all the signs, with homonyms, whose phonographic values he identified. Altogether he proffered twenty-five purely phonemic signs, comprising seven vowels and eighteen consonants, which, he remarked, nicely coincided with a remark by Plutarch.[11] The "general table" included at the end of the *Précis* contains explanations of the 450 signs or sign sequences reproduced in the accompanying plates. These he divided into nine groups: grammatical forms; phonetic, figurative, and symbolic names of divinities; proper names of Egyptian sovereigns; common Egyptian names (both phonetic, "phonetico-symbolic," and "totally symbolic"); Greco-Roman phonetic proper names; common names expressed by single or grouped hieroglyphic signs in any of the three methods; and qualifiers and titles. For each entry Champollion provided what he took to be the Coptic equivalent. Although he refused to hazard a guess as to the total number of hieroglyphic signs given "the present state of our knowledge," Champollion claimed to have identified 864 distinct characters. Since the phonetic signs, including the considerable number of homonyms, comprised but a fraction of this number, the vast majority of signs fell into his figurative and symbolic categories.

For all its detail, the *Précis* was not a definitive treatment giving Champollion the last word on the decipherment of Egyptian hieroglyphs. He offered virtually no argument in support of his "general table," which left room for criticism, as we shall see. More damaging was the lack of a transliteration of even a single passage in any of the three Egyptian scripts, and this despite the ready availability of several such passages, including the large papyrus reproduced in the *Description*. We now know that Champollion had essayed a transliteration of the Rosetta demotic into Coptic, but this undated work was never published.[12] This omission posed a significant problem since Coptic, after all, underpinned his system, and he remained certain that Coptic hardly differed from ancient Egyptian. Much of the criticism leveled against Champollion in the four years following the publication of the *Précis* orbited about the lack of such a transliteration and the lability of the meanings that he had assigned to his many "symbolic" hieroglyphs. As we will see, Champollion's earliest critics easily located these weaknesses and found many ways to press the advantage.

CHAPTER 33

THE RECEPTION
OF THE *PRÉCIS*

Reactions to the *Précis* were slow to arrive. Reviewers had good reason to hesitate, for the *Précis* was a demanding work in which complicated arguments concerning this or that Egyptian sign's meaning or pronunciation were embedded in an intricate theory designed to uphold Champollion's claims to originality. For months after the *Précis* reached print, knowledge of Champollion's work derived principally from earlier sources—the Dacier letter and reports regarding it. To some extent the lack of rapid response may also reflect, in Britain at least, disdain for Egypt of the sort that we have already uncovered in Young's disgust with Egyptian zoomorphic idolatry. Barrow's scathing report on the Dacier letter, for example, rejected not only Champollion's work but much else concerning ancient Egypt, at least for periods before the Greco-Roman domination.[1] The chief exception to the British silence was Young's own *Account* in early 1823, which provided little that was new beyond a carefully phrased discussion of priorities and attacked Champollion's position, expressed at the conclusion of the Dacier letter, that phonetic hieroglyphs extended beyond Greco-Roman royal names.[2] But even Young noted that "the discovery of the general import of the hieroglyphics has by no means excited any great sensation in this country."[3] British engagement escalated only when the priority question began to preoccupy Young's friends and supporters after his death in 1829. In France, the initial burst of enthusiasm prompted by the Dacier letter soon waned, and only a small group of philologists, historians, and travelers remained interested in Egyptian writing.

It was none other than de Sacy who produced the first detailed French review of the *Précis*. Published in March 1825, de Sacy's review covered the original Dacier letter, the *Précis*, and Young's *Account*. The broad remit required de Sacy to exercise care in order to avoid unduly offending the English savant while giving sufficient credit to a compatriot whose moral qualities de Sacy had years before cast in doubt, but who now had royal support. De Sacy took the safest route: he temporized. Though some "light points of contact" obtained between Champollion and Young, nevertheless their "manners of proceeding are essentially different." Noting that Young's method, about which he gave few details, would lead "in a false direction," de Sacy repeated his request, made privately years before, for clarification as to how Young had rendered the Rosetta demotic. Young's work had nevertheless "suggested to our compatriot [Champollion] some of the ideas that, after many trials and errors, finally led him to adopt certain principles that were fruitful of happy results." Continuing to apportion praise and blame, de Sacy remarked that "one knows that a clever person can draw some light even from the errors of those who preceded," which were kind words indeed, coming from him. Given de Sacy's own work on grammar, it seems likely that he had in mind Champollion's insistence that phonetic signs had to be essentially alphabetic in a way that allowed combinations of stripped signs. These could not be syllabic, a point that countered Young's easy movement between the two possibilities.[4]

De Sacy's armchair quarterbacking about priority did not obscure a deeper question as to whether phonetic hieroglyphs had been used in ancient Egypt. Agreeing with both Young and Champollion, as well as just about everyone else in this regard, de Sacy acknowledged that phonography had first been recognized in "words of foreign origin." The question, as usual, was whether it applied to hieroglyphs of much greater antiquity. If such signs had been re-purposed in a manner similar to the one by which foreign words were written in China, then the ancient Egyptians originally must have had only a "pure ideographic system without any [phonetic] admixture." Casting Champollion's path to the *Précis* in terms of a progressive enlightenment, de Sacy claimed, accurately as it turns out, that Champollion had at first regarded phonetics as "an auxiliary" to a fundamentally ideographic system.[5] Further "application of his phonetic alphabet" had led Champollion to conclude that it was "an essential part, necessary and inseparable, of hieroglyphic writing, in a word *the very soul of the system*." Noting that the *Précis* did not completely demonstrate the claim, de Sacy reported that Champollion had decided to publish despite this incompleteness because "powerful considerations" pressed upon him. De Sacy tactfully did not specify these "considerations," but few in his circle would have been unaware of the threat that Young had posed to Champollion with his *Account*, not to mention the equally pressing need to justify the patronage of Blacas and the king, especially in light of the questions concerning originality that might easily arise given what Young had written.[6]

De Sacy condensed Champollion's argument into four central assertions: first, that his alphabet applied to names of all eras; second, that the alphabet was "the true key" to the entire writing system; third, that the alphabet had always been used to represent the sounds of the language; and fourth, that all hieroglyphic inscriptions were composed "in major part" of phonetic signs.

De Sacy somewhat resisted the third and fourth claims, which of course meant that he also hesitated concerning the second. "Perhaps," he wrote, "it's permissible to doubt that the hope conceived by M. Champollion . . . is entirely realized," for it seemed to him that difficulties of interpretation might continue to defy Champollion's efforts and that of "those who will walk the path that he first opened." For, de Sacy thought, no one knew "in all its extent the ancient language of the Egyptians," a point that cast doubt on Champollion's claim for Coptic as essentially the unaltered original language. Change, after all, likely occurred to the language "in its grammatical structure, its nomenclature," and in the several dialects. Court de Gébelin, whom de Sacy admired in his *Principes de grammaire générale*, had doubted the possibility that both ideographic and phonetic writing could have existed simultaneously as working representations of any language, whereas Champollion was arguing for precisely such a mixture, even if he was certain that phonetic signs predominated.[7]

De Sacy discerned the central weakness of Champollion's system in its tripartite division of hieroglyphs into figurative, symbolic, and phonetic signs. "Perhaps this proposition," he wrote, "[which is] altogether contrary to what one is inclined to suppose, is not to this point completely proven," and may even be doubtful, for the following reason. Champollion's phonetic hieroglyphs were, de Sacy noted, similar to Hebrew, Phoenician, and Syriac scripts in their suppression of medial vowels. However, unlike the signs of those writing systems, Champollion's phonetic hieroglyphs were all images of objects, the words for which in Egyptian (i.e., Coptic) began with the appropriate articulations. This, to de Sacy, seemed to demonstrate the anteriority—perhaps the very great anteriority—of ideography over phonography, in which case the antiquity of phonetic representation remained in question. Moreover, de Sacy could not identify a method for determining the meanings of Champollion's so-called symbolic signs, of which there seemed to be a great many, more perhaps than phonetic characters, which themselves might sometimes be figurative. Further, the system necessitated a considerable number of homonyms, some of which Champollion had even tried to identify. But the only sure way to determine which signs could substitute for one another phonetically required knowing enough of the original language to divine the words for the objects that the signs represented. This, de Sacy opined, "might seem to be the weakest part of his system" since it depended critically on the antiquity of the present form of Coptic, which de Sacy doubted.[8] He concluded with extensive quotations from the *Précis* and praised the "fruits" the system had so far yielded for historical knowledge of Greco-Roman times in Egypt.

Writing to his brother from Italy, Champollion rather optimistically declared that de Sacy's review constituted a "complete profession of faith" in his work. The review had "perfectly satisfied" him, despite de Sacy's skeptical remarks about nonphonetic signs. "It's true," he admitted, "that that part of my system is not clearly enough developed." Such developments awaited the arrival of more evidence, particularly monuments bearing nonphonetic signs. These things "do not exist," or at least not in collections to which Champollion presently had access, and so, he wrote, "I postpone to another time the demonstration of my remarks on that subject."[9] Favorable and evidently pleasing enough to Champollion, de Sacy's review nevertheless constituted a warning: others less well inclined might fasten on the questions raised by Champollion's interpretations of signs. That is indeed precisely what occurred as scholars assimilated the arguments of the *Précis*.

Of the full studies of hieroglyphs to reach print alongside the *Précis* in 1824, the Russian Ivan Aleksandrovich Goulianoff's *Opuscules archéologiques* constituted the most direct immediate response to the work.[10] In the *Opuscules*, published under the alias "Th. Ausonioli," Goulianoff praised Champollion while taking a strikingly different approach to the nature of phonetic signs. In Paris three years later, we shall see, the talented but combative linguist Julius Klaproth would take up arms against Champollion while mounting a defense of Goulianoff.[11] Meanwhile, a work edited and published at Leipzig by Gustav Seyffarth presented a study of the Rosetta inscriptions completed by the classical philologist Friedrich Spohn that remained unpublished at the time of Spohn's death; inspired by his reading of Spohn, Seyffarth would later produce a critical response of his own to the *Précis*. In what follows, we survey these works, which, taken together, constitute the *Précis*'s initial reception.

*

Trained at Leipzig, Seyffarth had prepared for a theological professorship by immersing himself in "all the languages into which the Holy Bible had once been translated." Convinced that the sounds of Greek and Hebrew letters differed from the ones commonly accepted, his investigation of this question led to his first publication in 1824, followed soon thereafter by his appointment as professor of philosophy at Leipzig. There Spohn had died in 1824, leaving in raw form a treatise on Egyptian demotic that involved extensive work with Coptic. As Seyffarth was the only scholar at Leipzig who knew anything at all about Coptic, he was asked to edit Spohn's work, which led reasonably quickly to publication in Latin of elements drawn from Spohn's manuscripts in 1825, and the consequent production of Seyffarth's critique of Champollion's work.[12]

Spohn's material included a translation into Latin of the Rosetta demotic together with a series of lines from hieroglyphic papyri for which Spohn had left both Coptic and Latin correspondences. Publishing these manuscripts, Seyffarth did not provide the scheme that Spohn had used, perhaps due to the state of the manuscripts. (Seyffarth certainly knew this scheme, as we will see in a moment.) The work's publication prompted Johann Kosegarten, professor of oriental languages at Jena, to write a few words to Thomas Young, whom he felt had been slighted by the new work. The principal aim of the letter, written in late December 1825, was to inform Young about a defense Kosegarten had launched in a recently published critique of Spohn and Seyffarth. In it, he explained, "I tried to save your rights with respect to the explanation of the enchorial script, and to show that Spohn has not said anything true that you haven't said and read before him, and that Spohn only added a mass of errors." Seyffarth had nevertheless continued "with the most extravagant flattery" of the deceased scholar. Moreover, there were in Germany some who "pretended to be able to read everything like Spohn and Seyffarth." Indeed, Kosegarten complained, so strong were sentiments favorable to the departed Spohn, who had (unspecified) "great friends here," that Kosegarten was even warned not to "say anything against" him.[13]

When Kosegarten wrote his letter, he and Young had already been in touch for some time. In the late fall of 1824 Kosegarten had published a lengthy review of Young's and Champollion's works. The review avoided comparative judgments, though Kosegarten ended by noting that Champollion had not as yet provided renderings of anything beyond names and a few words.[14] He had written Young that summer after reading the *Account* and the letters Young had published in the *Museum Criticum*. "It's certainly to you," he wrote, "that Europe owes the first certain knowledge with respect to this subject, as new as it is interesting; and it's above all in your 'Account' that you communicated important pieces of this kind of literature." Kosegarten was particularly interested in the "enchorial" papyrus that Young had obtained from Grey and had translated, and wanted very much to see the original. Young's reply is unknown, but we do know that he had sent along a copy of his *Hieroglyphics*.

Kosegarten's admonition may have stemmed as much from a desire to preserve the departed Spohn's reputation as from any conviction regarding the accuracy of what Spohn and Seyffarth had done. Before his death, Spohn had occupied a peculiar position in the world of German letters. He was respected for his brief work on the *Iliad* and the *Odyssey*, as well as for an edition of Hesiod's *Works and Days* created for school use.[15] Among Spohn's admirers was Wilhelm von Humboldt, who was also a friend of Young's. In the fall of 1814, while serving as a diplomat in London, Humboldt had met Young, to whom he confessed that he was "a complete stranger to" Egyptian matters. By 1821, he had begun to remedy the ignorance to which he had confessed, writing to Young specifically about "four Egyptian, lion-headed statues in the Royal collection."[16] Between 1820 and 1824, Humboldt published four books on the nature of language, including the relationship between speech and writing. To Champollion, whose *Précis* he had recently read, he wrote on June 26, 1824, for assistance with two hieroglyphic manuscripts.[17] In view especially of Champollion's claim that Coptic was essentially unaltered ancient Egyptian, Humboldt wrote Young in late August 1825 that, since his sojourn in London, he had studied the language "with care" and had "occupied [himself] with the decipherment of hieroglyphs without pretention [himself] to make discoveries, but simply to know yours and those of M. Champollion."[18]

Humboldt informed Young that the posthumous first volume of a work on the "demotic alphabet" had appeared—this, of course, was the volume of Spohn's work that Seyffarth had edited. It included translations of the Rosetta demotic and "many other Egyptian texts," but not the alphabet that Spohn had used to make his translations. Humboldt nevertheless judged Spohn's "discovery" to be "true" (*véritable*) and adduced "the solid erudition and character of Spohn" to support his view. Humboldt had not yet compared Spohn's translation with Young's on demotic (presumably from the *Museum Criticum*), while Champollion had not provided any substantial applications of his alphabet to a demotic text. Young responded a month later. Despite his best efforts with Spohn, he had "looked in vain for any one addition to what even Mr. Åkerblad had made out, more than thirty years ago." Nothing in the text could "justify the pomp and ceremony with which Professor Seyffarth's *Prodromus* is issued to the world." Spohn's "mode of reading the words, by an alphabet, which, the newspapers tell us, is like the Armenian" had failed badly. Young attributed this failure to Spohn's exclusive use of proper names, "originally Egyptian, not one of which has been read by Professor Spohn in any way at all approaching the truth." Of course, at this time neither Young nor anyone besides Seyffarth had a clear idea of Spohn's methods, about which Seyffarth's volume had given only hints.

Seyffarth, meanwhile, had been fully captured by Spohn's efforts. He decided to examine "all the Egyptian museums of Europe" where he intended to copy "the principal papyri and inscriptions," an enterprise he undertook with extraordinary vigor from 1826 to 1828. In late 1826, he published a critique of Champollion's system in which he advanced an understanding of hieroglyphs that departed completely from those of both Young and Champollion. The extent of his reliance on Spohn's work is unclear, for Spohn had apparently limited his study to demotic. Seyffarth nevertheless seems to have accepted the alphabetic system that informed Spohn's translations.[19]

Although Seyffarth and Spohn agreed that Coptic descended in some fashion from ancient Egyptian, Seyffarth rejected the idea that hieroglyphs might be strictly alphabetic while insisting that they were, nevertheless, phonetic. Specifically, he thought the signs formed an unusual syllabary that not merely pervaded Egyptian writing but constituted the entire system. In his view, every one of the hieroglyphs for royal names represented two or three of the consonants contained in the ancient Coptic name of the figure represented by the sign. Ancient Coptic, on his view, not only differed from its descendant but was also closely related to Hebrew, which Seyffarth, like many others, considered the origin of all languages. Take, for example, the hieroglyph that figured a bee. According to Seyffarth, this was a "queen bee," and as such the name of the sign contained the sounds for "queen." Since "the language of the ancient Egyptians was a Hebrew dialect," Seyffarth produced *melik* as the spoken name of the sign, from which the three consonants *mlk* were extracted to generate what he took to be the sign's actual vocalization. According to Seyffarth, ancient Egyptian was a "Hebrew dialect" of an idealized ur-language with a script derived from an even more original "Noachian" alphabet of "25 letters."[20] As such, Egyptian writing had always been peculiarly phonographic. What remained to be explained was this: Given the extreme variety of characters, not to mention the multiplicity of scripts, why had ancient Egyptians so greatly multiplied the signs of this primordial Noachian alphabet? Apparently because they needed to compress their inscriptions in order to conserve space: "The Egyptians and their predecessors were compelled to excogitate syllabic signs for the purpose of comprehending, in the same spaces, a greater number of words and phrases [than was] possible by single letters."[21]

Young obtained another view of Seyffarth from William Gell, who had been living in Rome since 1820 and who fancied himself an aficionado of the Egyptian scripts.[22] In the spring of 1821 he had written to Young about hieroglyphs, explaining that he intended to head for Egypt "as soon as circumstances permit" and interspersed his thoughts on the script with tidbits of new material for Young's consideration and breezy reports of local events.[23] In another diverting letter to Young, written in early August 1826, Gell described a contretemps between Champollion and Seyffarth. Champollion was "in full swing at Rome," Gell reported, "where he was preceded about ten days by Mr. Professor Seyffarth, his antagonist." The German scholar had not impressed Gell, who referred to him as "Mr. Sighpoop." With his friend and collaborator Antonio Nibby, Gell arranged for Champollion and Seyffarth to meet at the residence of the Russian minister in Rome (whom Gell, in his inimitable way, dubbed "Italinsky") where Gell had mischievously suggested a duel with Champollion using "two obelisks for swords and the labrum of Monte Cavallo for a shield."

Gell reported that the following day, while "at the French minister's," Champollion asked Seyffarth "into what language he translated his hieroglyphics, to which Seyffarth replied, 'Coptic.' Then says Champollion, 'I will not say there is no sentence, but there is no word of Coptic in your translation.' 'Oh,' says Seyffarth, 'it is a more ancient Coptic than that of the books.' C. 'Where did you learn it?' S. 'In the Rosetta inscription.' C. 'In the two lines you have published?' S. 'Yes.' C. 'Then give me leave to say that as you have published them they are so falsely copied that they give no idea of the real figures, and that no ten figures together are correct.'" Seyffarth did not respond except to murmur to Nibby that "he thought it was better to be silent, as Champollion was so violent." But Gell had sensed no such vehemence, at least not from Champollion. "All I can say," Gell concluded, "is even the Germans did not support him, and that as all his figures may mean any letter of the alphabet according to their position, if his scheme be true it cannot be worth learning."[24]

So much for "Sighpoop." Champollion may not have been "violent" at Nibby's, but he took up metaphorical cudgels against Seyffarth in a letter written shortly thereafter to Blacas. Dated September 7, 1826, the letter was printed in Florence and Paris, ensuring its wide circulation. The substance of the letter was characteristic: Faced with an alternative system, Champollion typically insisted that such schemes amounted to baseless speculations, whereas his own was founded on examination of the facts. This letter was no exception. "Seyffarth, or M. Spohn" had "abandoned themselves to illusions that only the study of original monuments can dissolve," he wrote. Theirs was an "arbitrary system," devoid of "positive facts." According to Champollion, his system differed from Seyffarth's principally because Seyffarth did not recognize the existence of either "figurative" or "symbolic" hieroglyphs, believing all such signs to be "character-letters," that is, phonetic signs, making hieroglyphs exclusively phonetic.

Whereas Champollion's examinations had uncovered no more than "eight or nine hundred" distinct hieroglyphs, Seyffarth had produced a phonography of some six thousand characters. What evidence, Champollion wondered, did Seyffarth have for this extraordinary result? Worse, Seyffarth's scheme strictly disallowed signs with fixed values. According to Seyffarth, a sign could be rendered by up to six different letters depending on context. This could not have been further from Champollion's view, which limited the number of different signs having the same sound. He did not explain the basis of Seyffarth's system in its construction of phonography on the conjectural basis of the spoken name of a sign—nor did he feel obliged to do so, since the system had no evidentiary basis whatsoever. Why had the "two German savants" been so misled? To explain this, Champollion engaged in a characteristic move, shifting blame from the dangerous cliff of moral virtue to the blameless shoals of available evidence. The Germans had access only to inferior "designs and engravings of inscriptions," principally from the *Description*, whereas Champollion worked insofar as possible from exactingly produced copies.[25]

A second challenger to Champollion emerged in the otherwise obscure figure of Francesco Ricardi, also known as Carlo di Oneglia, who in 1824 produced a ten-page brochure in which he used an ideographic interpretation of hieroglyphs to establish a "history, chronology of the cult of all peoples, ancient and modern."[26] Champollion brought Ricardi to his brother's attention in August of that year, just two months after arriving in Turin. Forwarding the brochure "by the brave Ricardi," Champollion advised his brother to shelve the piece, and not only figuratively, "next to the one by Goulianow [*sic*]," whose *Opuscules* had just been published.[27] In the *Précis* Champollion had mocked an 1821 piece by Ricardi (whom he dubbed "the new Oedipus") on the Barberini obelisk in which Ricardi had insisted that the hieroglyphs referred to the biblical pharaoh Sesac who "pillaged Jerusalem," providing what Champollion derided as a "triumph against the impious."[28] Champollion's opinion of Ricardi only worsened upon meeting him personally in October. "Ricardi is ever himself," Champollion reported to his brother, "but through a forest of stupidities his new production [the brochure] offers several elements of hemlock and wolfs-bane: he should have taken as epigraph *stupidity* and *wickedness*." To this note Champollion attached a "new, dwarf [*bamboche*] piece" also authored by Ricardi. This new offering Champollion judged "as stupid as the first, but more malicious. He denounces me, me and my makers, to the Holy Inquisition."[29]

Champollion considered Ricardi something of a Kircher redivivus, but without Kircher's linguistic talent and with a taste for religious dogmatism. Ricardi's other work amply justified this view. Among Ricardi's early publications was a meditation on the proper way to "read and understand Hebrew" that combined these qualities in exemplary fashion. The original presence of Hebrew vowel signs, he explained, had been lost in antiquity. Though imperfectly reconstructed by the Masoretes, these signs could be restored by looking to grammar, context, and sound. In so doing, he insisted, "one must not diverge from grammatical rules and the best dictionaries, and above all such an explication must conform to the infallible truths of the Church."[30] Ricardi had been engaged in efforts to establish such so-called truths for more than a decade, and in 1811 he had published an application of his system to the "theological, political, and moral maxims of Solomon." In a telling "letter of the author" appended to the work, Ricardi claimed that idolatry had first arisen among "the Arabs" who resorted to "contemplation of the heavens" as a result of unbearable daytime heat:

[T]he Arabs, a people subjected as a result of their topography to the scorching rays of the sun, could not in the hottest hours of the day attend to their business, were nearly constrained to pass such times in rest, while the cool air of the night, during which they remained vigilant, gave them ease to contemplate the many marvels to be seen in the heavens: Sun, Moon, Stars, Planets, Meteors, whose real or supposed influence on our globe gave rise in them to the idea that these celestial bodies might be animate, and therefore to attribute to the Sun, as the most beautiful, and as the most influential on our well-being, the supremacy, with the Moon as companion, over the twelve signs of the Zodiac, through whose passage the Sun was given principal title, and the rest of the celestial bodies as secondaries, for less was their influence.[31]

To the extent that it can be extracted from the surrounding verbiage, Ricardi's main idea seemed to be this: Writing developed from the need to establish an organized society, while worship of the heavens led to the production of hieroglyphs as teraphim, or cult objects representing influential celestial objects, as well as terrestrial ones, with particular importance to organized social life. "This, my opinion," he continued, "is also confirmed by the superstitious Egyptian monuments." Where Seyffarth would read every hieroglyph as a phonogram, Ricardi insisted on precisely the opposite, that every such sign had only a "cult" (i.e., religious) meaning.[32]

In 1826 Ricardi published a lengthy, chapter-by-chapter critique of the *Précis* in which he attempted to demolish Champollion's central claim about phoneticism. Ricardi's criticism depended for the most part on the lability of Champollion's phonetic assignations and his seemingly unsystematic determinations of when a sign might have a figurative or symbolic meaning. Founded, Ricardi wrote, on "arbitrary suppositions and many other inadmissible licenses," Champollion's scheme lacked consistency and evidentiary support. Why, using his method and applying it to a hundred hieroglyphic signs, one might as easily read "the names of the Kings of France." Damagingly, Ricardi pointed out that Champollion had never produced a rendition of a single hieroglyphic sentence. Seyffarth had criticized Champollion on precisely this point, claiming that he had been "repeatedly challenged to verify his theory by a translation of the Rosetta Stone, but failed in his attempts."[33] Summing up, Ricardi dismissed Champollion's work as madness: "All novelty dazzles, but in the end the illusion vanishes, and certainly posterity will have trouble believing that scholars and scientific establishments could have lent an ear, could possibly have applauded anything like such a delirium."[34]

Visiting Rome in 1826, Champollion sensed danger all around. He had the waspish Gell to contend with, not to mention the hostile Seyffarth, but his anxieties were likely further exacerbated by his long-standing and vocal dislike of the clergy. He wrote to Costanzo Gazzera, a member of the Turin Royal Academy of Science whom he'd recently befriended, that he'd heard of a "terrible conspiracy" forming against him, one aimed at demonstrating that he "*knew nothing about* hieroglyphs, and, what's more, *that I don't even know* Coptic." These conspirators "surround me daily," he complained, and they were sneaky, pretending friendliness before foreign ministers who praised Champollion while undermining him elsewhere.[35]

Among these so-called pretenders was Michelangelo Lanci, a prominent orientalist. Lanci was incensed by the danger to sacred chronology posed by the use of Champollion's system for purposes of dating Egyptian artifacts. In a pamphlet written in early May, Lanci had claimed that Champollion could not know the true pronunciation of the Egyptian signs because Coptic must have been contaminated by Greek and Latin. The only proper route to the original sounds had to be through Moses, who spoke to the ancient Hebrews with the sounds of Egypt. Egyptian writing could therefore only be known through Moses himself, dispelling any fear "that the new hieroglyphic system might ever overshadow in any way that History which alone merits universal veneration," by which he meant, of course, the Bible.[36]

In a letter to his brother of June 17, Champollion protested that Lanci, motivated by "pure jealousy," had published "impertinences about my system."[37] But before Champollion could counter Lanci's attack, he had first to deal with Pope Leo XII, Annibale della Genga, with whom he had an audience two days before. This formidably reactionary pope would, among other repressive endeavors, require Roman Jews to sell their property; on November 26, 1826, he forced their return to the ghetto. Uprisings and revolts throughout the papal states were the eventual consequences of Leo's policies. Not content with merely persecuting Jews, Leo also rejected tolerance for "Deism and Naturalism," which he had attacked in an encyclical entitled *Ubi Primum* that had been issued on May 5, 1824. As intolerable as Rome's Jews were, so too the "flood of evil books that are intrinsically hostile to religion."[38] Among these infernal publications would certainly have been any that undercut sacred chronology, such as ones that claimed great antiquity for the Egyptian zodiacs.

Champollion's papal audience was arranged by Gian Domenico Testa, one of the earliest and most ardent defenders against the threat posed to sacred chronology by the French savants' dating of the zodiacs.[39] Testa, who had spent nearly his entire life in the church, had carefully, if incompletely, prepared Leo for his meeting with Champollion. He greeted Champollion enthusiastically, praising him three times for the "beautiful, great and good service" he had rendered to the church in showing, by means of his identification of "autocrator," that the zodiacs did not contradict biblical chronology. To the French ambassador, Leo praised Champollion for having "abased and confounded the arrogance of that philosophy which pretended to have discovered in the zodiac of Dendera a chronology anterior to that of the sacred Scriptures."[40] When he offered the married Champollion a cardinalship, the reluctant savior of biblical chronology demurred, remarking in a tongue-in-cheek reference to his wife and daughter that "two ladies wouldn't agree."

Having emerged intact from his papal adventure, Champollion turned to finishing off Lanci. This he did in a published letter to Stanislaw Kossakowsky, an influential Polish diplomat and Egyptian amateur who was among the "foreign ministers" who had expressed interest in Champollion's work. Known as the "Letter to 'Z,'" the piece opened on a somewhat palliative note, expressing a sneering gratitude to the "amiable" Lanci, against whom no one could possibly take offense when the estimable critic had only the most "praiseworthy intentions."[41] Having dispensed with these minimal niceties, Champollion went in for the kill. "This scholar," he continued, "begins by demonstrating how easy it is to write about Egyptian antiquities and even to say something *new* about this branch of archaeology without knowing even its first elements." For example, Champollion criticized Lanci's remarks about a Phoenician bas-relief at Carpentras done in Egyptian style, discussed long before by Barthélemy, in which Lanci had mistaken five elongated fingers for a "flame," absurdly concluding that the female figure in question must be a vestal virgin since such an unmarried woman would have had "hands that burn." Elsewhere in the same bas-relief Lanci mistook a whip for a water sprinkler. In response to these missteps, Champollion could only shake his head. The work, he averred, contained an "infinite number of [such] misunderstandings."

Champollion was incensed as well by Lanci's assertion that Champollion's efforts, however misguided, were really just a codicil to what Young had previously accomplished: "So here I am, despoiled as proprietor by fiat of the discovery of the hieroglyphic system's alphabet." Even so, Champollion continued, his "severe corrector" proposed to help the misled decipherer by "magisterially" proposing a reform that would correctly determine the sounds represented by hieroglyphs. Wholly innocent of Coptic, Lanci nevertheless proposed to use the language he did know, Hebrew—the original of languages, so Lanci believed—to uncover the sounds of Egyptian via Moses. "Surely!" Champollion exclaimed, "one scarcely expected to see Moses in this matter." This, he opined, would be like trying to recover French pronunciation from French words badly quoted in English books. Tongue planted firmly in cheek, Champollion struggled to understand how Lanci could possibly have concluded that "my ideas on the hieroglyphic system lead nowhere," since "Mr. Lanci, *as he himself admits, has understood absolutely nothing of four or five hieroglyphs* cut on the bas[-]relief. He doesn't know how to translate this little inscription, he even doubts whether it is complete, therefore my hieroglyphic system leads nowhere . . . a *nice conclusion, and worthy of a beginning.*" Why, "the sword of Damocles is suspended over our heads, and this double-edged sword" was none other than "Mr. Lanci." Well, he concluded, Lanci certainly had the right to depreciate all of Champollion's results "if he could," but "he has passed the outer limits of literary criticism" in claiming that "the results of my system . . . tend to contradict the history presented to us by the sacred texts. I declare this accusation of Mr. Lanci's entirely false." And that, indeed, was the danger Champollion had to avoid since, were such an accusation to stick, he would have had the church arrayed against him just when that nemesis seemed to have been neutralized.

Lanci responded by distributing libelous attacks against Champollion. Lanci's behavior so displeased Pope Leo—who, after all, had just offered the Frenchman a cardinalship—that Leo apparently decided to eject Lanci from the Vatican, where he was employed. According to Hartleben, who cites no source, Champollion intervened to save Lanci's position.[42] Nevertheless Lanci's animadversions continued, at least in Rome. A year later, Champollion reported to Blacas that Lanci's slanders, "though devoid of any foundation, produce an impression," and worried that the resulting damage to his reputation would oblige him "to engage in a deep discussion of the subject" in order to demonstrate Lanci's "bad faith and evil intentions."[43]

Ricardi and Lanci may have presented Champollion with tactical problems, but their efforts were comparatively easy to counter since neither had anything like a solid grasp of the available evidence, much less of Coptic. Considerably more dangerous to Champollion's scholarly reputation was the critique and alternative system proposed by the German expatriate Julius Klaproth, a scholar of languages from Berlin who had been working successfully for some years in Paris.

Klaproth's original interest, in Chinese, attracted the attention of Count Jan Potocki, a colorful Polish aristocrat and author who took him along on a mission to Peking.[44] With Potocki's monetary support, Klaproth later found his way to Paris, where the presence of Abel-Rémusat, and the ability of Parisian print shops to handle foreign scripts, assured a favorable context for his work. Klaproth had previously met the brothers Humboldt, who assisted with his appointment while still in Paris as professor of Asian languages at the recently established University of Bonn.[45] As the materials he needed could only be found in Paris, Klaproth nevertheless wished to remain there and was granted an ongoing stipend by the Prussian government enabling him to do so.[46] A founding member of the *Journal Asiatique*, he was also its coeditor when in 1823 the journal published its anonymous report on the Dacier letter, which may have been written by Klaproth himself.

Klaproth was at first quite taken with Champollion's work. "Your works on the language and the monuments of Egypt," Klaproth wrote to him in 1823, "give us the hope finally to dissipate the darkness that has for fifteen centuries covered the patrimony of the pharaohs." Anticipating the trouble to come, Klaproth in the next breath bracketed Champollion with Young: "It is to your indefatigable researches and those of Doctor Young in England that we owe the explanation of hieroglyphic inscriptions which, since the regeneration of letters, has been the despair of antiquaries and philologists."[47] What gripped Klaproth was not the issue of the Egyptian scripts per se but Champollion's reliance on Coptic as the language of ancient Egypt. This had stimulated him to wonder whether Coptic might be affiliated with the languages "north of Asia and the north-east of Europe" in which he was expert. To that end he provided a list of putatively parallel words, such as for "bone," which in Coptic is "kas" (ⲕⲁⲥ), in Slavic "kost," in Circassian "kouchha," and so on. Many parallels were rather forced.

Klaproth's admiration proved fleeting. His allegiance shifted to Goulianoff, the Russian who, in 1824, had pseudonymously published the *Opuscules*, which, as we noted above, was the fullest and most direct immediate response to the *Précis*. In the next few years, as Klaproth absorbed and adapted Goulianoff's views concerning hieroglyphs, his opinion of Champollion darkened. By 1827 Klaproth, ever a harsh critic, was overtly antagonistic, and this antagonism eventuated in a published attack, framed as a "letter" to Goulianoff, who had recently published an article on the hieroglyphs of Horapollo that further developed the theory which he had adumbrated three years earlier in the *Opuscules*.

Goulianoff's *Opuscules* was directed at the ninth chapter of the *Précis* where Champollion developed the "first elements" of his scheme. Since, Goulianoff wrote, Champollion insisted that his system was fully in accord with the words of Clement of Alexandria, he would first examine those words, critiquing a commentary provided by Letronne to Champollion which had been used in the *Précis*. He would then investigate whether Champollion's theory did accord with Clement's remarks when the latter were properly understood, and, in a final section, he would examine the details of the theory itself. Goulianoff concluded that "the primitive alphabet of the Egyptians was composed of only EIGHT *characters*, which would make precisely *half* of the quantity admitted by the commentator [Champollion]" who had misconstrued the correct sense of Clement's words.[48] In particular, Champollion had read Clement's remark in the *Stromata* concerning "anaglyphs" (loosely, carved images), as referring only to allegorical carvings or bas-reliefs and not to actual writing. In which case, the *Précis* argued, Horapollo's work, with its provision of meanings for many hieroglyphic signs, applies only to such things and not to "hieroglyphic writing, properly speaking."[49] Goulianoff disagreed, asserting that Champollion had thereby "excluded from the hieroglyphic system of the Egyptians the elements that constitute its essence, the fundamental instruments of all the sacerdotal mysteries."

Goulianoff found it impossible to believe that some ancient Egyptian had taken the word for an object represented by a given image and reduced it to its initial consonantal sound. Why, the very schematic character of hieroglyphic signs seemed to militate against such a procedure. For what would then be left of the original connection, and how could such a process have proceeded as rapidly and early as Champollion insisted that it had? "Let's for example patiently calculate," he sarcastically remarked, "the time needed imperceptibly to decompose the image of a lion traced in all its details to remove its skin, then its flesh, then to dismantle bit by bit its skeleton, finally leaving only an unarticulated trace of *its dorsal spine* in order to serve the representation of the consonants L or R!"[50] And why would such a person, having so easily accomplished such a task, not have gone further by substantially reducing the large numbers of phonetic homonyms that Champollion claimed to have uncovered if the point was to reduce the set of symbols to form a compact, alphabet-like system? Goulianoff did not develop an alternative, but the gist of his critique implied that the hieroglyphs had always remained closely associated with the word for the object or situation represented, that Champollion's radical stripping had incorrectly obliterated that continuing, close association. If Champollion's system were truly worthy, then instead of advancing a theory to justify it along with the interpretation of a scattering of words and names, he would have done much better to have "given an explication of some text, of some papyrus, an explication one had the right to expect from an investigator who consecrated ten years to his hieroglyphic researches." In the end, Goulianoff concluded, the scheme had not gone much beyond what Warburton had essayed in the previous century. Indeed, "*the theory* of M. Champollion is a *faithful abbreviation* of Warburton's." Twisting the knife, he added that Champollion's "*discovery* is just the *perfection* of Young's."[51]

The *Opuscules* ended there, promising a "second part" that would go further by working through an examination of the applications that Champollion had provided. This does not seem to have been printed, but something like it surely existed, at least in manuscript, because in 1827 Klaproth, stimulated by Goulianoff's critique, published what seems to have been the Russian's system with a large number of examples. Klaproth framed the account as a direct response to Champollion's remark in the *Précis* that the values of the "symbolic" hieroglyphs presented an "obstacle" to the completion of the scheme.[52] Goulianoff, Klaproth enthused, had surpassed Champollion by overcoming precisely that obstacle. According to Champollion, the purely "symbolic" signs were "happily . . . the least numerous of all" even though, he had admitted, "it is precisely with these that the Greeks were the most preoccupied," citing, among others, Horapollo. Such signs, Champollion wrote, were "very complicated" and for the most part concerned "religious ideas." When they were employed in such a fashion, they expressed concepts that, in hieratic, were instead given by "a group of two, or three or four characters." Situations in which a hieratic equivalent was available would, Champollion hoped, eventually lead to fixing the proper meanings of the symbolic signs, comparatively limited in use though, he insisted, they were.[53]

As for Horapollo, whose work had often been taken to provide the proper meaning for hieroglyphs, and in particular for the very symbolic signs whose ubiquity and significance Champollion downgraded, he was certainly an "invaluable" source, but useful just for establishing a suggestive set of connotations, not the specific meanings, of many signs.[54] Though Horapollo had "until now" been thought to shed "great light" on "*hieroglyphic writing*, properly speaking," the *Précis* averred, nevertheless study of the *Hieroglyphica* had "given birth only to vain theories, and the examination of Egyptian inscriptions, his book to hand, has produced only feeble results. Does not this prove that most of the signs described and explained by Horapollo do not really form part of what we call *Hieroglyphic writing*, and principally hold in an altogether other system of representation of thought?"[55] Champollion's strong demotion of Horapollo's importance and that of his "symbolic" signs, formed a crucial part of his brief for the new system.

Klaproth and Goulianoff thoroughly disagreed regarding the superfluity of symbolic signs, but not because they thought hieroglyphs did not represent sounds. In their view, hieroglyphic writing was vastly *more* phonographic than Champollion thought, with Horapollo providing an excellent guide to uncovering true phonography through the use of Coptic. How so, when Horapollo had usually been deployed to assign broad conceptual meanings to hieroglyphs with no direct link to sounds? As an example, Klaproth adduced Horapollo's interpretation for the sign resembling a camel as representing "a *feeble* man on his legs, so that he could only walk with difficulty."[56] In Coptic, according to Klaproth, the word for camel is pronounced "djamoul," while the word for feeble is pronounced "djhasè." Both of these words begin with the same "let-ter," namely, "dj." This coincidence hinted at the correct way to understand hieroglyphs by Klaproth's lights. Each sign's phonetic value derived from the Coptic words for both the object depicted and for a quality beginning with the same initial sound as that object.[57] As a result, vastly more hieroglyphic signs have the same sound than Champollion had counted in his homonyms, and, indeed, the number of phonographic hieroglyphs must greatly exceed Champollion's limited set. Only the "magic circle of the cartouches," limited to Greco-Roman times, contained the letters found by Young and Champollion.[58]

Significantly, neither Goulianoff nor Klaproth pressed on any of Champol-lion's interpretations that might have attested to a weakness in his system. For instance, Goulianoff questioned Champollion's interpretation of the characters that figured a lion. Both he and Klaproth would have been well aware that Champollion's phonetic reading of the sign as figuring an L or an R comported not at all with the Coptic word for lion, which, as a variant of ⲘⲞⲨⲓ, begins with an M sound, seemingly contravening Champollion's insistence on an-cient Egyptian's essential equivalence to Coptic. The reason for this apparent inconsistency, Champollion had insisted, was that the lion signified "force or courage in the idea of all people who have known this superb quadruped," and so it was used in the royal cartouches of the foreign rulers despite the inconsis-tency with Coptic. Which, no doubt, is precisely why neither Goulianoff nor Klaproth pressed the point, given their conviction that Champollion's system simply did not apply beyond foreign names. In some cases, it seemed, a sign was chosen whose initial sound worked within the language of the foreign name, provided that the represented object possessed, through metonymy or synecdoche, appropriate connections to an aspect of the person in question. Put to such limited purposes, Champollion's scheme did square with the way in which Goulianoff and Klaproth thought ancient Egyptians handled phonogra-phy, namely, by taking the initial sound of the word for the represented object.

Champollion did not take immediate public notice of Goulianoff's *Opus-cules*. In early 1826, Champollion wrote his brother that he would have "pre-ferred not to speak of [Goulianoff] at all, because I am not at all obliged to defend my work against the first imbecile who will want to attack it." By this time a friend of Klaproth's, Baron Andreas Adolph von Merian, had apparently signed his name to a pamphlet against Champollion whose author was in fact Klaproth himself.[59] Still, Champollion put nothing in print until the following year, when he finally published a response.[60]

To begin, Champollion reminded readers that anyone who made "sym-bolic" characters "the essential elements of hieroglyphic writing," as Klaproth had, contradicted "the evidence of the most esteemed authors of all antiquity." This reminder overlooked, deliberately or otherwise, Goulianoff and Klaproth's claim that the most important of these ancient authors, Clement of Alexandria, had been misunderstood in what he meant by "symbolic" hiero-glyphs.[61] This misapprehension broadened to a general critique as Champollion began to consider the evidence in detail. If his antagonists were correct about symbolic characters, why had not Horapollo, for instance, simply effected the requisite association by pointing out that the image of, say, a vulture signified "mother," and so spared himself the recitation of "natural facts" or "popular traditions"? In any case, Champollion added in a note, the two words, mother and vulture, do not begin with the same letter in Coptic.[62] Horapollo's assignations always involved "similarities" between the object depicted and an idea. Consequently, Champollion insisted, they really were useful only for grasping situations involving characters that were "truly *symbolic* or *ideo-graphic*," and not for phonography, reducing Klaproth's suggested alternative to "singular suppositions" that were utterly vacuous. Such blatant errors could only have arisen, Champollion averred, "from a very superficial knowledge of the Coptic language." Anyone who claimed "to overturn or modify doctrines that rest on known facts" should after all be "obligated" to study "Coptic with the greatest care."

Champollion romped through examples of failure, all based on Klaproth's errors with Coptic. He gave Klaproth a taste of his own medicine by faulting him for having only applied his system to a small subset of Horapollo's designations. Given the unacceptability of the ones that he had done, Champollion remarked, "what would happen if he tried" on the rest? In response to Klaproth's limitation of Champollion's and Young's phonetics to the "magic circle of the cartouches," Champollion insisted that he had made "numerous applications" of his system beyond those presented in the *Précis*, albeit not to an extended inscription. Klaproth and Goulianoff had, in the end, just replaced "facts" with "suppositions."[63]

Klaproth did not take Champollion's criticism well. In a "second letter" to Goulianoff, Klaproth complained, with some justice, that "I never spoke of him [Champollion], in that piece, without joining to his name respectful words." He accused Champollion of territoriality: "M. Champollion does not like one to speak of Egypt without his permission and above all he dislikes mention of anyone who had been occupied with it before him: that's an impermissible crime." The territory extended beyond scholarship to national identity. "Champollion seems to believe that supposing someone other than himself could anticipate this part of his literary career is to attack French honor," he grumbled.[64] Such an "unreflective" attack obliged the wounded Klaproth to defend himself, and so he tried.

Pillorying Klaproth's mishaps with Coptic, Champollion particularly faulted his use of Kircher's Copto-Arabic dictionary of 1644, which contains many printing errors. Klaproth admitted that he had indeed used Kircher's dictionary, but not before checking the published version against an original manuscript held at the king's library. According to Klaproth, Champollion's diatribe proved only that "his knowledge of Coptic and Arabic is much less profound than one had thought."[65] Item by item for fifty pages, Klaproth disputed Champollion's critiques, often by reconstruing something in Horapollo to retrieve an appropriate vocalization. No doubt, Klaproth admitted, Champollion was correct that he, Klaproth, was no expert in Coptic, but his limited effort had borne fruit, enabling Klaproth to point out errors in the work of "the man who passes, in the eyes of some people, for having discovered and made known at the deepest level and beyond the idiom of the Pharaohs." He hoped in a third "more extensive" publication to show that Young and Champollion were "far from leading us to knowledge of the inscribed monuments remaining from ancient Egypt."[66]

Writing to Gazzera in mid-September 1827, Champollion vowed to respond neither to Seyffarth nor Klaproth.[67] This restraint availed briefly, but two years later, Klaproth returned to the attack. After insisting that Young was "without contradiction" the first to "discover" the "hieroglyphic alphabet," Klaproth immediately adduced Champollion's abbreviated 1821 piece from Grenoble to prove that he had not then thought the scripts to be alphabetic. Worse, in his 1824 *Précis* Champollion wrote "only in passing of his obligations to M. Young, to whom nevertheless he owed the first idea of what he termed *his discovery*." In any case, Klaproth continued, Champollion's effort "applied only to a fairly limited number of hieroglyphic signs" because he had truly read only proper names in phonetic form. All of these were, moreover, "royal names," and anyone alerted to their occurrence in cartouches could have reached "the same result as M. Champollion." This criticism touched on the point at which Champollion's system was the most vulnerable. Even de Sacy had noticed the vulnerability, as Klaproth pointed out. After all, "up to now one has still not read a single phrase on the monuments *with certainty*, not a single proposition." Coptic, he insisted, which underpinned Champollion's work, could not be a reliable guide since the Egyptian language must have changed radically over the years, particularly during the period of Greco-Roman domination. In his *Précis*, and especially in the following publications, Champollion had merely piled "conjecture on conjecture and contradiction on contradiction." According to Klaproth, Champollion's obdurate insistence on the importance of Coptic had "forced [him] to construct a new *Egyptian Mythology* [*sic*] which is hypothetical and founded on nothing," which Klaproth worked hard to demonstrate with page after page of examples.[68]

Klaproth, of course, had his and Goulianoff's sign-switching alternative in mind, and he was a biting and often angry critic.[69] Still, he had a point, for Champollion after all had not published a rendition of even the Rosetta inscription using his phonetic signs. Champollion had not done so because he was acutely conscious that the array of so-called symbolic signs, including determinatives, required more data than he had accumulated in Paris or even in Turin, evidence that he hoped to acquire on visiting Egypt. Writing anonymously in a lengthy 1829 review entitled "Hieroglyphics" in the *Foreign Quarterly*, the same year in which Klaproth's second critique appeared, the Scottish author, journalist, and historian James Browne dismissed Seyffarth's and Klaproth's efforts and praised Young as the originator of the true understanding of the Egyptian scripts. Browne regarded Champollion as the elaborator of Young's original work. He looked forward to the results of the Frenchman's upcoming Egyptian expedition. He was however "no stranger to the infirmities and defects of M. Champollion. We know his presumption, his vanity, his intolerance of rivalry, and his habitual tendency to be unjust to the labours and researches of others." Nevertheless "we also know and appreciate his great merits; his enthusiastic zeal in the cause of Egyptian Archaeology, his indefatigable perseverance, his ingenuity, and his skill in applying to the monuments the key which Dr. Young discovered, and which he himself has so greatly improved."[70]

Champollion again kept silent, replying neither to Klaproth nor to Browne. There was little to gain in attacking those he considered his inferiors in Egyptian matters, and certainly not without an urgent rationale for doing so. Irritating as it was, Klaproth's attack—and indeed most anyone else's at the time—posed little danger. Following years of work in relative obscurity, Champollion had secured the support of powerful friends, and his research was bearing substantial fruit. For once, Champollion could choose his battles.

CHAPTER 34

"HOLD YOUR LAUGHTER, FRIENDS!"

Louis XVIII died on September 16, 1824, just two months after Champollion arrived in Turin. The king's successor, Charles X, was much less inclined than Louis to tolerate opposition. Although this development might once have boded ill for Champollion, he could now count on the protections afforded by his patron Blacas and his widening reputation, undeserved as it was, as a defender of the faith. The demands of travel and fresh administrative responsibilities drove Champollion ever further from political activities. While in Rome Champollion was asked to catalog the papyri held at the Vatican. That June he visited Florence to examine a collection bought by the grand duke of Tuscany, Leopold II, from Giuseppe Nizzoli, formerly the Austrian consul in Alexandria. The following month he traveled to Livorno, where one of Henry Salt's collections was in play. In February the king charged Champollion with the task of estimating the collection's value, allocating 5,000 francs for the mission to do so. After negotiations for its purchase by the British Museum stalled, Champollion succeeded in acquiring the collection for the French Crown, and it arrived in Paris at the end of 1826.[1]

Pushed forward by Blacas and his ever-vigilant brother, Champollion's career advanced rapidly despite continuing opposition from Saint-Martin and others. Although this opposition blocked Champollion's election to the Académie des inscriptions et belles-lettres in February of 1827, the setback proved fairly minor in comparison to his substantial progress on other fronts. In the summer of 1827, he proposed an expedition to Egypt to the grand duke of Tuscany. By the end of the year, he was installed as the first curator of the Egyptian museum at the Louvre, named the Musée Charles X after the king. Champollion obtained support for a joint Franco-Tuscan expedition from the French monarchy in April 1828, and he left for Egypt on July 31.

Before leaving, he met with Thomas Young, who traveled to Paris following his election as a foreign member of the Institut d'Égypte. Their meeting was cordial. Young spent "seven whole hours" with Champollion perusing "his papers and the magnificent collection which is committed to his care." To Gurney he observed that "as [Champollion] feels his own importance more, he feels less occasion to be tenacious of any trifling claims which may justly be denied him; and in this spirit he has borne my criticisms with perfect good humour."[2] Soon heading to Egypt, Champollion agreed to let Young "have the use of all his collections and his notes relating to the enchorial character." The "enchorial" was ever Young's main concern, and in the final months of his life he completed the *Rudiments of an Egyptian Dictionary in the Ancient Enchorial Character*. The book was published two years after Young's death on May 10, 1829; he had worked on proofs even after falling ill the previous February. To the end, Young maintained that Champollion's "system of phonetic characters" was at best "of use in assisting the memory," that it could only be trusted in those cases where "the same kind of evidence that had been employed before its invention" provided support—and there were few enough of those. Still, Young took care to acknowledge those of Champollion's "manuscript communications [to him]" wherever they "furnished many valuable additions" to the *Rudiments*.[3] The "enchorial alphabet" included in the book remained essentially what it had been in Young's *Britannica* piece a decade before, while his "dictionary" consisted principally of sequences of characters identified as whole words or phrases, with the occasional Coptic analog indicated but usually followed by a question mark to indicate uncertainty.

Champollion arrived in Alexandria on August 18. He remained in Egypt for a year and a half, returning to France on December 6, 1829, with a hundred pieces acquired for the museum. These still bear the marks he added for identification. He also copied a large number of inscriptions, some of which his brother included in the posthumously published volumes of the *Monuments of Egypt and Nubia*.[4] These drawings greatly expanded Champollion's collection of signs, providing him with enough further evidence to draft two new works, his *Grammar* and *Dictionary*, both of which were also published posthumously and edited by Champollion-Figeac.[5] All three publications were assembled from the mass of drawings and partial drafts that Champollion left at his death. The published *Grammar* and *Dictionary* consisted of sentences in French, hand-drawn by Champollion-Figeac from his brother's notes, together with line drawings of the hieroglyphs, also drawn by Champollion-Figeac. These were engraved on copper plates for publication, requiring extraordinary expense and time. They formed principal sources for an emerging coterie of scholars who began to constitute the discipline of Egyptology.

For Champollion Egypt had been an all-consuming interest pursued from early youth. After years of adversity, he had secured the assistance of an unexpected patron by dint of having read "autocrator" on the Dendera zodiac.

Though Blacas was interested in antiquities, as a conservative Royalist he was unlikely to have so strongly supported the politically unsavory Champollion in the absence of that achievement. Having done it, the distasteful radical became palatable to altar and throne. With this apparent rescue of biblical chronology, Champollion had inadvertently countered a threat to religious orthodoxy just when such things were particularly contentious. The zodiac, now in Paris, eventually formed part of the museum's Egyptian collections. Little wonder, then, that Champollion made a special effort to reach Dendera, where he could see firsthand the object's original home and the lamentable depredations of Jean Lelorrain and Sebastien Saulnier.

Champollion and his entourage arrived at Dendera on November 16, 1828, at four in the afternoon. Alone, without guides, but "armed to the teeth," the party set out immediately for the temple.[6] Singing marching songs from the latest operas, the troupe encountered a man who, terrified by their appearance, tried to flee; they chased him down and persuaded him to lead them to the temple. This "walking mummy," as Champollion called him, not unsympathetically, guided them until the monument at last appeared. "I would not even try to describe the impression" the sight made, Champollion wrote his brother, except to say that it was "grace and majesty united in the highest degree." They sat for two hours in "ecstatic" contemplation of the temple under "magnificent moonlight."[7] The next morning Champollion canvassed the temple's columned interior, enthralled by the architecture but unimpressed by the bas-reliefs, which reflected the "decadent" period of their carving, as he saw it. Although the temple had itself been constructed during this same era, nevertheless its architecture, Champollion averred, being based on a "numerical art," had better resisted the vagaries of fashion.

Ascending to the temple's roof, he entered the small room that had housed the circular zodiac until it was ripped from its millenial-old home. There, on the margins of the void created by Lelorrain, Champollion could examine the carvings that had been left behind, most especially the cartouche in which he had so influentially read the word "autocrator." Gazing at the ruined ceiling, Champollion must have frozen in shock. "Risum teneatis, amici" (Hold your laughter, friends!), he wrote his brother.[8] For "the piece of the famous circular zodiac that carried the cartouche is still in place, and that same cartouche is *empty*" (fig. 34.1). As indeed are all the others in the temple's interior. Champollion decided that "the members of the *Commission*" had "added the word *autocrator* to their drawing, thinking they had forgotten to draw an inscription that doesn't exist," a self-defeating move that Champollion described as "carrying whips to beat yourself." Denon's original drawing, with its vacant cartouches, was correct after all, whereas a hieroglyph for "autocrator" had

FIGURE 34.1. The empty cartouches at Dendera. Photo by JZB.

somehow crept into the purportedly more accurate rendering by the *polytech-niciens* Jollois and Devilliers, who never wrote a word about it, though they surely knew of the mistake by 1845.

Champollion told only his brother about the empty cartouches. In 1832, having at last gained admission to the Académie des inscriptions et belles-lettres, and by then in failing health, Champollion collapsed in what was described as a fit of fatal "apoplexy." He was just forty-two. The tax authorities evaluated his modest estate, the records for which remain available. His possessions included his copy of the *Description*, which was assessed at 300 francs, with the entire estate amounting to 11,700.[9] Champollion-Figeac became the custodian of his memory, which he tended carefully. Nineteen of Champollion's letters to his brother had been printed as they arrived from Egypt in the widely read *Moniteur*. The year after his death they were collected and reprinted in Paris. The seventh letter, dated November 24 and written from Thebes, contained a version of the visit to Dendera, but the entire passage concerning the zodiac had been excised, replaced with an anodyne remark that the cartouches throughout the temple were all empty.[10] By 1845, Champollion-Figeac was ready to be more forthright. In that year, the fourth volume of drawings done during Champollion's expedition was printed. It contained an illustration done under Champollion's auspices that shows the empty cartouches. The caption, written by Champollion-Figeac, reads:

Planisphere or zodiac of Dendera. This plate was revised from the original monument, whether at Paris, where the zodiac, properly speaking, is, or at Dendera, where the part of the monument on which the figure of a woman, placed between two vertical columns of hieroglyphs, still exists. The two cartouches, at the base of these two inscriptions, are empty, and do not contain the Roman name that was inserted by error. This plate is the most faithful representation of all those that have been published of this celebrated monument.

And so, one might say, the events that would soon give rise to the development of the new discipline of Egyptology owe as much to absence as to presence. Were it not for his early reading of "autocrator," Champollion might not have acquired Blacas's patronage or his own position at the Louvre, and the Franco-Egyptian expedition was unlikely to have occurred either. Champollion might well have languished in obscurity, bereft of support and position, had it not been for the empty space mistakenly introduced into an engraving. "Risum teneatis, amici," indeed. Yet these absences jostle with other, more obvious, and less ironic contingencies—a worldly older brother who was generous with allies and advice; a competitor whose focus was often elsewhere, split between natural philosophy and his medical practice; and fortunate pecuniary accidents that enabled long periods of uninterrupted work, including mornings spent sketching antiquities in galleries frequented by the wealthy and powerful. The contingencies of character have their roles to play as well. Champollion's vast energy, keen insight, and remarkably flexible and sympathetic imagination propelled him beyond his competitor, Thomas Young, whose interests and attitudes limited his progress in ways that even he, perspicacious as he was, did not apprehend.

ACKNOWLEDGMENTS

This book has taken nearly a decade to write, and in that time, we've accumulated many debts. For their advice and assistance we thank Karine Chemla, Ivan Eubanks, Mordechai Feingold, Kristine Haugen, Gwen Kordonowy, Linda Josefowicz, Michael Josefowicz, George Landow, Ruth Macktez Landow, Susan Brind Morrow, James Pasto, Mac Pigman, Jean-Laurent Rosenthal, David Shawn, Fran Tise, and Victoria Spitz. In Grenoble, Dominique Raynaud provided invaluable practical assistance, and Karine Madrigal offered information about archival materials that were otherwise unavailable. We also wish to thank the three anonymous reviewers for their thoughtful comments on a late draft of the manuscript. At Princeton University Press, Rob Tempio and Matt Rohal kept the entire project on track while Sara Lerner shepherded the manuscript through production and Cathy Slovensky, our copyeditor, saved us from many blunders. We're especially grateful to Diana Kormos Buchwald and to Matthew and Jane Josefowicz for their support of this project from inception to conclusion and whose sacrifices on our behalf defy enumeration and description.

NOTES

INTRODUCTION

1. Lach-Szyrma 2009, 204.
2. The extensive literature on Young and Champollion is cited below and listed in the bibliography.
3. The principal exponent of this view was Athanasius Kircher; see Stolzenberg 2013.
4. Leitch 1855, 434.
5. Madrigal 2016, 37.
6. On these paintings, see Omer 1996 and Jasper 1998.
7. The only biography of Champollion-Figeac is Carbonell 1984.
8. Athanasius Kircher made a careful study of Coptic, which he considered the descendant of ancient Egyptian. His work on the language, though defective in several respects, remained the most complete in Europe for well over a century (Stolzenberg 2013, 94). Kircher was not the first to link Coptic to ancient Egyptian. The tenth-century alchemist and historian Ahmad bin Wahshihya had also done so, and in addition he considered at least some hieroglyphs to be true phonetic signs (El-Daly 2005, 57–73).
9. In what follows, we have tried to be faithful to our subjects' different nomenclatural preferences, for example, by using "enchorial" when discussing Young's views and "demotic" when discussing Champollion's. When such a stratagem proved impossible, we resorted to neutral expressions (e.g., "the intermediate script) to avoid the appearance of favoring one perspective.
10. Henry Brougham's 1803 attack on his ideas about optics nevertheless did provoke an unavoidable reaction that has often been discussed by historians of science. This was the second attack upon his reputation in natural philosophy. As we shall see, it followed an earlier conflict in which his uncle's friends intervened to stave off disrepute.
11. As the more fragile of these materials were often poorly preserved, reproducing them posed difficulties for engravers. In what follows, we often touch upon the different material cultures of printing in Britain and France.
12. For a recent study of the development of Egyptology, see Thompson 2015.
13. Hostilities between England and France did not in any case obliterate the social and cultural links between them. See, most recently, the relevant chapters in Cornick and Kelly 2013.

CHAPTER 1. DINNER AT LONGMAN'S

1. Thornbury 1880, 274. On Longman's as it appeared from the street, see Murray 1844, 78.
2. Thornbury 1880, 275. On Longman's business interests, see Briggs 2008, 72–73. The middle-class rage for the antique in early nineteenth-century Britain is explored in Coltman 2006.
3. On Edinburgh's rise, see Duncan 2005, 48–54.
4. Over the next decade and a half, Longman also published Scott's *Guy Mannering* (1815), *The Antiquary* (1816), and *Rob Roy* (1817).
5. Irving in Campbell and Beattie 1849, 1:433.
6. Kinghorn 1991, 8n5.
7. Campbell and Beattie 1849, 1:433.

8. Leersen 2004, 110–11. The arrival in Britain of the new philology was slow and irregular; see Jose-fowicz 2016.

9. The literature documenting the sources and manifestations of eighteenth-century British neoclas-sicism is enormous. See Coltman 2006 for a useful overview. For a brief but suggestive portrait of the political engine behind the fashion, see Mitchell 2007, 28–29.

10. Peacock 1855b, 128–31. Brougham's review was likely motivated by revenge for, some years earlier, Young had published a set of critical remarks on mathematical paper by Brougham. In an extended critique of the use of what Young regarded as impenetrable algebraic methods when geometric would do, he criticized "a young gentleman in Edinburgh" (i.e., Brougham) for doing just that.

11. Milburn 1890, 674.

12. Peacock 1855b, 192–215. Young's response to Brougham sold but a single copy, and Young's first authoritative biographer called Brougham's attack a "poison" that "sank deep into the public mind, and found no antidote in reclamations of other journals of coordinate influence and authority." Ibid., 184. Nevertheless, recent scholarship has shown that Young's work on light was not universally ignored, that Brougham and Young differed substantially over methodology, and that in any case the *Edinburgh Review* was too new and provincial to have a major effect on the reading public, though Brougham's attack certainly had a major effect on Young. On Brougham and Young's differences, see Cantor 1970 and 1978. In addition, Brougham did not understand Young's new principles, though he was competent enough otherwise; Kipnis 1991, 154.

13. The physician John Ring, for example, produced a pamphlet entitled *Beauties of the Edinburgh Review, alias the Stinkpot of Literature*, which compared the journal's reviewers to "the Malays, Lascars, and other Indian savages, who sally forth in a state of intoxication," neatly bringing Scotland under the umbrella of English colonial prejudice. Ring, who was adept at Latin (and later amended Dryden's translation of Virgil's *Georgics* to some praise), defended Young and blamed the *Review*'s tone on its object "to deprecate whatever tends to elevate or support this country [England]" (1807, 16). Young almost certainly knew Ring, who was a member of the Royal College of Surgeons. Politics may have played a part in all of this, because the *Quarterly Review* was strongly associated with liberal conservatives linked to George Canning (one of the *Review*'s founders, together with Walter Scott and John Murray), whereas the *Edinburgh Review* had a Whiggish cast. Extensive information on the journal is available in Cutmore 2017.

14. The argument rested on a broad consideration of various ancient sources, most notably on tenth-century manuscripts containing versions of the Alexandrian originals, the so-called Venice scholia, published in 1788 by Jean Baptiste Gaspard d'Ansse Villoisin.

15. Grafton, Most, and Zetzel 1985, 209.

16. On this topic, see Finley 1980, chapter 3.

17. Wolf's methods were similar to those of Johann Eichhorn, Göttingen professor of oriental languages, and almost certainly modeled on them (Grafton, Most, and Zetzel 1985, 20).

18. The wish for purity was particularly intense among aficionados of ancient Greek. Late eighteenth-century British "Grecians," as they were known—like the classicist Richard Porson, who Young knew well, and who we will encounter again—reciprocally idealized the ancient Greeks and their language, believing that the formal, logical syntax of ancient Greek mirrored the unique precision of thought for which the Greeks were also credited.

19. Modern scholars have greatly tempered this romanticism. Suzanne Marchard, for instance, suggests that the *Prolegomena* might best be considered "an introduction to the mental world of the academic philologist" for whom Homer "figures in the work only as a distant voice, speaking from the murky depths of an illiterate—and hence irrecoverable—age" (1996, 20).

20. Beattie 1849, 36–37.

21. Haugen 1998, 311.

22. It also indirectly supported a program to convince the public of the need for an autonomous Scots militia. Sher 1982, 58.

23. Haugen 1998, 311–12; Macpherson 1760. Considerable skepticism greeted Macpherson's claims. Samuel Johnson, for instance, asserted that the Gaelic original never existed (1775, 273). For the reception of Ossian in Scotland and England, see Moore 2004, who writes that for some in England,

Ossian was "a symbol [not only] of Scottish ambition and pretentiousness, but also of laughable bad taste" (24).

24. Macpherson may indeed have collected Highland sagas and not have altogether forged them, as some did suspect at the time. See, for example, Gaskill 1991. But see also Curley 2009, 8, who argues for "*Ossian*'s largely fabricated makeup." See also Stafford 1988. The *Ossian* controversy has generated an immense literature in which positions remain hotly contested more than two and a half centuries later.

25. Trumpener traces the literary influence of Ossian through the appearance, in 1816, of Walter Scott's *The Antiquary.* See Trumpener 1997, 115–27.

26. Johnson 1775, 274–76. For Johnson on Scotland, see Trumpener 1997, 87–92, especially chapter 2.

27. On Johnson and Ossian, see Curley 2009.

28. Trumpener 1997, 74–82.

29. Boswell 1904, 149; Burnett 1773.

30. John Locke, for example, considered writing to be a subsidiary, purely practical development.

31. Aarsleff asserts that Tooke was so influential in England that "for thirty years [his ideas] kept England immune to the new philology until the results and methods finally had to be imported from the continent in the 1830s, and even then they met strong opposition" (1983, 73). This somewhat over-states the case. If the new philology were completely unavailable, it is difficult to understand how Wolf's *Prolegomena* became the subject of an argument over dinner at Longman's in 1803. There is evidence that the *Prolegomena* made other incursions into Britain as well. See Josefowicz 2016.

32. For a concise account of Monboddo's conceptions, see Aarsleff 1983, 36–43. Monboddo was familiar with Locke's ideas but claimed not to have read Condillac.

33. Monboddo 1773–87, 12, 473.

34. "Johnson, like Locke," writes Elizabeth Hedrick, "believes in the close relationship between mental and linguistic processes; he believes in the irreducibility of simple ideas, though he manages to define them efficiently; and, most importantly, he sees complex ideas as created in the mind out of combinable parts" (1987, 600).

35. Monboddo 1773–87, 341. "I say," Monboddo wrote, "that instead of being short, and consisting of monosyllables, [these words] were of great length: and this too is a consequence of those languages being derived from natural cries; for such cries of almost all animals have a certain tract and exten-sion, such as the lowing of an ox."

36. Ibid., 4:15–6.

37. Johnson 1775, 23.

38. Ibid., 267.

39. Trumpener 1997. Johnson's interactions with Monboddo are discussed in Nash 2003, 142–44, on which we draw; see 144 for the remark to Piozzi.

40. Boswell 1904, 383.

41. On the connections between speed of thought and abbreviation in the linguistic philosophy of Tooke and others, see Aarsleff 1983, 46–48.

42. Ibid., 48.

43. Hazlitt 1825, 116, from an essay, "The Late Work of Horne Tooke," quoted in Aarsleff 1983, 71. Even Young's nemesis Brougham shared this opinion, believing Tooke's methods to be "cognate" with those of "inductive investigation" (1856, 104); quoted in Aarsleff 1983, 92. Ibid., 89–95, on evolution to efficiency.

44. Beattie 1849, 434.

CHAPTER 2. IN THE CLASSROOM OF NATURE

1. *Gulliver's Travels* particularly impressed him. As a small child in the company of adults who must have seemed quite ignorant, he may have seen himself as a variant on the eponymous hero.

2. The first biography of Young, by the surgeon Thomas Pettigrew (or "Mummy" Pettigrew as he be-came known for his work on embalming and Egyptian mummies), appeared in his *Medical Portrait*

Gallery ten years after Young's death in 1829. Peacock published translated extracts from a Latin autobiography by Young written late in life to the age of fourteen (Peacock 1855b, 3–8). The original is now lost. Young had, however, written another, in English, intended for the *Encyclopaedia Britannica*, which his friend Hudson Gurney apparently used to write his own reminiscence of Young (ibid., p. ix on Gurney 1831), eventually included as a preface to Young's never-completed Egyptian dictionary (Young 1831). The dictionary was first printed, without Gurney's reminiscence, as an appendix (Tattam 1830). Young's original was lost until 1978 when the historian Victor L. Hilts discovered it in the papers of Francis Galton (Hilts 1978). Although this memoir is less detailed than the Latin version translated by Peacock, it contains a valuable year-by-year list of Young's reading from 1775 to 1797. That list includes many items not given in the (Latin) fragment, but the fragment also mentions several important items that are not in the longer list. These together comprise what we have of Young's own thoughts on his life. Peacock's and Gurney's materials were used by Alexander Wood and (completing Wood) Frank Oldham to produce the first comprehensive biographies of Young (Oldham 1933; Wood 1954). The most recent, and discerning, account that has used Hilts's discovery is Robinson 2006.

3. Goldsmith 1770, 2–3, 15–16.
4. Peacock 1855b, 4.
5. Gay 1772, 2–4.
6. Peacock 1855b, 3.
7. Maturinus Corderius was also known as Mathurin Cordier.
8. Telescope [pseud.] 1794, 120–21. Young would likely have used the 1779 edition. Some later editions assign several of the quoted lines to a schoolmarmish character named Mrs. Mentor rather than to Tom.
9. This was almost certainly the extravagantly titled *Complete Dictionary of Arts and Sciences in which the Whole Circle of Human Learning is Explained* by Croker, Thomas Williams, and Clark, 1766.
10. Peacock 1855b, 5.
11. Peacock 1855a, 5–6.
12. "Oriental literature . . ." (Peacock 1855b, 6).
13. Most likely in the edition by Camden and Westminster 1754.
14. Martin 1771; Ryland 1772. Section 5 of Simpson and Nourse 1752 is titled "A short account of the nature and first principles of fluxions." A good deal of Martin's account was developed in smaller type at the bottom of the pages, where he used Newtonian fluxional calculus. Nevertheless, the main text could be understood without that apparatus. The section does not move on to fluents (integrals).
15. Dilworth 1765; Ewing 1779.
16. Vyse 1772; Ward 1724.
17. Peacock 1855b, 6–7.
18. Priestley 1775. On Priestley, see Hiebert et al. 1980. There's no evidence that Young then or later also read Priestley's treatise on speech (Priestley 1781).
19. Priestley 1775, ix.
20. Toulmin eventually switched from the Baptist denomination to Unitarianism, and in 1804 became pastor at the Unitarian congregation in Birmingham that Priestley had been forced to leave by mob violence. On April 22, 1804, Toulmin gave an address in memory of Priestley, who had died on February 6. See "Joshua Toulmin" in Rose 1857, 267.
21. Sharpe, shortly followed by Warburton (see below), produced one of the many eighteenth-century conjectural histories of language and writing. They differ from seventeenth-century reductions of language and writing to systems of arbitrary signs and from consequent attempts to generate perfected schemes by projecting a complex series of cultural developments. Sharpe differed as well in considering that the original linguistic and graphical systems were connected to natural vocalizations. On the seventeenth century, see Hudson 1994, 32–54; for England, see Aarsleff 1983.
22. Mordechai Feingold (private communication). Note that Hogarth apparently did not realize that Newton's telescope was a reflector and not the pictured refractor.
23. Hutchinson 1749, 101–2: "Do the Quakers employ or lay out their money with a Christian, if they can find a Quaker . . . and yet everyone knows they are ready to spend that money Christians put into their hands to establish blasphemy, and root out Christianity."

24. Peacock 1855b, 8; Sharpe 1751, ix–x.

25. Sharpe 1751, xi.

26. On conceptions of the first language, see Eco 1997.

27. Sharpe 1751, 1–2. Condillac's work on the origin of knowledge was printed in 1746, five years before Sharpe's.

28. Ibid., 6. Sharpe thought that Chinese and Egyptian had "an amazing conformity" to each other. What he had in mind for Egyptian is not clear, but he perhaps meant Coptic since the Jesuit Athanasius Kircher had, in the previous century, suggested the connection. John Webb, whom Sharpe discussed, averred that Chinese was the original language, that the Chinese were descended from Noah, and that, as Sharpe put it, "they now talk in China the language of Paradise" (ibid.); cf. Webb 1669.

29. See Olender 1992.

30. Sharpe claimed that with an alphabet of twenty-two letters, taking two at a time yields 506 words, three yields 11,154, and four produces 245,410. The oddity of these results may have caught Young's attention; it certainly demands explanation. Sharpe must have counted a word consisting of two of the same letters twice to produce the two-letter count (otherwise 462 if the letters are different or 484, allowing two of the same letters to form a word but disallowing a double count); adding another letter and allowing for a word consisting of three of the same letters to count twice produces the result for a three-letter count, while adding another and allowing for a word consisting of four of the same letters produces the four-letter count on the same basis. Each additional letter in such a procedure adds 22 times the previous count plus 22. Sharpe provided no reasoning here, but he may have had in mind that a word consisting, for example, of two of the same letters might have different meanings depending on whether the first or the second of these two same letters is emphasized, and so forth. Sharpe 1751, 14.

31. Ibid., 17.

32. Ibid., 40–41.

33. Ibid., 56.

34. Ibid.

35. Warburton 1742, 67. On Warburton's attack on Newton's attempt to revise chronology in the *Divine Legation*, see Buchwald and Feingold 2012, 394–97. For short accounts of his views on writing, see Iversen 1993, 103–5, and Pope 1999, 48–53. A particularly informative discussion of Warburton's views for our purposes here is Hudson 1994, 59–71. See also David 1965, 95–99, and Rosenfeld 2001, 38–40.

36. Warburton on hieroglyphs was translated into French; see Warburton 1744. Warburton was not the first to offer an evolutionary account of writing; Giambattista Vico had done so much earlier. Warburton's developmental discussion has traditionally been linked to Neo-Epicureanism, but recent work indicates that it was more likely embedded in an ongoing historiographical tradition grounded in philology and dating to Isaac Casaubon and Joseph Scaliger (1540–1609) (Levitin 2015, 724–25).

37. For the several graphological distinctions used here, see Powell 2009. We have used "sign" to refer to written signs throughout this book, following our protagonists' customary ways of talking about matters relating to the Egyptian scripts. The modern understanding of "sign" is considerably broader, including gestures and vocalizations, to name a few.

38. Hudson 1994, 61.

39. "[A]s we learn from an old saying of Heraclitus: that the king, whose oracle is at Delphi, neither speaks nor keeps silent, but reveals himself by signs; a plain proof that speaking by actions was once the common mode of information" (Warburton 1742, 86–87).

40. Warburton 1742, 140.

41. Kircher had, however, suggested an alphabetic component to the "vulgar" or (as Champollion termed it) demotic, Egyptian script written by Herodotus.

42. Sharpe 1751, 57.

43. Ibid., 60.

44. Bayley 1742, ii; his emphasis.

45. Moreover, without vowel points there lurked the danger that the resulting vagueness "opens a flood-gate to Popery." Because the absence of points "naturally reduces the sense of the Hebrew scriptures," either the Old and New Testaments could no longer be granted divine authority, or allowance had to be made for "the perpetual residence of an infallible spirit somewhere in the church of Christ; to which recourse might always be had, for solving hard questions and dubious occurrences." Bayley 1782, ii, viii.

46. Jones 1771, 16. Five years later, in a February 1786 address to the Asiatick Society, Jones further asserted that Sanskrit, Greek, and Latin derived from a common ancestor.

47. The Gurney family (then in Norwich) descended from Hugh de Gournay who had arrived with William the Conqueror. Their bank eventually merged with others run by Quakers, including the one controlled by the Barclays and their relatives, to form the present-day Barclay's Bank.

48. All accounts of Young's years in Hertfordshire are based on two sources: Peacock (1855b, 9–26), and the account published by Gurney (1831, 10–15).

49. Peacock 1855b, 9.

50. Peacock 1855a, 17.

51. The Euclid was Simpson 1756; the algebra, Bonnycastle 1782.

52. Simpson 1776.

53. Peacock 1855b, 10.

54. TH 1891, 63.

55. Sharpe 1751, 98.

56. The English edition of Arnauld and Launcelot's *Abrégé de la nouvelle méthode pour apprendre facilement la langue greque* (1655) that Young likely used was Nugent 1773; there was also a 1748 edition.

57. The book is Scheller 1779; cf. Peacock 1855b, 14.

58. Peacock 1855b, 16.

59. Peacock 1855a, 19–20.

60. Gibbon and Sheffield 1827, 1:228–29. The "comma," which appears in the Latin Vulgate in John 5:7–8, is a phrase that can be construed as supporting Trinitarianism. In the eighteenth century, the sign's authenticity was attacked by, among others, Gibbon, to whom Porson responded.

61. Porson and Kidd 1815, 120.

62. Ibid., 112–13, for Porson's extract from Knight and his critique.

CHAPTER 3. AN ERRAND IN THE CITY

1. London's markets tended to be differentiated according to the products bought and sold at each. Besides the central market at Smithfield, markets for beef and its by-products were held at Newgate, Farringdon, Fleet, and Eastcheap. See Smith 1999, 110–39; see also Sheppard 1971, 189–90.

2. Forshaw and Bergström 1980, 20.

3. Dodd 1856, 244; quoted in Forshaw and Bergström 1980, 54.

4. Wynter 1868, 214. This is a reprint of an essay that appeared in the *Quarterly Review* in 1854.

5. Dickens 1837, 203.

6. Dickens 1860–61, 189.

7. Forshaw and Bergström 1980, 102.

8. Adams 1817, 29.

9. Ibid., 39–40; Ottley 1839, 31.

10. Gloyne 1950, 53.

11. Hunter 1786.

12. Musschenbroek 1745, 394–403.

13. Peacock 1855b, 385. Pemberton 1719 is the original source. Peacock asserts that Young "had not studied [Pemberton's] writings."

14. See Pettigrew 1839, 4–5, for the timing of Young's studies in 1792–93; see also Peacock 1855b, Wood 1954, and Robinson 2006.

15. Hunter died of a heart attack on October 16, 1793, in the midst of a heated argument about the admission of students to St. George's Hospital. The publication of Young's paper on ocular accommodation preceded Hunter's death by several months. On Hunter's death, see Kobler 1960, 281–302.

16. Young 1793, 172. Young asserted that he had "discovered . . . a structure which appears to remove all the difficulties with which this branch of optics has long been obscured"—quite a claim from a barely twenty-year-old neophyte.

17. Kobler 1960, 280; the quote is from Young 1793, 172.

18. Young 1793, 173.

19. Ibid., 176.

20. Ibid., 173.

21. Smith 1738.

22. Hilts 1978, 255.

23. Young 1793, 174.

24. He cited Simpson's *Fluxions*. See Simpson 1776.

25. Holmes 2008, 306.

26. Hunter and Home 1794. Established in 1701 as a result of a bequest from William Croone, the Croonian Lectures were by this time jointly presented by the Royal Society and the Royal College of Physicians. Two lectures were given, one before the society and the other before the college; the society lecture had to concern muscular behavior.

27. Peacock 1855b, 39.

28. McIntyre 2003.

29. Getman 1937.

30. Ibid. In Cavendish's case, the contest concerned "dephlogisticated air." Blagden 1787 made much of phlogisticated alkalis and galls.

31. Peacock 1855b, 39, and Robinson 2006, 38.

32. Hunter and Home 1794.

33. The fibers are called the "zonules of Zinn" after their discoverer, Johann Gottfried Zinn.

34. As it turns out, Young had the explanation precisely backward. Years later, the physician Robert Knox fastened on a thin structure, quite difficult to observe, known as the "annulus albus," which he described as the "external prolongation of the choroid" (the region between the retina and the eye's outer layer). He strongly suspected that the annulus was involved in accommodation, but just how remained unclear. In 1855 the German physician and physicist Hermann von Helmholtz proposed that the region in question was itself muscular and acted to change the lens's shape, albeit in a counterintuitive manner. When this circular muscle is relaxed, and therefore not contracted, the fibers connecting it to the lens pull the lens out of its otherwise more spherical shape. The relaxed muscle consequently accommodates for distant objects. When the fibers loosen, the lens springs back to become more convex, thereby accommodating for near vision. Muscular contraction rounds the lens, while muscular relaxation flattens it, with the lens's unstressed state being convex. Helmholtz's proposal was utterly counterintuitive, hence its power and novelty. See Helmholtz 1855, 1–74; Knox 1826–27, 110–16; and Zinn 1780.

35. Hunter 1786.

36. Young 1793.

37. Foot 1794.

38. Ibid.

39. Adams 1817. He excused Hunter's poor grammar by attributing it to an excess of maternal influence in his early life, and opposed to this a stronger grammatical command, which he suggested was due to "paternal inheritance."

40. Hunter and Home 1794.

41. In a footnote, Adams (63) cites Hunter's *Treatise on the Blood* (1794, 16) and Tooke's *Diversions of Purley*, vol. 2, chapter 8, 1805 (the first London edition).

42. Foot 1794; our emphasis.

43. On language and politics in late Georgian England, see Smith 1986.

44. On Port-Royal, see Aarsleff 1983.

45. Smith 1986, 3.

46. Thelwall 1795.
47. Smith 1986, 37. According to Olivia Smith, Burke's success among radicals was due to his use of homely metaphors, sensational vocabulary, and other markers of common sensibility—direct language on which Hunter also notably relied.
48. Ibid., 31.
49. Quoted in ibid., 33.
50. Ibid., 13.
51. See Harris 1771 for his views. On Tooke and on Harris, see Manly 2007.
52. Johnson, quoted in Smith 1986, 13.
53. Peacock 1855b, 21.
54. Peacock 1855a, 26.

CHAPTER 4. THE VOCAL CIRCLE

1. At the time of Young's nomination, Weston was known for his books on ancient and modern Rome, on Greek poetry, and on interpretations of biblical passages. In later years he translated works in Arabic, Chinese, and Persian; in 1808, he penned a brief memoir of Porson. Elected to the Royal Society in 1791, Townley, a wealthy Catholic, had collected the "Townley Marbles," a Greco-Roman assortment that he displayed, in 1778, in a house constructed expressly for that purpose in London's West End. The remainder of the list consisted primarily of medical men and important society members.
2. Peacock 1855b, 45.
3. Peacock 1855a, 10, 46. The title page added to a new edition in 1807 announced that Young had formulated "the method" used.
4. Hodgkin 1794.
5. Ibid., 10.
6. Watson 1861, 362.
7. Full membership after a year's candidacy to fellowship in the college had previously been restricted to Oxford and Cambridge graduates. The regulations had also allowed those who had taken their degree at another university to become members of the college with the rank of licentiate. After seven to ten years licentiate holders could become fellows.
8. Peacock 1855a, 88–89; Wood 1954, 47, 53–55.
9. Peacock 1855a, 58–60; Robinson 2006, 46.
10. Peacock 1855a, 43–44, 53; Robinson 2006, 44–45.
11. Peacock 1855a, 55–56.
12. For example, Young claimed that one of Homer's words had "about the same sense and sound as the Gothic *scathe*" (Dalzel 1802, 33n144, in the *Notae in Homerum* section).
13. Peacock 1855b, 89–90. Although the copy preserved at the British Library lacks these pages, they are present in the one held at the library in Göttingen. The elements of the system were later included in the extensive "Catalogue of Works" published in the second volume of Young 1807, 276–77.
14. Newton 1704, 114–17.
15. Young 1796.
16. Ibid., 117.
17. Young 1807, 2:276.
18. Hudson 1994, 83–85; cf. Helmont 2007, 105–8; and Wachter 1752.
19. Young apparently never met Johann Eichhorn, who arrived at Göttingen from Jena in 1788. At Göttingen Eichhorn taught Near Eastern (then called oriental) languages and developed a critical approach to the Old Testament in which he rejected miracles as ancient fables and tried to place the biblical accounts in the frameworks of their originators.
20. Young's logical orientation poorly suited the complexity and particularism of Heyne's historicist conception. On Heyne, see Grafton 2010, 436–37, and for his Philological Seminar, Leventhal 1986, 243–60.

21. Peacock 1855b, 81–84, and Dalzel 1862, 120–21, 39, for quotation. Fiorillo was at the time working on the first of the five volumes of a comprehensive art history, *Geschichte der Zeichnenden Künste von Ihrer Wiederauflebung bis auf die Neuesten Zeiten* (1798–1808).

22. Herder is remembered primarily for his philosophy of history, according to which all aspects of a culture had to be understood on the culture's own terms. The fourth and final volume of his *Outlines of a Philosophy of History of Man* had been published only several years before Young's visit, and there, too, Herder insisted on examining events and beliefs in the context of their production. Herder 1792–94, translated as Herder 1800. On Herder's philosophy of history, about which a great deal has been written, see especially Beiser 2011, Forster 2010, and Nisbet 1970.

23. Peacock 1855b, 101.

24. Herder 1772; translated in Forster 2002, 65–164. See Beiser 2011, 120–27, and Forster 2010, 55–130.

25. Forster 2002, 98.

26. Ibid., 69 and 71.

27. Ibid., 54 and 62, the former from 1767–68.

28. Herder 1800, 104.

29. Ibid., 104 and 2.

30. Peacock 1855b, 124. On Brocklesby's life, see Curran 1962. The value of Brocklesby's bequest is equivalent in purchasing power to more than three-quarters of a million pounds.

CHAPTER 5. LECTURER AND PHYSICIAN

1. Peacock 1855b, 124.

2. Langdon-Davies 1952, 106. Even the well-connected Brocklesby had never enjoyed a regular hospital connection though he had a yearlong appointment to London Lying-In Hospital and was later nominated for, but did not get, a post as physician to St. Bartholemew's.

3. Brocklesby's set included the third duke of Richmond as well as Burke and Johnson. See Curran 1962, 509–21.

4. On van Butchell, see Kent 1954, 123, and Thompson 1928, 321.

5. Langdon-Davies 1952, 78. While their ingredients were touted as closely held secrets, most patent medicines contained mainly herbs and spirits, if also, occasionally, arsenic; see Thompson 1928, 345.

6. Gould and Uttley 1997, 5. On Queen Anne's fondness for Sir Walter Raleigh's Cordial, see Southey 1837, 4:379.

7. Thompson 1928, 344.

8. Ibid., 270–72.

9. Timbs 1855, 200.

10. See Milburn 1890, 670.

11. Its holdings included "the curious library of [Thomas] Astle, the antiquary," author of *The Origin and Progress of Writing* (1803) and longtime fellow of the Society of Antiquaries. Timbs 1855, 655 and 464.

12. On the Royal Institution, see Berman 1978 and James 2002. We have drawn our details from these accounts and others, as noted below.

13. Feltham 1803, 284.

14. Wood 1954, 119.

15. Brown 1967, 37.

16. Peacock 1855b, 134.

17. Rumford to Sir Thomas Bernard, May 13, 1798. Quoted in Brown 1967, 119.

18. Wood 1954, 120.

19. Rumford to Banks, September 21, 1801; Banks and Chambers 2007, 110.

20. Feltham 1803, 296–98.

21. Young, quoted in Wood 1954, 125.

22. Peacock 1855b, 135–36.

23. Young, quoted in Peacock 1855a, 191–92.

24. Bagot 1909, 165.

25. Ibid., 164.
26. James Gillray, "Scientific Researches! New Discoveries in Pneumaticks! Or Experimental Lecture on Powers of Air," published by H. Humphrey, May 23, 1802.
27. Ilchester 1908, 61.
28. Wood 1954, 124.
29. When the costs of colonial adventuring and wars began to exceed funds raised by traditional customs and excise taxes, Pitt implemented an income tax in 1799. This unpopular measure was repealed in 1816. O'Gorman 2016, 252.
30. Feltham 1803, 285.
31. Cantor 1970, 87–112, 95. Brougham did, however, praise Davy for what Brougham saw as his imperviousness to the Royal Institution's corrupting influence.
32. Young 1855, 480.
33. Wolcot had served as personal physician to the governor of Jamaica in the late 1760s and early 1770s. On moving to London in 1780 Wolcot turned his attention to writing and publishing satirical portraits of prominent figures.
34. Young 1855, 479.
35. Peacock 1855a, 189.
36. Chard 1977, 138–44, 41.
37. Lockhart 2005, 239.
38. Peacock 1855b, 188.
39. Peacock 1855a, 211–18.
40. Ibid., 220–21.
41. Langdon-Davies 1952, 32–34.
42. Gould and Uttley 1997, 3.
43. Knight 1851, 783.
44. Gould and Uttley 1997, 3–4.
45. Wood 1954, 74.
46. Gould and Uttley 1997, 5; Gould 1997, 5.
47. The troubles with the physical plant at St. George's are described in Gould 1997, 3–6.
48. Wood 1954, 7.
49. Brodie's remarks are reprinted in ibid., 78–79.
50. Young 1815, vii.

CHAPTER 6. THE HERCULANEUM PAPYRI

1. Moser 2006, 84–91. As early as 1773, Wedgwood's catalog had included ceramics with Egyptian motifs such as sphinxes and hieroglyphics. Home furnishings of this period were similarly adorned.
2. Carrott 1978, 31–32. Carrott notes British architects' substantial debt to Piranesi. We note as well the widely circulated drawings of the artist Dominique-Vivant Denon, who accompanied Napoleon on the expedition.
3. Ibid., 34–35.
4. Hayter was a graduate of Eton and King's College, Cambridge, where he had won the Browne gold medal for his rendition of a Greek ode. The college gave him the rectory of Hepworth, and at some point Hayter became chaplain in ordinary to the Prince of Wales.
5. Hayter 1811, 52. The precise figure would soon come into dispute, as we will see.
6. It is unclear whether Hayter kidnapped the girl or rescued her from a forced marriage.
7. See Harris 2007, 139–45, for a recent account of these matters. For details on Hayter, see Auricchio 1980. The most detailed nineteenth-century account of the materials at Herculaneum remains Comparetti and Petra 1883.
8. Peacock 1855b, 508, from Young 1813b, 18. Although it's not clear when Young was asked to work on the scrolls, he may have examined them as early as 1804, the year they arrived at the prince's residence, Carlton House (Lloyd 1830, 249). It was during Young's years at the Royal Institution

that his expertise in "mechanics and the operations of chemistry" became more broadly known. He later wrote an account of these matters for the *Britannica*; Young 1819a, 624–37.

9. Peacock 1855b, 119. Young's interest in the properties of sound—and from there of light as well—had developed as a consequence of his work at Göttingen on the sonic properties of the human voice.

10. Drummond and Walpole 1810.

11. Walpole and Charles 1805.

12. Anonymous 1806, 164; Drummond 1805. The quotation is from Drummond's *Academical Questions* (ibid., x).

13. The papyri were found in a villa that had probably belonged to Lucius Calpurnius Piso, the father-in-law of Julius Caesar. The fragment presented by Drummond was by the minor Epicurean philosopher Philodemus.

14. Drummond and Walpole 1810, 123 and 44.

15. Blomfield 1810; ibid., 379–80. The review was anonymous, but Peacock was informed at thirdhand that the author was Blomfield. When he wrote his biography of Young, Peacock did not know that the review had been published (1855a, 237).

16. Again according to Peacock, Blomfeld desired to stop the publication (ibid.).

17. Young 1810, 1–6. Peacock 1855a reproduces Young's review but omits its quotation from the *Herculanensia* on the scrolls' history (488).

18. See the account in Haugen 2012, 14–15.

19. Foster 1773, xviii. Foster formally resigned his Eton post in 1773 to take up a canonry at Windsor.

20. Young 1810, 6. Musical notation did not however express absolute temporal duration, which Young thought should be remedied (1807, 2:571—this "Essay on music" originally appeared in October 1800 in the *British Magazine*).

21. For Young's merging of music, sound, and light, see especially Pesic 2013.

22. "The same constitution of the human mind, which fits it for the perception of harmony, appears also to be the cause of the love of rhythm, or of a regular succession of any impressions whatever, at equal intervals of time" (Young 1807, 1:392).

23. Peacock 1855b, 236–37.

24. Young 1810, 10. When compared with the original apograph, two of the omitted letters (τν) are above the line (των), one below it (o), and two apparently in-line (αν). Young included the *Herculanensia*'s σω at the end of his reconstructed line in order to give sense to the line.

25. Peacock 1855b, 232.

26. Casanova's and Orazj's apographs in lead pencil, as well as engraved plates made from them and intended for publication by Hayter, which never took place, were presented by the Prince of Wales to Oxford in 1810 (Scott 1885, 4). The "fragment" presented by Drummond was the first of the last twelve columns of what is now Pap. 1428, cols. 356–67 in the Bodleian Library, Oxford collection of apographs. On the first of these (no. 356) we read "svolto da D. Giuseppe Paderni, disegnato da D. Giuseppe Casanova." Paderni was one of the unfolders (Hayter 1811, 55). For Young's re-creation, see Young 1810, 9.

27. Young 1810, 15–16.

28. Hayter 1810, 6.

29. Ibid., 7.

30. Young's reconstituted text reads in his translation: "He has made no vows to the Gods, thinking it a folly for one, who has no distinct conception respecting them, to give himself trouble on their account." Young 1810, 12; Hayter 1810, 6.

31. Drummond characterized Young's critique as showing "both ingenuity and learning," though he considered it "sometimes too hasty" (Hayter 1811, 18).

32. Peacock 1855b, 234–35. Reviews were customarily unsigned. Peacock's views of both Drummond and Hayter were likely colored by his theological commitments. Peacock, who in 1839 became dean of Ely Cathedral, generally disapproved of Drummond, criticizing him for being "somewhat apologetic and deprecatory," the latter of which is something of an exaggeration. Drummond may have offended Peacock by privately printing a book, the *Oedipus Judaicus* (an allusion to Kircher's *Oedipus Aegyptiacus*) that treated the Old Testament as a series of imaginative fables and subsequently

generated much offended criticism (Drummond 1811; D'Oyly 1818). As for Hayter, his career did not quite come to an end at this point, Peacock's assessment notwithstanding; several years later he again became involved in Herculaneum matters when he made an unsuccessful attempt to unroll the six scrolls given to the Prince of Wales (Anonymous 1817, 5). Though Young did not directly answer Hayter's defense, he took the opportunity in 1819 to attack his competence: "It is difficult to understand by what test the merits of such a scholar were appreciated, when he was appointed to superintend the operations at Portici" (Young 1819a, 628).

33. Adelung 1806–12. After Adelung's death in 1806, the second and third volumes were continued by the philologist and theologian Johann Severin Vater, who from 1809 to 1820 was professor at Königsberg, before and after that at Halle.

34. Young 1813a and 1815.

35. Young 1813a, 252.

36. Chamberlayne 1715.

37. Young 1813a. The view that Chinese was "principally monosyllabic" remained common until the publication by Abel-Rémusat of his major work on Chinese (1822).

38. Marshman 1809. A Baptist missionary in India, Marshman translated Indian works into English and the Bible into several Indian languages and Chinese. Although the review of Marshman's *Chinese* was unsigned, there are persuasive reasons for assigning the essay to Barrow and Staunton, on which see Cutmore 2017, 147.

39. The self-educated son of a tanner from the village of Dragley Beck in Cumbria, Barrow eventually became second secretary of the Admiralty and, in 1830, a founder of the Royal Geographical Society. His account of his two years in China was widely read (Barrow 1804; Proudfoot 1861).

40. Young never met Staunton, who was still in China when the review appeared. Barrow, on the other hand, would eventually draw close to Young, supporting him vigorously against Champollion (Barrow 1849, 162–73) and later writing an admiring biography.

41. Peacock 1855b, 272; Barrow 1816.

42. Barrow and Staunton 1811, 377.

43. Ibid., 380–92.

44. Though *mooching* seems unavoidable.

45. Barrow and Staunton 1811, 391.

46. According to Barrow and Staunton, phonography's eventual development in China did not displace the semantic structure of "imitative" writing, which persisted as a result of the force of tradition (ibid., 399). In general they suggested polysyllabic languages were originally represented orthographically by sets of characters representing entire words or word phrases, and that the development of an "alphabetic" script for such a language would occur in a manner similar to the repurposing of character sounds in the Chinese script to represent new or foreign words. They claimed Egyptian hieroglyphics provided the impetus for the development of alphabets when the same "simple and natural" process unfolded in the West (ibid., 392). However, because the words of a polysllabic language are more complex than those of a monosyllabic one, the binding between a character's meaning and its vocalization may be looser. That, in Barrow's opinion, made the replacement of an original "imitative" script by an alphabetic one more likely.

47. Barrow 1804, 355.

48. Barrow's *Travels* does not appear among the books listed in Young's brief autobiographical sketch, written after 1826 (Hilts 1978, 256). However, Young likely did know that Barrow and Staunton were the authors of the review discussed above in the *Quarterly*, and in his 1823 *Account* designed to support his claims to priority in hieroglyphs, Young praised Barrow's "clear and concise explanation of the peculiar nature of the Chinese characters" (Young 1823b, 7). Consequently it's likely that Young was aware of what Barrow had written (Barrow 1804, 236). For Barrow as an exemplar of Europeans judging other cultures entirely on the basis of whether or not they had developed machines and practiced something like a science along European lines, see Adas 2014, 178–83.

49. Barrow 1804, 168.

CHAPTER 7. WORDS FROM EGYPT'S PAST?

1. Solé and Valbelle 1999, 2.
2. Anonymous 1799 [29 Fructidor, An VII]–Sept. 15. Quoted in Solé and Valbelle 1999, 8.
3. Ibid. The article is translated in Solé and Valbelle 1999, 7–8. According to the British Museum online, the Rosetta Stone is 112.3 centimeters (approximately 3.68 feet) long, 75.7 centimeters (2.48 feet) wide, and 28.4 centimeters (nearly 1 foot) thick, and weighs 760 kilos (1,675 pounds); https://research.britishmuseum.org/research/collection_online/collection_object_details.aspx?objectId=117631&partId=1.
4. Gillispie and Dewachter 1987, 21.
5. Reybaud et al. 1830, 36 and 432–33. Syriac derived from Aramaic and in its middle form had been broadly used for liturgical and literary purposes among Christians, which is perhaps why the stone's discoverers, having decided that the middle script must be alphabetic, initially associated it with Syriac.
6. Clarke 1814, 270. Clarke's *Travels* appeared in six volumes between 1810–23 and was reprinted many times thereafter. Clarke's account of his meetings with Menou appeared originally in the second volume (1812) of the first edition. His account also appeared in several British publications before 1820, including Anonymous 1814, 623–24.
7. Clarke 1814, 372.
8. Ibid., 367–68.
9. Turner 1812, 213.
10. Ibid., 214. Turner's account raises several questions concerning just who was involved in the affair and when, and bears comparison with Clarke's, which was first published in his *Travels* at nearly the same time as Turner's in the *Archaeologia*. Turner, it seems, overplayed a number of points to his own advantage. For a detailed account of what likely transpired, see Downs 2008, 165–84.
11. Turner 1812, 212. The antiquities seized from the French and sent by Turner to England are listed in Walsh 1803, 136–37. The manifest was signed by both Fourier and Turner. The Rosetta Stone, item number 8 of 17, is described as "of black granite, with three inscriptions, hieroglyphic, Cophitic, and Greek." Walsh was a captain in the 93rd Foot Regiment and aide-de-camp to Major General Sir Eyre Coote. The extensive list of subscribers to Walsh's account included various luninaries and royalty—and Thomas Young.
12. Ray 2007, 36–37.
13. Turner 1812, 213.
14. Leitch 1855, 3:60.
15. Turner's letter was read to the Society of Antiquaries on June 8, 1810. Clarke's account was printed in 1812, as was volume 16 of the society's memoirs, the *Archaeologia*. Apart from the two we discuss, notices about the Rosetta Stone were rare and brief, for example, Clarke 1805, 24.
16. Anonymous 1802b, 3.
17. Anonymous 1801, 1194. Dugua apparently had two proofs, although the later French account refers to only one of them; one was given to Dugua by Marcel, whereas the other was a Conté etching—cf. Reybaud et al., 1830–36, 6:444. Ameilhon, who was among the first in France to comment on the Greek inscription, noted that Dugua had brought back "two copies," which he had given to the Institut (Ameilhon 1803, 1). De Sacy made clear in 1802 that one of the two that Dugua had was indeed the etching. De Sacy 1802, 7.
18. Gough (writing as "D.H.") 1802, 726. Gough was an expert in British antiquities, and while director of the society had pressed unsuccessfully for publications beyond the *Vetusta*, which makes his concern about the printing delay understandable. The next year he printed his own translation of the Greek text (1803).
19. Additional copies were sent to the Vatican, the Imperial Libraries at Vienna and St. Petersburg, the Berlin Academy, Cardinal Borgia in Rome, the Philosophical Society at Philadelphia, and the National Library in Paris, among others.
20. The scholar here is de Sacy. The full set of plates are in Anonymous 1803 (published 1815). At the turn of the century, the Egyptologist Wallace Budge asserted that all three inscriptions were printed

by the Society of Antiquaries and "issued to the public in three plates on July 8, 1802" (1904, 33). However, there is no evidence that anything beyond the Greek plate was issued at that time. The other two plates are dated April 23, 1803, nearly a year later.

21. Penn 1795 and 1802; Anonymous 1802b. Penn had previously written a pamphlet warning of the dangers arising to England from the dissemination of "French principles."

22. Evans 1956, 199, remarks that Weston "was perpetually producing scrappy papers in shaky writing on such subjects as 'Men of Great and Little Stature among the Ancients.'"

23. Heyne's Latin translation was printed in a Göttingen journal in 1804. Weston's English and Heyne's Latin translations of the Greek inscription were printed in the *Archaeologia* in 1812, together with Turner's reminiscences (Heyne 1812; Turner, ibid.; Weston, ibid.).

24. Plumptre 1802, 1106–8. Plumptre hastened to print because he feared that "a copy of it [his translation] has already got into hands which it was not designed for, and that too not by honourable means, or for very fair purposes."

25. Gough 1803.

26. In the 1812 *Archaeologia*, Matthew Raper remarked that both Weston and Porson had reconstructed the missing Greek, but that Porson's was the version printed (1812, 210).

27. The only significant *Vetusta* engravings of an ancient script before the Rosetta were printed in 1789. Delineated by Adam de Cardonell (d. 1820), but engraved by Basire, they portrayed the Ruthwell Cross with its runic inscriptions. These are plates 54 and 55 in vol. 2 of the *Vetusta*.

28. On the antiquarian aesthetic of the *Vetusta*, see Lolla 1999, 24.

29. On antiquarianism, see, for example, Ceserani 2013 and Schnapp 1997. Cruikshank's print is from *Scourge, or, Monthly Expositor of Imposture and Folly* (June 1812), 431.

30. Savoy 2009, 808, is a useful account of the early reception in Paris.

31. Anonymous 1800, 254–56. Millin became a center of diffusion for knowledge about the scripts. Between 1800 and 1810, more than three hundred pages concerning the Rosetta inscriptions were printed in his *Magasin*. Savoy 2009, 809–13.

32. Jollois and Lefèvre-Pontalis 1904, 25–26.

33. Denon 1802a.

34. Jollois and Lefèvre-Pontalis 1904, 26–27.

35. See Gillispie and Dewachter 1987, 23–29.

36. On Jomard's editorial direction of the *Description*, see Laissus 2004, 98–106.

37. The publication history of the first edition of the *Description* is complex. See Gillispie and Dewachter 1987, and especially Bednarski 2005, on the *Description*'s comparatively small impact in Great Britain.

38. The plates are nos. 52, 53, and 54 in Jomard 1822.

39. Ameilhon 1803.

40. De Sacy received Marcel's lithographs from Chaptal, the minister in charge, who must have obtained them from Dugua. He also had a print from the Conté etching, the latter of which is what Visconti most likely had since there were no other copies of the lithographs except one possessed by Marcel himself.

41. Savoy 2009, 813; cf. Winckler 1802, 66–67.

42. This, the *Vetusta* engraving, had been sent to the National Library in Paris shortly after its printing on July 8.

43. Ameilhon 1803, 116–17. There is some confusion as to what precisely Ameilhon received. Millin, who favorably reviewed Ameilhon's Greek, asserted that Ameilhon's English facsimile, which he found inadequate, was an earlier version, and that in the meantime the Society of Antiquaries had printed an improved version. Although Millin claimed that the inadequate print that Ameilhon saw was taken "from the stone itself," there's no indication of any other Greek plate than the one printed on July 8, and the problems that Ameilhon pointed out are visible on that plate.

44. Ibid., 117. The extant English engraving (the one printed on July 8, 1802) unambiguously has the words ΑΣΠΙΔΟΕΡΔΩΝ, ΤΡΙΑΝΑΔΑ, whereas Ameilhon had read ΑΣΠΙΔΟΕΙΔΩΝ, ΤΡΙΑΚΑΔΑ, because his lithograph did not clearly distinguish a P from I in the first word, or the N from a K in the second, and because only the second readings made any sense at all. On Ameilhon, see also Ray 2007, 38–40.

45. Anonymous 1802b, 572.
46. Millin 1802b, 504. Savoy sees the allusion to Visconti as an implicit critique that he was monopoliz-ing the inscription (2009); Visconti seems never to have done anything with it. For more on the hapless Visconti, whose hopes to oversee the Musée Napoleon, with its collection of purloined art and artifacts, were dashed when Napoleon chose Denon for the post, see Buchwald and Josefowicz 2010.
47. Millin had reprinted Penn's transcription in his *Magasin Encyclopédique*; Millin 1802b, 504–12.

CHAPTER 8. THE SOUNDS OF AN ANCIENT LANGUAGE

1. Original translation (here altered) in Solé and Valbelle 2002, 10, from Marcel 1799, 293–94.
2. Herodotus 1958, book 2, chapter 36, 2:106–7.
3. Emphasis ours; Weston 1812, 224. Plumptre translated the passage as "in hieroglyphics, the vulgar tongue (*the Coptic*), and in Greek characters" (1802, 1108). The parenthetical was Plumptre's addition.
4. Young's transliteration brought "enchorial" into English.
5. Kircher's first work, the *Prodromus Aegypticus* (1636), provided a Coptic grammar and dictionary based on Pietro della Valle's Coptic-Arabic dictionary. A second work, his *Lingua Aegyptiaca Resti-tuta* (1644), provided the first argument intended to establish Coptic as a modern, degraded form of the ancient language. Here, Kircher claimed that the ancient city of Coptus was home to a special monastic tradition that preserved the ancient language, which was everywhere else altered by Arabic at the time of the Arab conquest. The *Lingua Aegyptica Restituta* presented della Valle's dictionary directly, along with ten additional original essays and a bibliography of manuscript sources. A third Egyptological treatise, the *Obeliscus Pamphilius* (1650), inventoried the obelisks of Rome; it also developed the claim that hieroglyphs were purely symbolic, and that knowledge of them was limited to an exclusive priesthood. Finally, his *Oedipus Aegyptiacus* (1652–54) offered an imaginative deci-pherment of hieroglyphs on the Bembine tablet. Kircher also first advanced the view that Chinese was a descendant of Egyptian hieroglyphics, a position that made China chronologically younger than Egypt, and by doing so allowed Kircher to insert himself in one of the major controversies of the time, the dating of the Flood.
6. La Croze and Scholtz, 1775. Students relied on Kircher's flawed Coptic dictionary "often to their consternation" during these years, although some apparently had access to La Croze 's version in manuscript (Stolzenberg 2013, 228; for details, see Hamilton 2006). La Croze used Coptic texts provided to him by David Wilkins to produce his Coptic dictionary. Though not published at the time, the dictionary was copied by Carl Gottfried Woide, known in England as Charles Godfrey Woide, in 1750. Later Woide, who was well known in London by Joseph Banks's circle, also worked an improved version of La Croze's dictionary with patronage from George III, which was the one published in 1775 and used by Young.
7. Stolzenberg 2013, 233.
8. Rigord 1704, 985–90; cf. Pope 1999, 43–44.
9. Several of Rigord's and Montfaucon's plates are reprinted in Pope 1999, 44–45; cf. the multivolume Montfaucon 1722–24.
10. Bandini 1750, 16. Bandini wrote in Italian with facing Latin.
11. Ibid., 17. Because hieroglyphs were not intended to represent the "the sounds comprising the voice," Bandini held that "such figures were not signs of simple sounds" but of "joined voices [*voci insieme*] of things," that is, spoken words (ibid., 18–19).
12. This interest extended to his own practice, and he became a knowledgeable craftsman in his own right. His skillful engravings distinguished his *Receuil* from such comparable works as the multivol-ume compendium of images issued by Montfaucon. See Callatay 2011, 1999. The craft methods of the past were of particular interest to Caylus, who wrote a noted memoir on encaustic painting, a technique mentioned by Pliny the Elder. "Caylus," notes Alain Schnapp, "put the greatest importance on observation, on placing the object within the process of its manufacture" (1997, 242). Notably, the philosophes despised his erudition; Caylus reciprocated their disdain. After Caylus was cremated,

Denis Diderot proposed the following epigraph: "Here lies a short and cantankerous antiquarian; how nicely this Etruscan urn houses him" (cited and translated in Callatay 2011, 1999). On Caylus, see the nineteenth-century biography by Rocheblave (1889) and the more recent one by Rees (2006).

13. Caylus 1752–67, 5:79 (1762).

14. Ibid.

15. Ibid., 69–71 (1752).

16. Ibid., 5:78 (1762).

17. Ibid., 74 (1752).

18. Ibid., 72–74 (1752). Like others in France, Caylus entertained the view that both the "vulgar" and "sacerdotal" scripts were likely alphabetic, and perhaps even the original source of alphabetic scripts transmitted to the Phoenicians.

19. On Barthélemy, see the early *éloge* by Sainte-Croix 1798 and the introduction to his collected works by Villenave 1821, vol. 1, followed in the same volume by Barthélemy's reminiscences. On Barthélemy and the Palmyrene script, see Pope 1999, 95–97, and Daniels 1988, 430–36.

20. The reading of Palmyrene was accomplished nearly simultaneously by Barthélemy and John Swinton of Christ Church, Oxford. Barthélemy's paper was read to the Académie des inscriptions on February 12, 1754; Swinton's was read to the Royal Society in London on June 20 of that year. Barthélemy's was not printed until five years later (1759), while Swinton's was soon available in the *Philosophical Transactions* (1753–54). Barthélemy's work on Phoenician was printed six years after it was read at the Académie as Barthélemy (1764).

21. Barthélemy 1797, 390.

22. J-N [Jourdain] 1857.

23. Michaud 1843, 18:126–27.

24. Barthélemy 1797, 398. On de Guignes's imaginative—and, for a time, influential—excursion from Egypt to China, see Allen 1960, 536–40.

25. Napoleon's soldiers first called them "cartouches" for their resemblance to cartridge casings, and the name has stuck. Zoëga 1797, 465: "On the Egyptian monuments are observed here and there some figures, egg-shaped or elliptical, sitting on a flat base [or pedestal], which emphatically include certain arrangements of signs, *either to express the proper names of persons or to designate sacred formulas*" [Conspiciuntur autem passim in Aegyptiis monumentis schemata quaedam ovate sive elliptica planae basi insidentia, quae emphatica ratione includunt certa notarum syntagmata, *sive ad propria personarum nomina exprimenda sive ad sacratiores formulas designandas*]; emphasis added. On Zoëga, see Jørgensen 1881; the brief remarks in Allen 1960, 546–47; Curran, Grafton, Long, and Weiss 2009, 222–27; and Buzi 2015. Zoëga was trained at Göttingen by Heyne. We thank Mac Pigman for the translation.

26. Including, at the age of twenty-three, a set of notes on a Syriac version of Kings 4, said to be a third-century translation by Origen from the Septuagint.

27. For near contemporary accounts of de Sacy, see Daunou 1839 and Reinaud 1838. Edward Said identified de Sacy as the founder of orientalism and characterized him as having generated that entire field of study from textual "fragments," an approach that, Said claimed, was inherently expressive of European arrogance (2004, 128). Subsequent scholarship has identified weaknesses in Said's characterization and has somewhat rehabilitated de Sacy's reputation as a scholar of Arabic. Although Irwin 2006 is flawed by polemicizing, see the useful discussion on 141–50, and see, most recently, Espagne, Lafi, and Rabault-Feuerhahn 2014.

28. Marcel apparently had a third, even better copy, but it was destroyed in the fire that consumed his possessions, including his antiquities collection, in Egypt.

29. De Sacy 1802, 3.

30. Like Rousseau, Gébelin believed the first object of imitation was always nature. "Man invents nothing," Gébelin opined, "but he imitates and perfects. From these faculties writing was born." The connection between Rousseau and Gébelin was drawn in LeGros 1785.

31. Gébelin 1776, 107–12.

32. De Sacy 1799, v–vi.

33. De Sacy 1802, 8.

34. He left no instructions for making such comparisons, but he seems to have followed a procedure of this sort: Place one point of a compass at the beginning of the Greek inscription and the other at the beginning of a line that contains a particular word, say, the proper name Alexander or Alexandria, which is repeated in the Greek. Without changing the position of the compass ends, use the compass to mark points on a ruled paper. Do the same to mark off the distance from the beginning of the line containing the word to the word itself. Finally, use the compass to mark a line that measures the entire height of the Greek inscription and another that measures the entire length of the line containing the word. Measure off these lengths to find the proportions of line position to inscription height and of word position to line length. Next use the compass to measure the height of the "Egyptian" text and mark the resulting distance on the ruled paper. From that determine a length that has the same proportion to this height that the position of the Greek line had to the height of its text and then use the compass to find the corresponding line in the Egyptian. Do the same for the position in that line of the Egyptian word that should correspond to the Greek.

35. De Sacy to Chaptal: "[Since Coptic] is incontestably formed from the debris of ancient Egyptian, I could reasonably hope to push my discovery further, and to recompose at least in part this Egyptian inscription" (1802, 4).

36. The third line of signs appears in the second line in the *Description*'s plate of the Rosetta Egyptian script.

37. Ibid., 12–15.

38. As noted in Solé and Valbelle 2002, 46–49; cf. de Sacy 1802, 41–42.

39. Zoëga 1810. On Zoëga, see Picchi 2010.

40. The details of Åkerblad's involvement with the Rosetta inscription are carefully discussed in Thomasson 2013 on which we have relied. See also Solé and Valbelle 2002, 49–52.

41. Åkerblad 1802a and de Sacy 1803. Citing Åkerblad in support, de Sacy applauded Barthélemy's effort while denigrating Swinton's.

42. Swinton 1764, 133.

43. Ibid., 141; Pococke 1745, 213; Swinton 1764, 414; Åkerblad 1802a.

44. Barthélemy 1764, 423; the translation of Åkerblad's Latin was printed in Gough 1803, 28, which also printed the English version of Swinton's Latin (27–28).

45. One English reviewer noted as much: Anonymous 1804, 543.

46. Quoted and translated in Thomasson 2013, 230–31, who notes that de Sacy's subsequent treatment of Åkerblad makes his words here less than convincing.

47. Ibid., 233, referring to MS N 71 (Åkerblad's Coptic dictionary) held at the National Library in Stockholm.

48. Åkerblad 1802b, 40.

49. Ibid., 14–16. The following sign is the *A*, which could also be found in the analog for the name *Arsinoë*. The second is an *L*, which he found in the analog for *Ptolemy*. The third is a *K*, but it is placed *above* a sign equivalent to the Coptic *shei* (ϣ) that represents the analog of Greek Ξ, which must not otherwise have existed in Egyptian. The fifth letter, again an *A*, is also placed above another, this latter signifying an *N*.

50. Ibid., 4.

51. As reproduced in Anonymous 1802a, 553, where Young may have first seen Åkerblad's alphabet.

52. Åkerblad 1802b, 69.

53. Åkerblad asserted that, in light of his understanding of the Egyptian, the Greek inscription might be missing final words, and he offered a tentative reconstruction (1802b, 62–63). Ameilhon thought this particular attempt was unjustified by any supporting indication in the Greek text, in fact, quite the contrary (1803, 120). Specifically, Åkerblad proposed that a missing segment on the last line of the Greek should be reconstituted as "in each of these temples of the first, second, and third order in which the statue of the king would be erected." Interestingly, a recent rendition of the last line of the demotic reads "in the first-class temples, the second-class temples and the third-class temples, next to the statue of the King, living forever," which is not altogether far from Åkerblad's version (Simpson 1996, 271).

54. Millin 1802b. This was volume 30.

55. Villoisin 1803, 84, followed by a supplement on 378–79. Of course the subject of the report was a particular Greek passage and not Åkerblad's accomplishments with the Egyptian. On Villoisin and Åkerblad, see Thomasson 2013, especially 239–40.
56. Gough (writing as "D.H.") 1802, 725.
57. Gough (writing as "D.H.") 1803 and Anonymous 1802b.
58. Millin 1802a, 142.
59. Consulted by his friend the duke of Richmond concerning a possible visit of his two grand-nephews to France to learn the language, Young offered to accompany the party to Rouen to assist since the tutor engaged did not know French. Placing the children with a French family, he then visited Paris (Peacock 1855b, 211).

CHAPTER 9. PARIS ATMOSPHERES

1. Kettel 1990 provides a detailed list of Champollion's earliest works.
2. Hartleben 1906, 37–44.
3. Hartleben 1983. For the description of the diligence, see Planta 1827, 36–37.
4. For details about Champollion's early life, see Hartleben 1906.
5. The work was published in précis in Champollion 1811a, followed by a full treatment in Champollion 1814.
6. Rostaing 1945.
7. The street still provides a shortcut between rue de Rivoli and avenue de l'Opéra. On the persistence of medieval elements of Parisian streets, see Papayanis 2004.
8. Hillairet 1953, 69.
9. Ibid. and de Gaulle 1839, 576.
10. Jones 2005.
11. Pugin 1831, 2.
12. Examples abounded, from street names that referred to Egyptian cities to architectural elements inspired by Denon's drawings of Egypt to Egyptian-themed productions in the popular theaters, not to mention the obelisks that punctuated the streets of central Paris. See Carrott 1978, Pao 1998, and Buchwald and Josefowicz 2010, 222–67.
13. On Langlès as the "Tatar," see Hartleben 1983. For Champollion's Parisian curriculum, see ibid., which has substantially provided the basis for all subsequent accounts.
14. "One wonders who could possibly live here and what happens after dark," Balzac wrote with gloomy, noirish excitement, "when the alley fills with cut-throats and the vices of Paris, under cover of night, have free rein" (1976–81, 5–451); quoted material appears on 99–100. Although published at midcentury, the novel featured characters who remembered and continually referred to life in Paris under Napoleon. Balzac's descriptions of Napoleonic Paris are echoed by contemporary travel guides and accounts, including Champollion's. See Planta 1827, 124–25. Planta began to publish his popular guidebook in 1814. On Planta's and similar contemporary guidebooks, see Clark 1997, 8–29.
15. Lacouture 1988, 115.
16. The quoted material is from a letter from Champollion to his brother reproduced fragmentarily in La Brière 1897, 50, and reprinted in full in Lacouture 1988, 117. The emphasis is Champollion's in the manuscript, according to Lacouture.
17. Madame Vauquer and her establishment are described in Balzac 1976–81, 52–54.
18. All these quotes are reproduced in Lacouture 1988, 115–16, and scattered as well throughout the letters from Paris quoted by La Brière 1897.
19. On the unfolding of this demographic shift in nineteenth-century France, see Coller 2005, 206–14; Coller 2006, 433–56; and Coller 2011. We are indebted to Coller for much of what we know about this aspect of Champollion's student years.
20. Coller 2006, 433–56.
21. Lacouture 1988, 114.
22. Tadié 1995, 65.

23. On Chiftichi, see Louca 2011 and Louca 1988, 209–25. On Dom Raphaël, see Tadié 1995, 61–65. On Champollion's Coptic studies, see Bourguet 1982, 62–75.

24. Although the walkway still exists, the bell tower does not. Having fallen into ruin, it was dismantled in 1875, and the bells vouchsafed to Sacré-Coeur. See Hillairet 1953, 45.

25. Champollion-Figeac 1880–89, 2:390.

26. Ibid., 402.

27. Ibid., 395.

28. Champollion describes Dom Raphaël's affectionate attitude in ibid., 395. On Champollion's difficult father, see Hartleben 1909.

29. Louca 2011.

30. Tadié 1995, 62.

31. Sacy 1802, 4.

32. Champollion 1811a.

33. Herder 1772. We owe the useful phrase "central linguistic dogma" to Turner 2015, 127.

34. Turner 2015, 135–36; Humboldt 1823.

35. Turner 2015, 129. Hamilton was first cousin to the American founder of the same name.

36. Ibid., 130–31. The project was taken up, refined, and extended in the next decade by others, notably Franz Bopp.

37. The parallel here with Chiftichi and Dom Raphaël, also speakers of an ancient, liturgical, and (presumed to be) dead language, is worth pointing out, though it is hard to know how far to press the comparison.

38. Pollock 2009, 938–39.

39. Jones claimed that Sanskrit shared a common ancestor with Latin, Greek, and Celtic; on this point, and Arzu, see Vasunia 2013, 17. On the "classicizing" of Sanskrit, see Trautmann 1997. The selective uptake of Sanskrit also shows in the elements that were left wholly to one side. For instance, unlike the Europeans, who were ever more interested in the historical dimensions of language, Panini proposed an entirely synchronic system, considered linguistic development as "foreign to the tradition," and treated unexpected or incongruous elements as variations rather than novelties. See Kiparsky 1993, 2918–23.

40. Quoted in Hartleben 1909. According to Hartleben, Champollion was skeptical of the comparative philology developing in Paris around Hamilton and his patron, Langlès (99), but this skepticism did not prevent him from making his own investigations along these lines (126). Citations for the 1808 and 1813 essays are given in Kettel 1990, 2 and 5. Hartleben alludes to Champollion's early comparative work on ancient Egyptian and Indian philosophy and religion (126) but does not provide a citation.

41. On the movement of library materials under Napoleon, see Cuccia 2013, 66–74, and Maras 1994, 567–78. On the Vatican Archives, see Blouin 1998. For catalogs of Coptic materials still held at the Vatican, see Hebbelnyk and Lantschoot 1937 and Lantschoot 1947.

42. On Colbert and the acquisition of manuscripts from Egypt, see Thompson 2015, 67–69. On Vansleb, see Hamilton 2006, 142–51; Hamilton 2009, 142–51; Delahaye 2009; and Martin 1997. The *Bulletin* in which Martin 1997 appears was formerly the *Bulletin Monumental*.

43. A list of Coptic sources available at the Bibliothèque nationale through the end of the nineteenth century appears in Amelineau 1888–89. For an overview of the collection, see Chabot 1906, 5–21.

44. The library fire is mentioned in Silvestre, Champollion-Figeac, and Champollion 1849, 169; apparently manuscripts were still held at the library at the time of the fire. For more about this remarkable library, see the relevant section in Buringh 2010.

45. Hamilton 1993, 225–49.

46. Although knowledge of Coptic was rare in Europe before the seventeenth century, Fabri de Peiresc and Pietro della Valle played major roles in focusing attention on the language in the early modern period. See Allen 1960, 527–47.

47. Ibid.

48. Bunsen 1848, 259–69. Bunsen's account of the late eighteenth-century European reception of Coptic manuscript sources draws on Quatremère 1808.

49. Buzi 2015, 216–24.

50. Champollion 1814.

51. Champollion made special mention of Marcel's reports in Champollion 1811a, 18.
52. Ibid., 22.
53. Åkerblad 1802b, 19.
54. Åkerblad 1801, 490–94; 93.
55. Ibid., 492.
56. Åkerblad 1802b, 19–21; 21.
57. Thomasson 2013, 254–55.
58. Åkerblad 1801, 494.
59. Champollion 1811b, 31. This article was also published as an offprint.
60. Ibid., 34.
61. Champollion 1811a, 3.The first italicized phrase is our emphasis; the second one is Champollion's. This point was repeated verbatim in the 1814 publication.
62. Saint-Martin 1810–11, 79 recto; emphasis his. This undated letter appears to have been written during the late spring or summer of 1811, ahead of the July 30, 1811, publication of Saint-Martin's review of Champollion's "Introduction" in the *Moniteur Universel* (1811, 807–8).
63. Champollion 1814, 1:xiv.
64. Arnauld and Nicole 1662, 95–97.
65. Dumarsais 1730, 26–30, subsequently reissued as *Traité des tropes*.
66. Turgot 1751–72, 6:98–111; 99.
67. Ibid., 100.
68. Ibid., 104.
69. Ibid., 108.
70. Bergier 1837 [1764], 14–15.
71. De Brosses, IX (1800), viii.
72. Gébelin 1776, 26.
73. Ibid., 40.
74. Ibid., 41–43.
75. Barthélemy 1768, 212. The report was published in 1768 in the society's *Mémoires*.
76. Rocher 1968. The connections between the philologists of eighteenth-century India and the West, mediated by representatives of colonial institutions such as the East India Company, have recently attracted attention from Anglophone scholars, though much remains to be uncovered. See Pollock 2009, 938.
77. Kiparsky 1993, 2922.
78. One of several Sanskrit riddle hymns appearing in the *Rig Veda* nicely illustrates the governance of semantics by syntax in Sanskrit. In this riddle hymn, speech appears as a divine force, "the perfected speech in the heart of the poet," and is figured as a formidable bull-like creature: "Four horns, three feet, has he; two heads and seven hands has he. Bound threefold, the bull bellows. The great god has entered mortals" (verse 4.58.3; O'Flaherty 1981, 126–27). According to traditional interpretations of the riddle, the four horns represent the four parts of speech in Sanskrit; the three feet refer to the three main tenses, past, present, and future; the two heads refer to the conventional and etymological meaning of the word; the seven hands refer to the seven grammatical cases; the thrice-bound bull refers to the three resonant places of the body, the chest, neck, and head; and the "great god" is the word, which moves through the person, shaping the person's form and gesture according to the specific context of utterance. This interpretation of the riddle is traditional according to Moebus 2011, 476–77.
79. De Sacy 1836, 146–55; 8. The review appeared in the March issue.
80. Kircher's contributions to early modern scholarship have lately been reconsidered and his reputation somewhat restored. See Stolzenberg 2013.
81. The letter is reprinted in Vaillant 1984, 17.
82. Ibid., 18; our emphasis.
83. The chair was itself a style of furnishing that had been revived under the Directory.
84. Hartleben 1983.
85. Ibid.
86. The quoted material appears in Lacouture 1988, 117.

CHAPTER 10. ROOTED IN PLACE

1. For our descriptions of the Dauphiné landscape, the Lyon coach, and the children of the Place Grenette, we are indebted to the classic account by Durrell 1969, 378–88. The quoted line about the Alps and the city's sightlines appears on 379.

2. For a careful analysis of the question of the elder brother's influence on the younger, see Champollion-Figeac 1887, 78–109.

3. His precise words were "que moi c'est toi." Ibid., 90.

4. The story is related in Robb 2007, 203–4.

5. Of these ideas, the classic treatment remains Rostaing 1945.

6. "The names attached to localities tend to be extremely persistent and to resist replacement even when the language spoken in the area is itself replaced. This resistance to replacement is particularly marked in the case of important topographical features such as large rivers and mountains." Bynon 1977, 273–74.

7. The quote is taken from the first page of the essay "The Longevity of Toponyms," which appears in Sobin 1999, 112–20.

8. Champollion 1811a, 24–30.

9. Ibid., 24–25.

10. Champollion 1814, 1:xiv, note 3.

11. Sanskrit was believed to possess the same eternal quality. See Turner 2015, 129, and Alger 1885–1900, 134.

12. Champollion 1811a, 24.

13. Champollion 1814, 23.

14. Following conventions of his time, Champollion recognized three Coptic dialects: Memphitic (later known as Bohairic), Theban (later known as Sahidic), and Bashmuric (later known as Fayyumic). The sequence of development of these dialects was a matter of some interest, since it was believed that the oldest dialect would retain the clearest and most reliable links to ancient Egyptian. Champollion knew of, and commented on, recent work on the dialects of Coptic, especially that of A. A. Georgi, who, in his preface to *De Miraculis Sancti Coluthi* (1793), claimed that the most ancient Coptic dialect was Bohairic/Memphitic.

15. Jomard 1809a.

16. Champollion 1815, 7.

17. Champollion 1811–12b, verso 62, recto 63.

18. *Diligere*'s object is an estimable one. As a British Latin teacher of the next generation explained, "*Amare* denotes the affection, whether instinctive, as that of parents to offspring, or excited by a rational conviction of the amiableness of the object; *Diligere*, love towards an object, as being preferable to others." Crombie 1838, 2:34–35.

19. First scrap: In the original: "Ce que nous appellons les racines ne sont autre chose de mots inventès ~~arbitrairement und d'apres~~ [illegible] pour ~~exprimer~~ representer nos ideas & nos pensées." Champollion 1811–12a, 31.

20. Second scrap: In the original: "Nous appellerons Racine le mot representatif d'une idée [ou] abstraction faite de tout rapport avec les personnes ou les choses." Ibid., 31.

21. Third scrap: In the original: "Cette exposition [illegible] ajouter aux Racines, leur donne la valeur de l'Infinitif Latin." Ibid., 36–37.

22. Throughout his *Pharaons* work, he did note and analyze the presence of particles and other small grammatical units that could be mistaken for monosyllables representing roots.

23. Champollion 1814, xv.

24. Champollion referred to the Denon manuscript as "Ms 138" (Vaillant 1984, 8). Denon had received this manuscript from "Amelin"—probably Ameilhon—and had published it as plate 138 in volume 2 of his *Voyage*. He described it in detail in Denon 1802a, 265–66, vol. 3.

25. Vaillant 1984, 13.

CHAPTER 11. *HIER POUR DEMAIN*

1. Germaine de Staël's *De L'Allemagne*, written in 1810 but not published until 1813, is usually cited as the first account of this phenomenon. See also Leventhal 1986. Other general overviews include Turner 2015, 127–36, and Irwin 2006, 150–57.
2. Hartleben 1906, 69.
3. Champollion-Figeac 1810, 12–13.
4. Sweet 1978, 238.
5. Bolter 1980, 84.
6. The *Magasin* was founded in 1795 (Revolutionary Year 3) and ceased publication in 1816.
7. Espagne 2005.
8. Hartleben 1906, 82.
9. Kotzebue 1804, 310.
10. Ibid., 311. On the Wednesday salons, see the relevant articles in Espagne 2005.
11. Champollion-Figeac 1880–89, 2:393; Faure 2004, 118.
12. Parsis-Barubé, 2011, 21.
13. Posner 2000, 536–37.
14. We have taken this phrase, and the title of this chapter, from *Hier pour demain: Arts, traditions et patrimoine, catalogue d'exposition du Grand Palais* (1980), published to commemorate an exhibition exploring the study of folklore and popular culture in France.
15. Gerson 2003, 33.
16. Pomian 1987, 288, and Parsis-Barubé 2011, 32.
17. Gerson 2003, 1–19.
18. Leersen 2004, 111.
19. On the transmission of ancient objects to museums, where those objects were recontextualized as patrimony rather than aestheticized, see Blix 2009, 60–71.
20. The Académie celtique published the *Mémoires de l'Académie celtique* until the society dissolved in 1812; it was reconstituted in 1814 as the Société des antiquaires de France. Senn 1981, 24.
21. Ibid.
22. Guiomar 1992, 63–85; 64.
23. Belmont 1980, 54–60; 55.
24. Ibid., 57.
25. Anonymous 1807, 62–63.
26. Ibid., 383; quoted in the original in Belmont 1980, 57–58.
27. On Lenoir and his milieu, see Spieth 2007, 62–110.
28. Hartleben 1983. On Cambry, see Guiomar 1992.
29. Hartleben 1983; Champollion-Figeac 1880–89, 2:396.
30. Champollion-Figeac 1880–89, 2:396.
31. Ibid., 2:397–98.
32. Grafton, Most, and Zetzel 1985, 99.
33. Sweet 1978, 209.
34. Moore 2008, 28–32. Moore's thesis develops earlier work by Bercé and Furet on the founding of the École des chartes in relation to the Charter of 1814 and the (ultimately fruitless) efforts undertaken to expand Louis XVIII's constitutional monarchy to a truly parliamentary government, and it departs substantially from earlier studies (e.g., by Mathiez) that have viewed the school's mission as aligned with ultra-Royalist ideologies of the Bourbon Restoration, a perspective that remains close to the time of the school's official founding (1821) but does not account for aspects both of the school and the context of its founding, including Degérando's long-term interest in the school and the influential position held there by Champollion-Figeac, who was hardly an Ultra.
35. Champollion-Figeac 1807, 43.
36. Champollion-Figeac had apparently even written an essay, now lost, on the toponyms of the Dauphiné as evidence of Celtic settlement (Faure 2004, 289).
37. Champollion-Figeac 1807, 122.

38. Ibid., 65–66.

39. Champollion-Figeac 1880–89, 2:290. The letter was written in response to Fourier's of 20 Vendémiaire Year 12.

40. Ibid., 2:390.

41. Ibid., 2:392.

42. Champollion-Figeac 1807, 8.

43. Letter reprinted in Champollion-Figeac 1880–89, 2:290.

44. Champollion-Figeac 1804.

45. Champollion-Figeac 1814, 2.

46. Ovand was martyred in the sixth century toward the end of the reign of Clovis.

47. Champollion-Figeac 1804, 3–4.

48. Ibid., 5.

49. Ibid., 7.

50. Ibid., 9.

51. Ibid., 15. Compare with a letter he received from Eloi Johanneau on 22 Prairial: "L'etymologie du mot *Drac*, dont vous parlez . . . éclairera beacoup mes idées. Les mots *Drac*, riviere, le Drac, nom du diable, dans les Languedoc, viennent du breton *Deve* or *Drove*, le malin, le mauvais, de là le latin Draco et le français drague, draguer, etc. Ce mot existe aussi en gallois et dans quantité de noms ou monuments celtiques, attribués au diable par nos ancêtres, par opposition à d'autres monuments attribués aux fées ou *fades* des Bretons, *fad*, les bonnes" (emphases in the original). The letter is printed in Champollion-Figeac 1880–89, 1:68–69. The idiom "mettrer Grenoblo un savon" is hard to translate but, like the familiar sense of dressing-down, conveyed by phrases like *passer un savon* and *prendre un savon*, it may suggest that the rivers are forever teaching the city's inhabitants a harsh lesson.

52. Clockwise from left: Champollion 1816–19, 2; Braun and Hogenberg 1581–88, vol. 3; Boisseau 1648.

53. The doodle may be found in Champollion 1816–19, 244.

CHAPTER 12. *L'AFFAIRE POLYCARPE*

1. Champollion's biographers offer broadly similar accounts of this conflict. See Lacouture 1988; Champollion-Figeac 1880–89, 2:406–7; Hartleben 1906; Lacouture 1988, 157–60; Chassagnard 2001; and Faure 2004, 155–58. "L'affaire Polycarpe" appears to be Lacouture's phrase, from Lacouture 1988, 158.

2. On the circumstances leading to the execution of Quatremère's father, see Leclerc 1909, 299–319. On the Quatremère family, see Saint-Hilaire 1861, 4, and Sadjedi 1983.

3. Sadjedi 1983, 19–32. Toward the end of 1809, Quatremère was called as professor of Greek to the university at Rouen, where he remained for a little more than a year and a half before abruptly returning to Paris in August 1811.

4. Champollion-Figeac 1880–89, 2:406.

5. Backus 1997, 1:xxii. Historically minded theologians and patristic scholars who privileged documentary evidence over traditional forms of religious authority threatened the power of the church.

6. Champollion-Figeac 1880–89, 2:406.

7. Hartleben 1906.

8. Guillaume 1891–1958, 259.

9. Millin 1796, 45–46.

10. Cointreau 1803, 22.

11. Quatremère 1808. The announcement of the work in progress appears in the preface on p. x.

12. Hartleben 1983; Lacouture 1988, 158.

13. Quatremère 1808, 25.

14. In May 1811, de Sacy reported that the manuscript had arrived at the Institut and that he'd read it to the Institut's membership in September 1810 (1811, 196).

15. Quoted in Thomasson 2013, 256, who translates from Hartleben, German edition, 1:149. The order of words here is rearranged. Hartleben does not give the date.

16. On Millin's travels, see d'Achille 2011.

17. The phrase appears in a letter to Saint-Martin, February 20, 1811; see Saint-Martin 1810–11, 83 recto. The phrase also appears in Lacouture 1988, 158.

18. "He has no idea": Letter to Saint-Martin, February 20, 1811; Saint-Martin 1810–11, 83 recto.

19. Champollion-Figeac 1880–89, 2:406.

20. Faure 2004, 155. On the search for a typeface, see Champollion-Figeac 1880–89, 1:406. Champollion jeered at Peyronard in a letter to Champollion-Figeac dated October 4, 1811; quoted in Faure 2004, 115.

21. Kettel 1990, 3–4. Kettel lists three 1811 publications, on Thebes, Memphis, and "The Nile and Its Egyptian Names," all of which correspond to extensive, similarly titled sections in the 1814 *Pharaons* volume. According to Hartleben, Champollion read an essay on the names of the Nile to the Académie at Grenoble on August 13, 1811, but there is no mention of publication (1906). In his recent biography of Åkerblad, Thomasson claims that Champollion in 1811 published only two of these articles, those on Thebes and Memphis. See Thomasson 2013, 256–57; 7n21.

22. Jomard's letter of June 1, 1811, to Champollion-Figeac is quoted in Lacouture 1988, 158.

23. De Sacy 1811, 201.

24. Ibid., 200. By 1812, Quatremère had seen at least Zoëga's catalog of the manuscripts, if not the manuscripts themselves; he wrote approvingly of Zoëga's work as "a treasure of erudition, and the most important work that has yet appeared on the Egyptian language" (1812, 2).

25. Champollion to Saint-Martin, January 3, 1811, and Saint-Martin 1810–11, 81 recto, verso. De Sacy finally published Åkerblad's piece in 1834. On de Sacy's admission, decades later, that he had failed to make the manuscript available to others immediately, see Thomasson 2013, 257.

26. Champollion 1811a, 3.

27. De Sacy 1811, 198. In their popular history of the decipherment, Adkins and Adkins write that "a whispering campaign" ensued, in which Champollion was accused of plagiarism, which would be most interesting if correct. No source is given, however, and we have been unable to verify it (2000, 98).

28. Hartleben 1983.

29. Letter from January 3, 1811, Saint-Martin 1810–11, 81 recto. Although the letter is addressed merely to "Msr.," de Sacy may be inferred as the addressee because the first paragraph thanks the addressee for his review of Champollion's *Introduction*, and—apart from a review by Saint-Martin, discussed below, which repeats much of this memo verbatim—de Sacy was the only reviewer of this work. It is not clear how Saint-Martin obtained this letter. We agree with Lacouture and Faure that it was likely written in the weeks after de Sacy's review, that is, late May or early June 1811. See Lacouture 1988, 159, and Faure 2004, 156. Hartleben does not give a date but notes that the letter alludes to Champollion's recent recovery from a serious illness, the details of which are not specified; the letter is partly reproduced in Champollion-Figeac 1880–89, 4:407.

30. The letter is preserved in Saint-Martin 1810–11, 79 recto.

31. Ibid., 79 verso. According to Adkins and Adkins, de Sacy responded indignantly: "I have not even thought of suspecting M. Champollion of plagiarism." Quoted in Adkins and Adkins 2000, 98. We have been unable to locate this letter.

32. Saint-Martin 1811, 807–8; emphasis ours. The quoted material appears on 808.

33. Ibid.

34. Faure 2004, 156.

35. Quatremère 1812, 3.

36. Ibid., 4.

37. Ibid., 8.

38. For instance, he made much of the manner in which mountains were named by Herodotus, Champollion, and Zoëga, focusing on the importance of a prefix letter in differentiating between a mountain's name and the adjectival form of it (ibid., 8–10). He also claimed that the locations of mountains in Upper Egypt were not difficult to determine and noted that Champollion had nevertheless omitted many of them from his account (17).

39. Quatremère 1808, 26–28.

40. Quatremère 1812, 3.

41. Significantly, Quatremère's insistence on the absence of cultural remnants in Egyptian toponyms allowed him to avoid consideration of the religious life of ancient Egypt, for example, its zoomorphism, which he found distasteful. Moreover, by failing to investigate Coptic etymologies, he was able to avoid any appearance of distancing himself from or otherwise undermining de Sacy's commitments to Arabic source material.

CHAPTER 13. AN EGYPTIAN GEOGRAPHY OF EGYPT

1. The same root, Champollion continued ominously, "appears as well in ⲥⲁⲛⲕⲣⲟϥ, *sanchrof* [the power to do] evil."
2. Champollion 1814, 1:165; our emphasis.
3. For an example of Quatremère's skepticism, see Quatremère 1808, 25–26.
4. Champollion 1814, 1:138.
5. Ibid., 1:340–41.
6. Ibid., 216.
7. Ibid., 216–17; emphasis his. Unless otherwise noted, capitalization, underlining, and italics are reproduced as they appear in the original sources.
8. Ibid., 217.
9. Puns and other forms of wordplay abound in ancient Egyptian sources; Loprieno 2000, 3–20.
10. These towns were particularly interesting because, though very far to the south, they were not frontier towns like Philae. As sturdily Egyptian settlements rather than disputed border territories, their toponyms were less exposed to outside influence than others.
11. Champollion 1814, 1:172.
12. Ibid., 171.
13. Ibid., 169–71.
14. Ibid., 171.
15. On Panopolis, see ibid., 261–63; on Hermopolis-Magna, see ibid., 291–92.
16. Ibid., 263.
17. Champollion 1808–10, 288–89. Although the precise date of this material is uncertain, it is surely from his early career and likely precedes the 1814 publication of *Pharaons*. He was then working intensively with Coptic materials, and the notes include a great deal about Zoëga on whom Champollion relied heavily in *Pharaons*. The peculiar, uneven quality of these notes, boldly assertive but at times quite derivative, suggests a mind at work on material not yet fully understood, and contrasts markedly with the striking confidence of Champollion's later publications. Finally, the file includes an official notice dated May 23, 1810, of an event to be held at the Grenoble Academy, which suggests that the surrounding material may be of the same period.
18. Champollion 1808–10, 288.
19. Ibid.
20. Ibid.
21. Ibid., 288 verso.
22. Ibid., 288–89.
23. The lotus had long been considered a symbol of Horus's power of ongoingness, of "continuance of life in the world to come," while Horus, emerging from the flower's cup, was expressing his basic nature as the first principle of generativity. See Spence 1990 [1915], 299; cf. Gardiner 1964 [1961], 39.
24. Champollion 1814, 1:261–62.
25. See the first chapters of Rosanvallon 1985 and of Moore 2008.
26. Chassagnard 2001, 90.
27. Lacouture 1988, 164.
28. Champollion-Figeac 1887, 85, and Chassagnard 2001, 9.
29. On the history of the struggle to publish *L'Égypte sous les Pharaons*, see Champollion-Figeac 1887, 82–84.

CHAPTER 14. INDICATIONS

1. Leersen 2004, 111. Viccy Coltman has observed that the publications of antiquarians—the lavishly illustrated folio editions and subscription-only catalogs of antiquities seen in other books and on the grand tour—were also important objects of delectation and not merely proxies for the objects described between their covers. Coltman 2006, chapter 2.
2. Stendhal's tongue-in-cheek attribution of "hic et nunc" to Grenoble's governing body suggests as much. Interestingly, the notion of a concretizing literary language emerged at about this time. Works in this mode were embedded in the biographical circumstances of their authors. Stendhal registered this shift in his *Life of Henri Brulard*, a novel written in 1835–36 (but not published until 1890) based on the author's experiences at the turn of the century in Grenoble. Stendhal not only referred repeatedly to Laclos's notorious roman à clef, *Liaisons dangereuses*, as a touchstone for writing that sticks close to reality, but he also pointed directly to that reality by naming the actual Grenoblois women who, according to Stendhal, provided the inspiration for the schemers of Laclos's novel. These observations are drawn from remarks in Levin 1963, especially 24–31 and 84–149.
3. Buchwald and Josefowicz 2010.
4. Champollion-Figeac 1806, 8.
5. Hartleben 1983, 121. A facsimile of a letter in which Champollion recounts this discovery was published in an appendix to Kettel 1996.
6. Hartleben 1983; Buchwald and Josefowicz 2010, 312–13.
7. Quoted in Hartleben 1983, 121.
8. Original: "Tu vas la clouer, la crucifier" (viii). On the context for *Néologie* and Mercier's motives for writing it, see Rosenberg 2003.
9. Gattel 1803.
10. Paquet 1973, 35. A follower of Condillac, Gattel believed that alphabetical characters originated in Egypt. It is unlikely that Champollion agreed with Gattel on this point.
11. Robb 2007, 62.
12. Ibid., 62–63.
13. Champollion-Figeac 1809, 2–3.
14. Quoted in Hartleben 1906, 178.
15. Faure 2004, 180.
16. Ibid., 178.
17. Ibid., 177.
18. Leitch 1855, 62–64. Champollion wrote that he worked from the engraving "that is to form part of the third delivery of the *Description de l'Égypte*." Although the plate for the Egyptian script that eventually appeared in volume 5 was printed around 1812, the plate for the hieroglyphs had to await Jomard's return from London the following March 1815 after he had made his own copy directly from the artifact.
19. A long-standing member of the Antiquaries and a trustee of the British Museum, Banks had considerable antiquarian and Egyptological interests. In 1804 he had even written an unpublished piece on an ibis mummy. On Banks's antiquarianism at this time, see Gascoigne 1994, 125–29.
20. Champollion 1811–12b, 62–63.
21. Champollion-Figeac 1844, 216–17.
22. A draft of this proposal, preserved in the Champollion archives in Paris, provides the basis for our discussion here. Champollion 1815, 286–90.
23. Ibid., 289 recto, verso.
24. For a survey of the history of Coptic dictionaries and grammars in Europe from Kircher to Champollion, see Hamilton 2006, 229–49.
25. Champollion 1815, 289 recto.
26. Ibid.
27. Ibid., 288 recto.
28. Champollion 1809–12, 299 recto.

29. Hartleben 1983, 145, and Faure 2004, 249; an offhand remark by Aimé Champollion-Figeac, that the dictionary was eventually "redone in the ordinary format" (1887, 86), suggests that Champollion's original organization, by roots, remained untenable even after the decipherment.

30. De Sacy's response seems intended to head off claims that hieroglyphs might themselves be phonetic. Given this report from de Sacy, it's at least conceivable that Champollion's eventual, complete abandonment of the alphabet, which occurred when he became convinced of the identity of all three scripts, may have reflected not only his sense of the broader import of de Sacy's remark but also, perhaps, how to circumvent it.

31. Champollion-Figeac 1887, 87.

32. Vaillant 1984, 86–87.

33. Alexander 1991, 17.

34. Both quotations are from Lacouture 1988, 178. The capitalization is Champollion's.

35. Champollion-Figeac 1887, 88.

36. "C'est un nouveau malheur." This letter of July 21, 1815, appears in ibid.

37. Lacouture 1988, 180.

38. Alexander 1991, 240.

39. Faure 2004, 146; Alexander 1991, 240; Lacouture 1988, 181–87.

40. Alexander 1991, 240.

41. Champollion-Figeac 1887, 88.

42. Ibid.

CHAPTER 15. SUMMER AT WORTHING

1. Leitch 1855, 264.

2. See chapter 7n12. Despite Young's disavowal here, it is difficult to say precisely when Young became interested in the Egyptian scripts. Given his early exposures, it seems prudent to take Young's later recollection with a grain of salt. Robinson addresses Young's claim from a slightly different point of view (2012, 85).

3. Early editions were Denon 1802b. The plate in question is no. 137. "Manifestly compendious imitations": Boughton 1817 (1814), 60.

4. Leitch 1855, 264, in Young 1823b, xiv–xv.

5. Boughton 1817 (1814), 64.

6. Peacock 1855b, 214.

7. Ibid., 252. This refusal notwithstanding, Young would eventually contribute a spectrum of articles to the *Britannica*. Though he had been engaged in scientific matters for some time—Young had, of course, been the society's foreign secretary since 1802—Napier's request for a series of articles apparently went too far.

8. Young's reluctance to comment publicly on Boughton's papyri is otherwise odd since he had published on the Herculaneum papyri in 1810, and this would have been a single, short publication. Perhaps not coincidentally, Boughton was elected to the Royal Society that same May at the age of twenty-six.

9. Boughton 1817 (1814), 64. The full volume (18) of the *Archaeologia* that contained these efforts did not circulate until 1817, though offprints were available earlier, likely as early as the spring of 1815.

10. The May 1816 *Museum Criticum* version (no. 6), with letters to de Sacy and from Åkerblad and signed "A.B.C.D.," included Porson's modification of a translation of the Greek contained in a book on the Seleucid kings (Gough 1803). It was reprinted over a decade later in a collection of *Criticum* articles (Young 1816a). Young wrote Gurney in January 1816 that he had sent the manuscript for printing at Cambridge the previous August (1815) but that it was still not ready (Peacock 1855b, 277). The *Criticum* was edited at Trinity College, Cambridge, by James Henry Monk, who followed Porson as regius professor, and by Charles Blomfield, whom we encountered as one of the reviewers of the *Herculanensia*. The *Criticum* was published from 1813 to 1826; articles from the first four issues were published in a separate volume in 1814. In 1826, when the *Criticum* ceased publication, Monk and

Blomfield published a two-volume set of *Criticum* articles, reprinting the 1814 collection in the first volume and articles from issues 5 through 8 in the second (1826). On the dating of the issues and volumes, and on the *Criticum* in general, see Stray 2004.

11. Signed "I.J.," Young's article, completed in the spring of 1818, ran to thirty-seven pages in small print with five detailed plates.

12. Young 1813a, 290.

13. Young to de Sacy, October 3, 1814; translated from the French in Leitch 1855, 18; original in BL Add 21026, 13.

14. Boughton 1817 (1814), 60.

15. Ibid., 62.

16. Although Zoëga saw hieroglyphs as principally allegorical, in an unpublished manuscript he nevertheless developed an extensive and carefully organized list of signs to which he assigned meanings, occasionally via similar-looking signs in Coptic. See Frandsen 2015.

17. Compare with Horapollo 1993.

18. Young 1814b, 2.

19. A Prussian, Wilkins had moved to England and became a prelate in the Church of England. Wilkins's Coptic version of the New Testament, the first such to be produced in Europe, was based on a manuscript held at the Bodleian, together with manuscripts at the Vatican and in Paris; Wilkins 1716.

20. Young 1814b, 2. Although the *Vetusta* volume had not yet been assembled, the plates had been printed years before and would have been made available to Young by the Society of Antiquaries.

21. Ibid., 2–3.

22. Ibid., 3.

23. The derivatives embrace king, kingship, queen, and reign. Ameilhon had published the Greek in lowercase words in 1803 (27–107), together with extensive discussions and followed by a French translation. Nine years later, Raper's publication of the early Weston (English, 1803) and Heyne (Latin, 1802) translations included a lowercase line-by-line version that incorporated Porson's suggestions for the missing portions (1812, 215–19).

24. Young 1814b, 3.

25. Ibid., 11. Four years later, in his article on Egypt, Young essentially reiterated this point. His process, he remarked, did not differ altogether from de Sacy's or Åkerblad's, though he had relied entirely on frequency counts and not on any link to Coptic; Young 1819b, 53.

26. Ibid. "The second inscription," Young wrote in his 1819 *Britannica* article, "will be safest to call by the Greek name *enchorial*, signifying merely the 'characters of the country.'"

27. De Sacy 1802, plate 2, no. 5; Young 1814b, 92.

28. In the Greek inscription, a Ptolemy variant directly follows a variant for βασιλειας in the following lines: 9 (twice), 37, 38, 41 (twice), and 49. In line 8, a Ptolemy variant is separated from a variant for βασιλειας by one word.

29. Young 1814b, 3, verso.

30. The Greek phrase in question is αιωγοβιου ηγαπηυενου υπο του φθα, which first occurs in line 4 of the Greek and whose corresponding initial instance Young found in line 2 of the Egyptian. The phrase repeats in line 8, with the next corresponding instance in Egyptian line 5, where Young also attended to the succeeding phrase (namely, θεον επιφανουσ ευχαπιστον), which also followed elsewhere.

31. Young 1814b, 6, verso.

32. Ibid., 5 and 8.

33. In a supplemental letter to de Sacy sent October 21, 1814, Young gave an extensive list of Egyptian sign sequences with what he thought to be the corresponding words in Coptic, Greek, and Latin—not because he had himself worked with Coptic analogs, but because both de Sacy and especially Åkerblad had. Leitch 1855, 21–29.

34. Young 1814b, 82.

35. Ibid., 57–58.

36. Simpson 1996, 243. Budge rendered the phrase as "lord of the diadems" (1904, 2:100).

37. Young 1814b, 6 and 57.
38. Young explained his reasoning on this point in a letter to de Sacy dated October 3, 1814 (Leitch 1855, 18–21).
39. Peacock 1855b, 261–62.
40. Young 1814b, 100.
41. Boughton 1817 (1814), 70–72. The hieroglyphic "interpretation" was certainly not ready by the late spring of 1815, though the Egyptian "translation" likely was, at least in major part, as we shall see in chapter 16.
42. Peacock 1855a, 100; vol. 1 from "An Essay of Cycloids" printed in 1800. This should not be taken too far. Young does employ analysis extensively in many of his articles. He preferred, however, to use Newtonian fluxional rather than Leibnizian methods, and he was wont to complain about analyses he considered overly elaborate.
43. Young 1819b, 223. The quote appears in his *Britannica* article, "Languages."
44. Young 1823b, 6–7.
45. From Young's "Languages," *Britannica Supplement* (1824; see Leitch 1855, 554).

CHAPTER 16. LETTERS FROM PARIS

1. Contemporary biographies of de Sacy include Daunou 1839, Derenbourg 1905, and Reinaud 1838, the last translated in Reinaud 1842.
2. Portions of Young's correspondence concerning Egyptian matters were printed in Leitch 1855; several other letters, particularly to Hudson Gurney, were included in part by Peacock 1855b. Many letters printed by Leitch are extant in the British Library collections. Although he transcribed the letters with little editorializing, in his notes Leitch attempted to rescue Young's priority from what Leitch perceived to be Champollion's depredations. Young included some of these letters in Young 1816a.
3. On which, see Mansel 2003.
4. Young's letter and de Sacy's reply are in Leitch 1855, 16–18, with the original of de Sacy's reply in Young 1814–29, 11–12.
5. Leitch 1855, 30–35, and Young 1814–29, 1–10. Only Åkerblad's second letter is extant in the original.
6. Leitch 1855, 18–21; 18.
7. Ibid. and Young 1814–29, 27–28 of PDF.
8. Leitch 1855, 21–29. The original, written as usual in French, is not in the British Library collection; the extant version is Young's translation of his own letter as printed originally in the *Museum Criticum*.
9. It is hard to know what to make of this apparently offhand remark. Perhaps Young wished to remind Champollion of all he still held in common with the antiquarians who, as we have seen, were perceived as irrelevant if not ridiculous, in contrast to London's scientists who were doing so much celebrated work. The rest of his letter suggests that Young suspected he was dealing not with a naïf or a pretender but with a potentially serious adversary. Ibid., 64; Young 1814–29, 16. Young's reply to Champollion is undated both in his *Miscellaneous Works* and on the draft he wrote directly on Champollion's letter. It was certainly written after he received his first letter from Åkerblad because the latter was dated December 15, and Young's reply to Champollion mentioned results presented by Åkerblad in that letter. One of Champollion's biographers dates Young's reply to March 15, 1815, but does not provide evidence for the dating (see note 11 below), whereas his first major biographer dates it to March 10, also without providing the source (Hartleben 1983, 139). In any case Young's reply was certainly not written before March.
10. Young did not send Champollion a copy of his Latin translation. The *Archaeologia* piece (Boughton 1817 [1814]) that contained Young's English translation likely became available in February or March. At which point, Young sent this material to de Sacy and mentioned having done so in his reply to Champollion.
11. Leitch 1855, 64. Lacouture dates the letter to March 15, 1815—scarcely two weeks after Napoleon's escape from Elba—and has interpreted this remark and the absence of any mention of political

events as a cutting and typically dismissive British response (1988, 180). Possibly so, but all that Young seems to be saying is that anyone—and not just Young himself—would determine both the *Vetusta* and the Marcel lithographic reproductions to be "sufficiently exact." Pace Leitch, the failure to mention political events might just as likely be evidence of polite reticence on Young's part to raise additional—potentially divisive—issues in a scholarly correspondence in which Young already signaled a disagreement, implicitly at least, by noting that he had not relied at all on Coptic for his results.

12. Champollion 1811–12b, 58; Young 1814–29, 16; Young 1814b, 5.

13. See, in particular, Champollion 1811–12b, 63v.

14. Not long after receiving the *Pharaons* from Gurney in late fall 1814, Young told him that he had, at that point, "only spent literally five minutes in looking over Champollion . . . He follows Åkerblad blindly with scarcely any acknowledgment. But he certainly has picked out the sense of a few passages in the inscription by means of Åkerblad's investigations—although in four or five Coptic words which he pretends to have found in it, he is wrong in all but one—and that is a very short and a very obvious one" (Peacock 1855b, 264).

15. Champollion 1814, 47 and 23, respectively.

16. In the draft reply to Champollion (see fig. 16.1, *top*), Young identified several signs with "Vulcan" (Ηφαιστος), which occurs in line 2 of the Rosetta Greek. His identification of the latter had evolved since the previous summer, for at that time Young had isolated only a portion of the corresponding signs.

17. He was living at 73, rue de Lille. See Faure 2004, 231–32.

18. In August de Sacy was appointed as one of the five members of Louix XVIII's newly constituted, and powerful, Commission of Public Instruction.

19. "I have the honour to transmit to you, Sir, a copy of my conjectural translation of the Egyptian Inscription of Rosetta" (Leitch 1855, 18, and Young 1814–29, 13).

20. The copy that Young must have transcribed for de Sacy remains in the manuscripts held at the British Library. Young 1814b.

21. Leitch 1855, 49; the original is Young 1814–29, 24.

22. The Latin translation that Young included in his October 3 letter may have been accompanied by a discussion of this method. De Sacy asked about it, but until his article "Egypt" in the *Britannica* Young had not elsewhere published how he had gone about his work. Neither a copy of the Latin translation nor any detailed remarks about Young's method are present in the MS copy of the letter. Although Leitch's translated version concludes in a way that militates against further remarks having been present in the original, it is possible that Young included such remarks in his follow-up letter of October 21, but the latter no longer exists in MS form at the British Library. Perhaps there was a third communication from Young to de Sacy that no longer exists, but this seems unlikely given that the next known response from de Sacy, dated July 20, 1815, begins by referring to the "Latin translation" that Young had sent, that is, the one enclosed with Young's letter of October 3.

23. "[T]here are only a very few cases, in which I have been able to find similar words in Coptic, at all capable of representing the sense of the Inscription" (Leitch 1855, 21).

24. Young 1814–15, 11; 1814b, 11.

25. "[M]ost of the proper [Greco-Roman] names seemed to exhibit a tolerable agreement with the forms of letters indicated by Mr Åkerblad; a coincidence, indeed, which might be found in the Chinese, or in any other character not alphabetical, if they employed words of the simplest sounds for writing compound proper names" (Young 1819b, 54).

26. Leitch 1855, 49–52.

27. The entire first Greek line on the Stone reads: "Βασιλενονος του νεου και μαραλαβοντος την βασιλειαν παρα του ματρος κυριου βαγιλειων μεγαλοδοξου του την Αιγυπτον καταγτνσαμενον και τα προς τους." Gough's English translation "as amended by Porson," gives: "In the reign of the young prince, who received the kingdom from his father, Lord of 'kings,' highly glorious, who settled the affairs of Egypt." Gough's translation was printed by Edward Clarke (1809, 58–65). Ameilhon's 1803 Latin of the first Greek line begins "Regnante (rege) juvene et successor patris in regnum" (While the young man and successor of his father to the kingdom was ruling) (1803, 11), while Heyne's of

1811 begins "Regnante novo, et regnum a patre acceptum tenente" (While a new man, one holding the kingdom received from his father) (Raper 1812, 231). The sixth line of the Greek ends with
"Ξανδικου, τετραδι Αιγυπτιων δε Μεξειτ οκτωκαιδεκατη." The most recent English version of the
demotic continues to include Young's addition, still in brackets "Year 9, Xandikos day 4" (Parkinson
1999, 198; translation by R. S. Simpson). Champollion would later differ with Young concerning the
"decimo octavo" date, despite its appearance in the sixth line of the Greek. We thank Mac Pigman
for the translations.

28. Leitch 1855, 50; Young 1814–29, 24; Ameilhon 1803, 38, for the sixth line.
29. Leitch 1855, 52.
30. "I doubt whether the alphabet which Mr. Åkerblad has given us can be of much further utility than
in enabling us to decipher the proper names," and "the inscription contains at least a hundred different characters, which it is impossible to explain by means of [Åkerblad's] alphabet, ingenious as
it is, at least without long and laborious study" (ibid., 16 and 20).
31. Ibid., 52; dated July 20, 1815.
32. Ibid., 51.
33. In his letter to Young on May 19, 1815, Champollion wrote that he had not as yet seen "your memoir
on the Egyptian part and the hieroglyphic text of the Rosetta inscription" and asked that a copy be
sent to his brother in Paris (Leitch 1855, 66; Young 1814–29, 22v). In the event, de Sacy lent his copy
of the *Archaeologia* piece to Champollion, as we see from his July 20 letter to Young (Leitch 1855, 49).
In it de Sacy remarked that Champollion-Figeac had asked him to convey the piece to his brother
"according to a letter he received from you." Champollion returned the copy shortly before de Sacy
wrote. Six months later de Sacy decided to wash his hands altogether of Champollion, principally
because of political actions (ibid., 59).
34. The original, which was in French, is not extant in the Young MS at the British Library, but the
translation is in Leitch 1855, 53–56.
35. Young 1815.
36. Young 1816a, containing Young's edited versions of letters through August 1815 together with the
English "translation" of the Egyptian script that had also just been printed, with the Boughton letter,
in the *Archaeologia*.
37. Leitch 1855, 53.
38. Ibid., 34; the original French is in Young 1814–29, 1–10. The version printed by Leitch was taken
directly from Young's own "translation" by way of an "extract"—in fact, an extensive recension, as
we shall see—of the letter from Åkerblad that was printed in Young 1816a, 180–93. Until August 1815,
a year later, Young had written Åkerblad only the one (now missing) letter on August 21, at which
point he may not have had his Latin version. He had included enough in this first letter to give a few
of his results, but, to judge from Åkerblad's response months later, not much more than that.
39. ΗΕΛΙΟΣ occurs three times in the Greek inscription: once at the very end of its second line and
twice toward the last part of its third.
40. Leitch 1855, 53; the letter from Young to de Sacy is dated August 3, 1815.
41. From Young's notebook (1814b, sequentially 97, 57, 82, etc.).
42. Young 1819b, 38.
43. Leitch 1855, 54; emphasis added.
44. Ibid., 53; emphasis added.
45. Åkerblad to Young on January 31, 1815: "I received, about two months ago, the letter which you
did me the honour to write on the 21st of August" (ibid., 30), which includes Young's translation of
Åkerblad's French as published in Young 1816b, 180–93. The original in Young 1814–29, 3–13, now
lacks Åkerblad's Coptic version that Young included in print.
46. Leitch 1855, 3–44; Young 1814–29, 1–10. That Åkerblad had the list is evident from the original letters,
in which he refers to a word "N. 46." Åkerblad's long reply is dated January 31, 1815.
47. Young did not include the Egyptian sequences in the highly compressed extract of the letter he
published in the *Criticum*. The manuscript of the letter in the British Library lacks Åkerblad's French
and Coptic translations, which were certainly on separate sheets, but it does contain his commentary
with the several Egyptian sequences that he had included (Young 1814–29, 3–13 of PDF).

48. Leitch 1855, 44.
49. Young's reply, ibid., 44–49. Young simply altered what was (according to his reply) Åkerblad's original for the date in the first line, giving "the eighteenth of the Egyptian month Mechir" (Young 1814b, 96–97).
50. Leitch 1855, 44; emphasis added.
51. Young 1816c.
52. Ibid., 104–5.

CHAPTER 17. THE PAPYRI OF THE *DESCRIPTION DE L'ÉGYPTE*

1. At this time, the *Criticum* was still unfinished, so Jomard cannot be referring to it. De Sacy may have lent him Young's *Archaeologia* piece, for in his April 29 letter Jomard remarked that Young had "himself" noted that the question of the hieroglyphs was still very much alive even "after so much effort and research done without foundation" (Leitch 1855, 67; Young 1814–29, 20–21).
2. Young had twice gone to see Jomard but, as Jomard explained in his letter (whose first paragraph was not printed in the *Miscellaneous Works*), "a servant who had not fulfilled his duty" was to blame for the missed connection. Young had apparently left letters at Jomard's residence to be delivered to Laplace and to Champollion, which Jomard claimed to have done; the Champollion letter was no doubt the one that Young had written in his capacity as secretary of the Royal Society (Leitch 1855, 64–65; the original is not in the British Library, and Leitch's reproduction is undated and incomplete). Jomard might have visited Young, for he had obtained Young's address from Young's friend William Wollaston, but he had received "letters that obliged [him] to leave," perhaps due to Napoleon's return from Elba. As for his own letter to Young, Jomard did not specify what accompanied it, but, given his statement that it is "nearly entirely descriptive" to permit "judging the resources we possess," it was likely a draft of Jomard 1818, which concerns numerical signs and hieroglyphs. A more speculative version printed separately the following year contains references to Young, likely the result of their correspondence (Jomard 1819). The latter was read at both the Académie des inscriptions and at the Académie des sciences, respectively, on September 3 and 6, 1819.
3. Jomard 1809b.
4. Jomard took this position despite the extensive, and contrary, discussion of the point by Caylus, whom Jomard did not mention.
5. "Take care nevertheless not to conclude that the two classes of letters have shapes that are entirely different and that have no relation to one another" (Jomard 1809b, 371–72).
6. Ibid.
7. The papyrus in question was printed as plates 60–65 of vol. 2 of the *Description* (1812).
8. Godlewska 1999, 138. On Jomard's attachment to measurement, see Godlewska 1999, 132–37.
9. Jomard 1818.
10. Leitch 1855, 67–68.
11. Young did produce such an index, likely begun before he learned about Jomard's prospective table; Young 1814–15.
12. Scott 1816, 198. On Restoration culture and politics, see Mansel 2003.
13. The latter had in fact been found by Philippe-Joseph-Marie Caristie, an engineer (École polytechnique 1795) on the expedition.
14. Leitch 1855, 69–70, does not include this full remark. The original is in Young 1814–29, 26.
15. Leitch 1855, 70.
16. Ibid., 71; Young 1814–29, 18–19. Jomard's catalog of hieroglyphs would eventually be included in the third and final delivery of the *Description*: Jomard 1822, plates 50 and 51, respectively.
17. These months had indeed been busy. De Sacy had joined the Commission of Public Instruction, renamed the Royal Council of Public Instruction after Napoleon's expulsion to St. Helena. The organization's financial arrangements had fallen to de Sacy, who proved a reluctant accountant.
18. Leitch 1855, 56–59; Young 1807, 27–28.

19. Young 1816b, 194.
20. Leitch 1855, 59–60.
21. Weston 1812, 60.
22. Young 1842b, 57, 59, 60, 100, 82, 93, 92, and 98. He translated the word as "conspicuous" or "illustrious." Gough had left it as "Epiphanes."
23. The Rosetta plates could not be bought until the fifth volume (1822), though the two for the Greek and the Egyptian scripts had been printed—and circulated to several scholars and libraries—years before. The hieroglyph plate meanwhile awaited Jomard's return from England with his cast of the stone.
24. Jomard 1814.
25. The 1810 volumes could be bought in "thin paper" for either 750 or 800 francs, with the less expensive containing four color plates and the other sixteen. The vellum cost 1,200 francs. A few sets of the latter cost an extra 150 francs because their color plates had been retouched by hand. The second set cost either 1,200 or 1,800 francs. Champollion, we shall see, would a few years later be paid a (modest) salary of 750 francs per year in Grenoble, so clearly the *Description* was out of reach for all but the wealthy.
26. The 1810 set also included a volume containing Fourier's preface, together with an explanation of the antiquities plates, a volume of memoirs on ancient Egypt, and plates and text concerning Egypt's current state and natural history.
27. These plates came in several sizes, all rather large. Most plates had a standard height of 27.3 inches; all were available in three widths (21, 42, and 52.5 inches). The largest plates measured 44.1 by 31.5 inches.
28. Bednarski 2005, 24–33, provides these details.
29. Hamilton may have bought the *Description* at Payne's and lent it to Young; the price was high but by 1816 Hamilton would certainly have had the wherewithal to purchase a set. Hamilton's account of his Egyptian sojourn of 1801–2 appeared in Hamilton 1809; cf. Peacock 1855b, 269, note d. On Hamilton, see Downs 2008.
30. Plates 74, 62, and 71 in vol. 2 from the *Description de l'Égypte*; Young 1816a, 2, 35, and 38.
31. To de Sacy on August 15, 1815; Leitch 1855, 54.
32. In the case of "Epiphanes"—Young's "illustrious"—the additional enchorial sign can be seen in figure 17.1 as what looks like a "2" with a pair of underscores.
33. Young had concluded that the "cursive" script must be the sacerdotal (hieratic) of Clement of Alexandria.
34. Leitch 1855, 76–78.
35. Young remained skeptical that all of Åkerblad's assignments could be generally correct, though as we have seen, he did himself add to them. Young explained his skepticism to Åkerblad in a letter written on August 12 (1816), which was subsequently included in the supplement to the original *Criticum* piece.
36. Young 1819b, 62.
37. Ibid.
38. The hieroglyph sequence that Young read as Berenice appears twice in plate 50, in volume 3 of the *Description de l'Égypte*.
39. Young 1814b, 8 and 92.
40. Young 1819b, 62–63.
41. Ibid., 62. The remark Young quotes is a translation of the maxim, "Du sublime au ridicule, il n'y a qu'un pas," which appeared with some frequency in the British periodical literature of Young's day and was usually attributed to Napoleon. Readers could have seen it used this way in, for instance, Anonymous 1815, 526, a critical review of a self-excusing account by the archbisop of Mechlin, Dufour de Pradt, of his behavior as Napoleon's ambassador to Warsaw in 1812.
42. Young 1819b, 54; Leitch 1855, 135. Young's "Egypt" was printed in 1819, in the fourth volume of the *Supplement* to the *Encyclopaedia Britannica*. Volumes of the *Supplement* were issued at intervals over many years to supplement the *Britannica*'s fourth edition (1810). The final volume, printed in 1824, covered the fifth (1817) and sixth (1823) editions of the *Britannica* as well.

43. Young 1819b, 71. Young's attempted identification of a sign sequence for "Arsinoe" (based on an inscription at the temple of Kom Ombo found by William Hamilton) later proved to be the word for "autocrator," as Champollion would remark. For "Cleopatrides" Young provided no details, beyond writing that, at the Kom Ombo temple, there was a Greek inscription that mentioned both Ptolemy and Cleopatra (ibid., 63). Young's sequence does not, in the event, signify "Cleopatrides."

44. Ibid., 60.

45. Leitch 1855, 79; our emphasis.

46. Idolatry was especially repugnant to Quakers.

47. Peacock 1855b, 467; italics in the original.

48. Gange 2013, 55.

49. Turner's was hardly the only exemplar. The popular painter and illustrator John Martin, for instance, took up similar themes, including an 1823 rendition of the seventh plague. Of the large literature on biblical themes and typology in visual art of this period, we have found Landow 1980 and Omer 1996 particularly useful.

50. Anonymous 1821, 269, cited in Gange 2013, 76.

51. On which see Buchwald and Josefowicz 2010. Both Young, and especially Champollion, had their say on the issue, to which we will return below.

52. Halma 1822, 12. This was a common, and antique, calumny. In the images to which Halma referred there is no evidence whatsoever for human sacrifice.

53. See, for example, Curran, Grafton, Long, and Weiss 2009, and the classic Iversen 1993.

54. De Sacy 1808, 7–8.

CHAPTER 18. SEEKING UXELLODUNUM

1. Faure 2004, 265.

2. The remark is from Calmon et al., *Histoire de Figeac* (1998), quoted in Chassagnard 2001, 104.

3. Faure 2004, 271.

4. Hartleben 1983, 150; Faure 2004, 275.

5. Chassagnard 2001, 99.

6. Ibid., 102.

7. Ibid., 98.

8. Cabrerets, Lot département, quoted in Robb 2007, 81.

9. Letter to Albert Stapfer, September 2, 1837, reprinted in Chambon 1907, 85–86. Interestingly, Mérimée was a cousin of the physicist Augustin-Jean Fresnel, with whom Young famously tussled for priority regarding the discovery of the wave theory of light. On the Young-Fresnel rivalry, see Buchwald 1989.

10. Chassagnard 2001, 109.

11. On the fate of this material, see Hartleben 1983, 159; Faure 2004, 304; Lacouture 1988, 216–17; and Madrigal 2016, 97. The shipment contained several copies of the Rosetta inscription, including a large copy with numbered lines, bundles of papers, and a notebook of demotic inscriptions.

12. The essay is described in Faure 2004, 279. The quoted material is Champollion, from Faure (our translation).

13. Scores of fédérés were rounded up and massacred during the upheavals of 1815; a full account of the slaughter, according to one historian, "would require several dismal volumes" (Alexander 1991, 224).

14. We have used Jean Kimber's translation of "La Vénus d'Ille" published in Mérimée 1995, 61–90, with minor alterations (noted below) based on the French text appearing in Mérimée 1842, 171–208; the quote here is taken from p. 66 of Kimber's translation.

15. Mérimée 1995, 65–66.

16. Ibid., 74, with minor alterations; see Mérimée 1842, 187–88.

17. Mérimée 1995, 75.

18. In a curious instance of life imitating art, Chaudruc de Crazannes did report the discovery of a Gauloise iron statuette to Mérimée, but this article appeared in 1847 in the *Revue Archéologique*, a decade after the first publication of "The Venus of Ille." See Chaudruc de Crazannes 1847, 809–12.

19. Champollion-Figeac 1820, 90n1.

20. Letter from Champollion to Champollion-Figeac, August 27, 1817, reprinted in Madrigal 2016, 97–99; 8.

21. All quotations are from a letter from Champollion to Champollion-Figeac of August 23, 1817; the letter is reprinted in Madrigal 2016, 97–99.

22. Merimée 1838, referring to Roman antiquities throughout the Auvergne.

23. Faure 2004, 288; Chassagnard 2001, 373n6. A letter from one Delbrel of August 1, 1810, to Champollion-Figeac conveyed the excitement surrounding even the most provincial archaeological discoveries. During the excavation of a nearby quarry, workers discovered a cache of ancient objects. "Among the items recovered was a miniature metal statue that made a local splash. Immediately I thought that this ancient piece might be worthy of a place in your collection" (quoted in Madrigal 2016, 38).

24. Carbonell 1984, 134.

25. Faure 2004, 288.

26. "Old carcasses"; see ibid. On the discovery, see Hartleben 1983, 154–55.

27. On Chaudruc de Crazannes, see Chassagnard 2001, 373n6. Among the other amateurs involved with the project was Causse, the mayor of Capdenac, and the engineer and inventor Louis Vicat.

28. Faure 2004, 288–89.

29. The project was of immense interest to Caesar who, at this time, was competing with his rival Pompey, then conquering territories in the east in a war against Mithridates. Our knowledge of Hirtius comes mainly by way of Cicero, who did not especially admire him. An equestrian of means and one of Caesar's consuls, Hirtius is remembered as "a notorious epicure [who] was also a fluent and reasonably painstaking writer" and who by 45 BCE had become a praetor and the governor of transalpine Gaul. See Daly 1951, 113–17.

30. For the main events of the siege of Uxellodunum, we rely on the summary in Rivet 1971.

31. Champollion-Figeac 1820, 12–13.

32. Ibid.; cf. also Faure 2004, 288–90.

33. Champollion-Figeac 1820, 56.

34. After dismissing claims for Puy d'Issolu, Luzech, and Cahors, Champollion-Figeac turned to Capdenac, which shared a border with Figeac and was located a few miles from the town. Here he made a curious geological observation. Noting that the summit of the proposed site at Capdenac was "located a hundred and twenty meters above the riverbed," he observed the speed of the river at the foot of this deep chasm as well as the steepness of the slope rising from the riverbank, though he did not explicitly link the river speed to the process of erosion. An isthmus separating the possible site of the town from the rest of the valley appeared to correspond to the single route between it and the fertile fields outside, as noted by Hirtius. A pile of large boulders at the foot of the cliff and an ancient spring completed the picture. Champollion-Figeac 1820, 66.

35. In the five-page summary of his spadework, Champollion-Figeac reported the discovery of "semi-Gothic debris," a coin from the time of Louis XIII, and sherds of glazed and painted earthenware. Digging deeper, he discovered Roman coins and pottery, including a large vase decorated with symbols and people in relief. Gallic remnants were also among the debris, along with a cache of human remains. Champollion-Figeac 1820, 70. Perhaps these were the "old carcasses" Champollion reported to Dufour. Faure 2004, 288.

36. Champollion-Figeac 1820, 72.

37. Ibid., 88–89.

38. Ibid., 104.

39. Ibid., 105–6. As for the true location of Uxellodunum, Champollion-Figeac's was hardly the last word. Some claim Puy d'Issolu as the location, though the site might just as easily be near the modern towns of Vayrac or Luzech; and Capdenac had its supporters until at least the 1970s. See Sors 1971 and "Uxellodunum" in Hammond and Scullard 1970, 1104.

40. "The true meaning of this word DUNUM, which enters into so many Celtic names, excited great debate in the Académie between Mr. Falconet & Mr. Fenel, on the one hand, & Mr. Fréret on the other." Falconet 1763, xviii.

41. Apart from the notice in the *Catalogue de la bibliothèque de feu*, we have been unable to find any record of Fenel's participation in the debate.

42. De Brosses, IX (1800), 107.

43. Ibid., 107–8.

44. Ibid., 108–11.

45. Champollion-Figeac noted that the word "dunum" or its derivatives appeared frequently "in so many names of people and places that were part of ancient Gaul, as well as in the British Isles and Italy." Fréret's contribution, as he saw it, was that he tried to show "this word has neither an imitative sense, nor an analogous one, apart from idioms derived from Celtic," a position that linked the Uxellodunum study to Champollion-Figeac's ongoing interest in Celtic France. Champollion-Figeac 1820, 112.

46. La Brière 1897, 34–35.

47. Jomard's commitment to effective public instruction of all children regardless of their background was an important plank in the liberal platform during and after the Hundred Days. Jomard had discovered the method of mutual instruction (*l'enseignement mutuel*), a form of peer-to-peer instruction also known as the Lancaster method, on his trip to England. This discovery spurred him to implement a program to improve public education for French schoolchildren. In October 1815, Jomard tapped Champollion-Figeac to develop these schools in Grenoble. Champollion-Figeac in turn recruited his brother to establish one such school in Figeac. Despite resistance from local authorities, Champollion opened his Figeac school to acclaim in July 1817. On Champollion's pedagogical activities, see Armand 1990, 25–36. Little is known about his views on the teaching of languages despite their obvious relevance to his ideas of language generally. Nevertheless Champollion did possess and likely also developed curricular materials related specifically to language instruction in the new schools (Hartleben 1983, 155–56). To our knowledge, these materials are not held in the Champollion archives at the Bibliothéque nationale; if they do indeed exist, it is possible that they were cached in the Champollion family archives at Grenoble, now undergoing digitization under the direction of Karine Madrigal.

48. Champollion requested a copy of Sicard's work in an undated letter, likely written in May 1817, to his brother, then in Grenoble, reprinted in Madrigal 2016, 64–66; the request appears on page 66. "Seul primitif" appears in Sicard 1808, 1:xii.

49. The quote appears in the same undated letter reprinted in Madrigal 2016, 64–66.

CHAPTER 19. THE MASTER OF CONDITIONS

1. Hartleben 1983, 157; Faure 2004, 299–302; and Lacouture 1988, 214–16.

2. This chapter relies on letters from Champollion at Figeac to his brother in 1817 after the latter had left for Paris. We have only Champollion's side of this correspondence, which Karine Madrigal has transcribed and published in a valuable edition. If Champollion-Figeac's side of this correspondence still exists, it is most likely in the Grenoble family archives, which unfortunately remain unavailable for consultation while the archive is being digitized.

3. Hartleben 1983, 44–45.

4. We know little about the elder Champollion's book business, but Robert Darnton's painstaking studies of the inventories of eighteenth-century French booksellers are suggestive. See, for instance, Darnton 1982, 122–47, and Darnton 2014. On the anticlerical and pornographic material, see Darnton 1995, 85–114; on Mercier, see ibid., 115–36.

5. July 18, 1817; Madrigal 2016, 85–86.

6. August 1, 1817; ibid., 89–91. Champollion obtained the relevant fragment during the course of the crisis; he also obtained a copy of his deceased mother's will.

7. July 23, 1817; Madrigal 2016, 87–88.

8. From an undated letter, most likely written between April 30, 1817, and May 6, 1817; Madrigal 2016, 64–66.

9. May 13, 1817; ibid., 69–70.

10. May 26, 1817; ibid., 69.

11. On the maternal *jardin*, see Faure 2004, 300.

12. June 16, 1817; Madrigal 2016, 71–72.

13. July 23, 1817; ibid., 87–88.

14. August 1, 1817; ibid., 89–91; our emphasis.

15. September 1, 1817; ibid., 100–101.

16. August 27, 1817; ibid., 97–99.

17. June 17, 1817; ibid., 74–75.

18. June 5, 1817; ibid., 74–76.

19. Robinson 2012, 115.

20. Leitch 1855, 49–51.

21. Ibid., 59; Young 1814–29, 28. The Zoroastrian Ahriman was the evil counterpoint to the beneficent Ahura Mazda. With this allusion, de Sacy presents Napoleon as the evil spirit countered and overcome by the power of good as he saw it embodied in Louis XVIII.

22. The original French is given in Lacouture 1988, 213, and translated a bit differently in Robinson 2012, 113.

23. Madrigal 2016, 87–88.

24. June 29, 1817; ibid., 79–81.

25. Wahl 1816, 217–24.

26. July 18, 1817; Madrigal 2016, 85–86.

27. August 27, 1817; ibid., 97–99.

28. August 11, 1817; ibid., 92–93.

29. August 18, 1817:ibid., 95–96.

30. June 16, 1817; ibid., 77–78.

31. According to Lacouture, Champollion was sent the review directly from London (Lacouture, 216).

32. Anonymous 1816, 463–74.

33. Barring admiration for zoomorphic idolatry, which the English generally abhorred. To avoid the problem, the *Monthly Review* split chronological hairs ("It does not appear that Aegypt was polytheistic in the time of Joseph") and semantic ones ("Astronomical emblems and zodiacal signs have repeatedly been mistaken for pagan divinities; and phallus-worship, however coarsely idolatrous, is rather allegorical than polytheistic"). Anonymous 1816, 464 and 467.

34. Ibid., 469.

35. Madrigal 2016, 89–91.

36. Ibid.

37. Anonymous 1816, 468.

38. Madrigal 2016, 89–91.

39. Ibid.

40. Anonymous 1816, 472.

41. Madrigal 2016, 89–91.

42. All quotations are from a letter from Champollion to Champollion-Figeac, dated August 27, 1817, in ibid., 97–99.

43. Ibid.

44. Ibid., 89–91.

45. Ibid., 97–99.

CHAPTER 20. ABANDONING THE ALPHABET AT GRENOBLE

1. Hartleben 1906, 1:150; Hartleben 1983, 112. For Champollion's letter to the Royal Society, handed to Young, see Leitch 1855, 62–64. Champollion had, we have seen, begun to think about demotic at least as early as 1808, while still in Paris. Writing his brother on August 30 of that year, Champollion remarked that he "took Åkerblad's alphabet. I compared it [with a papyrus MS of Denon] and found 16 absolutely parallel letters. I gave them the same value" (Champollion and Vaillant 1984, 8–9). The Denon papyrus was, however, written (as usual) in hieratic, not demotic, so Champollion's comparisons were in vain.

2. Vaillant 1975, 35–36. There was also a "large white paper" with "French and Greek" written below the Rosetta demotic, and a copy of Ameilhon's translation.

3. "Mine of the English engraving is sloppy in some places," hence the desire for a much better rendition (Champollion and Vaillant 1984, 42). According to Hartleben the members of the *Description* commission were not forthcoming because they regarded Champollion as a "thorn in the eye" for, apparently, pushing to be involved in the plate's publication. The trouble may have stemmed from Jomard, the editor in chief, who was himself busy with hieroglyphs, though apparently Champollion-Figeac was on fair terms with him (1906, 299). Hartleben's claim is disputed in Laissus 2004, 43. On March 22, Champollion had asked his brother to obtain the *Description* engraving for him; letter held at the Musée Champollion. For Jomard and Champollion-Figeac at this time, see Champollion-Figeac 1887, 28.

4. Champollion and Vaillant 1984, 37–54.

5. Champollion 1814–21b, fig. 20.21 on p. 10 (7 of microfilm), 36 (27 verso of microfilm), and 1 (17 verso of microfilm).

6. Champollion's manuscript consists of distinct parts labeled "travaux sur l'écriture démotique," written in a hand that is not Champollion's. The description reads: "1. Essays on the alphabet; 2. Essays of interpretation; 3. First essay on the hieroglyphs of the Rosetta inscription (by comparison with the demotic figures); 4. Essay of concordance of the hieroglyphic text with the demotic [of the Rosetta]; 5. Hieroglyphic signs compared to demotic; 6. First table of demotic signs whose value seems certain; 7. Synoptic table of demotic signs; 8. Analysis of 102 demotic groups." A sheet introduces 4 as "Rosetta inscription. Material analysis of the two texts placed in concordance. Work anterior to the discovery of alphabet, done at Grenoble" (109). Item 8 is preceded by a sheet that reads "Rosetta inscription. Demotic text. Work on the material of the text before the discovery of the alphabet. This was done at Grenoble." Another note, added apparently as an afterthought (since the hand, though the same, is much looser than the foregoing, and the pen's application less consistent) reads: "One cannot use this as finished work but only [as] noteworthy for the happiest conjectures and as [illegible] already the incomparable perspicuity of the author even though the bases of Egyptian studies changed after the discovery [illegible]" (158). Given the carefully phrased encomium, it seems most probable that Champollion-Figeac wrote these notes.

7. For instance, he asserted that "the hieroglyphs concur in forming the equivalent of a given word of the spoken language, [and] are placed indifferently above others [but] always rigorously conserving the sequence order from right to left or from left to right or from below to above" (ibid., 91).

8. Champollion's remark to his brother is provided, in German only and without a source, in Hartleben 1906, 1:199. It is translated into French, apparently from Hartleben's German, with a somewhat different rendering from the English provided here, in Hartleben 1983, 132.

9. Coptic is, in modern terminology, highly synthetic. Old Chinese, by contrast, is analytic, in that its words consist of isolated morphemes.

10. Champollion, 1814–21a, 3; Champollion, 1814–21b, 73.

11. See Champollion 1814–21b, 198 of the PDF for a critique of Young's *Criticum 1*.

12. *Criticum 1* contains the following claims and materials:

> 1. *August 1814*: Åkerblad's alphabet would at the very most be useful "to decipher the proper names," and that his "letters" might resemble "the syllabic sort of characters, by which the Chinese express the sounds of foreign languages" (to de Sacy).
> 2. *October 1814*: Åkerblad's translation as to sense agrees with Young's (identifying the meanings of sign groups) (to de Sacy).
> 3. *October 1814*: The Egyptian text is likely the original, with the Greek its translation (to de Sacy).
> 4. *October 1814*: Similar words as to sense in Coptic to the Egyptian are difficult to find, but Young admitted insufficient "acquaintance with the Coptic language." He nevertheless provided 86 Greek-Coptic-Egyptian semantic equivalents, three of these being due to de Sacy and sixteen to Åkerblad (to de Sacy).

5. *October 1814*: Young in 1814 modified Åkerblad's Coptic alphabetic equivalents to Egyptian signs and added thirteen signs to the list but does not indicate whether he had made extensive use of these correspondences in adding to de Sacy's and Åkerblad's semantic lists (to de Sacy).

6. *January 1815*: Åkerblad wrote Young that the "different combinations of Egyptian letters are so diversified, and at the same time so difficult to determine with precision, that we should be in continual danger of error, if we attempted to make an enumeration of them."

7. *January 1815*: Åkerblad sent Young what he regarded as full Coptic equivalents line by line to the first five lines of the Egyptian inscription, with discussion.

8. *August 1815*: Despite Young's own putative addition to Åkerblad's alphabetical equivalents, he was certain that Åkerblad's own attempt "even in the first five lines" of the Egyptian inscription would inevitably produce an "inextricable confusion of heterogeneous elements" (to Åkerblad).

9. *August 1815*: Young remained convinced "that the Coptic is very nearly identical with the ancient Egyptian" but attempts at rendering sense lead him to "always change" his opinion (to Åkerblad).

10. *August 1815*: Young could not discern the sequence for Egypt in the MS published by Denon where Champollion claimed to see it, and in general considered that Champollion "has been somewhat hasty in several others of his remarks upon the Inscription" (to de Sacy). Since none of this occurs in Champollion's two letters to Young before August 1815, Young must have found the claim respecting the Egyptian for "Egypt" in *L'Égypte sous les Pharaons* (Champollion 1814, 2:338). In the *Pharaons* Champollion did not however provide the Rosetta Egyptian graphic for "Egypt." Instead, he provided three Coptic versions, identified the first of the three with sign sequences on the Rosetta, and all of the Rosetta sequences with one on plate 136 in Denon's *Voyage*. Young did not spy the sequence on Denon's plate, and indeed it is not clearly identifiable there.

11. *August 1815*: Young had decided that the Egyptian (i.e., demotic or enchorial) signs are "imitations of the hieroglyphics, adopted as monograms or verbal characters, and mixed with the letters of the alphabet," yet "none of these characters could be reconciled, without inconceivable violence to the forms of any imaginable alphabet" (Aug. 3, 1815, to de Sacy).

Criticum 2 adds:

12. *August 1816*: Comparison of the hieroglyphs on the great papyrus in the *Description* with the columnar text "put an end to the idea of the alphabetical nature of any of them" (to Archduke John).

13. *August 1816*: Young claimed to have "fully demonstrated" that the Egyptian script derived from the hieroglyphic, and "was not disposed to place any great reliance on the alphabetical interpretation of any considerable part of the inscription," though "it can scarcely be denied that something like a syllabic alphabet may be discovered in all the proper names, which seem to agree with Mr. Åkerblad's hypothesis" (to Archduke John).

14. *August 1816*: Quatremère and Champollion had tacitly adopted Åkerblad's idea, Young wrote, "but I am persuaded that if they will take the trouble of making the comparisons which I shall point out in this letter, they will be fully convinced that we have both been attempting an impossibility," namely, "to read the inscription of Rosetta into Coptic" (to Åkerblad).

13. Young did not at this time distinguish significantly between hieratic and demotic, but his assertion that enchorial was a degraded form of hieroglyphs parallels a nearly contemporaneous claim, by Thomas Christian Tychsen, professor of theology at Göttingen, that hieratic was an abbreviated form of hieroglyphs. This claim was reported in a translation of Arnold Hermann Ludwig Heeren's essay on the monuments of Thebes and Egypt (1816, 287n1). Heeren was professor of history at Göttingen and Tychsen's friend. Later, Champollion would make a point of remarking the assertion, rendering Young's claim as unoriginal as his own (1824, 20).

14. This date, cited by Lacouture 1988, 252, from the *Bulletin de l'Académie delphinal* in 1922 (vol. 13, p. 1) conflicts with the date of August 19 given by Hartleben 1906, 302, but without a source.

15. Champollion 1814–21b, 78 of folio.
16. Of these, the bird figure later proved indeed to work as such (known as *aleph* and sounding like *a* in *hat*), as did the handlike sign (denominated *ayin* and sounding like a swallowed *a*), whereas the feather-like sign corresponds phonographically to a sound rather like *y* (denominated *yodh*). The Greek "Ptolemy" is Πτολεμαιω, hence the "A."
17. Champollion and Vaillant 1984, 41–42. Champollion's remark that he would "return" to Clement's views indicates that he had, up to this point, thought hieroglyphs represented speech.
18. Champollion 1814–21b, 115 of folio.
19. Ibid., 80 of folio. For Young's identifications of the signs in 1814, see Young 1814–29, 38.
20. Champollion, 1814–21b, 120, 11, 30 of folio, which are 32, 2, 43, respectively, of microfilm. "XXIII" specifies the twenty-third line of the demotic text.
21. Champollion 1814–21b, 136, 47, 45, 59 recto, 59 verso, which are, respectively, 50, 72, 68, 97, 98 of microfilm.
22. Young's Coptic identification had not been produced by way of the putative alphabet, even though he had added several "fragments" to Åkerblad's set. Rather, his interpretation of sequences that were not Greek proper names relied entirely on semantics to find the corresponding *word* with the same meaning in Coptic.
23. Young thought otherwise: "In the four or five hundred years which elapsed, between the state of the inscription, and that of the oldest Coptic books extant, the language appears to have changed much more, than those of Greece and Italy have changed in two thousand"; letter to de Sacy, October 21, 1814 (p. 171 of *Criticum 1*).
24. Champollion 1814–21b, 162 verso, 204 of microfilm.
25. Ibid., 165, 209 of microfilm.
26. Ibid., 164 verso, 86 verso, respectively, 208, 52 of microfilm.
27. Ibid., 186 verso, 252 of microfilm.
28. Champollion 1814–21a, 221–22 of folio.
29. The Egyptologist John Ray has noted of determinatives that Champollion's "solution to the problem of these signs is probably his greatest single achievement" (2007, 90).
30. Champollion 1820. This MS contains several distinct pieces, all but one of which must have been written around 1822 or after since they reflect his later views. One among them, however, drafts a presentation concerning his revised notion that demotic (and hieratic and hieroglyphs, of course) cannot be alphabetic (ibid., 126–43 of PDF, contemporaneously numbered 237–49 on upper right). Following PDF 131 (numbered 239), there is another of the printed invitations to the Société des sciences et des arts de Grenoble on the verso of which Champollion continued his draft. He again used one of these printed cards for the next page of notes.
31. Ibid., 132 of PDF, 240 numbered. The papyri in question are the same ones that Young had worked on, printed in the *Description*, plates 66–75 of volume 2.
32. "One cannot," he wrote, "stop oneself from remarking that, in effect, the order of the words of the Greek text are only slightly intervened with [in the demotic passage] and that this intervention is just what must be when one submits a phrase belonging to a language of inversion like Greek to the logical or natural order ordinarily followed by the propositions of a language formed of words devoid of termination, like the Egyptian language [i.e., in his view, Coptic]." Ibid., 130 of PDF, 230 numbered.
33. Champollion 1821a.
34. Ibid., 2.
35. Champollion 1821a.
36. In this regard, the table was reminiscent of the arresting coup d'oeil productions we have already discussed in connection with Millin, Fourier, and their associates.
37. Ginzburg 2012, 95 (chap. 6: "The Europeans Discover [or Rediscover] the Shamans").

CHAPTER 21. DEMONSTRATIONS

1. Leitch 1855, 51 and 9.
2. Skuy 2003, 177. For another account of the disturbance at Grenoble, see Lacretelle 1820.
3. Champollion 1820, 3.
4. Moore 2008, 35.
5. A contested and contradictory product of the Congress of Vienna, the Constitutional Charter of 1814 established a bicameral parliament and guaranteed equality before the law, due process, religious tolerance, and the provisions of the Napoleonic Code while affirming the absolute power of the king. See ibid., 30–32.
6. Champollion 1820, 4.
7. Ibid.
8. Ibid., 5. According to Lacretelle, the students who defied the restriction on assembly convened at the Aveugles, a café in the Place Grenette known as a haven for the Ultras, where they shouted "Vive la charte!" as the duke arrived. See Lacretelle 1820, 43.
9. Pao 1998, 124–27.
10. Ibid., 134.
11. The history of literary realism cannot be separated from the history of the nineteenth-century novel. The scholarship dealing with this intersection is large but the same ideas—about the role of history, the importance of a central character, the use of montage, the importance of serial publication, the rise of the fourth estate, and so on—tend to repeat. Most useful for our purposes have been the classic studies by Levin 1963 and Auerbach 1953. Carlo Ginzburg has recently suggested an important corrective to Auerbach's tendency to conflate his own historicism with the different views of history that Ginzburg identifies in his readings of Stendhal and Balzac. See Ginzburg 2012 (chap. 10: "The Bitter Truth: Stendhal's Challenge to Historians").
12. Levin 1963, 137. The idea of battle as an experience of personal dislocation was not new. The painter Watteau, for instance, tended to generate work by transposing figures drawn previously, in different contexts, onto a single canvas; in his battlefield paintings, this practice gave the effect of unrelated individual soldiers sitting in groups with their minds elsewhere; the battlefield was also someplace else, indicated perhaps by a distant plume of smoke (Wile 2016, 45). After Waterloo, concentration supplanted dispersion. The effect could be visual as well as textual. In his use of bird's-eye maps in *The Life of Henri Brulard*, Stendhal famously combined the two (Engberg-Pedersen 2015, 211–18). The use of maps to suggest multiple points of view within a single narrative also recalls Champollion's juxtaposed sketches of Saint-Laurent and the bird's-eye view of the same corner in Grenoble.
13. Though Balzac focused on the novel, his criticisms echoed complaints voiced by critics of the preceding two decades, notably by theater critics against the writers of plays based on military events. See Pao 1998, 130–36.
14. Balzac 1840, 58.
15. Ibid., 81–82.
16. We have drawn the facts of our understanding of Balzac's struggle from Engberg-Pedersen's recent account; see 2015, 185–89.
17. Levin 1963, 137.
18. Balzac 1840, 82; Balzac, untitled essay in *Revue Parisienne*, 82. Engberg-Pedersen gives a slightly different translation; see 2015, 198. Not everyone shared Balzac's delight in this feature of Stendhal's style. Albert Sorel, for instance, was unimpressed when Stendhal adapted this style to consular reports: "His processes of observation and description are particularly dangerous and illusory. Stendhal ascribes too much importance to anecdotes and generalizes too easily." Quoted in Levin 1963, 109.
19. Indeed, tachygraphy is everywhere in *Attention*, starting with the title. It was an elegant choice, for it was not in the least didactic; it does not *tell* the reader anything. It merely makes a plea for the reader's attention.
20. For the events of the period, see Mansel 2003.
21. Alexander 2003, 166. See the accounts in Champollion-Figeac 1887 and Champollion-Figeac, 1844.
22. Hartleben 1906, 334–44; Hartleben 1983, 186–92; and Lacouture 1988, 240–46.

CHAPTER 22. ICONOCLASM AT THE ACADÉMIE DES INSCRIPTIONS ET BELLES-LETTRES

1. Now 20, rue Mazarine. A plaque commemorates Champollion's period of residence at this address.
2. Faure 2004, 415.
3. For a description of the apartment, including the atelier, see ibid., 413–15.
4. According to the biography written by Champollion-Figeac's son, Aimé-Louis.
5. Champollion-Figeac 1887, 28–30.
6. According to Lacouture 1988, 282. They worked together until Dacier's death, in 1832, at the age of ninety-one.
7. Champollion-Figeac 1887, 31. The block was organized by Quatremère de Quincy, de Sacy's friend and relative of the Coptic specialist with whom, as we have seen, Champollion became embroiled in a priority fight a decade earlier, together with Désiré Raoul-Rochette, a specialist in Roman antiquities, and the sinologist Jean-Pierre Abel-Rémusat, who had (paradoxically) praised Champollion's earliest publications.
8. Champollion 1821b, 2.
9. Ibid., 60, 106 of microfilm.
10. The sections comprised: 1. "Opinions given on the type of writing of the Egyptian manuscripts," 2. "On the writing of the non-hieroglyphics manuscripts," 3. "Nature of the writing of the papyri said to be non-alphabetic," 4. "Names given by the ancients to the type of Egyptian writing," 5. "Theory of the forms of hieratic writing," which he divided into five subsections, 6. "Number of hieratic characters," 7. "Comparison of the grammatical course of signs in the two systems," 8. "General classification of signs."
11. "[T]he various uses of an Alphabet in civil business not permitting it to continue long a secret, when it ceased to be so, they would as naturally invent another alphabetic character for their *sacred* use: which from that appropriation was called HIEROGRAMMATICAL [*sic*]" (Warburton 1765, 3:155). Champollion used the French translation: Warburton 1744.
12. Champollion 1821b, 6, 9 of microfilm.
13. Ibid.
14. Montfaucon 1722–24.
15. Using a short title, Champollion correctly cites page 59 of Humboldt's *Vues des cordillères, et monumens des peuples indigènes de l'Amérique*, published in Paris in 1810.
16. Since "the increase in the number of Egyptian manuscripts having become more common after the Egyptian expedition may modify these diverse opinions concerning the manuscripts . . . the savants who discovered them also published them in the *Description de l'Égypte*."
17. Champollion 1821b, 19, 2 of microfilm.
18. Ibid., 69, 115 of microfilm.
19. Champollion 1814, 1:xvii.
20. Champollion 1821b, 11–2, 4–5 of microfilm.
21. Ibid.: "le but unique et constant est d'exprimer beaucoup de sons avec peu de signes, et donc l'une des conditions principaux est de reduire l'art de peindre les sons d'une langue parlée, à des elemens bornées le plus possible dans leur nombre en n'en créant d'indispensables."
22. Ibid.
23. Ibid.
24. Ibid.
25. Ibid., 14, 7 of microfilm.
26. Ibid., 19, 22 of microfilm.
27. The "anomalies" involved apparent shape deviations. Champollion attributed these to scribal skill defects: "[T]he hand that worked these hieratic texts was unskilled [*inhabile*] in the imitation of the forms of living nature." He further speculated that scribes were classed according to "their greater or lesser ability to trace this or that bit of the character." Ibid., 31, 4 of microfilm.
28. Ibid., 19, 22 of microfilm.
29. Ibid., 1, 4 of microfilm.

30. Ibid., 63 of PDF, 31 numbered.
31. Champollion's vocabulary evokes that of Clement of Alexandria who, in his *Stromata*, identified three types of "symbolic" written signs: a "mimetic" depiction, a "tropical" or "figurative" representation, and an "allegorical" one (book 5, chap. 4). By "tropical," Clement meant a sign that refers metonymically or synecdochally to what it signifies.
32. Champollion 1821b, 57–243 of PDF, 31–180 numbered.

CHAPTER 23. THE OBELISK FROM PHILAE

1. On Bankes, see Sebba 2004, Seyler 2015, and Usick 2002. Since 1817, Bankes had been among the subscribers to a project Young had conceived of publishing all extant and reliable Egyptian inscriptions.
2. Young to Henry Bankes, February 10, 1818, BM AN30361001; Young 1814a.
3. In April 1815, Young wrote Jomard about this duplicate. Though he knew from Clarke's report that the object was in poor shape, Young hoped that "one of the savants" might have essayed a drawing (Leitch 1855, 68). Clarke described the stone in the court of the Cairo Institute as follows: "a very large slab, covered with an inscription, in Hieroglyphic, Egyptian, and Greek characters, exactly similar to the famous triangular stone now in the British Museum" (1814, 47). Champollion retrieved the artifact years later but it was too damaged to be useful.
4. Peacock 1855b, 315.
5. Denon 1802a, 21.
6. The events described immediately below later generated considerable enmity between Champollion and Young's English proponents. Although we do not have Bankes's account of the matter, in 1825 Salt wrote a lengthy description of what Bankes found. This essay amounts in several ways to a brief for Young's priority in establishing a "phonetic system" for hieroglyphs and so must be read with that in mind. Nevertheless, it is unlikely that Young, Bankes, and Salt conspired to falsify what had taken place in Egypt. See Salt 1825, 8–10.
7. Salt later remarked that at Diospolis Parva, Bankes had "confront[ed] the supposed name of Ptolemy, as furnished to him from the Rosetta stone by Dr. Young, with the hieroglyphical designation over the male figure" (ibid., 9). The male designators are a square with a semicircle, while the female designators are an oval with a semicircle. Bankes might have divined as much by comparing the sign sequences for the names Ptolemy and Berenice that Young had conveyed in his letter (see fig. 23.1).
8. In figure 23.2, note the additional signs in the obelisk "Ptolemy" (second from left) as compared to the ones that Young sent to Bankes. Although these signs are present in the Rosetta cartouche, Young omitted them in his letter to Bankes, believing them irrelevant to a phonetic rendering of "Ptolemy." Bankes 1821 and Young 1814a.
9. On Belzoni, see Mayes 2006.
10. Hullmandel 1824.
11. Charles Crowther of the Center for the Study of Ancient Documents at Oxford has recently led a team to examine the obelisk signs by reflectance transformation imaging.
12. Bankes was then accompanied as translator by the Italian adventurer Giovanni Finati, whose autobiographical "life and adventures" Bankes later translated and published (Finati 1830, 2:86).
13. On Bankes's drawings, see Letronne 1823, 298. In 1815, the jeweler and mineralogist Frédéric Cailliaud met Bernardino Drovetti, who recruited him with the sculptor Jean-Jacques Rifaud to help with excavations in Egypt. Drovetti eventually introduced Cailliaud to Egypt's ruler, Mehmet Ali, who hired him to search for lost emerald mines. This job took Cailliaud to Philae where he carefully copied the base of the obelisk. On Cailliaud, see Bednarski 2005, 6–8.
14. Since 1801 Letronne, originally an artist in the studio of David, had devoted himself to historical geography. His interests were extraordinarily diverse, and he always sought to assemble a broad array of evidence to reconstruct not merely the physical geography but the life of a period. Nevertheless, Letronne was attuned to the limitations of ancient evidence and skeptical of any overarching hypothesis about the past unless it could synthesize a very wide range of data. See Godlewska 1999, 298.

15. Letronne 1821.
16. "[I]in the supposition that the monument were of the same period as its plinth and relates to the circumstance expressed by the Greek inscription, it would be little probable that it translates the request in hieroglyphs; these hieroglyphs were rather an expression of the thanksgivings addressed by the priests to the prince and to the goddess, in memory of a benefit obtained" (ibid., 673).
17. Peacock 1855b, 315; Bankes 1821.
18. On the copy received at the Académie des inscriptions, see Letronne 1822, 211.
19. Nine years later, having completed his mutation into a conservative devotee of Throne and Altar, Saint-Martin joined with Abel-Rémusat to found a journal, *L'Universel*, devoted to the divine rights of monarchy. Although Abel-Rémusat, among others, blocked Champollion-Figeac's election to the Académie des inscriptions, he had been on good terms with Champollion in the 1810s.
20. Champollion, undated letter to his brother from 1817, reprinted in Madrigal 2016, 64–65.
21. Hartleben 1906, 349, no source provided.
22. For the debate, see Buchwald and Josefowicz 2010.
23. Jomard 1817. Note the engraving's detailed depiction of hieroglyphs to the side of the nude goddess. In 1822, this drawing was available only in full size; it was very expensive to have printing done in 1817.
24. Saint-Martin 1822b, 9, 37, 43, 51.This article was translated into English and printed as Saint-Martin 1823.
25. Saint-Martin 1822b, 46–47. Saint-Martin found the "same name" in the *Description*, vol. 2, plate 7.
26. Letronne to Saint-Martin, uncovered in Pezin 1993, 25: "I send you, my dear colleague, the hieroglyphs of the Obelisk, Champollion the younger, who had come to see them, agrees with you, however he did not know that you recognized Ptolemy: he is also convinced that the hieroglyphic inscription agrees with the plinth." Letronne also urged Saint-Martin to provide his piece on the subject for the same journal (*Savans*) that contained Letronne's own discussion of it.
27. Saint-Martin 1822a, 563. Young's *Britannica* piece was certainly available in Paris by this time. Champollion had read it ahead of his 1821 talk on hieratic. Moreover, in his article on the Casati manuscript—the printed version of a talk given at the Académie des inscriptions on August 16, 1822—Saint-Martin referred to remarks concerning the difference between hieratic and the Egyptian (enchorial, demotic) script made by both Young and Champollion. We discuss this article in chapter 25.
28. The main evidence was apparently his identification of a sign meaning "the two saviors," which he concluded had to refer to "two princes." Principles of collective naming obscured the gender of the persons so named, so "two princes" could refer to a Ptolemy and his queen, as on the obelisk; Saint-Martin 1822c.
29. Ibid., 217.
30. Regarding the February access, we follow Hartleben 1906, 404. Lacouture 1988, 511, though without source, gives the date of access as January 22. See Pezin 1993 for a letter to Saint-Martin from Letronne dated February 10, 1822, in which he indicates that Champollion had seen Bankes's plates. The relevant parts of the letter are translated in chapter 23n6. Champollion later wrote that he had seen the plate "by the kindness of M. Letronne." It would accordingly have been this copy, in Letronne's hands, that may have had Bankes's scribbled notation of "Cleopatra" (Champollion 1824, 9).
31. Champollion 1822c, 513. Despite its publication date of 1822, the journal contains material from 1821.
32. On pairs of obelisks, see Curran, Grafton, Long, and Weiss 2009, 14.
33. Champollion 1822c, 512–13.
34. Letronne had previously separated the Greek text on the plinth from the hieroglyphs on the obelisk: "[I]n the supposition that the monument were of the same period as its plinth and relates to the circumstance expressed by the Greek inscription, it would be improbable that it translates the request in hieroglyphs; these hieroglyphs were rather an expression of the thanksgivings addressed by the priests to the prince and to the goddess, in memory of a benefit obtained" (1821, 673).
35. Champollion 1822c, 515.
36. Ibid., 516. In his *Britannica* article, Young had drawn hieroglyphs for the names Ptolemy, Berenice, and Arsinoe. In addition, he provided sign sequences that he took to signify Euergetes, Philopator, Epiphanes, Philometor, and Cleopatrides, though he did not link these last identifications to phonetics. Of the three names, only Ptolemy occurs on the Rosetta Stone due to its damage;

Young obtained at least Berenice from a plate in the *Description*. The signs for Berenice come from Karnak, he wrote, and the *Description* shows the cartouche in its plate 50, volume 3. The cartouche that Young took to represent Arsinoe appears on plate 75, no. 60. Elsewhere in the article Young tentatively identified the name in a "temple at Ombos" (63). The *Description*, vol. 1, plate 41, has a cartouche (no. 6) from Ombo that matches Young's description of the non-Arsinoe part ("a basilisk is followed by two feathers"), before which are other signs that do contain the bird and curved sign with underscript that lead in Young's Arsinoe cartouche (plate 75, no. 60).

37. Bankes, quoted in Usick 2002, 79. Bankes claimed it was the copy sent to Denon.
38. See Leitch 1855, 371, for a remark Young made years later concerning Bankes's anger with Champollion.
39. Bankes, recall, had identified the cartouche as such because of the sequences' appearance on a temple, but Champollion could not then have known this.

CHAPTER 24. A SINGULAR AND PUZZLING ARTIFACT

1. See Buchwald and Josefowicz 2010 for how the zodiac came to France, and for the challenge it posed to religious orthodoxy.
2. Noted in Faure 2004, 426–27. The remarks follow Saulnier's letter, dated October 4, announcing the zodiac's arrival at Marseilles, and Dacier's congratulatory letter, dated October 5. Although the remarks in question are signed "note of the editors," they do certainly sound like Champollion, who is one of the very few likely to have protested the zodiac's removal (Champollion 1821a, 470).
3. Anonymous 1823b.
4. Henry 1854, 167–68.
5. Halma 1822.
6. In addition to the three memoirs Chabrol produced for the *Description* during the empire, a lengthy fourth on the customs of contemporary Egyptians was printed in 1822, the same year the zodiac caused an uproar in Paris, to which were appended fulsome thanks to his friend Fourier for his information about Egypt. Chabrol also collaborated with Jomard in describing the area around Kom Ombo, where they noted a temple whose construction "confirms very well the tradition that attributes the invention of geometry to Egypt, and that gives Egypt the honor of having made the first geographic projections." Given his interests and experiences, Chabrol certainly would have known quite a bit about the issues raised by the zodiac, and like many others on the expedition, he was convinced that the Egyptians had millenia ago invented the foundations of geometry, a specialty linked closely to his own.
7. Hartleben 1906, 1:49.
8. Champollion 1822d, 232–33.
9. The engraving (reproduced in Biot 1844) was done by François Gau and was available (together with the brochure that Saulnier had written to push the sale of the circular) at 52, rue de Rivoli for five francs.
10. Ibid., 237.

CHAPTER 25. A MOMENTOUS CHANGE

1. Saint-Martin 1810–11, 560. Letronne discussed Saint-Martin's talk in a letter; Pezin 1993. How Casati came into possession of the papyrus rolls remains obscure. They were eventually purchased by Louis XVIII. The publication in which Saint-Martin's article appeared was successively known as *Journal de sçavans*, then *Journal de Savans*, and finally *Journal de Savants*.
2. Saint-Martin 1822a, 561.
3. Young 1823b, 31; Champollion 1822e, 34, 280 of microfilm.
4. "The name of Ptolemy is not difficult to recognize: it is written precisely as in the Rosetta inscription; only, below the sign that marks the T in this name, one sees a trait or a kind of diacritic point

that is not found on the first monument; the same trait is found in the name of Cleopatra, which contains the same letter. After the name of Ptolemy comes that of the queen Cleopatra, daughter of Ptolemy and of Cleopatra. This princess must be Cleopatra, sister of Soter II, widow of Ptolemy Alexander I" (Saint-Martin 1822a, 564). Saint-Martin did not further remark that other signs in the sequences for the two names were also similar, though he surely noticed the fact, for example, the reversed Y-shaped sign that Young in the *Britannica* (and Åkerblad years before) suggested could be read as "L," as well as the sign immediately preceding the angle bracket, which Young had as "P."

5. Champollion 1822e. The memoir is divided into two chapters, a conclusion, and three appendices. The first chapter is entitled "On the non-alphabetic nature of the intermediate text of the monument"; the lengthy second, "Essay on the theory of the system of Demotic writing," consists of nineteen subsections, ranging from the "form and disposition of the signs" through signs for kinds, numbers, articles, pronouns, and other words, and including one on "Non-Egyptian proper names represented in Demotic characters." The first appendix consists of Champollion's rendering of six demotic lines into French (fig. 25.2, *bottom*).

6. Champollion had, of course, seen Young's *Criticum* years before his return to Paris. Here he wrote that he had first read Young's long *Britannica* article shortly after arriving in Paris the previous September (1821). There is no known evidence that Champollion had seen Young's article earlier, though he had heard about it from his brother (Hartleben 1906, 1:328). Young had sent a copy to Jomard, who might have informed Champollion-Figeac about it; see Leitch 1855, 208. In a letter to Young dated September 16, 1819, Jomard wrote that he "lacks at the moment leisure to speak to you about the rest of your vocabulary, which must have cost you infinite trouble." Jomard could only be referring to the "vocabulary" that Young presented in the plates of his *Britannica* article. According to Hartleben, Champollion requested a copy on March 20, 1820, but did not receive it, Young having sent only his "conjectural translation" (Hartleben 1906, 1:328; no source provided). No such letter from Champollion to Young exists in either the Young archive at the British Library or in Leitch's edition of the *Miscellaneous Works*. However, Alexander von Humboldt wrote Young on October 26, 1819, that he had circulated Young's *Britannica* piece to the "two Academies" (Leitch 1855, 208–9), so it certainly was known at the time, and Champollion-Figeac had been in Paris since April 1817.

7. Champollion 1822e, 63 of microfilm, 35 numbered.

8. Ibid., 36, 65 of microfilm.

9. Ibid., 30, 56 of microfilm (Coptic) and 237, 83 of microfilm. The latter is a finely drawn copy of a roughly drawn version on PDF 280. This document seems to be among the ones stolen by Francesco Salvolini and recovered after Salvolini's death by Champollion-Figeac (see Champollion-Figeac 1842 for the theft and recovery).

10. Ibid., 70–71 of microfilm, following 38 numbered.

11. Champollion 1822e, 39, 72 of microfilm.

12. Ibid., 33 verso, 92 of microfilm, and (*right*) 120, 82 of microfilm.

13. Ibid., 96–97, 38–39 of microfilm.

14. Ibid., 95–96 of microfilm, following 37–38 numbered. This definition of "ideographic writing" is Champollion's.

15. Champollion wrote approvingly that the priests of ancient Egypt, who "held first rank in the state," were "persuaded that the welfare of the people was attached to the conservation of their customs, proven by experience and established, like those of other Orientals, according to the physical character of place, eminently contributed to preventing all communication between foreign nations and the Egyptians. This fundamental maxim of Egyptian politics is today conserved among the Chinese, and the disastrous events that followed annihilated forever the liberty of Egypt fully justify this opinion of the priests and confirms their fears" (1814, 1:4). In such a society the weight of tradition necessarily conserved the original character of a writing system.

16. Champollion 1822e, 122 of microfilm, following 63 numbered.

17. Ibid., 108–9 of microfilm, 50–51 numbered.

18. Ibid., 126 of microfilm, following 65 numbered.

19. Ibid., 108–9 of microfilm, 50–51 numbered.

20. Even three years later, in the lengthy *Précis* that Champollion wrote to detail his new theory, he continued to insist that the names of ancient pharaohs were occasionally written in that fashion, that is, nonphonetically, reflecting the residue of his earlier convictions (see chapter 33).

21. Champollion 1822e, 62 of microfilm, following 34 numbered.

22. Ibid., 72 of microfilm, 38 numbered.

23. Ibid., 101 of microfilm, 43 numbered.

24. Though there are none for hieroglyphs, Champollion insisted that the two scripts were fundamentally similar. Ibid., 96, verso and recto, 155–56 of microfilm.

25. Ibid., 156 of microfilm, following 35 numbered.

26. As, for example, in rendering "Ptolemy" Young wrote "an oblique line crossed standing for the body, and an erect line for the tail: this was probably read not LO but OLE; although in more modern Coptic OILI is translated a ream; we have also EIUL, a stag; an[d] the figure of the stag becomes, in the running hand [enchorial], something like this of the lion" (1819b, 62).

CHAPTER 26. WORDS AND SOUNDS

1. Champollion 1822e, 74 of PDF.

2. Rochas 1856, 212.

3. Ibid.

4. Thomas Carlyle is usually credited with originating the "great man" theory of history; less often noted is his view that its power derived from the psychologization of religious belief. In *On Heroes, Hero-Worship, and the Hero in History* (1841), Carlyle lionized the "great man" type; in *Sartor Resartus* (1834) he satirized those who, like Rochas and even Carlyle himself at times, uncritically sought to assimilate their biographical subjects to it. The trope proved enduring, particularly in scientific biography, and its persistence attests to the slowness with which habits of even a cast-off faith are relinquished. See Landow 1980, 15–63 and 143–76.

5. The issue has been additionally confused by Rochas's supposition that Champollion had long sought evidence for phonography in the scripts, and this supposition led him to conflate Champollion's long-standing effort to recover the native language with a desire to recover its *sounds*, which was not his goal. Three decades later, Champollion's nephew, Aimé Champollion-Figeac, repeated verbatim Rochas's account (Champollion-Figeac 1887, 57); it is possible that both Champollion-Figeac and Rochas heard the original story from J.-J. Champollion-Figeac. A decade later Hartleben reiterated the story in her biography, though she replaced Rochas's "mon affaire" with "l'affaire" (1906, 1:422). There has been no significant variation in any subsequent account, for example, Adkins and Adkins 2000, 181; Faure 2004, 428–30 (who places most weight on Rochas's version); Lacouture 1988, 297; Robinson 2012, 142; Solé and Valbelle 1999, 120; and Parkinson 1999, 35.

6. Champollion 1822b, 2.

7. Ibid., 1 and 6. In the original French: *peignant les idées et non les sons d'une langue.* We have translated "peignant" as "indicating" rather than the more literal "painting." The phrase had a history of its own, and in using it, Champollion was likely signaling his awareness of and participation in this tradition. In his letter to Chaptal, de Sacy had used a similar phrase: "le caractère hiéroglyphique, peinture des idées, et non des sons, n'appartient à aucune langue" (1802, 8).

8. Champollion 1822b, 1 and 6.

9. Ibid., 5. In his demotic lecture, Champollion remarked: "The need to express, in writing, the proper names of countries, of peoples, and of foreign individuals with which a nation in possession of a script composed of signs of ideas finds itself in direct contact, becomes, in effect, the sole motor that can determine this same nation to essay a *syllabic or alphabetic script*" (Champollion 1822e, 153 of PDF, 93 numbered).

10. Young 1819b, 62.

11. Champollion 1822b, 15.

12. Champollion 1822a, 4.

13. Ibid., 5.
14. Champollion 1822b, 6.
15. For the angle bracket as a female designator, see Champollion 1814–21b, 209 of PDF, and for the equivalence between semicircle and angle bracket, see Champollion 1821b, 405 of PDF. Champollion always referred to the semicircle as a "quarter-circle," an odd choice, perhaps, for a person who chose his words with great intentionality. His use of "quarter" for "semi" may reflect his distance from mathematical terminology; it may also be another of his subtle jokes, or an allusion we have yet to understand.
16. Young 1819b, 63 and 55, respectively.
17. Ibid., 62.
18. Champollion 1822b, 8–9; Bankes 1821.
19. Champollion 1822a, 11.
20. Jomard 1812, vol. 3, plate 38, no. 13.
21. Saint-Martin 1822c, 217.
22. Young 1823b, 31; Champollion 1821b, 99. The latter sign equivalences appear ten other times in the MS.
23. Cf. Jomard 1812, vol. 2, plates 66–75.
24. Of these, the first (Champollion's no. 48) is a variant of what is shown on plate 28 in volume 4 of the *Description*, and a misprint in the Dacier letter refers to plate 27 rather than 28.
25. As Champollion worked his analyses through Coptic, he relied on previous works by scholars of Coptic's various dialects. Seventeenth-century scholars possessed Coptic texts in Bohairic, Sahidic, and a small number in Fayyumic. Bishop Athanasius of Qus, eleventh or fourteenth century, had noted the existence of three Coptic dialects, the third of which he termed "Bashmuric." Scholars then identified the Fayyumic texts with this putative Bashmuric, which nicely matched the presumption of dialects corresponding to Upper, Middle, and Lower Egypt. Champollion recurred to the so-called Baschmourique texts of Coptic to note that, in them, the Λ often substituted for the P (Champollion 1822a, 24). In the cartouches containing his identification of the phonetic hieroglyphs for αυτοκρατορ, the sign for Λ sometimes replaced that for P.
26. Ibid., 25.
27. Hartleben 1906, 1:417. There seems to be no source beyond Hartleben for Jomard's immediate reactions, and she provided no references.
28. Anonymous 1822, 445.
29. Champollion 1822b, 40.
30. Jacob Burckhardt had observed the monument in 1813, and Bankes visited it in 1815. On Bankes at Abu Simbel, see Seyler 2015, 118–21.
31. Leitch 1855, 251.
32. Champollion 1824, 274.
33. Young 1828, 47; Jomard 1812, vol. 4, plate 37, no. 10; Leitch 1855, vol. 3, 245; and Jomard 1822, vol. 5, plate 52; Champollion 1824, plate 12.
34. Champollion also found it in an isolated cartouche in the *Description*, and again in the Abydos inscription once he eventually obtained the latter from Cailliaud, who copied it during the summer of 1822 but only published the copy much later. See Cailliaud 1826–27; cf. Bednarski 2014, 17.
35. Robinson 2012, 140–41, succinctly describes the relevant signs.
36. Leitch 1855, 244–45.
37. With respect to the two groups in Champollion's second column: the first appears in Abydos cartouches but is missing one of the crooks; the second also appears but with only one scepter.
38. As noted by Robinson 2012, 140.
39. Champollion 1824, 69–70.
40. "XVII" appears about a third of the way past the crook-plus-trident. These signs were identified as such in Young 1819b, plate 77, no. 198.
41. Ibid., plate 77, no. 178.
42. Champollion 1822b, 41–42.

CHAPTER 27. PARISIAN REACTIONS

1. Hartleben 1983, 232. The letter, at the Bibliothèque nationale, was not available to Hartleben but was inserted in the French edition.

2. After an extract of the Dacier talk appeared in the October issue of the *Journal des Savants*, the *Lettre à M. Dacier* was published in full on November 5. Neither of Champollion's talks on hieratic and demotic scripts ever appeared in print. Hartleben 1906, German edition, 2:442.

3. Anonymous 1823a, 25. Covering literature, science, and the arts, the *Revue* was founded in 1819 by Marc-Antoine Jullien de Paris, who was a journalist, a former friend of Robespierre, and briefly a member of the Napoleonic expedition to Egypt.

4. Champollion had not linked the Dendera temple to Nero but had only suggested that the "autocrator" sign could apply either to him or to Claudius (1822b, 26).

5. Anonymous 1823a, 25–26.

6. The journal's full title was *Nouvelles Annales des Voyages de la Géographie et de l'Histoire.*

7. Eryiès was also the anonymous translator of *Fantasmagoriana* (1812), the collection of ghost stories Mary Shelley credited for prompting the storytelling contest in which she conceived *Frankenstein*.

8. On Malte-Brun, see Godlewska 1999, 90. Malte-Brun may have been volatile but he was no Byron. His next major publication was *Traité de la légitimité, considérée comme base du droit public de l'Europe chrétienne* (1824), a monarchist apologia ornamented with a preface by Chateaubriand.

9. Champollion-Figeac's influence is evident in this review, especially in the presentation of Champollion's lectures on demotic and hieratic. Although these lectures never appeared in print, the author knew their organization and contents in far greater detail than Champollion could possibly have presented in real time to a live audience. Since no one besides Champollion's brother had close contact with his work at this early date, it is likely that the review's author was either Champollion-Figeac himself or perhaps one of the journal's editors, working from notes provided by Champollion-Figeac who, as we saw during the conflict with Quatremère, did provide such things to reviewers of his brother's work.

10. Anonymous 1822, 119.

11. Letronne's 1823 volume on Egyptian history under the Greeks and Romans also referred to Champollion's extension of phonetic readings to the names of other Greek rulers. The extension supported Letronne's view that many of the Egyptian monuments thought to be ancient were actually built or inscribed much later, under the Greeks or Romans (Letronne 1823, xxx–xxxiv). As Letronne added these remarks to his introduction considerably after readying the bulk of the work for publication, we see little more than simple opportunism in his response.

12. Published in mid-1823.

13. Lesur 1823, 849–50.

14. Anonymous 1823c. The author of the remark may have been Julius Heinrich Klaproth, coeditor of the *Journal Asiatique*, who in October published a brief "Letter to Champollion the Younger" that referred to "the indefatigable research of you and Doctor Young, in England, to whom we owe the explanation of the hieroglyphic inscriptions" (1823). Although Champollion-Figeac's work had appeared in its pages, the *Journal Asiatique*, founded in 1818 by the sinologist Abel-Rémusat, was not an especially welcoming venue. As we will see, Klaproth was shortly to become one of Champollion's major critics, and Abel-Rémusat later blocked Champollion-Figeac's advancement.

15. Hartleben 1983, 232.

CHAPTER 28. WORDS ACROSS THE CHANNEL

1. Leitch 1855, 220–23.

2. Ibid., 235. To his sister-in-law, Caroline Maxwell, Young wrote at the time that "Champollion has adopted all my interpretations, almost without alteration; but he has [had] the good fortune to discover several important documents which were unknown to me," namely, the Casati papyrus (Peacock 1855b, 460).

3. Leitch 1855, 239–40. The only "enchorial papyrus" for which Young would in the next several months have generated a full translation was the Casati, printed in Young 1823b, 31. While Young was in Paris, Champollion had provided a "tracing of the enchorial papyrus of Casati," as Young wrote to Bankes on October 21. In that letter Young included "a *translation* of such parts of it, as I can pick out without too much trouble" (Leitch 1855, 236). Young provided the same to Champollion, probably in his reply to the latter's letter of November 12, since we shall see in a moment that Champollion thanked him on November 23 for remarks made by Young concerning Champollion in his "letters to MM. Hamilton and Bankes." Ibid., 220–23.

4. Young noted to Bankes that "this name *was* annexed to the zodiac of Dendera, though the notable speculators [i.e., Saulnier and Lelorrain], who have been so well rewarded by the liberality of the French government, found it convenient to *saw off* this most important part of the stone, in order to make it portable." Still, remarked Young, "so true it is, that a copy, for the purposes of literature, may be incomparably better than an original transported" (Leitch 1855, 236).

5. Ibid., 243–47.

6. Ibid., 247–49, for Champollion's apology at not having sent a proper facsimile of the Casati and for his complaint about the "conservators," and 236 for the October 21 letter to Bankes in which Young noted having obtained "a tracing of the enchorial papyrus of Casati" from Champollion. Young must have told Champollion that the Bankes and Hamilton letters would be printed in his reply to Champollion's first letter, which is unfortunately truncated in Leitch 1855, 240. The letters were printed in January; Young 1823a, 255–61.

7. Leitch 1855, 249–51; Young 1823b, 31. The fifty-three subscribers included Bankes, Boughton, Gell, Gurney, Hamilton, Jomard, Payne Knight, Henry Salt, and the Duc de Blacas, who had recently become Champollion's major supporter, as we shall see.

8. Leitch 1855, 371.

9. Barrow 1822. The piece, which reads like Barrow's earlier ones, has been attributed to Barrow based on his other contributions to the *Quarterly Review* (cf. Cutmore 2017).

10. Barrow 1822, 188.

11. Ibid., 191. According to Barrow, "this discovery," namely, that all three scripts were fundamentally ideographic, was not "due to M. Champollion" but to de Sacy, whose study of the "enchorial or demotic" on the Rosetta led him to identify names using the Greek inscription. Following de Sacy, Åkerblad had "constructed a sort of alphabet, by the help of which several foreign names in the hieroglyphical inscriptions, and in the writing of papyri, were made out," following which Young "extended the catalogue." Champollion's contribution was therefore nugatory.

12. Ibid., 192.

13. Champollion 1822b, 15n2; Young 1819b, 62. Recall Young's denigration of "syllabic and alphabetical writing," mixed "in a manner not extremely unlike the ludicrous mixtures of words and things with which children are sometimes amused."

14. Barrow 1822, 196.

15. Although Barrow's responsibility for a number of articles in the *Quarterly Review* was likely well known in Britain, especially given his position as second secretary of the Admiralty, he may have been less recognizable in France.

16. Leitch 1855, 255–56. In an example of the enduring hostility that began with this exchange, Leitch, the editor of Young's hieroglyphic work, expostulated in 1855 that "nothing can exceed the effrontery of Champollion in thus complaining to Dr. Young, the author of the discoveries ignorantly attributed by the reviewer to De Sacy and Åkerblad, as if he himself were the person aggrieved." With respect to the scripts' identities, Leitch had a point. As we have seen, Champollion knew Young had made the connection. The issue of phonetic interpretation is more problematic, as Champollion had not only made further identifications but had also derived them from his novel syllable-stripping technique.

17. We have been unable to locate a copy of the London issue 55 of the *Quarterly Review*, in which the advertisement for forthcoming Murray publications appeared along with Barrow's article. Only the original London issue 55 could have contained the advertisement that Champollion saw. Both issue 55 (October 1822) and issue 56 (January 1823) were gathered together and printed in 1823 in London by John Murray in volume 28. A separate printing of issue 55 was produced in Boston in April 1823.

Neither of these subsequent publications contained an advertisement for Young's *Account*. On these matters, see Peacock 1855b, 234.

18. Leitch 1855, 256.

19. Ibid., 256–57.

CHAPTER 29. GREY'S BOX

1. Young 1823c, 55–56.

2. Leitch 1855, 234–38.

3. Young 1823c, 57. In the Dacier letter, Champollion read the Casati demotic as containing the names Alexander, Ptolemy, Berenice, and Arsinoe, written in signs "similar to those on the Rosetta monument," while also, and crucially, finding Cleopatra as well, even though this name did not, of course, appear on the Rosetta monument (1822a, 4). In addition, Champollion spied in the papyrus three other Greek names: Apollonius, Antiochus, and Antigone. Of these, only Ptolemy occurred in the accompanying Greek lines. The other Greek names were Lysimachus and the stone mason "Léon son of Ronnophris, son of Orus." See Saint-Martin 1822c, whose interpretation of the Casati is discussed above in chapter 25.

4. Leitch 1855, 262; Young 1823c, xi.

5. Young 1823c, 301.

6. Like Casati, Grey remains obscure except in connection to these events, though it seems clear that he was a typical well-off traveler of the period. We have found no records indicating the provenance of the materials he gave Young.

7. Leitch 1855, 302.

8. Young 1823c, 60.

9. See the letter from Raoul-Rochette to Young dated December 21, 1822; Young 1814–29, 60 in PDF. Once Young had the Dacier letter, he would have seen a note mentioning that Raoul-Rochette intended to publish the Casati (Champollion 1822a, 5). Raoul-Rochette excused himself for having been unable to send the copy per Young's request via Champollion, and he never did publish the Casati, perhaps because it was included in Young 1823b. Saint-Martin 1822a, 555–67.

10. Peacock 1855b, 326.

11. Young 1823c, 61.

12. Ibid., 51–53. Champollion's letter apprising Young of his "Ramesses" identification is dated November 23, 1822 (Leitch 1855, 243–47).

13. Young 1823c, 52–53.

14. Ibid., 48.

15. Manetho (early to mid-third-century BCE) listed seventeen kings following "Tethmosis, the king who expelled the 'shepherds' [Hyksos]." The penultimate is "Amenophis," followed by "his son Sethos, who is also called Rhamsses." This is the longest sequence in Manetho's list, and Champollion accordingly linked Bankes's Abydos cartouches to it, yielding "Ramesses" (Sethos) as the final one in the list (Verbrugge and Wickersham 2003, 159).

16. In his *Britannica* piece, Young had, we have seen, made some use of Coptic, but in the *Account* he wrote that "my intention in placing the Coptic names in my vocabulary of hieroglyphics, was to assist in tracing any such analogies that might suggest themselves" rather than relying on Coptic for ab initio deductions (1823c, 48).

17. Leitch 1855, 209. Young had met Humboldt, Arago, Cuvier, Biot, and Gay-Lussac in England before his 1817 visit to Paris, where he met them again (Gurney 1831, 32).

18. Leitch 1855, 208–9. In his letter, Humboldt particularly praised what he termed Young's "Vocabulaire Hiéroglyphique" and his "Dissertation sur la Chronologie, les Arts, et les Antiquités de l'Egypte." The latter is Young's *Britannica* article on Egypt. The only "hieroglyphic vocabulary" Young published at this time appeared in the plates accompanying this article.

19. The announcement did not appear for two years. When it did, de Sacy grouped it with Champollion's Dacier letter and the *Précis* (de Sacy 1825).

20. The *Recherches* is Letronne 1823, which concerns, among other things, the appropriate translation of the Greek on the plinth found originally by Cailliaud. Letronne's reading differed from Young's (presumably in the published letter to Bankes of October 22, 1822, that is, Young 1823a) on several points, and from Champollion's as well on a number of issues.
21. Young 1823c, 19.
22. Barrow 1804, 329. Elsewhere in the *Travels* Barrow disdained as pernicious the effect in China of "the paraphernalia and almost all the mummeries of the Romish church" (303). Barrow admired Confucius but believed him uninfluential.
23. Leitch 1855, 366–69.

CHAPTER 30. AN OPPORTUNE ENCOUNTER

1. Louis Philippe would take power eight years later, following the July Revolution of 1830. An 1823 volume of the society's journal contained brief remarks about Champollion's work (see above, chapter 25n14), likely by Klaproth.
2. Closely associated with the Bourbon pretender, Blacas returned to France in 1814 at the first Restoration as a member of the new government. Following Napoleon's final defeat in 1815, Blacas was made Comte de Blacas d'Aulps by Louis XVIII. Blacas spent the Bourbon Restoration mostly in Rome, with visits to Paris, until the removal of Louis's successor, Charles X, in July 1830. Blacas has not received a full biographical treatment. The material presented here was developed by the genealogist Pierre Nicolas and can be found at Nicolas 2017.
3. Guichard 2007. The collection had belonged to Sauveur-Fortuné Thédenat-Duvent (b. 1800), the son of Pierre-Paul Thédenat-Duvent, French vice-consul at Alexandria from 1815 to 1820.
4. No contemporary account of the encounter between Blacas and Champollion has survived. All subsequent accounts apparently derive from Hartleben 1906, 451, no source provided, though one element of the account appears in Champollion-Figeac 1887, 58. Hartleben may have heard the story from Champollion's nephew, Aimé-Louis Champollion-Figeac, before he died in 1894 (Hartleben 1906, xviii). According to the French Egyptologist Gaston Maspero, who wrote the introduction to Hartleben's biography, Hartleben interviewed both Aimé-Louis Champollion-Figeac's daughter and her husband Leon de La Brière, as well as Champollion's granddaughter (daughter of Zoraïde), who "related what her mother had related of her father" (vi). Other family members who produced stories of Champollion's life did not mention this encounter. La Brière was the author of *Champollion Inconnu* (1897), in which the story of the encounter with Blacas does not appear; La Brière's account is not perfectly trustworthy, however, as it seems intended primarily to shield Champollion from the accusation that he was antagonistic to religion and is based on selective quotations from letters for which he provided no source.
5. Hartleben 1906, 450.
6. The king's interest had been fixed on Egyptian matters, at least since Biot convinced him to buy the Dendera zodiac "for the glory of France" (1823, v).
7. Hartleben 1906, 452, for the inscription. On the jewel-encrusted gift, see Champollion-Figeac 1887, 94, as reported in the *Moniteur Universel*.
8. For a full account of the zodiac debates, see Buchwald and Josefowicz 2010.
9. Hartleben 1909, 230.
10. Lacouture 1988, 304; no source provided.
11. The book contained six colored plates together with two hundred others and ran to 450 pages of descriptive text (Champollion 1823b). The lithographs were produced by Léon Jean-Joseph Dubois who became a curator of the Egyptian collections at the Louvre following Champollion's death in 1832. It was printed, like the Dacier letter, by Firmin Didot and was available by the fall of 1823. The book was not paginated, as was typical of luxury editions, which were valued chiefly for the images they contained.
12. La Brière 1897, 184. La Brière did not indicate to whom Champollion so wrote, but it was not likely his brother, who would have known about the gift and who would certainly have warned him not

to sell it. Champollion's estimate of the gift's worth was vastly too high. At his death the (unbound) copy was valued at 300 francs.

13. Although Henry Salt had reportedly tried to purchase Drovetti's collection for the British Museum, Drovetti's asking price was too high. Two years later, in 1818, Louis-Nicolas, Comte de Forbin, director general of Royal Museums, made similar efforts, but nothing came of them. Meanwhile, the kingdom of Piedmont under Victor Emmanuel I emerged as a potential buyer. Negotiations for the collection were undertaken by Count Carlo Vidua, and after several mishaps (not least the death of Victor Emmanuel), the Piedmont offer was formally accepted and signed in Alexandria on March 24, 1823. The collection arrived in Turin ten months later. See Forbin 1819, 23, cited in Ridley 1998, 251, upon which we rely.

14. Ridley 1998, 257–59.

15. La Brière 1897, 72.

16. Hartleben 1906, 460. Few if any contemporary accounts remain of this period. Hartleben apparently relied upon the material she obtained directly from the Champollions' descendants.

17. On these months, see Faure 2004, 456–67. The trip's cost was partly born by a grant from the king and by the largesse of Blacas. In return, Champollion kept Blacas abreast of work on Drovetti's rich collection.

CHAPTER 31. THE "TRUE KEY" TO EGYPTIAN HIEROGLYPHS

1. Champollion 1836 and 1841. The *Egyptian Grammar* was published posthumously, as was the *Egyptian Dictionary*, both from manuscripts edited by Champollion-Figeac.

2. Champollion 1824, dedication. "The approbation with which Your Majesty honored my first efforts sustained and redoubled my zeal; able now to flatter myself for having reached the goal to which I had constantly directed my studies, may I be permitted to take pride in your beneficence, and to say that it is to the munificence of Your Majesty that I owe the publication of this new work which you have kindly deigned to accept, as an offering of my profound and respectful gratitude." These fawning words are unlikely to be Champollion's alone and probably reflect the editorial hand of his ever-vigilant brother. In the pages that followed, Champollion mentioned neither the king nor Blacas but only his competitors in decipherment, especially Young.

3. Here and elsewhere in this chapter, emphases are in the original unless otherwise noted.

4. Champollion 1824, dedication, 6–7.

5. Ibid., 7–9.

6. Ibid., 5–6.

7. Printed the following fall, the *Précis* was available by early 1824. According to Peacock, Young's *Account* appeared "very early in the year" of 1823 (Peacock 1855b, 324).

8. The time in question here was no earlier than 1819 since Champollion referred to Young's *Britannica* piece.

9. Champollion 1824, 17–18.

10. Champollion refers to Heeren 1816, 287. A professor of philosophy at Göttingen and the son-in-law of the classical scholar Christian Gottlob Heyne, Heeren had turned to historical work by 1790. In the article to which Champollion refers, Heeren noted that the Göttingen theologian and Arabic specialist Thomas Tychse had made the connection, principally as a counter to Zoëga's claim that hieratic was entirely a more elegant version of the "alphabetic script," that is, the Rosetta intermediary.

11. Champollion 1824, 24.

12. Ibid., 28–29.

13. Ibid., 37; emphases his.

14. Ibid., 42–43.

15. Ibid., 44.

16. Ibid., 48.

17. The Mountnorris papyrus is at British Museum, EA 10088.6.

18. Champollion 1824, 52. The "phonetic alphabet, formed on the basis of Greek and Roman proper names" is the one he had presented in the Dacier letter (fig. 26.4).

CHAPTER 32. THE SEMANTIC TRAP AVOIDED

1. Champollion 1824, 312.
2. Abel-Rémusat 1822, 3–4, who remarked that "most names of trees, plants, fish, birds and many other objects that it would otherwise be hard to represent are designated by characters of this story, which form at least half of the written language." Here Champollion learned that the "Chinese language consists of four hundred fifty *syllables*, which are brought to twelve hundred three by variation of accents, and which serve for the pronunciation of many thousands of characters" (Champollion 1824, 306). The author of the Chinese grammar noted, on the page cited by Champollion, that "certain syllables, more used than the others, serve for the pronunciation of 30 or 40 characters, and express, in consequence, up to 30 or 40 different ideas" (Abel-Rémusat 1822, 33). For Champollion this extreme homophony definitively distinguished Chinese from Egyptian.
3. Champollion 1824, 304–7.
4. Champollion 1822e, 169 of PDF.
5. Champollion 1824, 66–68.
6. Ibid., 82.
7. The assumption was that such zodiacs were drawn to represent the state of the sky at the time of their construction, enabling the deployment of astronomical calculations with results that could oppose biblical chronology. Champollion's reading, in the Dacier letter, of "autocrator" on the Dendera temple had gainsaid such calculations by placing Dendera squarely in Greco-Roman times.
8. Champollion 1824, 174–75. On the astronomical dating and resultant controversy, see Buchwald and Josefowicz 2010.
9. Apparently a good number of such vases were produced and widely distributed. They date between the early and mid-fifth century BCE. Champollion 1824, 175–79; Barrow 1822, 188. The vase in question was first described by Caylus (1752–67, 5:79–81; 1762) and contains, in addition to the cartouche, cuneiform characters in which, at Champollion's request, Saint-Martin recognized the name "Xerxes," the latter having been identified in Persian inscriptions by the German philologist Georg Friedrich Grotefend. A similar vase exists today in the Egyptian collections at the University of Pennsylvania. See Clay 1910. On Champollion's interaction with Saint-Martin concerning "Xerxes," see Hartleben 1906, 430, which seems to be the only extant account.
10. Champollion 1824, 249–51.
11. Ibid., 65. Plutarch considered the Egyptians to be inclined to geometrical patterning. Cf. *Isis and Osiris*, where he wrote, "Five makes a square of itself, as many as the letters of the Egyptian alphabet, and as many as the years of the life of the Apis."
12. Champollion did show his rendition of the Rosetta into Coptic to Sir William Gell in Rome, who urged him to publish it (Leitch 1855, 421). Champollion never did so, probably because the Coptic words did not thoroughly match his system.

CHAPTER 33. THE RECEPTION OF THE *PRÉCIS*

1. Barrow 1823, cited in Gange 2013, 55. According to Gange, British antagonism to ancient Egypt stemmed from "the perceived biblical imperative to slight and dismiss the achievements of the civilization that brutalized the Old Testament Israelites" in a Britain "suffused with Evangelical Revival, anti-revolutionary enthusiasm and apocalyptic speculation" (ibid.)
2. Young 1823c, 53.
3. Leitch 1855, 286; Young 1823c, 34.
4. De Sacy 1825, 141–42. Four of Young's fourteen phonetic hieroglyphs were fully syllabic.
5. Champollion did indeed use the term, though even at the time, his logic pressed him to find evidence for phonographic antiquity.
6. De Sacy 1825, 144.
7. Ibid., 145. On de Sacy and Court de Gébelin, see chapter 8.
8. Ibid., 147 and 51–52.

9. Hartleben 1909, 213.

10. A second 1824 work on the hieroglyphs, by the German astronomer Johann Wilhelm Pfaff, had no substantive effect on subsequent discourse. On Pfaff, see Oestmann 2005.

11. Goulianoff 1824. Goulianoff's pseudonym may refer to Decimus Ausonius, a fourth-century CE poet and *grammaticus* from Gaul who tutored the future emperor and scourge of paganism Gratian.

12. For Seyffarth's autobiographical sketch, see Seyffarth and Knortz 1886. Spohn 1825 also contains a biography. The conflict between Seyffarth and Champollion is discussed in Messling 2012, 70–94.

13. Leitch 1855, 386–88.

14. Kosegarten 1824, 288. The review otherwise avoided comparative judgments.

15. Sandys 1908, 3:106.

16. Humboldt 1821.

17. These were held in a collection bought by the Prussian king from Count Heinrich Menu von Minutoli, who had amassed a large collection of Egyptian antiquities during an expedition in 1820 and 1821. Partly lost during shipwreck, the collection was purchased by the king. On Humboldt and hieroglyphics, see Messling 2008, 157–78, and Richter 2015. In the spring of 1824, before he had seen the *Précis*, Humboldt sent Champollion an elaborate response to the Dacier letter (Messling 2008, 317–57).

18. Leitch 1855, 383–84.

19. Seyffarth 1826.

20. On the "languages of paradise," see Olender 1992.

21. Seyffarth and Knortz 1886, 57.

22. Though Gell was close to William Drummond, he nevertheless enjoyed a chummy acquaintance with Young despite the latter's acerbic critique of Drummond's *Herculanensia*.

23. Leitch 1855, 226.

24. Ibid., 394.

25. Champollion 1826. This remark would certainly not have pleased Jomard, since Champollion had long made no effort to conceal his criticism of the *Description*'s renditions.

26. Ricardi 1824. There seems to be no remaining copy of the brochure, though de Sacy evidently owned one (de Sacy 1846, 2:259). The gist of Ricardi's understanding of hieroglyphs can be discerned, vaguely, in the following passage from his critique of Champollion: "[T]he signification of the hieroglyphs was based either on the expression of the object figured, or on one of its properties, or on some truth dictated by God and by reason." Since the Egyptians thought the signs "imperishable," they often wrote them on "very solid monuments in order to transmit the dogmas of their religion to the most distant posterity" (Ricardi 1826, 147).

27. To his brother; Hartleben 1909, 1:41.

28. Champollion 1824, 370–71.

29. Hartleben 1909, 1:68–71. We have not located this denunciation.

30. Ricardi 1820, 23.

31. Ricardi 1811, 83–84.

32. Ibid., 85–93.

33. Ricardi 1826, ix–x. Seyffarth and Knortz 1886, 10. Referring to the Coptic rendition of the Rosetta demotic inscription as a "failure," Seyffarth overstates his case. Champollion held back the material because he remained uncertain about the interpretations of all the signs.

34. Ricardi 1826, 151.

35. Hartleben 1909, 1:216.

36. Lanci 1825, 27 and 47. Lanci published the addition on the Egyptian script together with the much longer piece about an Phoenician bas-relief. "The principal guide for properly determining the sounds, and the phonetic value of the signs, must be Moses," he wrote.

37. Champollion 1825. On Lanci, see Testa 2002.

38. Leo XII 1824.

39. On Testa and the zodiacs, see Buchwald and Josefowicz 2010.

40. Hartleben 1909, 1:226–29.

41. Champollion 1825.

42. Hartleben 1909, 233–34n.
43. Ibid., 321–22.
44. Potocki is now remembered primarily as the author of a novel, *Manuscrit trouvé à Saragosse* (1805).
45. Klaproth's introduction to the Humboldts possibly came through his father, Martin Heinrich, a well-known apothecary and chemist in Berlin who, as a follower of Lavoisier's chemical system, had identified the elements uranium and zirconium.
46. This was especially necessary since his patron, Potocki, had shot himself with a silver bullet in 1815 on becoming convinced he was turning into a werewolf. On Klaproth, see Walravens 2005.
47. Klaproth 1823.
48. Goulianoff 1824, 11.
49. "Wishing to express Sun in writing, they make a circle; and Moon, a figure like the Moon, like its proper shape. But in using the figurative style, by transposing and transferring, by changing and by transforming in many ways as suits them, they draw characters. In relating the praises of the kings in theological myths, they write in anaglyphs" (*Stromata*, book 5, chap. 4). Champollion 1824, 300–301; Goulianoff 1824, 19.
50. Goulianoff was referring to Champollion's homonyms for "L." One among them certainly did look very much like a lion, but the others stripped away all such connections, leaving what Goulianoff referred to as a "bar," his schematic for something like the creature's "dorsal spine."
51. Goulianoff 1824, 26–38.
52. Klaproth 1827a; quoting from Champollion 1824, 397.
53. Champollion 1824, 397–98.
54. Ibid., 200. Horapollo, for example, on the sign that refers to a sunset: "a crocodile hunched up. For it is a self-producing and lewd animal," presumably on the grounds of the vivid color of the sunset and the Egyptian belief in the sun's daily death and rebirth (1993, 72; book 1, no. 69).
55. Champollion 1824, 299–300.
56. Klaproth 1827a, 9.
57. Klaproth coined the neologism "acrologic" for a sign that represents the initial sound of a word, from the Greek ακροσ (extremity).
58. Klaproth 1827a, 2.
59. We have not located the pamphlet signed by Merian, although Hartleben refers to it; see Hartleben 1906, 2:10 and 2:136. However, Goulianoff considered Merian a close friend (1827, 16n), and in a letter of May 31, 1827, to the English classical scholar Samuel Butler, whom Merian had met at Cambridge, Merian remarked that "Champollion, vain and silly like a peacock, thinks himself sole emperor of the Nile, and the rest of mankind doomed to hear and believe his fictions. I hope Mr. Drummond has sent you the first attack upon Champollion; a much severer one is just peeping out of the press" (Butler 1827, 329). This "much severer" publication was probably Klaproth's second letter attacking Champollion (Klaproth 1827b).
60. Champollion 1827.
61. The argument in respect to Clement concerned the precise meaning of a phrase that, referring to the "curiologic" signs, Letronne and others read as "by means of letters," whereas Goulianoff emphasized that the literal meaning is "by means of first elements" (1824, 4–5). Goulianoff further claimed that Clement's reference, in the following line, to "symbolic" signs also connoted "by means of first elements." This assumption enabled him and Klaproth to compress every one of Clement's "hieroglyphics" into essentially a single type since Clement had subdivided the "symbolic" hieroglyphs into three further types, namely, curiologic, "tropical" (the sign signifies the object represented), and "allegorical." For Goulianoff all of these types were essentially the same, except that signs in the "allegorical" subdivision could be semantically extended via something like Horapollo's explanations; even so, they always reproduced the initial sound of a corresponding word.
62. Champollion 1827, 290. Horapollo provided an extensive discussion on the meaning of "what they mean by a vulture" based on the bird's characteristics in book 1, no. 11 (1993, 49–51). The Coptic for "vulture" begins either with a ϲ (ϲελειλεν) or an ⲛ (ⲛⲟⲩⲣⲉ), whereas "mother" is some variant of ⲙⲁⲁⲩ.
63. Champollion 1827, 299.

64. Klaproth 1827b, 5–6.
65. Ibid., 7–8.
66. Ibid., 44–45.
67. Hartleben 1909, 422: "I will reply to neither Seyffarth nor to Klaproth. My comments remain, and neither the one nor the other has answered anything that is worthy of my objections or of the facts I opposed to them" (ibid., 1:422). Seyffarth had replied to Champollion's objections in Seyffarth 1827.
68. Klaproth and Dorow 1829, 1–6, followed by thirty-four pages critiquing Champollion's assignations.
69. On Klaproth as a critic: "Klaproth has inherited the reputation of a severe and relentless critic who always smelled charlatanry . . . In many cases this was certainly unnecessary, at least from today's point of view" (Walravens 2005, 184–85).
70. Browne 1829, 468–69.

CHAPTER 34. "HOLD YOUR LAUGHTER, FRIENDS!"

1. The first detailed account of the period from 1824 through Champollion's death in 1832 is Hartleben 1906, vol. 2, on which subsequent work principally relies. For later accounts, see Faure 2004, 553–758; Lacouture 1988, 380–492. The most readable and concise discussion is Robinson 2012, 167–252. Salt, who was poorly paid as a British representative in Cairo, expanded his income through the acquisition and sale of Egyptian antiquities, in which he competed with the French representative, Drovetti. On Salt, see Manley 2001 and Halls 1834.
2. Peacock 1855b, 341–42.
3. Young 1831, vi.
4. Champollion and Champollion-Figeac 1835–45. The volumes are now rare but were reprinted in toto in Champollion and Champollion-Figeac 1970.
5. Champollion 1836 and 1841.
6. Champollion 1833, 89.
7. Ibid., 89–91.
8. The phrase is from Horace's *Ars Poetica*, 5.
9. Mean labor income at the time was about 450 francs per year, so this was about ten times a mean yearly wage. Overall the estate was worth between 90 percent and 95 percent of contemporary French estate values at death. We thank Jean-Laurent Rosenthal for obtaining the estate information and for details concerning mean income and estate values. The Champollion estate record is dated August 9, 1832, CVI 807, "Inventaire après le décès dr Mr. Champollion."
10. Champollion 1833, 91–92. There's no doubt that Champollion-Figeac edited the volume of letters, despite the absence of his name from it. An edition reprinted more than forty years later and containing a few remarks by Champollion's daughter, Zoë, still omitted the passage, which only became available in 1887 when his nephew published the omitted portion (Chéronnet-Champollion 1868, 75, and Champollion-Figeac 1887, 174).

BIBLIOGRAPHY

Of the enormous number of articles and books on the decipherment of Egyptian hieroglyphics, many concentrate on who first obtained what Egyptologists now consider to be the correct value of which sign when. These priority-focused works tend to be older ones; by the end of the nineteenth century the controversy between Young and Champollion had cooled to some extent, and a degree of distance began to obtain in writing about the events.

Among the earliest detailed accounts is the Egyptologist E. A. Wallace Budge's 1904 *The Decrees of Memphis and Canopus: The Rosetta Stone*. It was followed in 1906 by Hermine Hartleben's two-volume biography, *Champollion: Sein Leben und sein Werk*. An Egyptologist herself, Hartleben had access to materials held by descendants of the family, some of which are no longer available so that subsequent work, including ours, has had in places to rely upon her presentation. Since Hartleben favored Champollion, care must be taken with her evaluations, though this work, as well as her 1909 *Lettres de Champollion*, remain invaluable sources. For Young we have the 1855 edition by John Leitch of his Egyptological researches in the third volume of the *Miscellaneous Works*. Written scarcely two decades after Young's and Champollion's deaths, Leitch fiercely defended Young's priority. Nevertheless, he did accurately, if incompletely, publish a considerable number of the relevant documents, confining his partisanship to his footnotes. Besides Peacock's, the biography of Young by Alexander Wood, and its extension by Frank Oldham, touch on the hieroglyphic affair. Since the turn of the nineteenth century, a nearly uncountable number of publications have followed. There are many gems in this material, but Jeannot Kettel's 1990 comprehensive catalog of materials concerning Champollion has proved the most indispensable for research. In recent years two major biographies of Champollion by Alain Faure and Jean Lacouture have appeared, both of which rely on Hartleben. The four most reliable and informative recent works in English on the decipherment of hieroglyphics are the Egyptologist John Ray's *The Rosetta Stone* (2007), Leslie and Roy Adkins' *The Keys to Egypt* (2000), and a pair of books by Andrew Robinson: a biography of Young, *The Last Man Who Knew Everything* (2006), and *Cracking the Egyptian Code* (2012), which focuses on Champollion. Many other secondary works appear in the following bibliography. The list includes all sources, primary and secondary, that we have used.

We have made extensive use of archival manuscript materials. The Young and the Champollion manuscripts are not uniformly, sequentially numbered. In all cases where we have cited a manuscript source, we have provided the shelfmark for that source in the bibliography and, in the notes, the number written on the page if such a number was available; for clarity, we often refer as well to the corresponding page in a PDF generated from the source. Where "MF" appears, it designates a microfilm source from which we have generally also produced a PDF.

Unless otherwise noted, all translations are our own, and emphases are as they appear in the original source.

PRIMARY SOURCES

Abel-Rémusat, J. P. (1822). *Eléments de la grammaire chinoise: Ou principes généaux du Kou-wen ou style antique, et du Kouan-Hoa, c'est-à-dire, de la langue commune généralement usitée dans l'empire chinois.* Paris: Imprimerie Royale.

Adelung, J. C. (1806–12). *Mithridates, oder Allgemeine Sprachenkunde.* Berlin: Vossische Buchhandlung.

Åkerblad, J. D. (1801). "Lettre au Silvestre de Sacy sur la découverte de l'écriture cursive copte." *Magasin Encyclopédique* 5:489–94. Paris.

———. (1802a). *Inscriptionis Phoeniciae Oxoniensis nova Interpretatio.* Paris: Typographia Republicae.

———. (1802b). *Lettre sur l'inscription égyptienne de Rosette, adressée au C.en Silvestre de Sacy, professeur de langue arabe à l'école spéciale des langues orientales vivantes, &c.* Paris: Imprimerie de la République.

Ameilhon. (1803). *Éclaircissements sur l'inscription grecque du monument trouvé à Rosette.* Paris: Baudouin.

Anonymous (1799 [29 Fructidor, An VII]). ". . . une pierre d'un très-beau granit noir . . ." *Courrier de l'Égypte* 3:3–4.

———. (1800). "Extrait du procès-verbal de la classe de littérature et beaux-arts: Séance du 3 Fructidor an 8." *Magasin Encyclopédique* 3, year 6:254–56.

———. (1801). ". . . a remarkable inscription . . ." *Gentleman's Magazine, and Historical Chronicle* 71:1194.

———. (1802a). "[Review of] *Letter on the Egyptian Inscription of Rosetta; addressed by Citizen Silvestre de Sacy, Professor of Arabic in the School for Living Oriental Languages. By J. D. Åkerblad.*" *Critical Review; or, Annals of Literature* 36:550–53.

———. (1802b). "The Greek Version of the Decree of the Egyptian Priests . . . From the Stone, Inscribed in the Sacred and Vulgar Egyptian, and the Greek Characters, Taken from the French at the Surrender of Alexandria." *Critical Review; or, Annals of Literature* 35:527–31.

———. (1803; published 1815). "Plates of the Rosetta Inscriptions." *Vetusta Monumenta: Quae ad rerum Britannicarum meoriam conservandam Societas antiquariorum Londini sumptu suo edenda curauit* 4.

———. (1804). "Inscriptionis Phoeniciae Oxoniensis nova Interpretatio Auctore J. D. Åkerblad." *Monthly Review; or, Literary Journal, Enlarged* 44:542–44.

———. (1806). "Review of Drummond's *Academical Questions.*" *Edinburgh Review* 7:163–85.

———. (1807). *Mémoires de l'Académie celtique.*

———. (1814). "Clarke's Travels." *Monthly Magazine.*

———. (1815). "Art. VII. Histoire de l'Ambassade &c. . . . by M. de Pradt." *Monthly Review* 78:517–28.

———. (1816). "*L'Égypte sous les Pharaons*, &c. . . . by M. Champollion, jun." *Monthly Review* 79:467–72.

———. (1817). *Herculaneum Rolls: Correspondence Relative to a Proposition Made by Dr. Sickler of Hildburghausen, upon the Subject of Their Development.* London: J. Barfield.

———. (1821). "Egyptian Antiquities." *London Literary Gazette and Journal of Belles Lettres* 223:268–69.

———. (1822). "Précis de mémoires relatifs aux écritures égyptiennes, rédigés par M. Champollion le jeune, et lus à l'Académie royales des inscriptions et belles-lettres." *Nouvelle Annales des Voyages de le Géographie et de l'Histoire* 16:108–19.

———. (1823a). "Notice sur les travaux scientifiques et littéraires, mentionnés dans la *Revue Encyclopédique*, en 1822." *Revue Encyclopédique* 17:15–28.

———. (1823b). "[Review of] *Notice sur le zodiaque de Denderah, par M. J. [Saint-]Martin.*" *North American Review* 41:233–42.

———. (1823c). "Les travaux de M. Champollion . . ." *Journal Asiatique* 2:61–62.

Arnauld, A., and P. Nicole. (1662). *La logique, ou L'art de penser.* Paris: Jean Guignart, Charles Savreux, and Jean de Lavney.

Balzac, H. de. (1840). "Lettres sur la littérature, le théâtre, et les arts." In *Revue Parisienne,* 1:47–98. Paris: Garnier Frères.

———. (1976–81). *Le Père Goriot: La comédie humaine.* Paris: Gallimard.

Bandini, A. M. (1750). *Dell'Obelisco di Cesare Augusto Scavato dalle Rovine del Campo Marzo, Commentario.* Rome: Marco Pagliarini.

Bankes, W. J. (1821). *Geometrical Elevation of an Obelisk . . . from the Island of Philæ, together with the Pedestal . . . First Discovered There by W. J. Bankes . . . in 1815: At Whose Suggestion & Expense, Both Have Been since Removed . . . for the Purpose of Being Erected at Kingston Hall in Dorsetshire.* London: John Murray.

Banks, J., and N. Chambers (2007). *Scientific Correspondence of Sir Joseph Banks, 1765–1820.* Vol. 5. London: Pickering & Chatto.

Barrow, J. (1804). *Travels in China.* London: T. Cadell and W. Davies.

———. (1816). "Review of *Brief View of the Baptist Missions, Clavis Sinica* and *A Dictionary of the Chinese Language.*" *Quarterly Review* 15:350–75.

———. (1822). "*Lettre à M. Dacier . . .* Par M. Champollion le Jeune." *Quarterly Review* 28:188–96.

———. (1823). "Egypt, Nubia, Berber, and Sennar." *Quarterly Review* 28:60–97.

———. (1849). *Sketches of the Royal Society and Royal Society Club.* London: John Murray.

Barrow, J., and G. T. Staunton. (1811). "Review of *A Dissertation on the Characters and Sounds of the Chinese Language.*" *Quarterly Review* 5:372–403.

Barthélemy, J.-J. (1759). "Réflexions sur l'alphabet et sur la langue dont on se servoit autrefois a Palmyre." In *Mémoires de l'Académie des inscriptions et belles-lettres,* 26:577–97.

———. (1764). "Réflexions sur quelques monuments phéniciens, et sur les alphabets qui en résultent." *Mémoires de littérature, tirés des registres de l'Académie royale des inscriptions et belles-lettres,* 30:405–27.

———. (1768). "Reflexions générales sur les rapports des langues égyptienne, phénicienne, et grecque." In *Mémoires de l'Académie des inscriptions et belles-lettres,* 32:212–33.

———. (1797). *Oeuvres Diverses de J. J. Barthélemy.* Part 2. Paris: H. J. Jansen.

Bayley, C. (1782). *An Entrance into the Sacred Language; Containing the necessary Rules of Hebrew Grammar in English.* London: R. Hindmarsh.

Beattie, W. (Ed.) (1849). *Life and Letters of Thomas Campbell.* London: Edward Moxon.

Bergier, N. S. (1837 [1764]). *Les élémens primitifs des langues.* Besançon: Lambert.

Biot, J.-B. (1823). *Recherches sur plusieurs points de l'astronomie égyptienne: Appliquées aux monuments astronomiques trouvés en Égypte.* Paris: F. Didot.

———. (1844). *Mémoire sur le zodiaque circulaire de Denderah.* Paris: Imprimerie Royale.

Blagden, C. (1787). "Some Observations on Ancient Inks, with the Proposal of a New Method of Recovering the Legibility of Decayed Writings." *Philosophical Transactions of the Royal Society of London* 77:451–57.

Blomfield, C. (1810). "Review of *Herculanensia.*" *Edinburgh Review* 16:368–84.

Bonnycastle, J. (1782). *An introduction to algebra with notes and observations.* London: Printed for J. Johnson.

Boswell, J. (1904). *Boswell's Life of Johnson.* London: Henry Frowde.

Boughton, W.E.R. (1817 [1814]). "A Letter from W. E. Rouse Boughton, Esq/F.R.S. to the Rev. Stephen Weston, B.D. respecting Some Egyptian Antiquities." *Archaeologia* 18:59–72.

Brougham, H. (1856). *Historical Sketches (Second Series).* London: Richard Griffin.

Browne, J. (1829). "Art. III [review of Spohn, Seyffarth, Champollion]." *Foreign Quarterly Review* 4:438–69.

Burnett, J. (1773). *Of the Origin and Progress of Language.* Edinburgh: A. Kincaid and W. Creech / London: T. Cadell.

Butler, S. (1827). *The Life and Letters of Dr. Samuel Butler: Jan. 30, 1774–March 1, 1831.* London: John Murray.

Cailliaud, F. (1826–27). *Voyage à Méroé, au Fleuve Blanc . . . fait dans les années 1819, 1820, 1821 et 1822.* Paris: Imprimerie Royale.

Camden, W., and S. Westminster. (1754). *Institutio graecae grammatices compendiaria, in usum regiae scholae Westmonasteriensis: Scientiarum ianitrix grammatica.* London: Excuderunt S. Buckley et J. Longman, Regii in Latinis, Graecis, et Hebraicis Typographi.

Campbell, T., and W. Beattie. (1849). *Life and Letters Edited by William Beattie.* London: E. Moxon.

Caylus, A.C.P. (1752–67). *Recueil d'antiquités égyptiennes, étrusques, grecques, romaines et gauloises.* Paris: Desaint & Saillant.

Chamberlayne, J. (1715). *Oratio Dominica in diversas omnium fere gentium linguas versa et propriis euiussque linguae characteribus expressa.* Amsterdam: G. & D. Goerei.

Champollion, J.-F. (1808–10). *NAF 20303.* Paris: Bibliothèque Nationale.

———. (1811a). *L'Égypte sous les Pharaons: Description geographique; Introduction.* Paris: Peyronard.

———. (1811b). *Observations sur le catalogue des manuscrits coptes du musée Borgia a Velletri, ouvrage postume de George Zoëga.* Paris: J. B. Sajou.

———. (1811–12a). *NAF 23142 (MF 22367).* Paris: Bibliothèque Nationale.

———. (1811–12b). *NAF 20372.* Paris: Bibliothèque Nationale.

———. (1814). *L'Égypte sous les pharaons, ou Recherches sur la géographie, la religion, la langue, les écritures et l'histoire de l'Égypte avant l'invasion de Cambyse.* Paris: De Bure Frères.

———. (1814–21a). *NAF 20389.* Paris: Bibliothèque Nationale.

———. (1814–21b). *NAF 20313.* Paris: Bibliothèque Nationale.

———. (1815). *NAF 20355 (MF 4854).* Paris: Bibliothèque Nationale.

———. (1816–19). *NAF 20317.* Paris: Bibliothèque Nationale.

———. (1820). *Attention.* Paris: Chez Corréard.

———. (1821a). *De l'écriture hiératique des anciens Égyptiens: Explication des Planches.* Grenoble: Baratier Frères.

———. (1821b). *NAF 20311.* Paris: Bibliothèque Nationale.

———. (1822a). *Lettre à M. Dacier, secrétaire perpetuel de l'Académie royale des inscriptions et belles-lettres, relative à l'alphabet des hiéroglyphes phonétiques employés par les Égyptiens pour inscrire sur leurs monuments les titres, les noms et les surnoms des souverains Grecs et Romains.* Paris: Firmin Didot Père et Fils.

———. (1822b). "Extrait d'un Mémoire relatif à l'alphabet des hiéroglyphes phonétiques égyptiens." *Journal des Savans,* 620–28.

———. (1822c). "De l'obélisque égyptien de l'Île de Philae." *Revue Encyclopédique* 13:512–21; printed originally in issue 39 for March.

———. (1822d). "Lettre à M. le Rédacteur de la Revue encyclopédique, relative au Zodiaque de Denderah." *Revue Encyclopédique,* 232–39.

———. (1822e). *NAF 20314.* Paris: Bibliothèque Nationale.

———. (1823). *Panthéon égyptien: Collection des personnages mythologiques de l'ancienne Égypte, d'après les monuments; avec un texte explicatif.* Paris: Firmin Didot.

———. (1824). *Précis du système hiéroglyphique des anciens Égyptiens, ou, Recherches sur les élémens premiers de cette écriture sacrée.* Paris: Imprimerie Royale.

———. (1825). "Lettre de M. Champollion le jeune, à M. Z***." *Bulletin Universel des Sciences et de l'Industrie* 7:3–10.

———. (1826). *Lettre à m. le duc de Blacas d'Aulps . . . sur le nouveau système hiéroglyphique de mm: Spohn et Seyffarth.* Florence: G. Piatti.

———. (1827). "[Review] of *Lettre sur la découverte des hiéroglyphes acrologiques, adressée à M.*" *Bulletin des Sciences Historiques, Antiquités, Philologie* 7:289–99.

———. (1833). *Lettres écrites d'Égypte et de Nubie en 1828 et 1829.* Paris: Firmin Didot Frères.

———. (1836). *Grammaire égyptienne, ou Principes généraux de l'écriture sacrée égyptienne.* Paris: Firmin Didot.

———. (1841). *Dictionnaire égyptien en écriture hiéroglyphique.* Paris: Firmin Didot.

Champollion, J.-F., and J. J. Champollion-Figeac. (1835–45). *Monuments de l'Égypte et de la Nubie d'après les dessins exécutés sur les lieux, sous la direction de Champollion le jeune, et les descriptions autographes qu'il en a rédigées; Publié sous les auspices de M. Guizot et de M. Thiers, Ministre de l'Instruction Publique et de l'Intérieur; par une commission spéciale; planches.* Paris: Firmin Didot.

———. (1970). *Monuments de l'Égypte et de la Nubie.* Genève: Éditions de Belles-Lettres.

Champollion, J.-F., and P. Vaillant (1984). *Jean-François Champollion: Lettres à son frère, 1804–1818.* Paris: L'Asiathèque.

Champollion-Figeac, A. L. (1880–89). *Chroniques dauphinoises et documents inédits relatifs au Dauphiné pendant la Révolution.* Vienne [Isère]: E.-J. Savigné.

———. (1887). *Les deux Champollion: Leur vie et leurs œuvres, leur correspondance archéologique relative au Dauphiné et à l'Égypt; Étude complète de biographie et de bibliographie, 1778–1867, d'après des documents inédits.* Grenoble: X. Drevet.

Champollion-Figeac, J.-J. (1804). *Dissertation sur un monument souterrain existant à Grenoble.*

———. (1807). *Antiquités de Grenoble.* Grenoble: Peyronard.

———. (1810). *Discours d'ouverture et programme du cours de littérature grecque.* Grenoble: Impr. de J.-H. Peyronard.

———. (1814). *Nouveaux éclaircissements sur la ville de Cularo, aujourd'hui Grenoble.* Paris: Impr. de J.-B. Sajou.

———. (1820). *Nouvelles recherches sur la ville gauloise d'Uxellodunum.* Paris: Imprimerie Royale.

———. (1842). *Notice sur les manuscrits autographes de Champollion le jeune perdus en l'année 1832, et retrouvés en 1840.* Paris: Firmin Didot.

———. (1844). *Fourier et Napoléon: L'Égypte et les cent jours; Mémoires et documents inédits.* Paris: Firmin Didot Frères.

Chaudruc de Crazannes, J.-C.-M.-A. (1847). "Lettre de M. le Baron Chaudruc de Crazannes a MP Mérimée, Membre de l'Institut, sur une statuette Gauloise en fer." *Revue Archéologique*, Year 4, no. 2 (Oct. 15, 1847 to March 15, 1848): 809–12.

Chéronnet-Champollion, Z. (1868). *Lettres écrites d'Égypte et de Nubie en 1828 et 1829.* Paris: Didier et Cie.

Clarke, E. D. (1805). *The Tomb of Alexander: A Dissertation on the Sarcophagus Brought from Alexandria and Now in the British Museum.* Cambridge: Cambridge University Press.

———. (1809). *Greek Marbles Brought from the Shores of the Euxine, Archipelago, and Mediterranean, and Deposited in the Vestibule of the Public Library of the University of Cambridge.* Cambridge: Cambridge University Press.

———. (1814). *Travels in Various Countries of Europe, Asia, and Africa: Part the Second; Greece, Egypt, and the Holy Land; Sections 1–3.* London: T. Cadell and W. Davies.

Cointreau, A.-L. (1803). *Tableau analytique de la réligion des anciens Égyptiens et des autres peuples originaires de l'Asie.* Paris: Ch. Pougens.

Croker, T. H., M. D. Thomas Williams, and S. Clark (1766). *The Complete Dictionary of Arts and Sciences in which the Whole Circle of Human Learning is Explained.* London: J. Wilson et al.

Crombie, A. (1838). *Gymnasium: Sive Symbola Critica: Intended to Assist the Classical Student in His Endeavors to Attain a Correct Latin Prose Style.* London: Simpkin, Marshall.

D. H. (1802). "Triple Inscription Lately Brought from Rosetta." *Gentleman's Magazine, and Historical Chronicle* 72:725–26.

———. (1803). "Inscription from Rosetta Illustrated by J. D. Åkerblad." *Gentleman's Magazine, and Historical Chronicle* 73:25–28.

d'Achille, A. M., Antonio Iacobini, Monica Preti-Hamard, Marina Righetti, and Gennaro Toscano (Eds.). (2011). *Voyages et conscience patrimoniale / Viaggi e coscienze patrimoniale: Aubin-Louis Millin (1759–1818).* Rome: Campisano Editore.

D'Oyly, G. (1818). *Remarks on Sir William Drummond's "Oedipus Judaicus."* London: W. Bulmer.

Dalzel, A. (1802). *Analecta Ellenika Meizona sive Collectanea graeca majora: Ad usum academice juventutis cum notis philologicis.* Edinburgh: Bell and Bradfute and C. Dickson / London: J. Robinson.

Dalzel, A. (1862). *History of the University of Edinburgh from Its Foundation*. Edinburgh: Edmonston and Douglas.

de Brosses, C. (IX [1800]). *Traité de la formation méchanique des langues et des principes physiques de l'étymologie*. Paris: Terrelonge.

de Gaulle, J. (1839). *Nouvelle histoire de Paris et de ses environs*. Paris: P. M. Pourrat Frères.

Denon, V. (1802a). *Voyage dans la basse et la haute Égypte, pendant les campagnes du général Bonaparte*. Paris: P. Didot l'Aîné.

———. (1802b). *Travels in Upper and Lower Egypt, during the Campaigns of General Bonaparte*. London: B. Crosby.

Dickens, C. (1837). *Oliver Twist*. London: Penguin.

———. (1860–61). *Great Expectations*. London: Penguin.

Dilworth, T. (1765). *The young book-keeper's assistant: shewing him . . . the Italian way of stating debtor and creditor . . . To which is annexed, a synopsis, or compendium of the whole art of stating debtor and creditor*. London: Printed by H. Kent.

Drummond, S. W., and R. Walpole. (1810). *Herculanensia, or, Archeological and Philological Dissertations: Containing a Manuscript Found among the Ruins of Herculaneum*. London: W. Bulmer.

Drummond, W. (1805). *Academical Questions*. London: Cadell & Davies.

———. (1811). *Oedipus Judaicus*. London: A. J. Valpy.

Dumarsais, C. C. (1730). *Des tropes, ou des différents sens dans lesquels on peut prendre un même mot dans une même langue*. Paris: J. B. Brocas.

Ewing, A. (1779). *A synopsis of practical mathematics. Containing plain trigonometry; mensuration of heights, distances, surfaces, and solids; surveying of land, gauging, navigation, and gunnery*. London: T. Cadell / Edinburgh: C. Elliot.

Falconet, C. (1763). *Catalogue de la bibliothèque de feu*. Paris: Jacques Barrois.

Feltham, J. (1803). *The Picture of London, for 1803*. London: R. Phillips.

Finati, G. (1830). *Narrative of the Life and Adventures of Giovanni Finati, Native of Ferrara*. London: John Murray.

Foot, J. (1794). *The Life of John Hunter*. London: T. Becket.

Forbin, C. de. (1819). *Travels in Egypt, Being a Continuation of the Travels in the Holy Land in 1817–1818*. London: Sir Richard Phillips.

Foster, J. (1773). *An Essay on the Different Nature of Accent and Quantity*. Eton: J. Pote.

Gattel, C.-M. (1803). *Nouveau dictionnaire portatif de la langue française*. Lyon: Bruyset.

Gay, J. (1772). *Gay's Fables*. Altenburgh: Gottlob Eman Richter.

Gébelin, A. C. de. (1776). *Histoire naturelle de la parole, ou Précis de l'origine du langage & de la grammaire universelle*. Paris: Boudet et al.

Gibbon, E., and J. H. Sheffield. (1827). *Memoirs of the Life and Writings of Edward Gibbon, Esq*. London: Printed for Hunt and Clarke.

Goldsmith, O. (1770). *The Deserted Village, A Poem*. London: W. Griffin.

Gough, R. (1802). "Triple Inscription Lately Brought from Rosetta." *Gentleman's Magazine and Historical Chronicle* 72:725–26.

———. (1803). *Coins of the Seleucidae, Kings of Syria*. London: J. Nichols and Son.

Goulianoff, I. A. (1827). *Essai sur les hiéroglyphes d'Horapollon et quelques mots sur la cabale*. Paris: P. Dufart.

Goulianoff, I. A. (1824). *Opuscules archéologiques*. Paris: P. Dufart.

Guillaume, J. (Ed.). (1891–1958). *Procès-verbaux du Comité d'instruction publique de la Convention nationale*. Paris: Imprimerie nationale.

Gurney, H. (1831). *Memoir of the Life of Thomas Young, M.D. F.R.S. . . . with a Catalogue of His Works and Essays*. London: John & Arthur Arch.

Halls, J. J. (1834). *The Life and Correspondence of Henry Salt*. London: Richard Bentley.

Halma, N.-B. (1822). *Commentaire de Théon d'Alexandrie sur le livre III de l'Almageste de Ptolemée: Tables manuelles des mouvemens des astres*. Paris: A. Bobée.

Hamilton, W. (1809). *Remarks on Several Parts of Turkey: Part I; Aegyptiaca, or Some Account of the Antient and Modern State of Egypt, as Obtained in the Years 1801, 1802*. London: Thomas Payne.

Harris, J. (1771). *Hermes, or, A philosophical inquiry concerning universal grammar*. London: Printed for J. Nourse and P. Vaillant.

Hartleben, H. (Ed.). (1909). *Lettres de Champollion le Jeune: Lettres écrites en Italie*. Berlin: Weidmann.

Hayter, J. (1810). *Observations upon a Review of the "Herculanensia," in the "Quarterly Review" of Last February, in a Letter to the Right Honourable Sir William Drummond*. London: W. Dulmer.

———. (1811). *A Report upon the Herculaneum Manuscripts: In a Second Letter, Addressed, by Permission, to His Royal Highness, the Prince Regent*. London: Richard Phillips.

Hazlitt, W. (1825). *The Spirit of the Age*. London: Henry Colburn.

Heeren, A. (1816). "Sur les monuments de Thèbes en Aegypte: Extrait de l'ouvrage allemand intitulé 'Idées sur la politique, les relations, et le commerce des peuples anciens,' Traduit par. M. Matter." *Magasin Encyclopédique* 2:241–90.

Helmholtz, H. (1855). "Ueber die Accommodation des Auges." *Archiv für Ophthalmologie* 1:1–74.

Henry, W. C. (Ed.). (1854). *Memoirs of the Life and Scientific Researches of John Dalton*. London: Cavendish Society.

Herder, J. G. (1772). *Abhandlung über den Ursprung der Sprache*. Berlin: C. F. Voss.

———. (1792–94). *Ideen zur philosophie der geschichte der menschheit*. Carlsruhe: C. G. Schmieder.

———. (1800). *Outlines of a Philosophy of the History of Man; Translated from the German of John Godfrey Herder, by T. Churchill*. London: Printed for J. Johnson, by Luke Hansard.

Herodotus. (1958). *The Histories*. New York: Heritage Press.

Heyne, C. G. (1804). "Commentatio in Inscriptionem Graecam Monumeni trinis insigniti titulis ex Aegypto Londinum apportati." *Commentationes Societatis Regiae Scientiarum Gottingensis* 15:260–80.

———. (1812). "Latin Version of the Trilinguar Stone, together with Illustrations of the Inscription." *Archaeologia* 16:229–46.

Hilts, V. L. (1978). "Thomas Young's 'Autobiographical Sketch.'" *Proceedings of the American Philosophical Society* 122:248–60.

Hodgkin, J. (1794). *Calligraphia Graeca et poecilographia Graeca*. London: Published October 1, 1794, by the author and sold by H. Ashby.

Horapollo. (1993). *The Hieroglyphics of Horapollo*. Translated by George Boas. Princeton: Princeton University Press.

Hullmandel, C. (1824). *The Art of Drawing on Stone*. London: C. Hullmandel.

Humboldt, W. von. (1821). *Über vier aegyptische, löwenköpfige Bildsäulen in den hiesigen königlichen Antikensammlungen*. Berlin: [s.n.].

———. (1823). *Über das Entstehen der grammatischen Formen und ihren Einfluss auf die Ideenentwicklung*. Berlin.

Hunter, J. (1786). *Observations on Certain Parts of the Animal Oeconomy*. London: Privately printed.

Hunter, J., and E. Home. (1794). "Some facts relative to the Late Mr. John Hunter's Preparation for the Croonian Lecture." *Philosophical Transaction of the Royal Society of London* 84:21–27.

Hutchinson, J. (1749). *The Religion of Satan, or Antichrist, Delineated, Supposed to have proceeded from Knowledge and Reasoning; But Proved to have proceeded from Want of Both*. London: James Hodges.

Ilchester, [G.S.H. Fox-Strangways] (Ed.). (1908). *The Journal of Elizabeth Lady Holland (1791–1811)*. London: Longmans, Green.

Johnson, S. (1775). *A Journey to the Western Islands of Scotland*. London: W. Strahan and T. Cadell.

Jollois, J.B.P., and P. Lefèvre-Pontalis. (1904). *Journal d'un ingénieur attaché à l'Expédition d'Égypte, 1798–1802*. Paris: E. Leroux.

Jomard, E.-F. (Ed.). (1809a). *Description de l'Égypte: Ou, Recueil des observations et des recherches qui ont été faites en Égypte pendant l'Expédition de l'armée française: Antiquités; Descriptions. T. 1*. Paris: Imprimerie Impériale.

———. (1809b). "De l'écriture des papyrus: De quelques symboles remarquables parmi les peintures des hypogées." In *Description de l'Égypte: Ou, Recueil des observations et des recherches qui ont été faites en Égypte pendant l'Expédition de l'armée française: Antiquités; Descriptions. T. 1.*, 369–81. Paris: Imprimerie Impériale.

———, (Ed.). (1812). *Description de l'Égypte: Ou, Recueil des observations et des recherches qui ont été faites en Égypte pendant l'Expédition de l'armée française: Antiquités, Planches*. Paris: Imprimerie Imperiale.

Jomard, E.-F.. (1814). *[Prospectus for] Description de l'Égypte.* Paris: Imprimerie Impériale.

———, (Ed.). (1817). *Description de l'Égypte: Ou, Recueil des observations et des recherches qui ont été faites en Égypte pendant l'Expédition de l'armée française: Antiquités, Planches. T. 4.* Paris: Imprimerie Royale.

———, (Ed.). (1818). *Description de l'Égypte: Ou, Recueil des observations et des recherches qui ont été faites en Égypte pendant l'Expédition de l'armée française: Antiquités; Mémoires. T. 2.* Paris: Imprimerie Royale.

———. (1819). *Notice sur les signes numériques des anciens Égyptiens, précédée du plan d'un ouvrage ayant pour titre: Observations et recherches nouvelles sur les hiéroglyphes accompagnées d'un tableau méthodique des signes.* Paris: Baudoin Frères.

———. (1822). *Description de l'Égypte: Ou, Recueil des observations et des recherches qui ont été faites en Égypte pendant l'Expédition de l'armée française: Antiquités, Planches, T. 5.* Paris: Imprimerie Imperiale.

Jones, W. (1771). *A Grammar of the Persian Language.* London: W. & J. Richardson.

Klaproth, J. H. (1823). *Lettre à M. Champollion le Jeune.* Paris: Dondey-Dupré Père et Fils.

———. (1827a). *Lettre sur la découverte des hiéroglyphes acrologiques.* Paris: J.-S. Merlin.

———. (1827b). *Seconde lettre sur la découverte des hiéroglyphes.* Paris: J.-S. Merlin.

Klaproth, J. H., and Dorow. (1829). *Collection d'antiquités égyptiennes recueillies par M. le chevalier de Palin.* Paris: Gide Fils.

Knight, C. (1851). *Knight's Cyclopaedia of London.* London: Charles Knight.

Knox, R. (1826–27). "Observations on the Comparative Anatomy of the Base of the Iris, and Its Mode of Union with the Cornea, Annulus albus, and Choroid Tunick." *Edinburgh Journal of Medical Science* 2:101–16.

Kosegarten, J. (1824). "Ueber die neuere Entrachtselung der aegyptischen hieroglyphen." *Hermes* 23:274–305.

Kotzebue, A. von. (1804). *Erinnerungen aus Paris im Jahre 1804.* Berlin: Heinrich Frölich.

La Brière, L. de. (1897). *Champollion inconnu: Lettres inédites.* Paris: Plon.

Lach-Szyrma, K. (2009). *London Observed: A Polish Philosopher at Large, 1820–24.* Edited by Mona Kedslie McLeod. Translated by Malgorzata Machnice and Agnieska Kiersztejn. Oxford: Signal Books.

La Croze, M. de, and C. Scholtz. (1775). *Lexicon Aegyptiaco-Latinum . . . Quod in compendium redegit . . . Christianus Scholtz.* Oxford: Clarendon Press.

Lacretelle, P.-L. (1820). "Les Conspirations." In *Panorama,* 39–48. Paris: Librarie Lacrette.

Lanci, M. (1825). *Osservazioni sul bassorilievo fenico-egizio che si conserva in Carpentrasso; Illustrazione di un Kil Anaglifo Copiato in Egitto da Sua Eccellenza Signor Barone D'Icskull.* Rome: Francesco Bourliè.

Langdon-Davies, J. (1952). *Westminster Hospital.* London: John Murray.

LeGros, C.-F. (1785). *Analyse des ouvrages de J. J. Rousseau, de Genève, et de M. Court de Gebelin.* Geneve: Barthélemy Chirol / Paris: La Veuve Duchesne.

Leitch, J. (Ed.). (1855). *Miscellaneous Works of the Late Thomas Young, M.D., F.R.S., &c. Vol. 3.* London: John Murray.

Leo XII. (1824). *Ubi Primum: On His Assuming the Pontificate.* Papal Encyclicals Online. https://www.papalencyclicals.net/leo12/l12ubipr.htm.

Lesur, C. L. (1823). *Annuaire historique universel.* Paris: Treuttel et Wurtz et al.

Letronne, J.-A. (1821). *Eclaircissements sur une inscription grecque contenant une pétition des prêtres d'Isis dans l'Île de Philae.* Paris.

———. (1822). "Supplément à l'explication de l'inscription grecque gravée sur le socle d'un obélisque égyptien trouvé dans l'Île de Philé." *Journal des Savans* (April 1822): 211–16.

———. (1823). *Recherches pour servir à l'histoire de l'Égypte pendant la domination des Grecs et des Romains tirées des inscriptions Grecques et Latines.* Paris: Boulland-Tardieu.

Lloyd, H. E. (1830). *George IV: Memoirs of His Life and Reign.* London: Treuttel and Würtz.

Lockhart, J. G. (2005). *Memoirs of the Life of Sir Walter Scott: Volume 1.* Whitefish, MT: Kessinger.

Macpherson, J. (1760). *Fragments of Ancient Poetry, Collected in the Highlands of Scotland, and Translated from the Galic or Erse Language.* Edinburgh: G. Hamilton and J. Bafour.

Madrigal, K. (2016). *Correspondances: Figeac et les Frères Champollion.* Figeac: Musée Champollion.

Marcel, J. J. (1799). "Précis des séances et des travaux de l'Institut d'Egypte, du 21 Messidor an 7 au 21 fructidor an 8 inclusivement." *La Décade Égyptienne, Journal Littéraire et d'Économie Politique* 3:293–94.

Marshman, J. (1809). *Dissertation on the Characters and Sounds of the Chinese Language: Including Tables of the Elementary Characters and of the Chinese Monosyllables.* Serampore, India: n.p.

Martin, B. (1771). *Philosophia Britannica, or, a new and comprehensive system of the Newtonian philosophy, astronomy, and geography.* London: Printed for W. Strahan; J. & F. Rivington; W. Johnston; Hawes and Co. T. Carnan and F. Newbery, jun. B. Collins, in Salisbury; W. Frederick, in Bath; and sold by the author.

Mérimée, P. (1838). *Notes d'un voyage en Auvergne.* Paris: H. Fournier.

———. (1842). "La Vénus d'Ille" (1837). In *Colomba: Suivi de la Mosaïque, et autres contes et nouvelles,* 171–208. Paris: Charpentier.

———. (1995). "The Venus of Ille" (1837). Translated by Jean Kimber. In *Demons in the Night: Tales of the Fantastic, Madness, and the Supernatural from the Nineteenth Century,* edited by J. C. Kessler, 63–90. Chicago: University of Chicago Press.

Millin, A. L. (1796). *Introduction à l'étude des médailles.* Paris: Magasin Encyclopédique.

———. (1802a). "Lettre sur l'inscription . . ." *Magasin Encyclopédique* 3, Year 8:141–44.

———. (1802b). "A Åkerblad . . ." *Magasin Encyclopédique* 3, Year 8:5.

Monboddo [Lord]. (1773–87). *On the Origin and Progress of Language.* Edinburgh: A. Kincaid & W. Creech.

Monk, J. H., and J. Blomfield. (Eds.). (1826). *Museum Criticum; or, Cambridge Classical Researches.* Cambridge: Cambridge University Press.

Montfaucon, B. (1722–24). *L'Antiquité expliquée et représentée en figures.* Paris: Dalaulne [u.a.].

Murray, J. F. (1844). *The World of London: Originally Published in "Blackwood's Magazine."* London: Thomas Tegg.

Musschenbroek, P. van. (1745). *Elementa physicae conscripta in usus academicos.* Venice: Joannem Baptistam.

Newton, I. (1704). *Opticks: or, A Treatise of the Reflexions, Refractions, Inflexions and Colours of Light.* London: S. Smith and B. Walford.

Nugent, T. (1773). *The Primitives of the Greek Tongue containing a complete Collection of all the Roots or Primitive Words, Together with the most considerable Derivatives of the Greek Language, as also a Treatise of Prepositions and other undeclinable Particles.* London: J. Nourse.

Peacock, G. (Ed.). (1855a). *Miscellaneous Works of the Late Thomas Young.* London: John Murray.

———. (1855b). *Life of Thomas Young.* London: John Murray.

Pemberton, H. (1719). "Diss. physico-medica inauguralis de facultate oculi, qua ad diversas rerum conspectarum distantias se accommodation."

Penn, G. (1795). *A warning voice to the people of England, on the true nature and effect of the two bills for the preservation of His Majesty's person and government, and for the prevention of seditious meetings.* London: Printed for Richard White.

———. (1802). *The Greek Version of the Decree of the Egyptian Priests, in Honor of Ptolemy the Fifth, Surnamed Epiphanes.* London: J. Nichols.

Planta, E. (1827). *A New Picture of Paris.* 5th ed. London: Samuel Leigh.

Plumptre, T. (1802). "Translation of the Greek Inscription from Rosetta." *Gentleman's Magazine, and Historical Chronicle* 72:1106–8.

Pococke, R. (1745). *A Description of the East, and some other Countries.* London: W. Bowyer.

Porson, R., and T. Kidd. (1815). *Tracts and Miscellaneous Criticisms of the Late Richard Porson, Esq., Regius Greek Professor in the University of Cambridge.* London: Printed by Richard and Arthur Taylor for Payne and Foss.

Priestley, J. (1775). *Experiments and observations on different kinds of air.* London: Printed for J. Johnson.

———. (1781). *A Course of Lectures on Oratory and Criticism.* Dublin: William Halhead.

Proudfoot, W. J. (1861). *"Barrow's Travels in China": An Investigation into the Origin and Authenticity of the "Facts and Observations."* London: George Philip and Son.

Pugin, A. (1831). *Paris and Its Environs.* London: Jennings and Chaplin.

Quatremère, É.-M. (1808). *Recherches critiques et historiques sur la langue et la littérature de l'Égypte.* Paris: Imprimerie Impériale.

———. (1812). *Observations sur Quelques Points de la Géographie de l'Égypte.* Paris: Schoell.

Raper, M. (1812). "An Account of the Rosetta Stone, in Three Languages, which Was Brought to England in the Year 1802." *Archaeologia* 16:208–11.

Reinaud. (1842). "Memoir, Historical and Literary, of the Late Baron de Sacy." *American Quarterly Register* 14:221–36.

Reybaud, L., et al. (1830–36). *Histoire scientifique et militaire de l'expédition française en Égypte.* Paris: A.-J. Denain.

Ricardi, F. (1811). *Massime teologiche politiche morali di Salomone ossia Versione latina e parafrasi italiana dell'Ecclesiaste.* Geneva: H. Bonaudo.

———. (1820). *Abrégé de la vraie méthode de lire et comprendre l'Hébreu, qui a été perdue pendant la dernière captivité des Juifs à Babylone, et maintenant recouvrée.* Geneva: H. Bonaudo.

———. (1824). *Découverte des hiéroglyphes domestiques phonétiques par lesquels, sans sortir de chez soi, on peut deviner l'histoire, la cronologie, le culte de tous les peuples anciens et modernes, de la même manière qu'on le fait en lisant les hiéroglyphes égyptiens selon la nouvelle méthode.* Turin: Bianco.

———. (1826). *Observations critiques sur le système hiéroglyphique des anciens Égyptiens de Mr. Champollion le Jeune.* Geneva: Yves Gravier.

Rigord, J.-P. (1704). "Lettre de Monsieur Rigord Commissaire de la Marine aux Journalistes de Trévoux sur une Ceinture de Toile trouvée en Égypte autour d'une Mumie." *Mémoires pour L'Histoire des Sciences et des Beaux-Arts* (April): 978–1000 (Art. 89).

Ring, J. (1807). *The Beauties of the Edinburgh Review, alias the Stinkpot of Literature.* London: H. D. Symonds.

Rochas, A. (1856). "Champollion le Jeune." In *Biographie du Dauphiné*, 1:205–16. Paris: Charavay.

Rose, H. J. (1857). *A New General Biographical Dictionary.* London: T. Fellowes.

Ryland, J. (1772). *An easy and pleasant introduction to Sir Isaac Newton's philosophy: containing the first principles of mechanics, trigonometry, optics, and astronomy.* London: Printed for E. and C. Dilly.

Sacy, S. de. (Ed.). (1799). *Principes de Grammaire Générale, Mis à la portée des enfans, et propres à servir d'introduction à l'étude de toutes les langues.* Paris: A. A. Lottin.

———. (1802). *Lettre au Citoyen Chaptal, ministre de l'Intérieur, membre de l'Institut national des sciences et arts, &c. au sujet de l'Inscription égyptienne du monument trouvé à Rosette.* Paris: Imprimerie de la République.

———. (1803). *Notice d'une dissertation de M.J.D. Åkerblad: Intitulée Inscriptionis Phoeniciae Oxoniensis nova interpretatio.* Paris: Didot Jeune.

———. (1808). *Notice de l'ouvrage intitulé: Recherches critiques et historiques sur la langue et la littérature de l'Égypte; par Ét. Quatremère.* Paris: J. B. Sajou.

———. (1811). "Untitled Review of *L'Égypte sous Les Pharaons.*" *Magasin Encyclopédique* 3:196–201.

———. (1825). *"Lettre à M. Dacier . . . Précis du système hiéroglyophique . . .* An Account of Some Recent Discoveries." *Journal des Savans* (March): 139–54.

———. (1836). "Untitled Review of *Lexicon linguae copticae* by Amadei Peyron." *Journal des Savants*, 146–55.

———. (1846). *Bibliothèque de M. le Baron Silvestre de Sacy.* Paris: Imprimerie Royale.

Saint-Martin, A. J. (1810–11). *NAF 9115.* Paris: Bibliothèque Nationale.

———. (1811). "Untitled Review of *L'Égypte sous les Pharaons.*" *Moniteur Universel* 211 (July 30, 1811): 807–8.

———. (1822a). "Notice sur quelques manuscrits grecs apportés récemment d'Égypte." *Journal des Savans* (January 1822): 535–67.

———. (1822b). *Notice sur le zodiaque de Denderah.* Paris: Delaunay.

———. (1822c). "Notice sur l'inscription hiéroglyphique de l'obélisque de Philae." *Journal des Savans* (April 1822): 216–20.

———. (1823). "Notice sur le zodiaque de Denderah." *North American Review* 41:233–42.

Sainte-Croix, de. (1798). *Éloge historique de J. J. Barthélemy (par le Baron de Ste Croix).* Paris.

Salt, H. (1825). *Essay on Dr. Young's and M. Champollion's Phonetic System of Hieroglyphics*. London: A. J. Valpy.

Scheller, I.J.G. (1779). *Imman. Io. Gerh. Schelleri Praecepta stili bene Latini in primis Ciceroniani, seu, Eloqventiae Romanae: qvatenus haec nostris temporibus in dicendo et scribendo vsvrpari potest. Svmma diligentia maximoqve perspicuitatis stvdio tradita et illvstrata*. Lipsiae: Svmptibus Caspari Fritsch.

Scott, J. (1816). *Paris Revisited in 1815, by Way of Brussels*. Boston: Wells and Lilly.

Seyffarth, G. (1826). *Rudimenta Hieroglyphices: Accedunt explicationes speciminum hieroglyphicorum glossarium atque alphabeta*. Leipzig: Ambros. Barth.

———. (1827). *Replique aux Objections de Mr. J. F. Champollion le Jeune contre Le Système Hiéroglyphique de Mm. F.A.G. Spohn et G. Seyffarth*. Leipzig: Ambros. Barth.

Sharpe, G. (1751). *A Dissertation upon the Origin, Construction, Division, and Relation of Language*. London: John Millan.

Sicard, R.A.C. (1808). *Théorie des signes pour l'instruction des sourds-muets*. Paris: A l'Imprimerie de l'Institution des Sourds-Muets.

Sichler, F.C.L. (1819). *Die Herculanischen Handschriften in England*. Leipzig: F. Brochaus.

Silvestre, J. B., J.-F. Champollion-Figeac, and A. Champollion. (1849). *Universal Palaeography: Or, Facsimiles of Writings of All Nations and Periods*. London.

Simpson, R. (1756). *The Elements of Euclid, viz. the first six books, together with the eleventh and twelfth: in this edition, the errors, by which Theon, or others, have long ago vitiated these books, are corrected, and some of Euclid's demonstrations are restored*. Glasgow: [s.n.].

Simpson, T. (1776). *The Doctrine and Application of Fluxions*. London: John Nourse.

Simpson, T., and J. Nourse. (1752). *Select exercises for young proficients in the mathematicks . . .* London: Printed for J. Nourse.

Smith, R. (1738). *A Compleat System of Opticks in Four Books, viz. A Popular, a Mathematical, a Mechanical, and a Philosophical Treatise*. Cambridge: Cornelius Crownfield.

Southey, R. (1837). *The Lives of the British Admirals*. London: Longman, Rees, Orme, Brown, Green & Longman.

Spohn, F. (1825). *De Lingua et Literis Veterum Aegyptiorum*. Leipzig: G. Reimer.

Swinton, J. (1753–54). "An explication of all the inscriptions in the Palmyrene language and character hitherto publish'd. In five letters from the Reverend Mr. John Swinton." *Philosophical Transactions of the Royal Society of London* 48:690–756.

———. (1764). "Farther remarks upon M. l'Abbé Barthelemy's Memoir on the Phoenician Letters, Containing His Reflections on Certain Phoenician Monuments, and the Alphabets Resulting from Them." *Philosophical Transactions of the Royal Society of London* 54:393–438.

Tattam, H. (1830). *A Compendious Grammar of the Egyptian Language . . . with an Appendix, Consisting of the Rudiments of a Dictionary of the Ancient Egyptian Language in the Enchorial Character; by Thomas Young*. London: John and Arthur Arch.

Telescope, T. [pseud.] (1779). *The Newtonian System of Philosophy. Adapted to the Capacities of Young Ladies and Gentlemen, And Familiarized and made entertaining, By Objects with which they are intimately acquainted*. London: T. Caman and F. Newberry.

TH (1891). *Dictionary of National Biography*. Ed. S. Lee. London: Smith, Elder.

Thelwall, J. (1795). *The Tribune, A Periodical Publication, Consisting Chiefly of the Lectures of J. Thelwall*. London: The Author.

Timbs, J. (1855). *The Curiosities of London*. London: D. Bogue.

Turgot, A.-R. (1751–72). "Etymologie." In *Encyclopédie, ou dictionnaire raisonné des sciences, des arts et des métiers*, edited by D. Diderot and J.L.R. d'Alembert, 6:98–111. Paris and Neuchâtel: Briasson.

Turner, S.T.H. (1812). "An Account of the Rosetta Stone: Letter to the Society of Antiquaries." *Archaeologia* 16:212–4.

Vaillant, P. (1975). "Deux lettres inédites de Champollion à propos des origines du déchiffrement des hiéroglyphes." *Revue Provence Historique* 25, fascicule 99:147–57.

———. (Ed.). (1984). *Jean-François Champollion, lettres à son frère, 1804–1818*. Paris: L'Asiathèque.

Verbrugge, G. P., and J. M. Wickersham. (2003). *Berossos and Manetho: Introduced and Translated; Native Traditions in Ancient Mesopotamia and Egypt*. Ann Arbor: University of Michigan Press.

Villoisin, D. A. de. (1803). "Lettre . . . à M. Åkerblad . . . sur un passage de l'Inscription grecque de Rosette." *Magasin Encyclopédique* 6:70–85.

Vyse, C. (1772). *The tutor's guide being a complete system of arithmetic; with various branches in the mathematics. In six parts.* London: Printed for Robinson and Roberts, No. 25 in Pater-Noster-Row.

Wachter, J. G. (1752). *Naturae et Scripturae Concordia. Commentatio de litteris ac numeris primaeuis aliisque rebus memorabilibus cum ortu litterarum coniunctis.* Lipsiae: Apud Viduam Gabriel / Hafniae: Christiani Rothe.

Wahl, G. (1816). "Entzifferung der egyptischen Buchstabenshrift." *Fundgraben des Orients* 5:217–24.

Walpole, R., and B. Charles (1805). *Comicorum Graecorum Fragmenta quaedam curavit et notas addidit R.W.* Cambridge: Cambridge University Press.

Walsh, T. (1803). *Journal of the Late Campaign in Egypt . . . with an Appendix; Containing Official Papers and Documents.* London: T. Cadell and W. Davies.

Warburton, W. (1742). *The Divine Legation of Moses Demonstrated, on the Principles of a Religious Deist, from the Omission of the Doctrine of a Future State of Reward and Punishment in the Jewish Dispensation.* 2nd ed. Vol. 2. London: Executor of the late Mr. Fletcher Gyles.

———. (1744). *Essai sur les Hiéroglyphes des Egyptiens, ou l'on voit l'Origine & le Progrès du Langage & de l'Ecriture, l'Antiquité des Sciences en Egypte, & l'Origine du culte des Animaux.* Paris: Hippolyte-Louis Guerin.

———. (1765). *The Divine Legation of Moses Demonstrated, in nine books.* London: A. Millar and J. and R. Tonson.

Ward, J. (1724). *A Compendium of Algebra: containing Plain, Easy, and Concise Rules in that Mysterious Science: exemplify'd by Variety of Problems, both in Arithmetick and Geometry, with the Solution of their Equations in Numbers [. . .].* London.

Watson, J. (1861). *The Life of Richard Porson, M.A.* London: Longman, Green, Longman, and Roberts.

Webb, J. (1669). *The Antiquity of China or an historical essay: endeavouring a probability that the language of the empire of China is the primitive language, spoken through the whole word, before the confusion of Babel.* London: Brook.

Weston, S. (1812). "Translation of the Inscription on the Rosetta Stone." *Archaeologia* 16:220–28.

Wilkins, D. (1716). *Hoc est Novum Testamentum Aegyptium vulgo Copticum Ex MSSS. Bodlejanis descripsit, cum Vaticanis et Parisiensibus contulit, et in Latinum Sermonem convertit.* Oxford: Oxford University Press.

Winckler, F. T. (1802). "Paris, den 14. September 1801." In *Der Neue Teutsche Merkur vom Jahre 1802*, by C. M. Wieland, 57–67. Weimar: [C.M. Wieland].

Young, T. (1793). "Observations on vision." *Philosophical Transaction of the Royal Society of London* (83): 169–81.

———. (1796). *De Corporis Humani Viribus Conservatricibus. Dissertatio.* Gorringae: Ioann. Christ. Dieterich.

———. (1807). *A Course of Lectures on Natural Philosophy and the Mechanical Arts.* London: Joseph Johnson.

———. (1810). "Herculanensia." *Quarterly Review* 3:1–20.

———. (1813a). "Review of *Mithridates, a General History of Languages*." *Quarterly Review* 10:250–92.

———. (1813b). *An Introduction to Medical Literature: Including a System of Practical Nosology, Intended as a Guide to Students, and an Assistant to Practitioners.* London: Printed by B. R. Howlett for Underwood and Blacks.

———. (1814a). *Letter to Bankes.* British Museum, 00030361001.

———. (1814b). *Inscriptio Rosettensis: Memorandums of an Attempt to Decypher the Egyptian Inscription of Rosetta.* British Library, ADD 27281.

———. (1814–15). *Lexicon Hieroglyphicum.* British Library, ADD 27282.

———. (1814–29). *Letters.* British Library, ADD 21026.

———. (1815). *A Practical and Historical Treatise on Consumpetive Diseases, Deduced from Original Observations, and Collected from Authors of All Ages.* London: Thomas Underwood.

———. (1816a). *Formulae Cultus Aegyptii.* British Library, ADD 27283.

———. (1816b). "Extracts of Letters and Papers Relating to the Egyptian Inscription of Rosetta." In *Museum Criticum; or, Cambridge Classical Researches* 6:155–204. Cambridge: Cambridge University Press.

———. (1816c). "Review of *Hermes Sythius* by John Jamieson and of *The Character of Moses* by Rev. John Townsend." *Quarterly Review* 14:96–112.

———. (1819a). "Herculaneum." In *Supplement to the Fourth, Fifth, and Sixth Editions of the "Encyclopaedia Britannica": With Preliminary Dissertations on the History of the Sciences* 4:624–32. Edinburgh: Archibald Constable / London: Hurst, Robinson.

———. (1819b). "Egypt." In *Supplement to the Fourth, Fifth, and Sixth Editions of the "Encyclopaedia Britannica": With Preliminary Dissertations on the History of the Sciences* 4: 38–74. Edinburgh: Archibald Constable / London: Hurst, Robinson.

———. (1823a). "Letters to Hamilton and Bankes." *Quarterly Journal of Science, Literature, and Art* 14:255–61.

———. (1823b). *Hieroglyphics, Collected by the Egyptian Society.* London: [Private Subscription].

———. (1823c). *An Account of Some Recent Discoveries in Hieroglyphical Literature, and Egyptian Antiquities: Including the Author's Original Alphabet, as Extended by Mr. Champollion, with a Translation of Five Unpublished Greek and Egyptian Manuscripts.* London: John Murray.

———. (1828). *Hieroglyphics, Continued by the Royal Society of Literature.* Vol. 2. London: Howlett and Brimmer.

———. (1831). *Rudiments of an Egyptian Dictionary in the Ancient Enchorial Character: Containing All the Words of which the Sense Has Been Ascertained; to which Are Prefixed a Memoir of the Author and Catalogue of His Works and Essays.* London: J. & A. Arch.

———. (1855). "Life of Sir Benjamin Thomson, Count Rumford." In *Miscellaneous Works of the Late Thomas Young*, edited by G. Peacock, 2:474–84. London: John Murray.

Zinn, J. G. (1780). *Descriptio Anatomica Oculi Humani Iconibus Illustrata.* Göttingen: Abraham Vandenhoeck.

Zoëga, G. (1797). *De origine et usu obeliscorum ad Pium Sextum, Pontificem Maximum.* Rome: Typis Lazzarinii.

———. (1810). *Catalogus codicum Copticorum manu scriptorum qui in Museo Borgiano Veletris adservantur; (opus posthumum) cum VII. tabulis aeneis.* Romae: Typis Sacrae Congregationis de Propaganda Fide.

SECONDARY SOURCES

Aarsleff, H. (1983). *The Study of Language in England, 1780–1860.* Minneapolis: University of Minnesota Press.

Adams, J. (1817). *Memoirs of the Life and Doctrines of the Late John Hunter, Esq. Founder of the Hunterian Museum, at the Royal College of Surgeons in London.* London: J. Callow, J. Hunter Sr., J. Ridgway.

Adas, M. (2014). *Machines as the Measure of Men: Science, Technology, and Ideologies of Western Dominance.* 2nd ed. Ithaca: Cornell University Press.

Adkins, L., and R. Adkins. (2000). *The Keys of Egypt: The Race to Read the Hieroglyphs.* New York: HarperCollins.

Alexander, R. S. (1991). *Bonapartism and Revolutionary Tradition in France: The Fédérés of 1815.* Cambridge: Cambridge University Press.

———. (2003). *Re-writing the French Revolutionary Tradition.* Cambridge: Cambridge University Press.

Alger, J. G. (1885–1900). "Hamilton, Alexander (1762–1824)." In *Dictionary of National Biography*, vol. 24, edited by L. Stephen. London: Smith, Elder.

Allen, D. C. (1960). "The Predecessors of Champollion." *Proceedings of the American Philosophical Society* 104:527–47.

Amelineau, E.-C. (1888–89). *Catalogue des manuscrits coptes.* Paris: Bibliothèque Nationale de France.

Armand, Y. (1990). "Les Champollions dans la Bataille pour l'enseignement mutuel." *Académie Delphinale* 3 (2): 25–36.

A.T.C. (1910). "A Vase of Xerxes." *Expedition Magazine* 1:6–7.

Auerbach, E. (1953). "In the Hôtel de la Mole." In *Mimesis: The Representation of Reality in Western Literature*, 454–92. Princeton: Princeton University Press.

Auricchio, F. L. (1980). "John Hayter nella officina dei papiri ercolanesi." In *Contributi alla storia della officina dei papiri ercolanesi*, by M. Giganti, 159–215. Napoli: Industria tipografica artistica.

Backus, I. (Ed.). (1997). *The Reception of the Church Fathers in the West: From the Carolingians to the Maurists*. Leiden: Brill.

Bagot, C. J. (Ed.). (1909). *George Canning and His Friends*. New York: E. P. Dutton.

Bednarski, A. (2005). *Holding Egypt: Tracing the Reception of the "Description de l'Égypte" in Nineteenth-Century Great Britain*. London: Golden House Publications.

———. (2014). *The Lost Manuscript of Frédéric Cailliaud*. New York: American University in Cairo Press.

Beiser, F. C. (2011). *The German Historicist Tradition*. New York: Oxford University Press.

Belmont, N. (1980). "L'Académie celtique." In *Hier pour demain: Arts, traditions et patrimoine, catalogue d'exposition du Grand Palais*, 54–60. Paris: Réunion des Musées Nationaux.

Berman, M. (1978). *Social Change and Scientific Organization: The Royal Institution, 1799–1844*. Ithaca: Cornell University Press.

Blix, G. (2009). *From Paris to Pompeii: French Romanticism and the Cultural Politics of Archaeology*. Philadelphia: University of Pennsylvania Press.

Blouin, F. X. (1998). *Vatican Archives: An Inventory and Guide to Historical Documents of the Holy See*. New York: Oxford University Press.

Boisseau, Jean. (1648). "L'ancienne ville de Grenoble capitale épiscopalle et siège du parlement de Dauphiné." In *Théatre des Citez, ou Recueil de plusieurs villes, dont les noms suivent par ordre alphabetique*. Paris: I. Boisseau.

Bolter, J. D. (1980). "Friedrich August Wolf and the Scientific Study of Antiquity." *Greek, Roman, and Byzantine Studies* 21:83–99.

Bourguet, P. du. (1982). "Champollion et les études coptes." *Bulletin de la Société Francaise d'Égyptologie*, 62–75.

Braun, G., and F. Hogenberg. (1581–88). "Gratianopolis Acusianoru Colonia." In *Civitates Orbis Terrarum*. Vol. 3. Köln: G. Kempen.

Briggs, A. (2008). *A History of Longmans and Their Books, 1724–1990: Longevity in Publishing*. London: British Library.

Brown, G. I. (1967). *Count Rumford: Scientist, Soldier, Statesman, Spy*. Oxford: Pergamon.

Buchwald, J. (1989). *The Rise of the Wave Theory of Light: Optical Theory and Experiment in the Early Nineteenth Century*. Chicago: University of Chicago Press.

Buchwald, J., and M. Feingold. (2012). *Isaac Newton and the Origin of Civilization*. Princeton: Princeton University Press.

Buchwald, J. Z., and D. G. Josefowicz. (2010). *The Zodiac of Paris*. Princeton: Princeton University Press.

Budge, E.A.W. (1904). *The Decrees of Memphis and Canopus: The Rosetta Stone*. New York: Oxford University Press.

Bunsen, C.K.J.F. von. (1848). *Egypt's Place in Universal History: An Historical Investigation*. London: Longman, Brown, Green and Longmans.

Buringh, E. (2010). *Medieval Manuscript Production in the Latin West: Explorations with a Global Database*. Leiden: Brill.

Buzi, P. (2015). "Genesis of a Masterpiece." In *The Forgotten Scholar: Georg Zoëga (1755–1809); At the Dawn of Egyptology*, edited by K. Ascani, Paola Buzi, and Daniela Picchi, 216–24. Leiden: Brill.

Bynon, T. (1977). *Historical Linguistics*. Cambridge: Cambridge University Press.

Callatay, F. (2011). "The Count of Caylus (1692–1765) and the Study of Ancient Coins." In *Proceedings of the XIVth International Numismatic Conference*, edited by N. Holmes, 1999–2003. Glasgow: International Numismatic Council.

Cantor, G. (1970). "Thomas Young's Lectures at the Royal Institution." *Notes and Records of the Royal Society of London* 25 (1): 87–112.

———. (1978). "The Historiography of Georgian Optics." *History of Science* 16:1–21.

Carbonell, C.-O. (1984). *L'autre Champollion: Jacques-Joseph Champollion-Figeac, 1778–1867*. Toulouse: Presses de l'Institut d'études politiques de Toulouse.

Carrott, R. G. (1978). *The Egyptian Revival: Its Sources, Monuments, and Meaning, 1808–1858*. Berkeley: University of California Press.

Ceserani, G. (2013). "Antiquarian Transformations in Eighteenth-Century Europe." In *World Antiquarianism: Comparative Perspectives*, edited by A. Schnapp, 317–42. Los Angeles: Getty Research Institute.

Chabot, J.-B. (1906). "Inventaire sommaire des manuscrits coptes de la Bibliothèque nationale." *Revue des Bibliothèques*.

Chambon, F. (Ed.). (1907). *Prosper Mérimée, l'homme, l'écrivain, l'artiste*. Paris: Journal des débats politiques et littéraires.

Chard, L. (1977). "Bookseller to Publisher: Joseph Johnson and the English Book Trade, 1760–1810." *Library* 32:138–54.

Chassagnard, G. (2001). *Les frères Champollion: De Figeac aux hiéroglyphes*. Figeac: Segnat Éditions.

Clay, A. T. (1916). "A Vase of Xerxes." *Art and Archaeology* 4, no. 1: 59–60.

Coller, I. (2005). "Egypte-sur-Seine: The Making of an Arabic Community in Paris, 1800–1830." *French History and Civilization* 1:206–14.

———. (2006). "Arab France: Mobility and Community in Early-Nineteenth-Century Paris and Marseille." *French Historical Studies* 29 (3): 433–56.

———. (2011). *Arab France: Islam and the Making of Modern Europe, 1798–1831*. Berkeley: University of California Press.

Coltman, V. (2006). *Fabricating the Antique: Neoclassicism in Britain, 1760–1800*. Chicago: University of Chicago Press.

Comparetti, D., and G. de Petra. (1883). *Villa ercolanese dei Pisoni: I suoi monumeneti e la sua biblioteca*. Torino: E. Loescher.

Cornick, M., and D. Kelly. (2013). *A History of the French in London: Liberty, Equality, Opportunity*. London: Institute of Historical Research.

Cuccia, P. (2013). "Controlling the Archives: The Requisition, Removal, and Return of the Vatican Archives during the Age of Napoleon." *Napoleonica* 2 (17): 66–74.

Curley, T. M. (2009). *Samuel Johnson, the Ossian Fraud and the Celtic Revival in Great Britain and Ireland*. Cambridge: Cambridge University Press.

Curran, B., A. Grafton, P. Long, and B. Weiss. (2009). *Obelisk: A History*. Cambridge, MA: Burndy Library.

Curran, W. S. (1962). "Dr. Brocklesby of London (1722–1797): An 18th-Century Physician and Reformer." *Journal of the History of Medicine* 17 (4): 509–21.

Cutmore, J. (2017). *Quarterly Review* Archive. http://www.rc.umd.edu/reference/qr/index.html.

Daly, L. W. (1951). "Aulus Hirtius and the Corpus Caesarianum." *Classical Weekly* 44 (8): 113–17.

Daniels, P. T. (1988). " 'Shewing of Hard Sentences and Dissolving of Doubts': The First Decipherment." *Journal of the Oriental Society* 108:419–36.

Darnton, R. (1982). *The Literary Underground of the Old Regime*. Cambridge, MA: Harvard University Press.

———. (1995). *The Forbidden Bestsellers of Pre-Revolutionary France*. New York: Norton.

———. (2014). "A Literary Tour de France." http://www.robertdarnton.org.

Daunou, P.C.F. (1839). "Notice historique sur la vie et les ouvrages de M. le baron Silvestre de Sacy." In *Mémoires de l'Académie des inscriptions et belles-lettres*, 12:507–40.

David, M. V. (1965). *Le débat sur les écritures et l'hiéroglyphe aux XVIIe et XVIIIe siècles, et l'application de la notion de déchiffrement aux écritures mortes*. Paris: S.E.V.P.E.N.

Delahaye, G. R. (2009). "Les Coptes vu par le voyageur J.-M. Vansleb au XVIIe siècle." In *En quête de la lumière: Mélanges in honorem Ashraf A. Sadek*, edited by A. A. Maravelia, 145–54. Oxford: Archeopress.

Derenbourg, H. (1905). *Esquisse biographique: Silvestre de Sacy (1758–1838)*, edited by G. Salmon, vii–lix. Cairo: L'Institut Français d'Archéologie Orientale.

Dodd, G. (1856). *The Food of London*. London: Longman, Brown, Green, and Longmans.

Downs, J. (2008). *Discovery at Rosetta*. New York: Skyhorse.

Duncan, I. (2005). "Edinburgh, Capital of the Nineteenth Century." In *Romantic Metropolis: Cultural Productions of the City, 1770–1840*, edited by J. Chandler and K. Gilmartin. Cambridge: Cambridge University Press.

Durrell, L. (1969). "The Three Roses of Grenoble." In *Spirit of Place: Letters and Essays on Travel*. London: Faber.

Eco, U. (1997). *The Search for the Perfect Language*. Translated by James Fentress. Oxford: Blackwell.

El-Daly, O. (2005). *Egyptology: The Missing Millennium: Ancient Egypt in Medieval Arabic Writings*. London: UCL Press.

Engberg-Pedersen, A. (2015). *Empire of Chance: The Napoleonic Wars and the Disorder of Things*. Cambridge, MA: Harvard University Press.

Espagne, G., and B. Savoy. (Eds.). (2005). *Aubin-Louis Millin et l'Allemagne*. New York: Georg Olms Verlag.

Espagne, M., N. Lafi, and P. Rabault-Feuerhahn. (Eds.). (2014). *Philologie, déchiffrement d'écritures et théorie des civilizations*. Paris: Les Éditions du Cerf.

Evans, J. (1956). *A History of the Society of Antiquaries*. Oxford: Oxford University Press.

Faure, A. (2004). *Champollion, le savant déchiffré*. Paris: Le Grand livre du mois.

Finley, G. E. (1980). *Landscapes of Memory: Turner as Illustrator to Scott*. London: Scolar.

Forshaw, A., and T. Bergström. (1980). *Smithfield: Past and Present*. London: Heinemann.

Forster, M. N. (Ed.). (2002). *Herder, Johann Gottfried: Philosophical Writings*. New York: Cambridge University Press.

———. (2010). *After Herder: Philosophy of Language in the German Tradition*. New York: Oxford University Press.

Frandsen, P. J. (2015). "A Concealed Attempt at Deciphering Hieroglyphs." In *The Forgotten Scholar: Georg Zoëga (1755–1809); At the Dawn of Egyptology and Coptic Studies*, edited by K. Ascani, P. Buzi, and D. Picchi, 160–73. Leiden: Brill.

Gange, D. (2013). *Dialogues with the Dead: Egyptology in British Culture and Religion, 1822–1922*. Oxford: Oxford University Press.

Gardiner, A. (1964 [1961]). *Egypt of the Pharaohs*. Oxford: Oxford University Press.

Gascoigne, J. (1994). *Joseph Banks and the English Enlightenment*. Cambridge: Cambridge University Press.

Gaskill, H. (1991). "Introduction in *Ossian Revisited*." Edinburgh: Edinburgh University Press.

Gerson, S. (2003). *The Pride of Place: Local Memories and Political Culture in Nineteenth-Century France*. Ithaca: Cornell University Press.

Getman, F. H. (1937). "Sir Charles Blagden, F.R.S." *Osiris* 3:69–87.

Gillispie, C. C., and M. Dewachter. (1987). *Monuments of Egypt: The Napoleonic Expedition; The Complete Archaeological Plates from "La Description de l'Égypte."* Princeton: Princeton Architectural Press in association with the Architectural League of New York and the J. Paul Getty Trust.

Ginzburg, C. (2012). *Threads and Traces*. Translated by Anne C. Tedeschi and John Tedeschi. Berkeley: University of California Press.

Gloyne, S. R. (1950). *John Hunter*. Edinburgh: E. & S. Livingstone.

Godlewska, A. (1995). "Map, Text and Image: The Mentality of Enlightened Conquerors; A New Look at the *Description de l'Égypte*." *Transactions of the Institute of British Geographers* 20:5–28.

———. (1999). *Geography Unbound: French Geographic Science from Cassini to Humbolt*. Chicago: University of Chicago Press.

Gould, T., and D. Uttley. (1997). *A Short History of St. George's Hospital and the Origin of Its Ward Names*. London: Athlone Press.

Grafton, A. (2010). "Christian Gottlob Heyne." In *The Classical Tradition*, edited by A. Grafton, G. W. Most, and S. Settis, 436–37. Cambridge, MA: Belknap Press of Harvard University Press.

Grafton, A., G. W. Most, and J.E.G. Zetzel. (Eds.). (1985). *F. A. Wolf: Prolegomena to Homer, 1795*. Princeton: Princeton University Press.

Guichard, S. (2007). "Une collection d'antiquités Égyptiennes méconnue." *Revue d'Égyptologie* 58:201–36.

Guiomar, J.-Y. (1992). "La Révolution française et les origines celtiques de la France." *Annales historiques de la Révolution française* 287:63–85.

Hamilton, A. (1993). "Eastern Churches and Western Scholarship." In *Rome Reborn: The Vatican Library and Renaissance Culture*, edited by A. Grafton, 225–49. Washington, DC: Library of Congress.

———. (2006). *The Copts and the West, 1439–1822: The European Discovery of the Egyptian Church*. Oxford: Oxford University Press.

Hammond, N.G.L., and H. H. Scullard. (Eds.). (1970). *The Oxford Classical Dictionary*. Oxford: Clarendon Press.

Harris, J. (2007). *Pompeii Awakened*. New York: I. B. Tauris.

Hartleben, H. (1906). *Champollion: Sein Leben und sein Werk*. Berlin: Weidmann.

———. (1983). *Champollion: Sa vie et son oeuvre, 1790–1832; Présentation de Christiane Desroches Noble- court*. Paris: Pygmalion.

Haugen, K. L. (1998). "Ossian and the Invention of Textual History." *Journal of the History of Ideas* 59:309–27.

———. (2012). "Hebrew Poetry Transformed, or, Scholarship Invincible between Renaissance and Enlightenment." *Journal of the Warburg and Courtauld Institutes* 75:1–29.

Hebbelnyk, A., and A. Van Lantschoot. (Eds.). (1937). *Codices coptici Vaticani, Barberiniani, Borgiani, Rossiani*. Vatican City: Biblioteca Apostolica Vaticana.

Hedrick, E. (1987). "Locke's Theory of Language and Johnson's *Dictionary*." *Eighteenth-Century Studies* 20:422–44.

Helmont, F. M. van. (2007). *The Alphabet of Nature*. Leiden: Brill.

Hiebert, E. N., A. J. Ihde, R. E. Schofield, L. Kieft, B. R. Willeford, and the American Chemical Society. (1980). *Joseph Priestley, Scientist, Theologian, and Metaphysician: A Symposium Celebrating the Two Hundredth Anniversary of the Discovery of Oxygen by Joseph Priestly in 1774*. Lewisburg, NJ: Bucknell University Press.

Hillairet, J. (1953). *Évocation du vieux Paris: Les Faubourgs*. Paris: Les Editions Minuit.

Holmes, R. (2008). *The Age of Wonder*. New York: Pantheon.

Hudson, N. (1994). *Writing and European Thought, 1600–1830*. Cambridge: Cambridge University Press.

Irwin, R. (2006). *Dangerous Knowledge: Orientalism and Its Discontents*. Woodstock: Overlook Press.

Iversen, E. (1993). *The Myth of Egypt and Its Hieroglyphs in European Tradition*. Princeton: Princeton University Press.

James, F.A.J.L. (Ed.). (2002). *"The Common Purposes of Life": Science and Society at the Royal Institution of Great Britain*. Hants: Ashgate.

Jasper, D. (1998). "Light and Darkness: J.M.W. Turner and the Bible." In *The Sacred and Secular Canon in Romanticism*. London: Palgrave Macmillan.

J-N [Jourdain]. (1857). "Joseph de Guignes." In *Biographie universelle ancienne et moderne: Nouvelle Édition*. Vol. 18, edited by M. Michaud, 126–29. Paris: Desplaces or Michaud.

Jones, C. (2005). *Paris: The Biography of a City*. New York: Viking.

Jørgensen, A. D. (1881). *Georg Zoëga: Et Mindeskrift*. Copenhagen: Bianco Lunos.

Josefowicz, D. G. (2016). "The Whig Interpretation of Homer: F. A. Wolf's *Prolegomena ad Homerum* in England." In *For the Sake of Learning: Essays in Honor of Anthony Grafton*, edited by A. Blair and A.-S. Goeing, 823–44. Leiden: Brill.

Kent, W. (1954). *London in the News through Three Centuries*. London: Staples Press.

Kettel, J. (1990). *Jean-François Champollion le Jeune: Répertoire de bibliographie analytique, 1806–1989*. Paris: Diffusion de Boccard.

———. (1996). "'Le grand noeud gordien délié?' L'énigme des canopes, Horus Apollon et les alphabets orientaux: Un lettre de Champollion à Antoine-Jean Saint-Martin." In *Poikila: Hommage à Othon Scholer; Études Classiques*, 8:103–36.

Kinghorn, A. M. (1991). "Two Scots Literary Historians: David Irving and John Merry Ross." *Studies in Scottish Literature* 28 (1): 78–89.

Kiparsky, P. (1993). "Paninian Linguistics." In *The Encyclopedia of Language and Linguistics*, edited by R. E. Asher and J.M.Y. Simpson, 6:2918–23. Oxford: Pergamon.

Kipnis, N. (1991). *History of the Principle of Interference of Light*. Basel: Birkhäuser Verlag.

Kobler, J. (1960). *The Reluctant Surgeon*. London: Heinemann.

Lacouture, J. (1988). *Champollion: Une vie de lumières*. Paris: Grasset.

Laissus, Y. (2004). *Jomard: Le dernier Égyptien*. Paris: Fayard.

Landow, George P. "Typology in the Visual Arts." In *Victorian Types, Victorian Shadows*, 119–42. London: Routledge & Kegan Paul, 1980.

Lantschoot, A. van. (Ed.). (1947). *Codices coptici Vaticani, Barberiniani, Borgiani, Rossiani: Codices Barberiani orientales 2 and 17, Borgiani coptici 1–108*. Vatican City: Biblioteca Apostolica Vaticana.

Leclerc, G. (1909). *La juridiction consulaire de Paris pendant la Révolution*. Paris: BnF-F.

Leersen, J. (2004). "Ossian and the Rise of Literary Historicism." In *The Reception of Ossian in Europe*, edited by H. Gaskill, 109–25. London: Thoemmes Continuum.

Leventhal, R. S. (1986). "The Emergence of Philological Discourse in the German States, 1770–1810." *Isis* 77:243–60.

Levin, H. (1963). *The Gates of Horn: A Study of Five French Realists*. New York: Oxford University Press.

Levitin, D. (2015). "Egyptology, the Limits of Antiquarianism, and the Origins of Conjectural History, c. 1680–1740: New Sources and Perspectives." *History of European Ideas* 41:99–727.

Lolla, M. G. (1999). "Ceci n'est pas un monument: *Vetusta Monumenta* and Antiquarian Aesthetics." In *Producing the Past: Aspects of Antiquarian Culture and Practice, 1700–1850*, edited by M. Myrone and L. Peltz, 15–34. Vermont: Ashgate.

Loprieno, A. (2000). "Puns and Word Play in Ancient Egyptian." In *Puns and Pundits: Word Play in the Hebrew Bible and Ancient Near Eastern Literature*, edited by S. B. Noegel, 3–20. Bethesda: CDL Press.

Louca, A. (1988). "Champollion entre Bartholdi et Chiftichi." In *Rivages et déserts: Hommage à Jacques Berque*, 209–25. Paris: Sindbad.

———. (2011). "Chiftichi, Yuhanna." In *Claremont Coptic Encyclopedia*, edited by A. Atiya et al. Claremont, CA: Claremont Colleges Digital Library.

Manley, D. (2001). *Henry Salt: Artist, Traveller, Diplomat, Egyptologist*. London: Libri.

Manly, S. (2007). *Language, Custom, and Nation in the 1790s: Locke, Tooke, Wordsworth, Edgeworth*. London: Ashgate.

Mansel, P. (2003). *Paris Between Empires*. New York: St. Martin's Press.

Maras, R. J. (1994). "Napoleon's Quest for a Super-Archival Center in Paris." In *The Consortium on Revolutionary Europe: Selected Papers, 1750–1850*, edited R. Caldwell et al., 567–78. Institute on Napoleon and the French Revolution, Florida State University, the Consortium on Revolutionary Europe.

Marchand, S. L. (1996). *Down from Olympus: Archaeology and Philhellenism in Germany, 1750–1970*. Princeton: Princeton University Press.

Martin, M. (1997). "Le journal de Vansleb en Égypte." *Bulletin de l'Institute Français d'Archéologie Orientale* 97:181–91.

Mayes, S. (2006). *The Great Belzoni*. New York: Taurus Parke Paperbacks.

McIntyre, I. (2003). *Joshua Reynolds: The Life and Times of the First President of the Royal Academy*. London: Allen Lane.

Messling, M. (2008). *Pariser Orientlektüren: Zu Wilhelm von Humboldts Theorie der Schrift*. Paderborn: Schöningh.

———. (2012). *Champollions Hieroglyphen: Philologie und Weltaneignung*. Berlin: Kulturverlag Kadmos.

Michaud, L. G. (Ed.). (1843). *Biographie universelle ancienne et moderne: Histoire par ordre alphabétique de la vie publique et privée de tous les hommes*. Paris: Desplaces.

Milburn, W. H. (1890). "Thomas Young, M.D., F.R.S." *Harper's New Monthly Magazine*, 80. New York: Harper & Brothers.

Mitchell, L. (2007). *The Whig World: 1760–1837*. London: Hambledon Continuum.

Moebus, O., and A. Wilke. (2011). *Sound and Communication: An Aesthetic Cultural History of Sanskrit Hinduism*. New York: Walter de Gruyter.

Moore, D. (2004). "The Reception of the Poems of Ossian in England and Scotland." In *The Reception of Ossian in Europe*, by H. Gaskill, 21–39. New York: Thoemmes Continuum.

Moore, L. J. (2008). *Restoring Order: The École des Chartes and the Organization of Archives and Libraries in France, 1820–1870*. Duluth, MN: Litwin Books.

Moser, S. (2006). *Wondrous Curiosities: Ancient Egypt at the British Museum*. Chicago: University of Chicago Press.

Nash, R. (2003). *Wild Enlightenment: The Borders of Human Identity in the Eighteenth Century*. Charlottesville: University of Virginia Press.

Nicolas, P. (2013). "Pierre Louis Jean Casimir de Blacas." http://www.comtedechambord.fr/entourage/le-duc-de-blacas.

Nisbet, H. B. (1970). *Herder and the Philosophy and History of Science*. Cambridge: Modern Humanities Research Association.

O'Flaherty, W. D. (Ed.). (1981). *The Rig Veda*. London: Penguin Books.

O'Gorman, F. (2016). *The Long Eighteenth Century: British Political and Social History, 1688–1832*. London: Bloomsbury Academic.

Oestmann, G. (2005). "J.W.A. Pfaff and the Rediscovery of Astrology in the Age of Romanticism." In *Horoscopes and Public Spheres: Essays on the History of Astrology*, edited by G. Oestmann, D. G. Rutkin, and K. von Stuckrad. New York: Walter de Gruyter.

Oldham, F. (1933). *Thomas Young, F.R.S.: Philosopher and Physician*. London: Edward Arnold.

Olender, M. (1992). *The Languages of Paradise: Race, Religion and Philology in the Nineteenth Century*. Translated by Arthur Goldhammer. Cambridge: Harvard University Press.

Omer, M. (1996). *J.M.W. Turner and the Romantic Vision of the Holy Land and the Bible*. Boston: McMullen Museum of Art.

Ottley, D. (1839). *The Life of John Hunter, FRS*. Philadelphia: Haswell, Barrington & Haswell.

Pao, A. (1998). *The Orient of the Boulevards: Exoticism, Empire, and Nineteenth-Century French Theater*. Philadelphia: University of Pennsylvania Press.

Papayanis, N. (2004). *Planning Paris before Haussmann*. Baltimore: Johns Hopkins University Press.

Paquet, J. (1973). "Les deux Champollion dans le milieu universitaire grenoblois." *Bulletin mensuel de l'Académie delphinale*.

Parkinson, R. (1999). *Cracking Codes: The Rosetta Stone and Decipherment*. London: British Museum Press.

Parsis-Barubé, O. (2011). *La province antiquaire: L'invention de l'histoire locale en France (1800–1870)*. Paris: Éditions de Comité des travaux historiques et scientifiques.

Pesic, P. (2013). "Thomas Young's Musical Optics: Translating Sound into Light." *Osiris* 28:15–39.

Pettigrew, T. J. (1839). "Thomas Young, M.D., F.R.S." In *Biographical Memoirs of the Most Celebrated Physicians, Surgeons, etc. etc.*, 1–24. London: Whitaker.

Pezin, M. (1993). "Un repère chronologique dans le déchiffrement des hiéroglyphes: Une lettre inédite de Letronne." *Cahiers du Musée Champollion: Histoire & Archéologie* 2:24–27.

Picchi, D. (2010). *Alle origini dell'egittologia: Le antichità egiziane di Bologna e di Venezia da un inedito di Georg Zoëga*. Imola (Bologna): La mandragora.

Pollock, S. (2009). "Future Philology? The Fate of a Soft Science in a Hard World." *Critical Inquiry* 35 (Summer): 931–61.

Pomian, K. (1987). *Collectionneurs, amateurs, et curieux*. Paris: Gallimard.

Pope, M. (1999). *The Story of Decipherment from Egyptian Hieroglyphs to Maya Script*. London: Thames and Hudson.

Posner, R. (2000). "Savigny, Holmes, and the Law and Economics of Possession." *Virginia Law Review* 86:535–67.

Powell, B. B. (2009). *Writing: Theory and History of the Technology of Civilization*. Chichester: UK / Malden, MA: Wiley-Blackwell.

Ray, J. (2007). *The Rosetta Stone*. Cambridge, MA: Harvard University Press.

Rees, J. (2006). *Die Kultur des Amateurs: Studien zu Leben und Werk von Anne Claude Philippe de Thubières, Comte de Caylus (1692–1765)*. Weimar: VDG.

Reinaud, M. (1838). *Notice historique et littéraire sur M. le Baron Silvestre de Sacy*. Paris: Librairie Orientale.

Richter, T. S. (2015). "Early Encounters: Egyptian-Coptic Studies and Comparative Linguistics in the Century from Schlegel to Finck." In *Egyptian-Coptic Linguistics in Typological Perspective*, edited by E. Grossman, M. Haspelmath, and T. S. Richter, 3–68. Berlin: Mouton De Gruyter.

Ridley, R. (1998). *Napoleon's Proconsul in Egypt: The Life and Times of Bernardino Drovetti*. London: Rubicon Press.

Rivet, A.L.F. (1971). "Hill-Forts in Action." In *The Iron Age and Its Hill Forts*, ed. M. Jesson and D. Hill, 189–202. Southampton: Southampton University Archaeological Society.

Robb, G. (2007). *The Discovery of France: A Historical Geography from the Revolution to the First World War*. New York: W. W. Norton.

Robinson, A. (2006). *The Last Man Who Knew Everything*. New York: Pi Press.

———. (2012). *Cracking the Egyptian Code: The Revolutionary Life of Jean-François Champollion*. New York: Oxford University Press.

Rocheblave, S. (1889). *Essai sur le Comte de Caylus*. Paris: Librairie Hachette.

Rocher, R. (1968). *Alexander Hamilton, 1762–1824: A Chapter in the Early History of Sanskrit Philology*. New Haven: American Oriental Society.

Rosanvallon, P. (1985). *Le moment Guizot*. Paris: Gallimard.

Rosenberg, D. (2003). "Louis-Sébastian Mercier's New Words." *Eighteenth-Century Studies* 36 (3): 367–86.

Rosenfeld, S. A. (2001). *A Revolution in Language: The Problem of Signs in Late Eighteenth-Century France*. Stanford, CA: Stanford University Press.

Rostaing, C. (1945). *Les noms de lieux*. Paris: Presses universitaires de France.

Sadjedi, T. (1983). "Bicentenaire de la naissance d'Étienne-Marc Quatremère." *Journal des Savants* (1–3): 19–32.

Said, E. W. (2004). *Orientalism: Western Conceptions of the Orient*. New York: Pantheon Books.

Saint-Hilaire, B. (1861). *Notice sur M. Étienne Quatremère*. Paris: Imprimerie Impériale.

Sandys, J. E. (1908). *A History of Classical Scholarship*. Cambridge: Cambridge University Press.

Savoy, B. (2009). "'Object d'observation et d'intelligence': La Pierre de Rosette entre paris, Londres, le Caire . . . et Göttingen (1799–1805)." *Études Germaniques* 64:799–819.

Schnapp, A. (1997). *The Discovery of the Past*. New York: Harry Abrams.

Scott, W. (1885). *Fragmenta Herculanensia: A Descriptive Catalogue of the Oxford Copies of the Herculanean Rolls*. Oxford: Clarendon Press.

Sebba, A. (2004). *The Exiled Collector*. London: John Murray.

Senn, H. (1981). "Folklore Beginnings in France, the Académie Celtique: 1804–1813." *Journal of the Folklore Institute* 18 (1): 23–33.

Seyffarth, G., and K. Knortz. (1886). *The Literary Life of Gustavus Seyffarth: An Auto-Biographical Sketch*. New York: E. Steiger.

Seyler, D. (2015). *The Obelisk and the Englishman*. New York: Prometheus Books.

Sheppard, F. (1971). *London: 1808–1870; The Infernal Wen*. Berkeley: University of California Press.

Sher, R. B. (1982). "'Those Scotch Imposters and Their Cabal': Ossian and the Scottish Enlightenment." *Man and Nature / L'homme et la nature* 1:55–63.

Simpson, R. S. (1996). *Demotic Grammar in the Ptolemaic Sacerdotal Decrees*. Oxford: Griffith Institute.

Skuy, D. (2003). *Assassination, Politics, and Miracles: France and the Royalist Reaction of 1820*. Montreal: McGill-Queen's University Press.

Smith, C. S. (1999). *The Market Place and the Market's Place in London, ca. 1660–1840*. PhD, University College, London.

Smith, O. (1986). *The Politics of Language, 1791–1819*. Oxford: Clarendon Press.

Sobin, G. (1999). *Luminous Debris: Reflecting on Vestige in Provence and Languedoc*. Berkeley: University of California Press.

Solé, R., and D. Valbelle. (1999). *La Pierre de Rosette*. Paris: Éditions du Seuil.

———. (2002). *The Rosetta Stone*. New York: Four Walls Eight Windows.

Sors, A. (1971). *L'Épopée gauloise en Quercy (Uxellodunum, cité martyre)*. Aurillac: Imprimerie Moderne.

Spence, L. (1990 [1915]). *Ancient Egyptian Myths and Legends*. New York: Dover.

Spieth, D. A. (2007). *Napoleon's Sorcerers*. Newark: University of Delaware Press.

Stafford, F. (1988). *The Sublime Savage: A Study of James Macpherson and the Poems of Ossian*. Edinburgh: Edinburgh University Press.

Stolzenberg, D. (2013). *Egyptian Oedipus: Athanasius Kircher and the Secrets of Antiquity*. Chicago: University of Chicago Press.

Stray, C. (2004). "From One Museum to Another: The 'Museum Criticum' (1813–26) and the 'Philological Museum' (1831–33)." *Victorian Periodical Review* 37:289–314.

Sweet, P. R. (1978). *Wilhelm von Humboldt: A Biography.* Columbus: Ohio State University Press.

Tadié, A. (1995). "Dom Raphaël de Monachis." In *Langues'O, 1795–1995: Deux siècles d'histoire de l'École des langues orientales,* edited by P. Labrousse, 61–65. Paris: Éditions Hervas.

Testa, A. M. del. (2002). *Michelangleo Lanci e l'interpretazione dei geroglifici.* Fano: Biblioteca Federiciana.

Thomasson, F. (2013). *The Life of J. D. Åkerblad: Egyptian Decipherment and Orientalism in Revolutionary Times.* Boston: Brill.

Thompson, C. J. (1928). *The Quacks of Old London.* New York: Brentano's.

Thompson, J. (2015). *Wonderful Things: A History of Egyptology; 1: From Antiquity to 1881.* New York: American University in Cairo Press.

Thornbury, W. (1880). *Old and New London: A Narrative of Its History, Its People, and Its Places.* London: Cassell, Petter, Galpin.

Trautmann, T. R. (1997). *Aryans and British India.* Berkeley: University of California Press.

Trumpener, K. (1997). *Bardic Nationalism: The Romantic Novel and the British Empire.* Princeton: Princeton University Press.

Turner, J. (2015). *Philology: The Forgotten Origins of the Modern Humanities.* Princeton: Princeton University Press.

Usick, P. (2002). *Adventures in Egypt and Nubia: The Travels of William John Bankes (1786–1855).* London: British Museum Press.

Vasunia, P. (2013). *The Classics and Colonial India.* New York: Oxford University Press.

Villenave, M.-G.-T. (1821). "Notice sur la vie et les ouvrages de J. J. Barthélemy." In *Oeuvres de J. J. Barthélemy,* 1:1–60. Paris: A. Belin.

Walravens, H. (2005). "Julius Klaproth: His Life and Works with Special Emphasis on Japan." Lecture given at Nichibunken, Kyoto in January.

Wile, A. (2016). *Watteau's Soldiers: Scenes of Military Life in Eighteenth-Century France.* New York: Frick Collection.

Wood, A. (1954). *Thomas Young: Natural Philosopher, 1773–1829.* Completed by Frank Oldham. Cambridge: Cambridge University Press.

Wynter, A. (1868). "The London Commissariat." In *Curiosities of London,* 201–44. London: R. Hardwicke.

NAME INDEX

SUBJECT INDEX